LABORATORY MANUAL
TEACHER EDITION

MERRILL
BIOLOGY
AN EVERYDAY EXPERIENCE

Authors
Albert Kaskel
Paul Hummer, Jr.
Lucy Daniel

GLENCOE
McGraw-Hill

New York, New York   Columbus, Ohio   Mission Hills, California   Peoria, Illinois

A MERRILL BIOLOGY PROGRAM

Biology: An Everyday Experience, *Student Edition*
Biology: An Everyday Experience, *Teacher Edition*
Biology: An Everyday Experience, *Teacher Resource Package*
Biology: An Everyday Experience, *Study Guide*
Biology: An Everyday Experience, *Transparency Package*
Biology: An Everyday Experience, *Laboratory Manual, Student Edition*
Biology: An Everyday Experience, *Laboratory Manual, Teacher Edition*
Biology: An Everyday Experience, *Computer Test Bank*
Biology: An Everyday Experience, *Cooperative Learning*
Biology: An Everyday Experience, *Spanish Resources*
Biology: An Everyday Experience, *Videodisc Correlations*
Biology: An Everyday Experience, *Lesson Plans*

Copyright © by the Glencoe Division of Macmillan/McGraw-Hill Publishing Company, previously copyrighted by Merrill Publishing Company.

All rights reserved. Permission is granted to reproduce the material contained herein on the condition that such materials be reproduced only for classroom use, be provided to students without charge, and be used solely in conjunction with the *Biology: An Everyday Experience* program. Any other reproduction, for sale or other use, is expressly prohibited.

Send all inquiries to:
Glencoe Publishing Company
936 Eastwind Drive
Westerville, OH 43081

Printed in the United States of America.

ISBN: 0-02-827306-0

3 4 5 6 7 8 9 10 POH 00 99 98 97 96

Table of Contents

To the Teacher	5T
Format	5T
Teaching Strategies	6T
Reading Features and Strategies	7T
Helping Your Students With Tables	8T
Helping Your Students With Graphs	9T
Materials for a Class of 30	10T–11T
Materials Lab by Lab	12T–17T
Preparation of Solutions	17T–19T
Making Slides (Temporary Mounts)	20T
Maintaining Living Materials	20T–21T
Setting Up Classroom Aquaria and Terraria	21T–22T
Living Things in the Laboratory	22T
SI Conversions	22T–23T
Suppliers	23T
Readings for the Teacher	24T

To the Teacher

The exercises in this laboratory manual are designed to give meaning to many biological concepts and enable students to use some of the tools a scientist uses. Students will recognize biology as something all around them and part of their lives. As new biological concepts and processes are presented in the text, students will have the opportunity for hands-on experiences in the lab to reinforce these concepts and processes. You may not be able to complete all 70 lab exercises in one school year. However, with the number and variety of exercises to choose from, you can do those labs you feel are most desirable for your students and your schedule.

The lab exercises are designed to require simple equipment, few solutions, and the purchase of few live specimens. Thus, preparation time in set up and tear down is minimal. The amount of equipment needed for an exercise is also minimal.

This laboratory manual can be used to develop skills in reading; following procedures; using technical and nontechnical vocabulary accurately; using diagrams, charts, and graphs; drawing conclusions; making judgments; and applying data to practical problems. The laboratory work also enables students to discover processes of science, such as observing, inferring, measuring, and comparing.

This laboratory manual provides your students with various types of laboratory work. Several exercises require students to do model building using simple household or laboratory materials. Building models makes certain concepts less abstract and thus aids student understanding. Some exercises require students to do simulation-type activities, such as counting populations and cells. Simulations allow students to see how biologists work in both the laboratory and the field.

This teacher annotated edition provides a variety of helpful information for both experienced and inexperienced biology teachers in the form of annotations and a complete teacher's guide. Overprinted in red on the pupil's copy are many teaching tips as well as answers to all questions. Sample data, drawings, places to obtain supplies, substitute supplies, and alternative procedures have been included. Additional biological information has been included to clarify or elaborate details. Throughout the manual, references to pages in the teacher guide provide quick access to other materials and information.

The teacher guide is located at the front of this teacher annotated edition. The teacher guide provides you with a wealth of information designed to aid in using *Biology: Laboratory Experiences* successfully. Helpful teaching strategies and strategies for aiding students' reading and comprehension are outlined in the teacher guide. Ways to help students work with and understand tables and graphs have been included. Care of living materials, use of equipment, and preparation of solutions are just a few of the topics that have been included to aid you in preparing for lab exercises. A list of suppliers has also been included to help you find certain laboratory supplies. At the end of the teacher guide is a list of readings for the teacher. The readings can furnish you with supplementary material, such as additional biological information, and different laboratory techniques.

Format

Most of the lab exercises in this laboratory manual are designed to be finished in one class period; only a few require several class periods to be completed. The longer exercises are centered around growth and development and life cycles. The format each laboratory exercise follows is outlined below.

Title: Each laboratory exercise is titled with a question. The question identifies the problem to be solved by students as they do the lab exercise.

Introduction: The introduction provides background information that students will need in solving the problem identified in the title.

Objectives: The objectives outline for the students exactly what the objectives of the lab exercise are. The objectives clearly define what the students will be doing and working to determine as they do the lab exercise.

Keywords: Important biology terms and boldfaced terms from the text are reviewed as students write definitions.

Materials: Living organisms, preserved specimens, solutions, media, and equipment are listed for each lab exercise. Number or amounts of each item are given per student or per team, depending on the exercise. These amounts may be adjusted according to the availability of materials and the number of students. In case of waste or breakage, extra materials and solutions should be on hand. Preparation directions for all solutions are given in the teacher's guide.

Procedure: Numbered step-by-step directions lead students through each lab exercise. Numbered steps make it easy for you, if you wish, to specify a point at which you want the entire class to stop before going on. Numbered steps also allow you to call attention to specific, important procedural steps that might otherwise go unnoticed. Incorporated with many procedural steps are tables to record results, observations, and other data. It is important that students understand what they are looking for in their results and observations so they know what to record in their tables.

Questions: Following the procedure steps are a number of thought provoking questions that guide students in analyzing data and drawing conclusions.

A glossary is provided at the back of this laboratory manual. It provides definitions to all of the keywords identified in each lab exercise. In addition, the glossary provides definitions to many other terms used in the laboratory manual. Many of these additional words are scientific terms with which students may not be familiar.

Teaching Strategies

An atmosphere of curiosity, interest, and enthusiasm for biology should be encouraged in the laboratory. It can be one of the most effective methods of teaching and learning. Students must be oriented to laboratory work and each exercise must appear to them real and worthy of study. It is important that students know what to do, what equipment to use, and how to use it. Even though this laboratory manual requires simple equipment and materials, the teacher should show the equipment to students and explain its use.

Students should be familiar with where the materials are for the laboratory work. You may wish to put materials on trays that can be moved around the room. It may be advisable to acquire laboratory assistants for your classes. A capable person who has successfully completed biology can be a great help in preparing solutions, passing out materials, and assisting with the laboratory work. If you do not have laboratory assistants, you may want to use some of the students in your class to prepare some of the materials in advance.

Some exercises require only paper and pencil. These exercises may be used in different ways. They can be started during class time and finished at home or they can be assigned totally as homework or as enrichment. Some paper and pencil activities lend themselves well to being used on the overhead projector. For example, in lab 1–2, the diagram could be placed on the overhead projector and as questions arise, the teacher can refer to the enlarged diagram from the overhead projector.

Maximum learning may be achieved by working in teams of two or four. A shy student may be stimulated to share duties and responsibilities. An extroverted student may assume leadership qualities not developed in individual work. Try to arrange groups so a strong reader is in each group. Team work in the laboratory provides social interaction that can lead to positive development and values. Many exercises in this laboratory manual are designed for team work.

As students are working in the lab, circulate around the room and help those who appear to be having trouble. A second explanation about the procedure can be very helpful and make the student more successful with the lab work.

Have students complete the keywords section before doing each exercise. Have students answer the questions after they finish each exercise. If there is not enough class time to do this, assign these sections as homework. Collect their written work and evaluate it. Return the corrected papers to students and conduct a post-lab session. A post-lab session will reinforce what they learned and also help those students who missed the lab understand what the exercise was about.

The need for safety in the laboratory should not be minimized. On the inside back cover is a list of safety and first aid rules. You should review these rules with your students and be sure students recognize the importance of the rules. Many of the exercises in this lab manual contain caution statements. You should be sure students notice and follow these cautions. The biology laboratory can be a safe place to conduct lab exercises if you and your students follow proper safety procedures.

To aid your students' reading and comprehension of the lab work, you may find, initially, a more teacher-oriented method of doing a lab helpful. Here are some suggestions:

1. Read the lab's introduction aloud to the class as they read along silently. Discuss the objectives listed for the lab before students begin. Relate how the lab correlates with the chapter being studied and the importance and worthiness of doing the lab.

2. Review any cautions that should be taken throughout the lab activity.

3. Have students work only a few steps at a time through the lab. To facilitate their work, prepare a transparency of the experiment and project it on a screen with an overhead projector. Mark and discuss the step where you wish all students to stop during the lab. After the entire class has reached the stopping point, mark again where students should proceed to. In this way, students will not be overwhelmed with too many directions at one time. Also, you will be better able to assist students.

4. Demonstrate to students how the equipment for a particular lab is to be used.

5. A number of labs may be done as class demonstrations. If this method is used, allow students to follow along in their lab books, recording needed observations and data.

6. Place students' data on the chalkboard or on an overhead projector at the conclusion of the experiment. Discuss which results may be in error and make sure students recognize sources of errors if errors exist.

7. You may want to occasionally discuss the answers to the lab questions and the application questions at the conclusion of each lab.

8. You may want to modify or delete parts of a lab to fit your time needs. In these cases, it might be beneficial for you to do part of the lab ahead of time before class. You might also designate a certain numbered procedural step as the stopping point for that class period if you anticipate running out of time.

9. Wherever necessary, you should adjust the labs to your students' needs.

10. Do not assume that students know or recall more than they do. Continually investigate what the student knows or does not know about laboratory procedures. For example, if it has been awhile since microscopes have been used, review the proper use of the microscope before students use them again.

Once students become accustomed to the direct step-by-step procedures, how to follow the diagrams, record their results, and answer the questions, they should be encouraged to work on their own. Students that need extra help often become dependent on the teacher and are reluctant to take the initiative in laboratory work.

See page 7T for reading features and strategies.

Reading Features and Strategies

Reading is an important tool in collecting, interpreting, and understanding data in the biology laboratory. Often a sense of frustration and lack of success with laboratory exercises is due to a basic inability to read and comprehend well. Student reading difficulties even can result in classroom management problems. *Biology: Laboratory Experiences* has many features incorporated to aid your students in reading and comprehending information presented. Math as well as reading skills are developed by some of the exercises.

In controlling the reading level of *Biology: Laboratory Experiences,* several things have been given careful attention. The vocabulary in the laboratory manual is consistent with the level of students for whom the text is designed. Words with large numbers of syllables have been avoided wherever possible. Important terms have been clearly defined and reinforced through use. In some cases, they are followed by phonetic spellings. Sentence construction has also been carefully controlled to aid in readability. Readability is also aided by use of many diagrams which tie directly to the printed words, in many cases showing the procedural steps. The typeface and type size in which the text is printed are other aids to readability.

The format of the exercises in *Biology: Laboratory Experiences* (See Format section in this teacher guide, page 5T) also can help students with difficulties in reading. Particular features, such as numbered procedural steps and vocabulary sections to review general as well as scientific vocabulary, can be easily used to facilitate laboratory success. A complete glossary of terms is included for student use.

Performance of laboratory exercises with students having reading and comprehension difficulties may require some adjustments from the usual approach of having students read the directions on their own and then work independently. Students who do not read well need positive reinforcement continually throughout the year. Some helpful suggestions follow.

1. You may want to have students follow along as you read aloud the title, introduction, and objectives of each exercise.

2. Show students how each exercise relates to material in the text chapter.

3. Have students read parts of the exercise aloud in a prelab session. Ask other students the meaning of what has been read.

4. You may want to have students write one complete sentence that clearly states what the exercise is about before it is carried out. Have students read their sentences aloud.

5. Introduce students to reading a laboratory exercise and learning its parts. Have students note that the problem to be addressed is the question title of the exercise.

6. Encourage students to use the phonetic spellings and the glossary. You may need to review how the pronunciation key and the glossary are used.

7. Students can be asked to copy words or sentences from the introduction or objectives. Make sure they know the meanings of the words they use and are not just copying letters. Words they do not understand or cannot pronounce can be discovered and easily reviewed in this way.

8. Make sure students understand the meanings of new words. Some words, such as balance and base, have more than one meaning in this laboratory manual. Make sure students are familiar with the meaning you are using in the particular exercise.

9. A transparency for the overhead projector can be helpful in reviewing the parts of the exercise with students. This procedure is particularly valuable at the start of the course but you may find it worthwhile to do all school year. Emphasize certain words and their meanings by underlining on the transparency as students follow along.

10. Go step by step through the procedure with students. Everyone can work together.

11. Show students *how* to use equipment or *how* to do procedural steps even after you have read aloud how it is done.

12. Repeat often. Reinforcement is essential.

13. Require students to spell words correctly, especially those words that appear in the exercise.

14. Students can be encouraged to develop a word list as terms are used in the exercises. Words can be kept on file cards and be used as a study guide.

15. Artwork in this lab manual is of two kinds, those used by students in the labs and those showing procedural steps. Have students match procedural steps to the appropriate diagrams.

16. Students can be encouraged to "read" through the diagrams in the exercise before reading through the words.

17. Encourage students to detail drawings they make in the exercises. Remind them that the pictures they draw "tell" what they saw. Drawing can be used in other ways, too, to help students communicate.

18. Encourage students to answer questions in complete sentences whenever that is asked for or whenever possible. You might want to review what parts a complete sentence must contain. Writing skills can also be aided this way.

19. Make sure students understand what is being asked in the questions. Ask students questions about the general vocabulary in the questions if you feel students do not understand what is being asked.

20. Make sure students can make and understand tables and graphs. See pages 8T and 9T for additional information in helping students with making tables and graphs of data.

Helping Your Students With Tables

Tables are used to record data in an orderly and concise manner for ease of interpretation and comparison. Many of the exercises in this laboratory manual will require students to complete tables. Poorly recorded data, incomplete data, and incorrectly recorded data will lead students to incorrect conclusions.

Before your students complete the first table in this laboratory manual, review with them what the major table components are. This review can be greatly enhanced by
1) using an overhead projector to show a transparency of a table.
2) providing each student with a ditto or photocopy of the table you project on the screen. (Students can then record information and have it available for future reference. This process also gets students involved as they record what you write on the projector transparency.)

The table on page 67 of this laboratory manual will be used here as an example of one you might use with students to review tables.

Table 2. Testing for Protein (A)

(B) Tube number and/or contents	Color with Biuret solution	Protein present?
Egg white (protein) + Biuret		
Biuret (control)		
1. Biuret		
2. Biuret + cottage cheese		(C)
3. Biuret + lard		
4. Biuret + tuna		
5. Biuret + water	(D)	

A is the table title. B is a column head. C is a column (section that runs up and down). D is a row (section that runs across). Have students identify these four main components. Have them record on their copies of the table what these components are and where they are located.

Ask students what is to be recorded in the column headed "Color with Biuret solution." Ask students what is to be recorded in the column headed "Protein present?" Then show students what a properly completed table might look like. (Note: If Biuret solution changes from blue to purple, protein is present. If no color change occurs, protein is not present.)

Remind students that all tables do not require the same kind of information. For example, the table shown here on bone lengths (from page 7 of this manual) requires numbers to be filled in. Prepare a copy of the Bone Lengths table from page 7 for your overhead projector. Also make copies of it for students.

Point out again the title, column heads, columns, and rows of this table. Have students notice that both main columns "Hand" and "Foot" have been divided into two subcolumns each. In this case, the subcolumn heads are "Millimeters" and "Centimeters."

Table 2. Bone Lengths

Bone	Hand		Foot	
	mm	cm	mm	cm
Bone A				
Bone B				
Bone C				
Bone D				
Bone E				
Thumb or big toe (F + G)				
Smallest finger or toe (H + I + J)				

Have students tell you where in the table each of the following would go:
1) Bone C of the foot in millimeters.
2) Big toe of the foot in centimeters.
3) Bone C of the hand in millimeters.

Many students have no trouble with locating information in tables or placing information in the proper places in tables that are given. However, when students are asked to organize information into their own table, problems often arise. It is important that you emphasize that more than one organization of data can be "correct."

Give students the paragraph below, which gives information on the types of seeds certain birds will eat. Have students make a table that clearly presents the information.

Cardinals, chickadees, and nuthatches like sunflower seeds to eat. Juncos, mourning doves, and pheasants eat oats and cracked corn. Sparrows usually will eat a variety of seeds. They have no special preferences.

The following table shows the information in the paragraph in a concise, organized form. Students may put the data in a "correct" table that is different than this one.

Birds	Seeds they eat		
	Sunflower	Oats	Corn
Cardinals	✓		
Chickadees	✓		
Nuthatches	✓		
Juncos		✓	✓
Mourning doves		✓	✓
Pheasants		✓	✓
Sparrows	✓	✓	✓

Helping Your Students With Graphs

Some of the exercises in this laboratory manual require your students to graph their data. Two types of graphs are used, bar graphs and line graphs. You may even want students to graph data when an exercise does not specifically call for graphing. If you do, it is recommended that you supply students with prenumbered and prenamed horizontal and vertical axes.

Bar graphs. Review the main components of a bar graph with students before they are asked to construct this type of graph. As with reviewing tables, the review will be greatly enhanced if it is conducted with a transparency on the overhead projector while students follow along on copies of the graph. In this way, students are directly involved in the review as they record information on their graphs.

Some of the main features of a bar graph are: A—vertical axis, B—title and numbers on vertical axis, C—horizontal axis, D—title and numbers on horizontal axis, E—horizontal lines within graph, and F—graph title. Have students identify where these parts are in copies of this graph. The letters of the parts go in the circles on the graph.

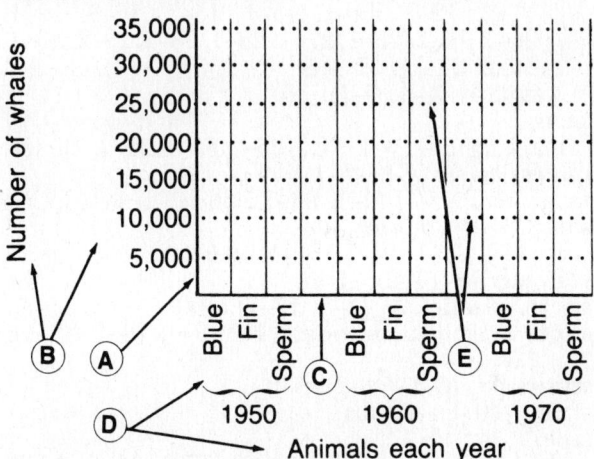

Once the main graph components are identified, show students how to put the following data onto the graph. Demonstrate on the overhead projector while students work on their copies.

Whale	Number of animals killed in		
	1950	1960	1970
Blue	7,000	2,000	0
Fin	25,000	32,000	5,000
Sperm	18,000	20,000	24,000

It is important that students realize that three sets of data are being compared for each year given. Data for three kinds of whales are given. Establish a pattern of shading such as solid for blue whales, stippled for fin whales, and striped for sperm whales. Students should be shown what to do with data that do not match exact numbers given on the vertical axis. For example, explain how the bar for blue whales killed in 1950 must be estimated because no line exists at the 7000 mark. Get students to tell the advantages of graphing.

Line graphs. Line graphs, although like bar graphs, differ in one main way. The data are plotted as points that are connected by a line. No bars are present. Review with your students how the following data would be made into a line graph. Again it is most advantageous to project the graph

Population of the United States

Year	Number	Year	Number
1800	10,000,000	1900	75,000,000
1850	25,000,000	1950	150,000,000

on a screen with the overhead projector and have students follow along on their own copies of the graph. Plot the data given above on the graph below while students work with you. Connect the points with a smooth continuous curve. Make sure students appropriately estimate the placement of points which are numbers not given on the vertical axis.

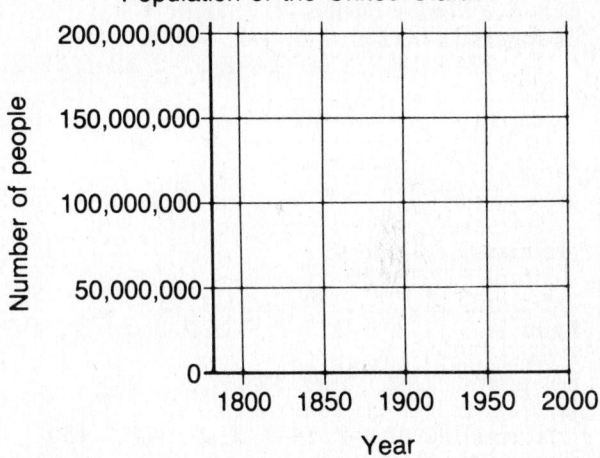

Show students two advantages that line graphs have over bar graphs. Namely, a line graph can estimate data points between those points plotted, and a line graph can be extended (extrapolated) in both directions. Students will see how the population of the United States can be estimated for the years 1840 or 1935, for example. Also the line can be extended to predict the population in the years 1990 or 2000, for example.

You may want to have students make a separate line graph of the following fly population data.

Number of flies in bottle	Time (days)	Number of flies in bottle	Time (days)
20	5	350	30
30	10	208	35
58	15	114	40
110	20	36	45
135	25	0	50

9T

Materials for a Class of 30

The materials on the following list are all you need for the 70 exercises in this book. The numbers after the items below indicate the exercises in which the items are used. Amounts listed are for one class of 30. Grocery-store items are marked with a large [G]. The pages that follow show supplies listed exercise by exercise. In this way, you have quick access to all that is needed for each lab exercise to aid you in planning.

Equipment
balances (31–1): 30
beakers
  50-mL (18–2): 60
  100-mL (18–1): 60
  250-mL (4–2, 19–2): 60
  400-mL (2–1, 32–2): 30
  500-mL (4–1, 19–2): 60
cardboard box, small (21–1): 30
clock/timer with second hand (16–1, 17–1, 17–2, 21–1, 32–2): 1
coins (27–2): 240
containers, large (31–1): 30
coverslips (2–2, 4–3, 5–1, 7–1, 23–1, 31–2, 32–2): 150
dish covers (17–1): 30
dishes
  petri (17–1, 18–2, 19–2, 31–3): 60
  small (7–1, 8–1): 30
dissecting needles (23–2): 30
dissecting pans (7–2, 7–3, 8–2, 8–3, 10–2, 16–3): 30
dropper bottles (7–1, 9–2, 18–1): 60
droppers (2–1, 2–2, 4–1, 4–2, 4–3, 5–1, 5–2, 9–1, 9–2, 12–2, 13–2, 18–1, 24–1, 31–2, 32–1, 32–2): 300
forceps (2–2, 5–1, 7–2, 7–3, 8–1, 8–3, 10–2, 16–1, 19–2, 32–1): 30
funnels (4–2): 60
glass slides (2–2, 4–3, 5–1, 6–1, 7–1, 23–1, 23–2, 24–1, 31–2, 32–2): 150
graduated cylinders
  10-mL (12–2): 30
  100-mL (2–1, 4–2, 18–1, 31–1): 30
hand lens (3–2, 6–2, 7–1, 7–2, 7–3, 23–2, 24–1, 31–3, 32–1): 30
hot plates (19–2, 32–2): 30
jar lids (2–1, 5–2): 130
jars, small (2–1, 5–2, 24–1, 31–3, 32–1): 130
metersticks (16–1): 30
models, chest cavity (13–1): 6
microscopes, compound light (1–1, 2–2, 3–2, 4–3, 5–1, 6–1, 7–1, 19–1, 20–2, 22–2, 23–1, 24–1, 27–1, 31–2, 32–2): 30
microscopes, stereoscopic (8–1, 8–3): 30
pennies (26–1): 60
probes (5–1, 7–1, 8–2): 30
scalpels (7–3, 10–2, 23–2, 24–1): 30
spoons
  metal (4–2, 5–2, 16–1, 18–2): 60
  plastic (16–1, 32–1): 30
stoppers, cork (4–1, 9–1): 240
test-tube holders (32–2): 30
test-tube racks (4–1, 9–1, 13–2, 32–2): 30
test tubes (4–1, 9–1, 9–2, 13–2, 32–1, 32–2): 300

Expendables
aluminum foil [G] (4–2, 19–2): 2 rolls
bags [G]
  brown (5–2): 25
  paper, small (30–2): 30
  plastic (13–2): 30
balls, polystyrene, 7.5 cm diameter (16–3): 15
blindfolds (16–1): 30
cheesecloth, 12 x 12 cm squares (18–2): 30
cloth strips, 2 x 8 cm (16–3): 60
cotton, balls [G] (13–2, 16–1, 31–2): 2 boxes
cotton swabs [G] (5–2): 1 package
cups [G]
  foam, insulated (21–1, 21–2): 120
  paper, 7 oz., (11–2, 16–1, 31–1): 360
  plastic, small, clear (12–2): 600
file cards (16–2, 17–2): 60
folders (17–2): 30
glue (11–2, 16–1, 22–1): 6 bottles
index cards (16–1): 30
labels (4–2, 9–1, 11–2, 18–1, 18–2, 24–1): 1000
paper
  black (17–1): 30 pieces
  brown (9–1): grocery bags cut into 30 pieces
  construction, various colors (22–1): 360 pieces
  filter (4–2): 30 pieces
  graph paper (30–2): 30 pieces
  heavy (11–2): 30 pieces
  lens (1–1): 100 booklets
  tracing (16–3, 17–2): 45 pieces
paper clips, large (16–3): 15
pencils
  colored (2–1, 4–3, 6–1, 7–2, 8–1, 10–2, 11–1, 11–2, 12–2, 14–1, 15–1, 15–2, 16–3, 19–1, 20–1, 20–2, 23–1, 27–1, 28–1, 28–2, 29–2, 30–1, 30–2, 31–1): 30 sets
  wax (2–1, 4–1, 9–1): 30
pens
  marking (5–1, 5–2, 16–2, 26–2, 32–1, 32–2): 30 sets
plates, paper [G] (16–1): 30
razor blades [G] (6–1, 23–1): 30
sand (31–1): 0.5 m^3
soil (4–2, 21–1, 21–2, 31–1): 0.5 m^3
soil, sterile (4–2): .1 m^3
sticks, plastic [G] (4–2, 17–1): 100
sticks or splints, wooden (4–1, 9–1, 18–1, 32–1): 210
straws [G] (6–1, 11–2): 330
string (10–1): 4 balls
tape
  clear (8–1, 10–1, 11–2, 16–3, 17–1, 17–2, 19–1, 20–1, 20–2, 24–3, 27–2): 12 rolls
  masking (26–1, 27–2): 1 rolls
thread (22–1): 3 spools
toothpicks [G] (16–1, 21–1, 22–1, 32–2): 5 boxes
twist ties [G] (18–2): 20
yarn (22–1): 1 skein

Chemical Supplies
For Solutions
antacid tablets, 2 kinds [G] (18–1): 1 bottle each
aspirin [G] (18–2): 1 bottle
Biuret solution, (9–1) purchase or make with potassium hydroxide: 100 g and copper sulfate: 0.75 g
bromthymol blue (4–1): 0.5 g
calcium chloride (5–1): 0.2 g
Congo red (18–1): 3 g
copper sulfate (32–1): 1 g
cow manure (31–3): 50 mL
fertilizer, liquid plant food (31–3, 32–1): 180 mL bottle
food coloring, blue, red, yellow, green [G] (2–1, 12–2, 13–2): 1 small bottle each color
glucose, or dextrose (4–2): 180 mL

hydrochloric acid, concentrated (18–1, 32–1): 36 mL
indophenol, 2, 6-dichlorophenol (9–2): 13 g
iodine (2–2, 19–2): 2 g
magnesium sulfate (5–1): 0.1 g
methylene blue (4–3, 32–2): 2 g
phenolphthalein powder (4–2): 1 g
potassium chloride (5–1): 1 g
potassium iodide (2–2, 19–2): 3.5 g
potassium phosphate, monobasic (5–1): 2 g
silver nitrate (13–2): 4 g
sodium bicarbonate (5–1, 18–2, 32–1): 3 g
sodium chloride (5–1, 13–2, 31–2): 37 g
sodium citrate (5–1): 3 g
sodium hydroxide (4–2): 0.4 g
sodium phosphate, monobasic (5–1): 2 g

Other Chemicals Needed
alcohol, ethyl (4–3, 18–2): 220 mL
alcohol, denatured (4–2): 50 mL
alcohol, isopropyl [G] (5–2, 19–2): 2 bottles
ammonia [G] (5–2): 1 bottle
bleach [G] (5–2, 7–1): 1 bottle
cleaner, household [G] (5–2): 1 bottle
cleaning powder, household [G] (5–1): 1 can
cobalt chloride paper (13–2): 120 pieces
oil, cooking [G] (9–1): 1 bottle
paper, pH 2-10 with color chart (32–1): 3 pkg.
paraffin, for dissecting pans [G]: 4 boxes
syrup, clear [G] (2–1): 1400 mL
vinegar, white [G] (2–1): 1400 mL
vitamin C, tablet (9–2): 250 mg
water
　distilled [G] (4–2, 5–1, 9–1, 9–2, 13–2, 18–1, 31–3, 32–1, 32–2): 3 jugs

Biological Supplies
bamboo shavings (2–2): 30
chewing gum, cinnamon [G] (16-1): 30 pieces
cigarettes [G] (18–2): 60
cork shavings (2–2): 30
food supplies [G]
　apples (9–1, 16–1): 4
　bacon (9–1): 6 strips
　bean pods, opened (29–1): 300
　bread (5–2): 15 slices
　broccoli (23–2): 30
　brussel sprouts (23–2): 30
　butter (9–1): 1 stick
　buttermilk (4–1): 1 cup
　cabbage juice, cooked (9–2): 200 mL
　cabbage juice, raw (9–2): 200 mL
　capers (23–2): 1 jar
　carrots (23–2): 30
　celery (6–1): 1 stalk
　cherry peppers (23–2): 30
　coffee, ground caffeinated (18–2): 1 small bag
　cottage cheese (4–1, 9–1): 12 oz.
　cucumbers (23–2): 2
　eggs, raw (2–1): 7
　egg whites
　　hard boiled (9–1): 6 eggs
　　raw (9–1): 6 eggs
　grapefruit juice, fresh (9–2): 1 grapefruit
　lard (9–1): 1 small can
　lemon juice, fresh (9–2): 1 lemon
　lettuce (6–1, 23–2): 2 heads
　lima beans (23–2): 30
　mayonnaise (9–1): 1 small jar
　milk (4–1): 1 cup
　okra (23–2): 15
　onions (2–2, 16–1, 23–2): 35
　orange juice, fresh (9–2): 1 juice orange
　peanut butter (9–1): 1 small jar
　peas (23–2): 30
　potatoes, instant (5–2): 1 pkg.
　potatoes, raw (16–1): 2
　raisins (5–2): 1 box
　shallots (23–2): 30
　sour cream (4–1): 4 oz.
　sweet potatoes (23–2): 30
　tomato juice, canned (9–2): 180 mL
　tomato juice, fresh (9–2): 1 tomato
　tuna fish (9–1): 1 small can
　veal, strained (9–1): 1 jar
　yogurt (4–1): 4 oz.
living specimens
　animals
　　Daphnia (24–1, 32–1): 1 culture
　　isopods (17–1): 180
　　termites (5–1): 90
　cultures
　　Euglena (24–1, 32–1): 1 culture
　　Gloeocapsa (4–3): 1 culture
　　Lyngbya (4–3): 1 culture
　　Paramecium (31–2): 1 culture
　planarian (24–1): 1 culture
　plants
　　bean, potted (19–2): 4
　　bean seedlings, germinated (21–1, 21–2): 180
　　Coleus, potted, (19–2): 15
　　duckweed (31–3): 270
　　elodea leaves (2–2): 30
　　flowers, various (23–1): 30
　　Mimosa, potted (21–1): 30
　　liverworts (6–1): 6
　　twigs, 3 kinds of conifers (6–2, 29–1): 60 twigs
　various living specimens, such as *Nostoc*, *Paramecium*, mushroom, fish, geranium, and earthworm (3–2)
　water samples, pond, lake, stream (5–1)
prepared slides: 30 each
　diatomaceous earth (5–1)
　frog blood (2–2)
　insect leg (1–1)
　leaf, dicot C.S. (19–1)
　normal and sickled red blood cells (27–1)
　onion root tip, L.S. (22–2)
　root, dicot (20–2)
preserved specimens:
　bedbug (8–1): 30
　flea (8–1): 30
　lamprey (8–2): 30
　louse (8–1): 30
　perch (8–2, 10–2): 30
　pig eye (16–3): 15
　sea anemone (7–3): 30
　shark (8–2): 30
　sponge
　　Grantia, *Spongia* (7–1): 1
　squid, frozen [G] (7–2): 30
　starfish (8–3): 30
　tick (8–1): 30
seeds
　beans, red kidney [G] (26–2): 900
　beans, pinto [G] (18–2, 26–2, 29–1, 30–1): 3530
　beans, white [G] (30–1): 390
　radish (18–2): 500
sheep hearts (11–1): 30
yeast [G] (32–2): 1 cake

Materials Lab by Lab

Exercise	Equipment	Expendables	Chemical Supplies	Biological Supplies
1–1	microscopes, compound light	paper, lens		prepared slides— insect leg
1–2	metric rulers			
2–1	beakers, 400 mL droppers jars, small jar lids graduated cylinders, 100 mL	colored pencils— blue, red, green pencils, wax	food coloring syrup, clear vinegar, white water, tap	eggs, raw
2–2	coverslips droppers forceps glass slides microscopes, compound light		iodine stain water, tap	bamboo shavings cork shavings elodea leaves onion skin prepared slides— frog blood
3–1	scissors	paper, white		
3–2	hand lens microscopes, compound light			specimens of living things
4–1	droppers stoppers, cork test-tube racks test tubes	pencils, wax wooden sticks or splints	bromothymol blue solution	dairy products— milk, buttermilk, yogurt, cottage cheese, sour cream
4–2	beakers, 500 mL breakers, 250 mL droppers funnels graduated cylinders, 100 mL teaspoons	aluminum foil filter paper labels sticks, plastic	alcohol, denatured glucose or dextrose phenolphthalein sodium hydroxide water, distilled	soil soil, sterile
4–3	coverslips droppers glass slides microscopes, compound light	pencils, green	alcohol, ethyl methylene blue stain	cultures— *Gloeocapsa*, *Lyngbya*
5–1	coverslips droppers forceps glass slides microscopes, compound light probes	pens, marking	buffer solution household cleaning powder	1–3 living termites prepared slides— diatomaceous earth water samples
5–2	droppers jars, small jar lids spoon	bags, brown cotton swabs pens, marking	alcohol, isopropyl ammonia bleach household cleaner	bread potatoes, instant raisins
6–1	glass slides microscopes, compound light razor blades scissors	colored pencils— green straws		celery lettuce sea lettuce
6–2	hand lens metric rulers			twigs of various conifers

Exercise	Equipment	Expendables	Chemical Supplies	Biological Supplies
7–1	coverslips dropper bottles glass slides hand lens microscopes petri dishes probes scissors		bleach water, tap	sponges— *Grantia, Spongia*
7–2	dissecting pans forceps hand lens scissors	colored pencils- red, green, gray, blue, yellow, purple		preserved squid, frozen
7–3	dissecting pans forceps hand lens scalpels scissors			preserved sea anemones
8–1	forceps metric rulers microscopes petri dishes scissors	tape, clear		preserved tick, flea, bedbug, louse
8–2	dissecting pans probes			preserved shark, lamprey, perch
8–3	dissecting pans forceps microscopes scissors			preserved starfish
9–1	droppers graduated cylinders, 100 mL stoppers, cork test-tube rack test tubes	labels or pencils, wax paper, brown sticks, plastic or wooden	Biuret solution water, tap	food samples— apples, bacon, butter, cooking oil, cottage cheese, egg white (hardboiled and raw), lard, mayonnaise, peanut butter, tuna fish, strained veal
9–2	droppers dropping bottles test tubes		indophenol, 2,6- dichlorophenol vitamin C tablets water, distilled water, tap	cabbage juice, raw cabbage juice, cooked grapefruit juice, fresh lemon juice, fresh orange juice, fresh tomato juice, fresh tomato juice, canned
10–1	metric rulers scissors	string tape, clear		
10–2	dissecting pans forceps metric rulers scalpels scissors	colored pencils— red, blue		preserved perch

Exercise	Equipment	Expendables	Chemical Supplies	Biological Supplies
11–1		colored pencils— red, blue		sheep hearts
11–2	metric rulers scissors	colored pencils— red, blue, purple cups, paper glue labels page of model parts paper, heavy straws tape, clear		

12–2	droppers
	graduated cylinders
	   10 mL
		colored pencils—
		   red, green, black
		cups, clear-plastic
			food coloring—
			   red, green	

13–1	models, chest cavity			
13–2	droppers test-tube racks test tubes	bags, plastic cotton, balls	cobalt chloride paper salt water silver nitrate solution "urine" water, distilled	
14–1		colored pencils— red, blue pencils, lead no. 2		
14–2	metric rulers			
15–1		colored pencils—red paper, white		

15–2		colored pencils—
		   red, green, blue,
		   gray, yellow
		pencils, lead no. 2		

16–1	clock/timer with
	   second hand
	forceps
	metersticks
	spoons, metal
		blindfolds
		cotton, balls
		cups, paper
		glue
		index cards
		plates, paper
		spoons, plastic
		toothpicks
		towels, paper
			chewing gum,
			   cinnamon
				apples
				onions
				potatoes, raw

16–2	metric rulers	file cards paper, blank pens, marking— red, yellow		

16–3	dissecting pan
	metric ruler
	scissors
		ball, polystyrene,
		   7.5 cm diameter
		cloth strips, 2 x 8 cm
		colored pencils—
		   blue, black
		paper, tracing
		paper clips, large
		pen, ball point
		tape, clear
				preserved pig eyes

Exercise	Equipment	Expendables	Chemical Supplies	Biological Supplies
17-1	clock/timer with second hand petri dishes scissors	paper, blank sticks, plastic tape, clear towels, paper		isopods
17-2	clock/timer with second hand metric rulers scissors	file cards folders paper, tracing tape, clear		
18-1	beakers, 100 mL droppers dropper bottles graduated cylinders, 100 mL	labels wooden sticks or splints	antacid tablets Congo red solution hydrochloric acid sodium bicarbonate solution	
18-2	beakers, 50 mL granduated cylinders, 100 mL petri dishes rulers scissors spoons	cheesecloth labels paper towels twist ties	alcohol, ethyl aspirin water, tap	cigarettes ground coffee, caffeinated
19-1	microscopes, compound light scissors	colored pencils paper, white tape, clear		prepared slides— leaf, dicot
19-2	beakers, 250 mL, 500 mL forceps hot plates petri dishes scissors	aluminum foil	iodine solution alcohol, isopropyl	plants—bean, *Coleus*
20-1	scissors	colored pencils paper, blank tape, clear		
20-2	microscopes, compound light scissors	colored pencils paper, blank tape, clear		prepared slides— root, dicot
21-1	cardboard boxes clock/timer with second hand metric rulers	cups—foam, insulated soil toothpicks	plants— bean seedlings, germinated *Mimosa*, potted	
21-2	metric rulers	cups—foam, insulated soil	water, tap	plants— bean seedlings, germinated
22-1	metric rulers scissors	paper, construction—4 colors glue thread toothpicks yarn		

Exercise	Equipment	Expendables	Chemical Supplies	Biological Supplies
22–2	microscopes, compound light			prepared slides— onion root tip, L.S.
23–1	coverslips glass slides microscopes, compound light razor blades	colored pencils	water, tap	flowers, various
23–2	dissecting needles glass slides hand lens scalpels			food samples— broccoli, brussel sprouts, capers, carrots, cherry peppers, cucumbers, lettuce, lima beans, okra, onions, peas, shallots, sweet potatoes
24–1	droppers glass slides hand lens jars, small metric rulers microscopes, compound light scalpels	labels towels, paper	water, tap, aged	cultures— *Euglena*, planarian, *Dephnia*/water flea
24–3	scissors	tape, clear		
25–1	metric rulers			
25–2	metric rulers			
26–1	pennies	tape, masking		
26–2		pen, marking		prepared bags of beans
27–1	microscopes, compound light	colored pencils		prepared slides— normal and sickled red blood cells
27–2	coins	pens, ball-point tape, masking		
28–1	scissors	colored pencils		
28–2		colored pencils		
29–1	metric rulers			twigs, conifer opened bean pods seeds—bean, pinto
29–2	metric rulers	colored pencils		
30–1	metric rulers	colored pencils		
30–2		bags—small, paper colored pencils graph paper		beans—brown, white
31–1	balances containers, large graduated cylinders, 100 mL	pencils cups, paper sand, dry soil, dry towels, paper		

Exercise	Equipment	Expendables	Chemical Supplies	Biological Supplies

31–2	coverslips
	droppers
	glass slides
	microscopes,
	  compound light
		cotton, balls
		towels, paper
			mineral salt
			  solutions—
			  0.5%, 2%, 5%
				culture,
				  Paramecium

31–3	hand lens
	jars, small
	petri dishes
			cow manure
			  solution
			liquid plant food
			  solution, 0.1%
			water, distilled
				plants, duckweed

32–1	droppers
	forceps
	hand lens
	jars, small
	spoons, plastic
	test-tube racks
	test tubes
		applicator stick
		pens, marking
			acid solution
			base solution
			chemical solution
			fertilizer solution
			pH paper with
			  pH color chart
			water, tap, aged
				cultures—
				  Daphnia/water
				  flee, *Euglena*

32–2	beakers, 400 mL
	clock/timer with
	  second hand
	coverslips
	droppers
	glass slides
	hot plates
	microscopes,
	  compound light
	test-tube holder
	test-tube racks
	test tubes
		pens, marking
		toothpicks
			methylene blue
			  stain
				yeast, suspension

No other materials except pencils are needed for exercise 12–1 and 24–2.

Preparation of Solutions

Solutions used in this laboratory manual are listed in order by the number of the lab exercise in which they are used. Preparation procedures, cautions, and amounts to make are also included. You may want to plan several weeks ahead so that you will have all the solutions prepared.

Add solvents to the solutes. If a specific order of adding is needed, it will be noted. Dissolve and mix thoroughly. Never add water directly to concentrated acid. Always add the acid to some of the water to be used and then continue diluting. Because the diluting produces heat, it is advised that you add the acid slowly down a stirring rod as you gently stir.

Always use distilled water in preparation of solutions where water is used. Using tap water may give erroneous results in the tests in which the chemicals are used.

Mix solutions in a beaker or flask of greater capacity than the amount you are making. Usually a container 100 to 300 mL works well. It is better to make a little more than the exact amount needed. Students may spill or waste some solutions.

It is more economical sometimes to buy chemicals in large quantities. Most chemicals can be stored for several years. Solutions, once prepared, can be stored in large screw-cap or stoppered bottles. These containers should be cleaned with a low sudsing detergent, and rinsed well in distilled water before use.

An excellent source of free bottles of all sizes is a local hospital or nursing home. Ask them to save screw-cap prescription bottles, serum bottles, and dextrose bottles for you. Many of these are 0.5 to 1 liter capacity.

If possible, ask your principal to schedule some student laboratory assistants to help you. These should be students who have already had a course in biology. These students can be of tremendous value. They can prepare solutions, set up equipment, culture many organisms, and serve as "extra hands" during laboratory instructions.

Exercise	Solution	Preparation	Cautions
2–2	Iodine stain	Dissolve 1.5 g potassium iodide and 0.3 g iodine in 1 L water. Store in dark room or brown bottle.	Iodine is poisonous and can burn the skin.
4–1	Bromothymol blue solution	Add 0.5 g bromothymol blue to 1 L distilled water. Bromothymol blue is sparingly soluble in water. (Makes enough for one class of 30.)	
4–2	Chemical A	Purchase from a biological supply or prepare as follows: Add 1 g phenolphthalein powder to 50 mL denatured alcohol. Add 50 mL distilled water. Place in dropping bottles.	
4–2	Chemical B	Add 0.4 g sodium hydroxide to 1000 mL distilled water. Place in dropping bottles.	
4–3	Methylene blue stain	Dissolve 1.5 g methylene blue in 100 mL ethyl alcohol. Dilute by adding 10 mL solution to 90 mL of water.	Alcohol is flammable; an eye irritant. Do not ingest. If body contact occurs, flush with water.
5–1	Trager's solution A	Add 1.169 g sodium chloride; 0.84 g sodium bicarbonate; 2.943 g sodium citrate; 0.69 g sodium phosphate, monobasic; 0.745 g potassium chloride; and 0.111 g calcium chloride to 1 L distilled water.	
5–1	Trager's solution U	Add 2.164 g sodium chloride; 0.773 g sodium bicarbonate; 1.509 g sodium citrate; 1.784 g potassium phosphate, monobasic; 0.083 g calcium chloride; and 0.048 g magnesium sulfate to 1 L distilled water.	
9–1	Biuret solution	Purchase from biological supply house or prepare as follows: Add 1 L 10% potassium hydroxide solution (100 g KOH in 1 L distilled water) with 25 mL of 3% copper sulfate solution (0.75 g $CuSo_4$ in 25 mL distilled water). (Makes enough for four classes of 30.)	Caustic. KOH is extremely caustic and corrosive. Do not get on bare skin. Flush with water if on skin. Do not ingest.
9–2	Blue testing chemical	Add 13 mg 2,6-dichlorophenol-indophenol to 100 mL distilled water. Place in dropping bottles. Shelf life—1 week. Refrigerate if possible.	
9–2	Vitamin C	Dissolve 250 mg vitamin C tablet in 200 mL water.	
13–2	Salt water	Add 18 g sodium chloride (table salt) to 200 mL distilled water. (Makes enough for one class of 30.)	
13–2	Silver nitrate solution	Add 4 g silver nitrate to 250 mL distilled water. (Makes enough for four classes of 30.)	Poisonous, caustic, and irritating to skin and mucous membranes. Flush with water if on skin.
13–2	"Urine"	Add 3 drops yellow food coloring and 9 g sodium chloride to 1 L distilled water. (Makes enough for one class of 30.)	

Exercise	Solution	Preparation	Cautions
18–1	Sodium bicarbonate solution	Add ¼ tsp. baking soda to 200 mL distilled water. (Makes enough for one class of 30.)	
18–1	Congo red solution	Add 3 g Congo red dye to 100 mL distilled water. (Makes enough for four classes of 30.)	May cause severe allergic reactions.
18–1	Antacids A & B	Crush solid antacid tablet. Add to 200 mL distilled water. Use different commercially available varieties for A and B. (Makes enough for one class of 30.)	
18–1	Hydrochloric acid solution	Add 35 mL concentrated hydrochloric acid to 500 mL distilled water. Place in dropping bottles. (Makes enough for four classes of 30.)	Warning: HCL is irritant and corrosive. Vapor may be harmful or irritating. Avoid spilling or contacting skin, eyes, or clothes. Flush with water.
19–2	Iodine solution	Dissolve 1 g iodine crystals and 2 g potassium iodide in 300 mL water.	Iodine is poisonous and can burn the skin.
31–2	Salt solution—0.5%	Add 0.5 g sodium chloride (table salt) to 100 mL water. (Makes enough for one class of 30.)	
31–2	Salt solution—2%	Add 2 g sodium chloride to 100 mL water. (Makes enough for one class of 30.)	
31–2	Salt solution—5%	Add 5 g sodium chloride to 100 mL water. (Makes enough for one class of 30.)	
31–3	Cow-manure solution	Add 50 mL dry manure to 1 L of distilled water. Drain off and use liquid. Discard sediment. Dry cow-manure is available in garden section of hardware store.	
31–3	Liquid plant food solution—0.1%	Add 20 drops of liquid plant food to 1 liter of distilled water. Stir to mix well. (Makes enough for one class of 30.)	
32–1	Fertilizer solution—1%	Add 20 drops of liquid plant food to 100 mL of distilled water. Stir to mix well. (Makes enough for one class of 30.)	
32–1	Chemical solution—1% copper sulfate	Add 1 g of copper sulfate to 99 mL of distilled water. Stir to dissolve. (Enough for one class of 30.)	
32–1	Acid solution—1% Hydrochloric acid	Add 1 mL of acid to 99 mL of distilled water. (Enough for one class of 30.)	Warning: HCL is irritant and corrosive. Vapor may be harmful or irritating. Avoid spilling or contacting skin, eyes, or clothes. Flush with water.
32–1	Base solution—1% sodium bicarbonate	Add 1 g of sodium bicarbonate to 99 mL of distilled water. Stir to dissolve. (Makes enough for one class of 30.)	
32–2	Methylene blue stain—0.001%	Add 1.0 g of methylene blue to 99 mL of distilled water. Dilute by adding 1 mL of this solution to 99mL of distilled water. Label 0.001% methylene blue.	
32–2	Yeast suspension	Add 1/4 yeast cake to 1 liter physiological saline solution. To prepare physiological saline solution, add 0.95 g NaCl to 99.5 mL distilled water.	

Making Slides (Temporary Mounts)

Temporary mounts can be made inexpensively in the lab and the slides can be used for several years. Before slides are made, make sure slides and coverslips are washed in hot soapy water, rinsed in warm water, and dried with a paper towel. Do not use plastic slides or coverslips.

To make a temporary mount, put the specimen part directly on the slide. Add one drop of polyvinyl acetate (PVA) to the slide. Place the coverslip on the PVA at an angle in order to avoid air bubbles. Gently lower the coverslip.

PVA is made by placing 5 g of PVA in 20 mL of 80% ethanol. Allow this material to set for 24 hours. Stir the material lightly with a stirring rod. Avoid putting air bubbles in the solution. PVA is a clear solution that can be stored in a small dropping bottle. Keep the bottle covered to prevent evaporation. PVA can be ordered from Shawnigan Resins Corp., Springfield, MA.

Prepared slides are needed for the following exercises in this laboratory manual. You may wish to make some of these slides as temporary mounts. Students may help you make them.

Slide	Exercise	Slide	Exercise
insect leg	1–1	root, dicot	20–2
frog blood	2–2	onion root tip	22–2
diatomaceous earth	5–1	normal blood cells	27–1
leaf	19–1	sickled blood cells	27–1

Maintaining Living Materials

Use of living materials has been kept to a minimum in this laboratory manual. However, you will need the following living things for certain exercises.

Animal or Protist	Exercise	Plant	Exercise
bacterial culture	4–2	various plants	19–2
termites	5–1	bean plants	21–1, 21–2
isopods	17–1	duckweed	31–3
planarian	24–1	Other living animals and protists (and some of those listed here) can be obtained from a biological supply house; other living plants can be obtained from a plant or grocery store or can be collected. It should not be necessary to maintain them in the lab.	
water flea/*Daphnia*	24–1, 32–1		
Paramecium	31–2		
Euglena	24–1, 32–1		

You may be able to collect the living things within your area or obtain them easily from a plant or pet store. However, if you have to order living organisms during the summer for the next school year, you may have to culture or maintain some living things throughout the school year. The living organisms used in this laboratory manual can be grown by you and student assistants for use in the classroom. Methods and materials listed here have been tried and proven by the authors. Easy techniques that take little time have been listed.

Isopods (Exercise 17–1). Isopods are easy to maintain. Punch air holes in the lid of a plastic container. Place several isopods in the container with a piece of raw potato for food.

Water Fleas/*Daphnia* (Exercises 24–1, 32–1. Water fleas are best grown in a four-liter jar or small aquarium. Fill the container with pasteurized spring or pond water. Tap water can be used if it is allowed to set for two to three days before use. Add several dozen water fleas and 500 mL of *Euglena* culture to the container. More Euglena can be added as the water fleas eat the Euglena.

***Paramecium* (Exercise 31–2).** Culture media for growing protozoa should be spring water or Chalkey's solution. A good source of spring water is a grocery or drugstore. If spring water is not available, use Chalkey's solution.

Chalkey's solution is a mineral salt solution that resembles the minerals found in pond water. Add the following chemicals to a liter of distilled water: 1 g sodium chloride, 0.04 g potassium chloride and 0.06 g calcium chloride. Stir the chemicals in the water until dissolved and label 10X Chalkey's solution. To use for protozoa cultures, dilute the Chalkey's solution by adding 10 mL of 10X Chalkey's solution to 90 mL of distilled water.

All glassware used in culturing protozoa should be cleaned in a low-sudsing soap. After washing, rinse the glassware in distilled water two times. Store dishes on a shelf lined with newspaper.

To start a new protozoan culture, fill a culture dish two thirds full of pasteurized spring or pond water, or Chalkey's solution. Add a dropperful of protozoa culture to the culture dish. For every 100 mL of culture liquid, add 2 grains of previously boiled split peas or rice. These are food for the bacteria, and the bacteria are food for the protozoa. Place *Paramecium* in an area where it gets light. Add more rice and split peas at two to three week intervals. Make new cultures from old cultures at least every four weeks.

***Euglena* (Exercise 32–1).** See precious section, *Paramecium*, for general information for culturing protozoa.

Various Plants (Exercises 19–2, 21–1, 21–2). Bean plants, *Coleus*, *Geranium*, and *Tradescantia* are just a few of the plants that can be grown easily in the lab. *Coleus*, *Geranium*, and *Tradescantia* can be successfully started from cuttings. Bean plants are easily germinated from seeds. Cuttings and seeds can be started in equal mixtures of sand and moistened topsoil. Empty milk cartons or plastic sandwich bags can be used as small pots and placed in well-lighted areas. As the plants become larger, they should be carefully transplanted into larger containers.

Many plants require different amounts of light, water, and minerals. You should be familiar with the requirements of each type of plant in the lab. Flowering plants can be induced to flower in the classroom with the proper conditions.

Duckweed (Exercise 31–3). Place several duckweed plants in a four-liter jar that is nearly two-thirds full of water. For each 100 mL of water in the jar, add 1 mL of 1% fertilizer solution. (To make 1% fertilizer solution, add 20 drops of liquid plant food to 100 mL of distilled water and stir.) Within a week, the population will double. Subculture duckweed when it covers the surface of the water.

Setting Up Classroom Aquaria and Terraria

Aquaria and terraria can provide an easy way to keep certain animals in the laboratory. Properly set up and maintained, they will furnish you with living organisms for study as well as for creating and/or stimulating student interest.

Aquaria. To redo a functioning aquarium or set one up from "scratch," certain procedures should be followed. A filled aquarium can be drained by siphoning. Immerse a piece of rubber tubing (about 1 meter long) in a container of water. When the tubing is full, close both ends with your fingers. Place one end into the full aquarium and the other end into a sink that is lower than the aquarium. Remove your fingers and water will flow from the aquarium into the sink.

Once an aquarium is empty, it should be washed with soapy water and rinsed thoroughly. Any trace of soap left can kill the organisms you put into the aquarium. When you are sure the aquarium is well rinsed, place it on a flat surface where it will not have to be moved again. Select a well-lighted area. If the aquarium is put in a dark area, an aquarium light (fluorescent) will have to be used. Do not place the aquarium near radiators or other heat sources.

You can check your aquarium for leaks after filling it with water. Small leaks can be sealed with an aquarium sealant. Aquarium sealants are available in hardware and pet stores. Leaky aquaria that cannot be sealed can be used for terraria. Do not attempt to move the aquarium after it is full of water. You may distort the frame and loosen the glass which will cause leaks.

Conditioned tap water can be used to fill the aquarium. Chlorine in tap water will kill living things. Therefore, allow the water to set out for two days before its use to "condition" it. A dechlorinating chemical can be purchased in pet stores to add to tap water if you want to add living things to the water sooner. Be sure to follow the directions on the chemical for its use.

Aquarium gravel for the bottom should be washed with hot running water until the water runs clear. If an undergravel filter is going to be used in the aquarium, the filter should be put in the aquarium before the gravel is added. The bottom of the aquarium should be covered with about 4 cm of gravel. If other larger rocks are being added, you may want to scrub them and rinse thoroughly to remove slime from previous use. Be careful what kinds of rocks you add; soluble rocks such as limestone are not appropriate.

A variety of aquatic plants may then be added to the aquarium. It is generally best to put rooted plants such as *Vallisneria* and *Sagitarria* near the back corners of the tank. *Vallisneria* should be spaced at least 5 cm apart to allow room for runners. *Elodea* can be put in the front or center or allowed to float. Anchor plants by pushing the roots into the gravel. Be careful not to heap the gravel above the roots. Duckweed can be floated on the top. Use plants sparingly. Too few plants are better than too many. Overstocking may cause all of them to die.

Goldfish and guppies are two common fish used in classroom aquaria. If other fish are used, make sure they are good in a community tank. Someone in your pet store can give you advice on what fish get along well with each other. Adding one 3-cm fish per gallon of water works well. Fish should be introduced to the tank slowly so that they can adjust to the temperature, other fish, and so on, more easily. Add some of the aquarium water to the fish first. Gradually add more, and finally add the fish to the aquarium. This process should take several hours. The aquarium water and the water from which the fish comes should be the same temperature. An aquarium heater can be used to maintain warm water for the fish. Be sure to find out the optimal temperature for the fish you are keeping. For example, guppies should be kept at 24°C to 27°C, goldfish at 20°C to 24°C. Fish should be fed when added to the aquarium. In this way they are distracted from each other temporarily.

If snails are added, add only one per gallon of water. They breed rapidly. If snails become too numerous, some should be removed.

For best results, an aquarium filter and aerator should be used. The charcoal for the filter should be rinsed thoroughly before use. Directions for setting up a filter are included with the filter.

A properly set up and balanced aquarium will need little cleaning. It is not advisable to replace all the water at one time. About one fourth of the water can be siphoned off and replaced with conditioned water periodically.

The aquarium should be covered to reduce evaporation and keep dust out. Use aquarium sealant to glue at least six pieces of cork to the top of the aquarium. A piece of glass set on the corks serves as a good lid. If one corner of the glass is cut away, feeding the fish is more convenient.

Guppies and goldfish eat a variety of food, both plant and animal. Give the fish a variety of foods by mixing different brands or giving a different one each day. Live food such as *Daphnia* (water fleas) can be fed occasionally. You should feed the fish a small amount twice daily, but one daily feeding will do. Avoid overfeeding. The snail population will increase and the water will become cloudy and polluted if overfeeding occurs.

An aquarium requires little daily care. Feed the fish, add conditioned water to replace evaporated water, remove excess plants and snails, and remove dead or diseased animals and plants. If a green scum appears or the water turns green, introduce *Daphnia* or an algae-eating fish to feed on the algae. Also, reduce the amount of light. Samples of the green water can be removed from the tank for certain laboratory activities.

Fish need dissolved oxygen to breathe. If the fish stay near the top of the water there is probably not enough oxygen in the tank. Low oxygen can result from too many animals in the tank or no aerator. Therefore, reducing the number of animals and using an aerator helps. If enough live plants are present and the aquarium is in the light, an aerator may not be needed.

Terraria. Any glass container with a glass plate cover can be used as a terrarium. The container should be at least four liters in size. If animals are to be kept in the terrarium, it should be 60 to 80 liter capacity. An old aquarium makes an excellent terrarium.

Wash and rinse the container thoroughly. Place 2 to 3 cm of washed gravel on the bottom of the container. Add 2 cm clean sand. Then add about 6 cm of moist soil. On top of the soil, place a layer of healthy, green moss. Then add small ferns, liverworts, and horsetails. Lichens also can be added. Cover the terrarium and place it in filtered light. The gravel layer allows for drainage. If water accumulates in the pebble layer, remove the cover of the terrarium for a time.

You may wish to set up a terrarium for plants and animals that live in a dry environment. Cacti and chameleons can live in a terrarium with dry conditions. Add a few branches or rocks for chameleons to climb. Place a thin wire cover on the terrarium. Keep the terrarium in a sunlit room but avoid direct sunlight. Feed the chameleons twice weekly with a varying diet of fruit flies, flies, and mealworms. Sprinkle a few drops of water on a branch or on the side of the container for the chameleons to drink.

Living Things in the Laboratory

The importance of living things in the lab should not be minimized. Only through working with and caring for live specimens can students truly develop a knowledge and respect for different kinds of life. Recognizing and filling the needs of living things is important. It is also important that certain procedures and precautions are followed.

Successfully growing different types of plants can help students expand their knowledge of living things as well as develop their responsibility and respect for life. However, plants that are used in the lab should be well known to the teacher. There are over 700 species of plants known to cause death or illness. The teacher should be familiar with the most common poisonous plants found in the local area.

Live animals in the lab are always a source of fascination for students. Before animals are brought into the lab, however, it is important that the policies of the local school district are checked concerning this issue. The overriding concern over live animals in the lab is that neither the students nor the animals will be harmed.

Many times, students wish to bring wild animals such as birds, raccoons, squirrels, turtles, and insects to class. For reasons of student safety as well as for proper caring for the animals, it is advised that you use thoughtful discretion in allowing wild animals to be brought in. Also it is recommended that you check state and local laws before animals are brought into the classroom. Any animals taken from a natural habitat should be returned after observation.

Students should develop a respect for life by applying humane principles in the educational use of living organisms. If animals are brought in, there should be adequate facilities and designated people to care for the animals. Students should learn the responsibilities of providing food, space, fresh water, and adequate ventilation for the animals. No activities should be conducted that will cause pain, hardship, or death to any of the animals.

SI Conversions

Scientists usually use the SI system of measurement. SI is a modern version of the older metric system. Some SI units are the same as metric units. SI comes from the French *Systeme Internationale*.

The laboratory exercises in this book use the SI system. If you do not have all the necessary lab equipment to accommodate the SI system, you may find it necessary to use some English equivalents. However, the SI system is used in scientific measurements and students should be discouraged from using or converting to the English system.

The charts on the next page will help you determine the English equivalents for many SI measurements if their use is needed.

Mass*

Ounces	Grams
¼ oz	7.09 g
½ oz	14.17 g
¾ oz	21.26 g
1 oz	28.35 g
3.527 oz	100 g
176.35 oz	500 g

Pounds	Kilograms
½ lb	0.227 kg
¾ lb	0.34 kg
1 lb	0.454 kg
2.2 lb	1 kg

Length

1 inch	2.54 cm
1 foot	30.48 cm
1 yard	91.44 cm
50 feet	15.24 m
100 feet	30.48 m
1 mile	1.609 km

*Equivalent measurements of mass and weight are given. Mass, however, is not the same as weight.

Volume

Spoonfuls	Milliliter
½ tsp	2.5 mL
¾ tsp	3.75 mL
1 tsp	5 mL
½ tbls	7.5 mL
¾ tbls	11.25 mL
1 tbls	15 mL

Cups	Milliliters
¼ c	59 mL
⅓ c	79 mL
½ c	118 mL
⅔ c	157 mL
¾ c	177 mL
1 c	236 mL

Ounces	Milliliters
¼ oz	7.5 mL
½ oz	15 mL
¾ oz	22.5 mL
1 oz	30 mL

Others	Milliliters
½ pint	236 mL
1 pint	473 mL
1 qt	946.3 mL
1 gal	3785 mL

Temperature

°F	°C
0	−17.78
10	−12.22
20	−6.67
30	−1.11
40	4.44
50	10

°F	°C
60	15.56
70	21.11
80	26.67
90	32.22
98.6	37
100	37.78
212	100

Suppliers

Cambosco Scientific Company, Inc., 342 Western Ave., Brighton Station, Boston, MA 02135

Carolina Biological Supply Company, Burlington, NC 27215

Central Scientific Company, 2600 S. Kostner Ave., Chicago, IL 60623

Difco Laboratories, Inc., 920 Henry, Detroit, MI 48201

Edmund Scientific Co., 101 E. Gloucester Pike, Barrington, NJ 08007

Learning Things, Inc., Littleton, MA 01460

Macmillan Science Co., Inc., 8200 S. Hoyne Ave., Chicago, IL 60620

Parco Scientific Co., P.O. Box 595, Vienna, OH 44473

Science Kit, Inc., 777 East Park Drive, Tonawanda, NY 14150

Ward's Natural Science Establishment, Inc., P.O. Box 1712, Rochester, NY 14603

Readings for the Teacher

Allen, Dorthea, *Biology Teacher's Desk Book*. West Nyack, Parker Publishing Co., 1979.

Allen, Robert D. and Michael B. Moll, "A Realistic Approach to Teaching Mendelian Genetics," *American Biology Teacher**, April, 1986, 227.

Bardell, David, "The Biological Nature of AIDS Virus," *American Biology Teacher*, February, 1986, 75.

Barman, C. R. and others, *Science and Societal Issues: A Guide for Science Teachers*. Science Activity Fund, Price Laboratory School, University of Northern Iowa, 1979.

Behringer, M. P., *Techniques and Materials in Biology*. New York, McGraw-Hill, 1981.

Biology Teacher's Handbook. New York, John Wiley and Sons, Inc., 1978.

Browning, R. F., "Custom Undergravel Filters Made From Scraps," *American Biology Teacher*, Vol. 42, (December 1980), 549.

Cobb, V., *Science Experiments You Can Eat*. New York, Lippincott, 1972.

———, *More Science Experiments You Can Eat*. New York, Lippincott, 1979.

Collette, Alfred T. and Eugene L. Chiappetta, *Science Instruction in the Middle and Secondary Schools*. St. Louis, Missouri, Times Mirror/Mosby College Publishing, 1984.

Corcos, A. F., "Tongue Rolling in the Classroom," *American Biology Teacher*, Vol. 42, (May 1980), 311.

Glasenapp, Douglas J., "The Nuts and Bolts of Classification," *American Biology Teacher*, September, 1986, 362.

Good, Ron and Mike Smith, "How Do We Make Students Better Problem Solvers?" *The Science Teacher*, Vol. 54, (April 1987), 31.

Hagquist, C. W., "Preparation and Care of Microscope Slides," *American Biology Teacher*, Vol. 36, (October 1974), 414.

Hickman, F. M. and J. B. Kahle, *New Directions In Biology Teaching*. Reston, Virginia: National Association of Biology Teachers, 1982.

Hounshell, P. B. and I. R. Trollinger, *Games for the Classroom*. Washington, DC, National Science Teachers Association, 1977.

Hummer, P. J., "Fish Ponds—Sources of Organisms for the Laboratory," *The Science Teacher*, Vol. 41, (May 1974), 55.

Hummer, P. J., "Life Science Culture Center—Teacher Designed and Student Maintained," *The Science Teacher*, Vol. 39, (September 1972), 43.

Jackson, J. R. and L. P. Thomas, "Building a Terraquarium," *American Biology Teacher*, Vol. 42, (December 1980), 551.

Lawson, Anton E. and William Hegebush, "A Survey of Casual Hypothesis Testing Strategies: K-12," *American Biology Teacher*, September, 1985, 348.

Lennox, J. E. and A. R Mikula, "The Sickled Cell—A Simulation Game," *American Biology Teacher*, Vol. 39, (April 1977), 238.

Lennox, John E., "Wavelength and Resolution in Microscopy, An Explanation by Analogy," *American Biology Teacher*, November/December, 1986, 487.

Lewis, Ralph W., "Teaching the Theories of Evolution," *American Biology Teacher*, September, 1986, 344.

Lunetta, V., A. Hofstein, and G. Giddings, "Evaluating Science Laboratory Skills," *The Science Teacher*, Vol. 43, No. 1 (1981), 22.

Madrazo, G. M. and C. A. Wood, "Playing the Cell Game," *American Biology Teacher*, Vol. 42, (December 1980), 554.

Morholt, E. P. et al, *A Sourcebook for the Biological Sciences*. New York, Harcourt Brace Jovanovich, 1966.

"NABT Guidelines for the Use of Animals at the Pre-university Level," *American Biology Teacher*, Vol. 42, No. 7 (1980), 426.

Novak, J. D., "Applying Psychology and Philosophy to the Improvement of Laboratory Teaching," *American Biology Teacher*, Vol. 41, No. 8 (1979), 466.

Orlands, F. B., *Animal Care from Protozoa to Small Mammals*. Menlo Park, CA, Addison-Wesley Pub. Co., 1977.

Rakow, Steven J. and Thomas C. Gee, "Test Science, Not Reading," *The Science Teacher*, Vol. 54, (February 1987), 28.

Safety First in Science Teaching. Raleigh, NC, North Carolina Department of Public Instruction, Division of Science, 1977.

Safety in the Secondary Science Classroom. Washington, DC, National Science Teachers Association, 1978.

Sestini, V. A., "Biology Culture Bank—A Growing Investment," *The Science Teacher*, Vol. 48, No. 2 (1981), 23.

Shepherd, D. L., *Comprehensive High School Reading*. Columbus, OH, Charles E. Merrill Pub. Co., 1982.

Stewart, J., J. Van Kirk, and R. Rowell, "Concept Maps: A Tool for Use in Biology Teaching," *American Biology Teacher*, Vol. 41, No. 3 (1979), 171.

Sund, R. B. and L. Trowbridge, *Teaching Science by Inquiry in the Secondary School*. Columbus, OH, Charles E. Merrill Pub. Co., 1973.

*Most issues of *American Biology Teacher* have good suggestions for laboratory work.

LABORATORY MANUAL
STUDENT EDITION

MERRILL
BIOLOGY
AN EVERYDAY EXPERIENCE

Authors
Albert Kaskel
Paul Hummer, Jr.
Lucy Daniel

GLENCOE
Macmillan/McGraw-Hill

New York, New York
Columbus, Ohio
Mission Hills, California
Peoria, Illinois

A MERRILL BIOLOGY PROGRAM

Biology: An Everyday Experience, *Student Edition*
Biology: An Everyday Experience, *Teacher Edition*
Biology: An Everyday Experience, *Teacher Resource Package*
Biology: An Everyday Experience, *Study Guide*
Biology: An Everyday Experience, *Transparency Package*
Biology: An Everyday Experience, *Laboratory Manual, Student Edition*
Biology: An Everyday Experience, *Laboratory Manual, Teacher Edition*
Biology: An Everyday Experience, *Computer Test Bank*
Biology: An Everyday Experience, *Cooperative Learning*
Biology: An Everyday Experience, *Spanish Resources*
Biology: An Everyday Experience, *Videodisc Correlations*
Biology: An Everyday Experience, *Lesson Plans*

Copyright © by the Glencoe Division of Macmillan/McGraw-Hill Publishing Company, previously copyrighted by Merrill Publishing Company.

All rights reserved. Printed in the United States of America. Except as permitted under the United States Copyrights Act of 1976, no part of this publication may be reproduced or distributed in any form or by any means, or stored in a database or retrieval system, without prior written permission of the publisher.

Send all inquiries to:
Glencoe Publishing Company
936 Eastwind Drive
Westerville, OH 43081

ISBN: 0-02-827305-2

Printed in the United States of America.

2 3 4 5 6 7 8 9 10 POH 00 99 98 97 96

To the Teacher

Biology: Laboratory Experiences presents readable, workable laboratory exercises that make working in the laboratory an enjoyable experience for students and teachers. Students will learn how to do the investigations, answer the questions, and also learn how to question what they do. Some of the main features are listed here.

- 70 exercises are included, most of which require simple, inexpensive equipment.
- Performance objectives follow the introduction to each exercise.
- Procedural steps are short, simple, and numbered.
- Many exercises use models and simulations.
- Few labs use living materials.
- Few labs carry overnight; most can be completed in one class period.
- Many illustrations are used. Some are to be used in the exercise; others show procedural steps.
- Three aspects of scientific literacy are developed: science vocabulary, familiarity with the scientific method, and working with graphs and tables.
- Each exercise includes a section to review scientific and general vocabulary used in the exercise.
- A full glossary at the end of the book gives students a place to find words that may be unfamiliar to them.

The 70 exercises in this laboratory manual are divided evenly among the chapters of *Biology: An Everyday Experience;* there are either two or three exercises for every text chapter. This feature provides full coverage of basic biology in the laboratory.

Biology: Laboratory Experiences is a completely self-contained laboratory manual. All observations, data, and answers are to be recorded in the spaces supplied in the manual. No extra paper or notebook is needed. These features were carefully designed to facilitate individualized learning.

To the Student

Biology is an everyday occurrence all around you. One way to learn more about biology around you is to experience biology in the laboratory. *Biology: Laboratory Experiences* is your tool to do just that.

You will get to experience biology in the laboratory in many different ways this school year. Some of the exercises in this book will show you how to organize data. In some exercises, you will take your own data. Sometimes you will work with living things. Some exercises will require you to build models to represent living things. In other exercises, you will work with chemicals. Let's look at some of the features of this book.

Each of the exercises in this book has these seven parts.

1. **Title**—The title of each exercise is a question. You will see from the title what you will investigate in the exercise. The number of the exercise is in the title. The number tells

you which chapter of *Biology: An Everyday Experience* has information to which the exercise relates. For example, Exercises 24–1 and 24–2 relate to information in Chapter 24 of the textbook.

2. **Introduction**—The introduction is in a box on the first page of each exercise. The paragraphs tell you about the laboratory exercise and give you background information.
3. **Objectives**—The objectives list just what you are expected to do during the exercise.
4. **Keywords**—This section helps you review scientific words as well as general words used in the exercise. You will find the vocabulary section a handy way to tell if you understand the words used in the exercise.
5. **Materials**—The materials section lists the supplies you will need to do the exercise.
6. **Procedure**—Procedure steps are numbered. They are simple, easy-to-understand steps telling you what to do. There also are many diagrams to show you what to do. Sometimes tables are included in this section.
7. **Questions**—Different kinds of questions covering the exercise are in this section. Some ask you to recall parts of the exercise. Some ask you to go a step further in analyzing what you did in the exercise.

In addition to the exercises, this laboratory manual has two other main features. It has a complete **glossary**. The glossary tells you how to pronounce words and what the words mean. The words in the glossary are the science words and other terms used in this book with which you may not be familiar. Get in the habit of looking up the words you do not understand.

Inside the back cover of the laboratory manual is information on **safety and first aid**. Safety is very important in the laboratory. Read the safety information before you begin. Many of the exercises have caution statements. Additionally, safety codes that appear next to the materials list show what precautions you must take when doing a lab. Familiarize yourself with what the safety codes stand for by studying the chart on page viii. Always be sure to note and follow directions to avoid injury to yourself and/or others.

Working in the laboratory will be a rewarding experience for you this year. Using this book, you will learn how to *question* and how to *answer questions*. Biology can become much more understandable.

Table of Contents

1-1	How Is the Light Microscope Used?..	1
1-2	How Are SI Length Measurements Made?..	5
2-1	What Are Diffusion and Osmosis?...	9
2-2	What Cell Parts Can You See With the Microscope?.......................	13
3-1	How Can Paper Objects Be Grouped?...	17
3-2	How Can Living Things Be Grouped?...	21
4-1	How Can You Test for Bacteria in Foods?.......................................	25
4-2	Does Soil Contain Bacteria?..	27
4-3	What Are the Traits of Blue-green Bacteria?...................................	31
5-1	Where Are Protists Found?..	33
5-2	What Do Molds Need in Order to Grow?..	37
6-1	What Are the Traits of Vascular and Nonvascular Plants?..............	41
6-2	How Can Conifers Be Identified?..	45
7-1	What Traits Does a Sponge Have for Living in Water?...................	49
7-2	What Are the Parts of the Squid?..	53
7-3	What Are the Parts of a Stinging-cell Animal?...............................	57
8-1	What Are the Traits of Certain Jointed-leg Animals?.....................	59
8-2	How Do Fish Classes Compare?...	63
8-3	What Are the Parts of a Starfish?..	67
9-1	What Are the Tests for Fats and Proteins?......................................	71
9-2	How Do You Test for Vitamin C in Foods?.....................................	75
10-1	How Do Digestive System Lengths Compare?...............................	79
10-2	How Does the Fish Digestive System Work?..................................	83
11-1	How Does the Heart Work?..	87
11-2	How Do Hearts of Different Animals Compare?............................	91
12-1	How Can Blood Diseases Be Identified?..	95
12-2	What Blood Types Can Be Mixed?...	99

13-1	How Does Breathing Occur?	103
13-2	What Chemical Wastes Are Removed by the Lungs, Skin, and Kidneys?	107
14-1	What Causes Sports Injuries?	111
14-2	How Do Male and Female Skeletons Differ?	115
15-1	Which Brain Side Is Dominant?	119
15-2	What Are the Functions of the Brain?	123
16-1	How Can You Test Your Senses?	127
16-2	How Are Your Senses Sometimes Fooled?	131
16-3	What Are the Parts of an Eye?	135
17-1	What Self-protecting Behaviors Do Isopods Show?	141
17-2	What Happens When You Learn Through Trial and Error?	145
18-1	Which Antacid Works Best?	149
18-2	How Do Drugs Affect the Ability of Seeds to Grow into Young Plants?	153
19-1	What Do the Inside Parts of Leaves Look Like?	157
19-2	Where in a Leaf Does Photosynthesis Take Place?	161
20-1	What Does a Woody Stem Look Like Inside?	165
20-2	What Do the Inside Parts of a Root Look Like?	169
21-1	What Tropisms Can Be Seen in Growing Plants?	173
21-2	Is Light an Important Growth Requirement for Plants?	177
22-1	What Happens When Cells Divide?	181
22-2	Are There More Dividing Cells or Resting Cells in a Root Tip?	185
23-1	What Are the Parts of a Flower?	189
23-2	What Plant Part Are You Eating?	193
24-1	How Do Some Animals Reproduce Asexually?	197
24-2	How Do Internal and External Reproduction Compare?	201
24-3	What Are the Stages in the Menstrual Cycle?	205
25-1	How Does a Human Fetus Change During Development?	211
25-2	What Changes Occur During Birth?	215
26-1	How Can the Genes of Offspring Be Predicted?	219
26-2	What Is a Test Cross?	223
27-1	What Do Normal and Sickled Cells Look Like?	227
27-2	How Are Traits on Sex Chromosomes Inherited?	231

28-1	How Does DNA Make Protein?	**235**
28-2	How Can a Mutation in DNA Affect an Organism?	**239**
29-1	How Do Some Living Things Vary?	**243**
29-2	How Do fossils Show Change?	**247**
30-1	What Are Some Parts of a Food Chain and a Food Web?	**251**
30-2	How Do Predator and Prey Populations Change?	**255**
31-1	How Much Water Will Soil Hold?	**259**
31-2	How Can a Nonliving Part of an Ecosystem Harm Living Things?	**263**
31-3	How Can a Nonliving Part of an Ecosystem Help Living Things?	**265**
32-1	How Do Chemical Pollutants Affect Living Things?	**267**
32-2	How Does Thermal Pollution Affect Living Things?	**271**
Glossary		**275**

Safety Symbols

The safety symbols shown here appear next to the materials lists in this laboratory manual. They also appear on the activities pages in your textbook. Many of the labs have one or more of these symbols. Each symbol represents the kind of caution you must take when doing a lab. Read the entire chart before doing any of the lab exercises. Each time you start a new lab, refer to the chart again and read the information for symbols that appear in the lab. For more information about safety and first aid in the laboratory, see the back cover of this laboratory manual.

	DISPOSAL ALERT This symbol appears when care must be taken to dispose of materials properly.		**ANIMAL SAFETY** This symbol appears whenever live animals are studied and the safety of the animals and the student must be ensured.
	ELECTRICAL ALERT This symbol appears when an electrical appliance should not be used.		**RADIOACTIVE SAFETY** This symbol appears when radioactive materials are used.
	OPEN FLAME ALERT This symbol appears when use of an open flame could cause a fire or an explosion.		**CLOTHING PROTECTION SAFETY** This symbol appears when substances used could stain or burn clothing.
	THERMAL SAFETY This symbol appears as a reminder to use caution when handling hot objects.		**FIRE SAFETY** This symbol appears when care should be taken around open flames.
	SHARP OBJECT SAFETY This symbol appears when a danger of cuts or puncture caused by the use of sharp objects exists.		**EXPLOSION SAFETY** This symbol appears when the misuse of chemicals could cause an explosion.
	FUME SAFETY This symbol appears when chemicals or chemical reactions could cause dangerous fumes.		**EYE SAFETY** This symbol appears when a danger to the eyes exists. Safety goggles should be worn when this symbol appears.
	ELECTRICAL SAFETY This symbol appears when care should be taken when using electrical equipment.		**POISON SAFETY** This symbol appears when poisonous substances are used.
	PLANT SAFETY This symbol appears when poisonous plants or plants with thorns are handled.		**CHEMICAL SAFETY** This symbol appears when chemicals used can cause burns or are poisonous if absorbed through the skin.

Name _____ Class _____ Period _____

1–1 How Is the Light Microscope Used?

A microscope is a tool used to look at very small things. "Micro" means small and "scope" means to look at. The microscopes that you will use in class have two or more lenses. A lens is a curved piece of glass. The lenses inside your microscope make the objects you look at appear larger. They are located in the eyepiece and in the objectives.

You may wonder how much larger your microscope can make something look. The magnifying power of a microscope is how many times larger a microscope makes something look. The eyepiece of your microscope probably makes things look ten times larger. If so, it has 10× written on it. Each objective lens also has a power written on it. To find the magnification for your microscope, multiply the eyepiece power by the power of the objective lens you are using.

EXPLORATION

OBJECTIVES
In this exercise, you will:
 a. learn the names and jobs of microscope parts.
 b. learn how to use and care for the microscope.
 c. determine the magnification of your microscope.

KEYWORDS
Define the following keywords:

compound light microscope _____

field of view _____

lens _____

stage _____

MATERIALS
light microscope   prepared slide of
lens paper   insect leg

PROCEDURE
1. The microscope should always be handled with care. Use one hand to hold the arm. Place the other hand under the base. Move the microscope to your table and gently set it down. (The arm should be toward you.)
2. Use of the microscope is easy if you know the parts. Find the parts listed in Table 1 on page 2 on your microscope.

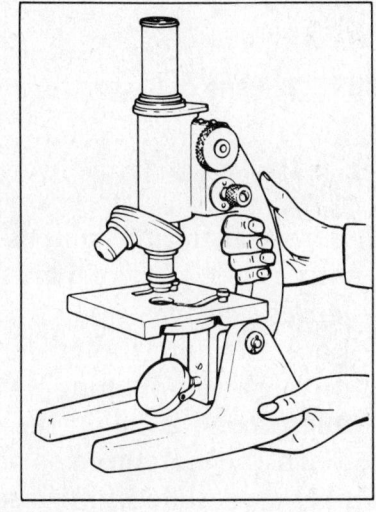

FIGURE 1. Carrying a microscope

Table 1. Microscope Parts and Their Jobs

Part	Name	Job
A	Eyepiece	Holds top lens, usually 10×
B	Body tube	Holds top lens certain distance from lower lenses
C	Arm	Supports body tube
D	Nosepiece	Holds lower lenses, turns to change objectives
E	High power objective	Contains 43× lens
F	Low power objective	Contains 10× lens
G	Coarse adjustment	Moves body tube up and down, brings objects into focus
H	Fine adjustment	Moves body tube up and down slightly, brings objects into focus
I	Stage	Supports slide
J	Stage clips	Holds slide in place
K	Diaphragm	Controls amount of light entering microscope
L	Light or mirror	Sends light through microscope
M	Base	Supports microscope

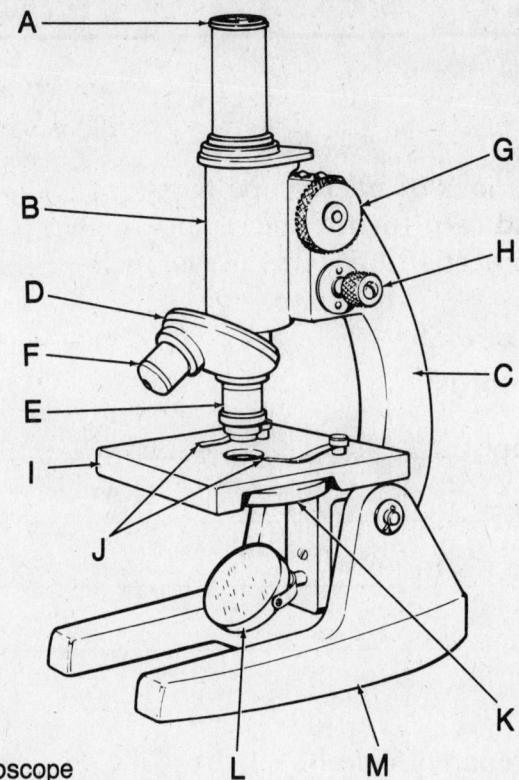

FIGURE 2. Parts of the microscope

3. Before using the microscope, make sure the lenses are clean. Use lens paper *only*. Any other kind of paper may scratch the lenses. Wipe the eyepiece and objective lenses gently.
4. Look through the eyepiece. Turn the diaphragm so that the most light comes through the opening in the stage. The circle of light that you see through the microscope is called the field of view.
5. Turn the nosepiece so that the low power (10×) objective is in place. Put a prepared slide of an insect leg on the stage under the clips. A prepared slide is a slide made to last a long time. Keep the slide clean by holding it by the edges.

Name _____ Class _____ Period _____

6. Always find an object first on low power. Move the slide until the leg is directly over the hole in the stage. Then use the coarse adjustment knob to make what you see clear. Look to the side of your microscope when turning the coarse adjustment to keep from hitting the slide with the objective. Turn the coarse adjustment slowly. When the object is clear, we say it is in focus.

7. Move the slide to the left. Which way does the leg move as you look through the microscope?

   _____

8. Move the slide away from you. Which way does the leg move as you look through the microscope? _____

FIGURE 3. Using coarse adjustment

9. Draw the insect leg in the circle in Figure 4 as it appears under low power. Then turn the nosepiece carefully until the high power objective clicks into place. Bring the object into focus by turning *only* the fine adjustment. Observe and draw the leg in the circle in Figure 4 as it appears under high power.

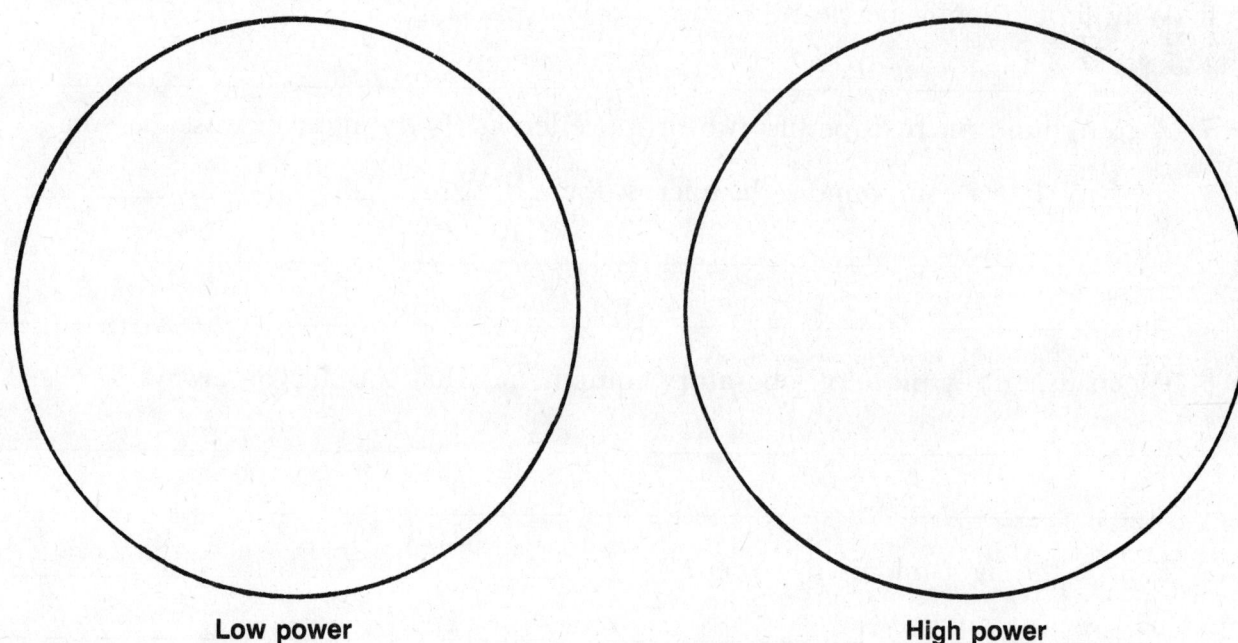

Low power **High power**

FIGURE 4.

10. Switch back to low power. Remove the slide and put it away. Answer the questions on the next page. Then put your microscope away.

3

QUESTIONS

1. Fill in the chart below to show the total magnification of your microscope on low and high power.

	Eyepiece magnification	× Objective magnification	= Total magnification
Low power			
High power			

2. How does the leg look under high power that differs from how it looks under low power? _____

3. When you moved the slide to the right, which way did the insect leg move? _____

4. Is the field of view brighter or dimmer under high power? _____

5. How should you carry a microscope? _____

6. Why should lenses be cleaned only with lens paper? _____

7. A compound microscope has two or more lenses. Is the light microscope you used in class a compound light microscope? _____ Explain. _____

8. When using any piece of laboratory equipment, what should you always do? _____

9. How is the light microscope used? _____

Name _____ Class _____ Period _____

1-2 How Are SI Length Measurements Made?

Often measurements are made to learn more about biological problems. The International System of Units, or SI, is a system of measurements you will become more familiar with this year. The measurements you will make in this exercise are SI measurements.

In this exercise, you will make length measurements. The basic unit of length is the meter. The meter is divided into one hundred smaller units called centimeters. Smaller measurements are made with millimeters. Ten millimeters equal one centimeter.

When measurements are made, you should write them down. Data are observations you record—in this case, the measurements you write down. The data will be written in a table to help you keep them organized.

INTERPRETATION

OBJECTIVES
In this exercise, you will:
  a. compare hand and foot bones.
  b. record your data in tables and draw conclusions.

KEYWORDS
Define the following keywords:

data _____

length _____

meter _____

SI measurements _____

MATERIALS
metric ruler

PROCEDURE
1. Look at the diagram of the hand in Figure 1 on page 6. Count the number of bones present in the thumb, fingers, palm, and wrist. (They are shaded in different ways in the diagram to help you.) Record your counts in Table 1.

Table 1. Bone Counts

Part		Part	
Thumb		Big toe	
Fingers		Other toes	
Palm of hand		Center of foot	
Wrist		Ankle and heel	

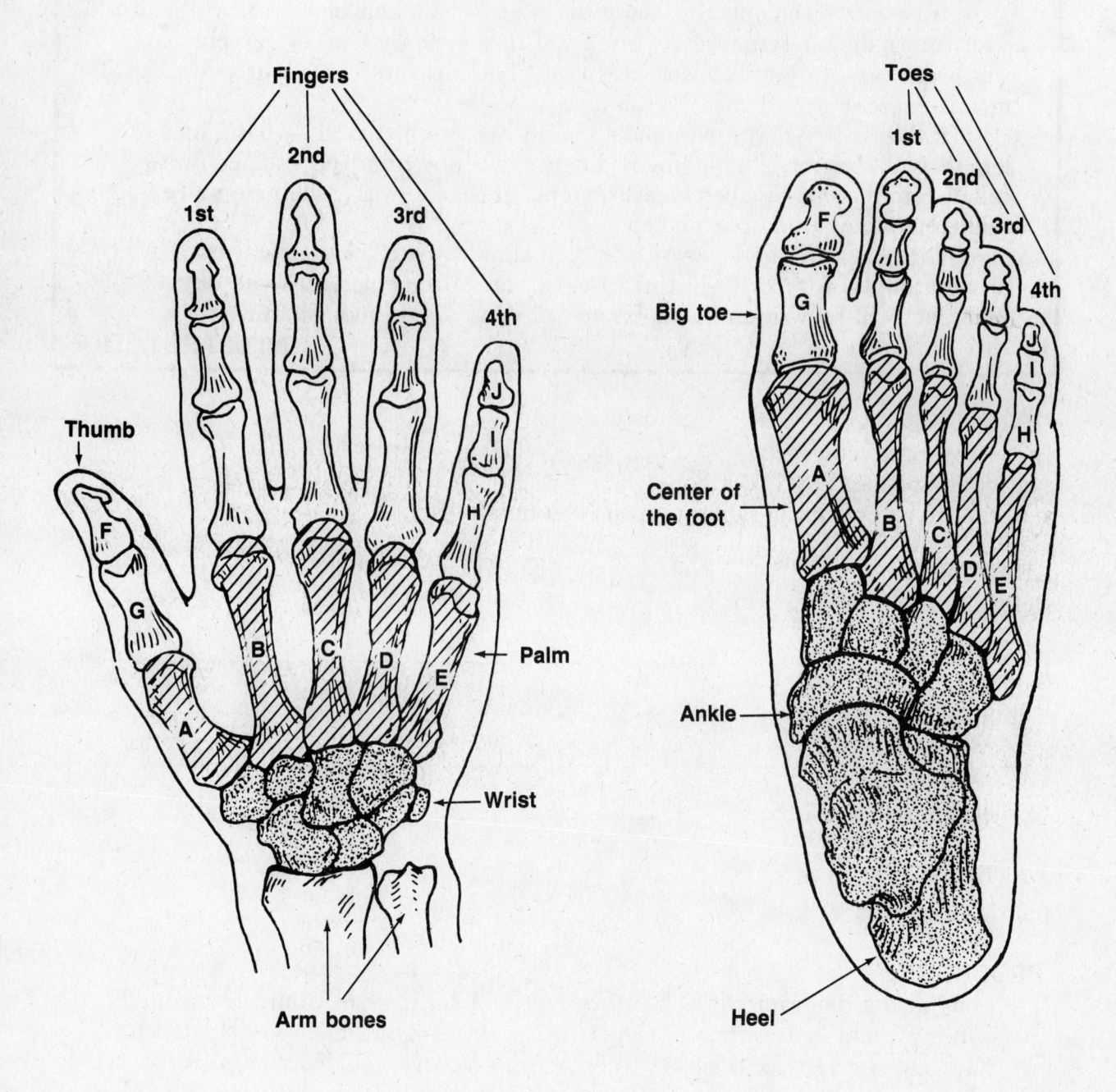

FIGURE 1.

Name _____ Class _____ Period _____

2. Measure in millimeters the lengths of the bones marked A, B, C, D, and E in the hand diagram. Record your measurements in Table 2.

3. Measure in millimeters the lengths of the bones marked A, B, C, D and E in the foot diagram. Record your measurements in Table 2.

4. Measure the length of the thumb (F + G) and record the number in the table.

5. Measure the length of the big toe (F + G) and record it in the table.

6. Measure the lengths of the smallest finger and toe (H + I + J). Record these data in the table.

7. Change all the millimeter measurements to centimeter measurements in the table. Recall that there are ten millimeters in one centimeter.

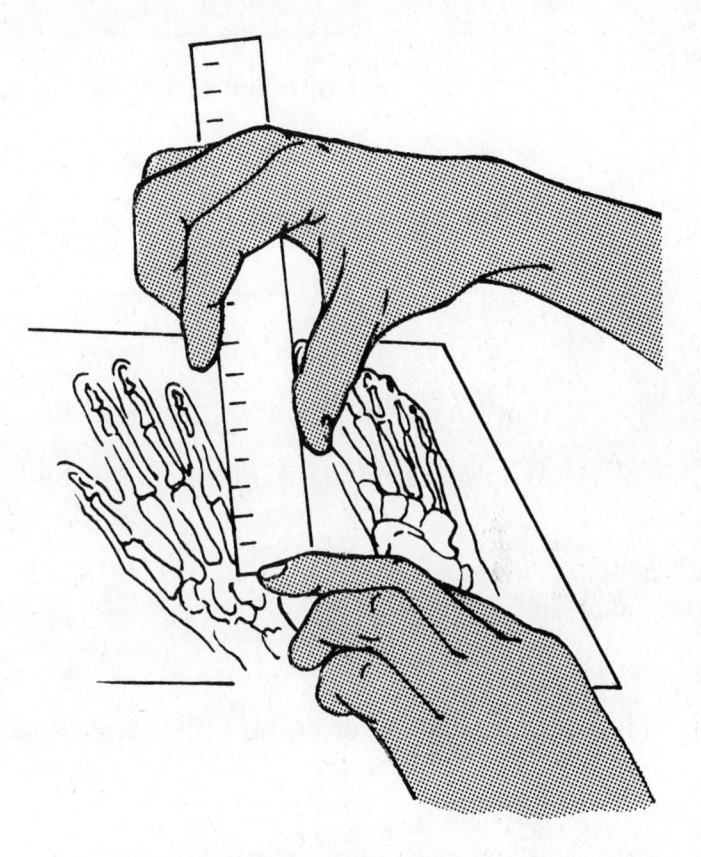

FIGURE 2. Measuring bone length

Table 2. Bone Lengths

Bone	Hand		Foot	
	Millimeters	Centimeters	Millimeters	Centimeters
Bone A				
Bone B				
Bone C				
Bone D				
Bone E				
Thumb or big toe bones (F + G)				
Smallest finger or toe bones (H + I + J)				

QUESTIONS

1. What is the total number of bones:

    a. in the hand? _____ b. in the foot? _____

2. How do the total number of bones in the hand and foot compare? _____
   _____

3. What is the total number of bones in the:

    a. palm? _____ b. center of foot? _____

4. How do the total number of bones in the palm and foot center compare? _____
   _____

5. What is the total length of:

    a. bone A in the hand? _____ b. bone A in the foot? _____

6. How much longer is bone A in the foot than bone A in the hand? _____
   _____

7. What is the total length of the:

    a. little finger? _____ b. little toe? _____

8. How much longer is the little finger than the little toe? _____
   _____

9. Describe the main differences between the lengths of the bones in the hand and the foot. _____
   _____
   _____

10. Why are data often kept in tables? _____
    _____

11. Suppose you were working in a department store. What unit of measurement (meter, centimeter, millimeter) would you use to measure the length and width of shoes and window curtains? _____
    _____

12. How are SI length measurements made? _____
    _____

Name _____ Class _____ Period _____

2–1 What Are Diffusion and Osmosis?

Cells have an outer covering called the cell membrane. The cell membrane controls what moves into and out of cells. Food and oxygen move into cells through the cell membrane. They move by diffusion. The movement of a substance from where there is a large amount to where there is a small amount is called diffusion.

The movement of water across the cell membrane is called osmosis. Osmosis is a special kind of diffusion.

INVESTIGATION

OBJECTIVES
In this exercise, you will:
   a. observe diffusion of food coloring in water.
   b. observe osmosis across the membrane of an egg.
   c. measure the amount of water that moves across the egg membrane.

KEYWORDS
Define the following keywords:

cell membrane _____

diffusion_____

_____

milliliter (mL) _____

osmosis _____

MATERIALS

beaker, 400 mL
water
dropper
food coloring
3 glass jars with lids, 500 mL

colored pencils: blue, red, and green
wax pencil
graduated cylinder, 100 mL
200 mL white vinegar
200 mL clear syrup
raw egg in shell

PROCEDURE
Part A. Diffusion
1. Add water to the beaker until it is three-fourths full. Let the beaker stand until the water is very still.
2. Carefully drop one drop of food coloring to the surface of the water (Figure 1). Observe what happens.

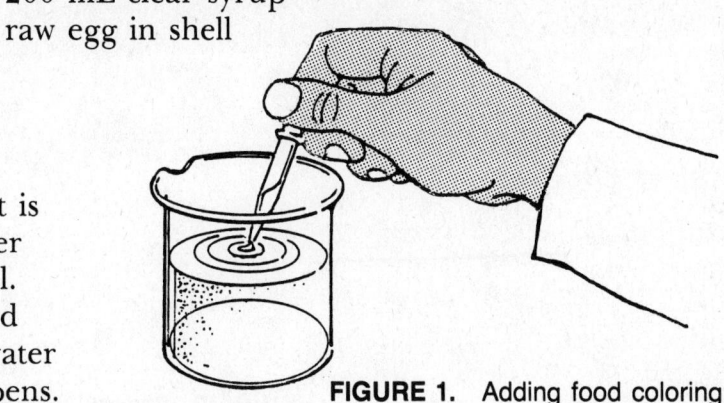

FIGURE 1. Adding food coloring

9

3. Draw what happens in the beakers in Figure 2.

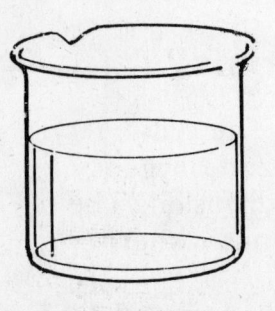

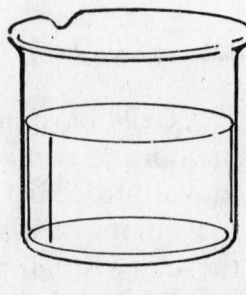

Early   Middle (after 10 minutes)   Late (after 1 hour)

FIGURE 2. Diffusion in beakers

4. Empty the water from the beakers into the sink.

Part B. Osmosis

1. Label the glass jars with the wax pencil. Write *vinegar* on one jar, *syrup* on the second jar, and *water* on the third.
2. Look at the raw egg. Note the appearance of the eggshell. Use your pencil and trace the egg in the space provided.
3. Use the graduated cylinder to measure 200 mL of vinegar. Put the vinegar in the vinegar jar. Place the raw egg into the vinegar jar. The vinegar should cover the egg. Cover the jar with the lid. Leave it undisturbed for two days (Figure 3).
4. After two days, observe what has happened. Write your results in Table 1.

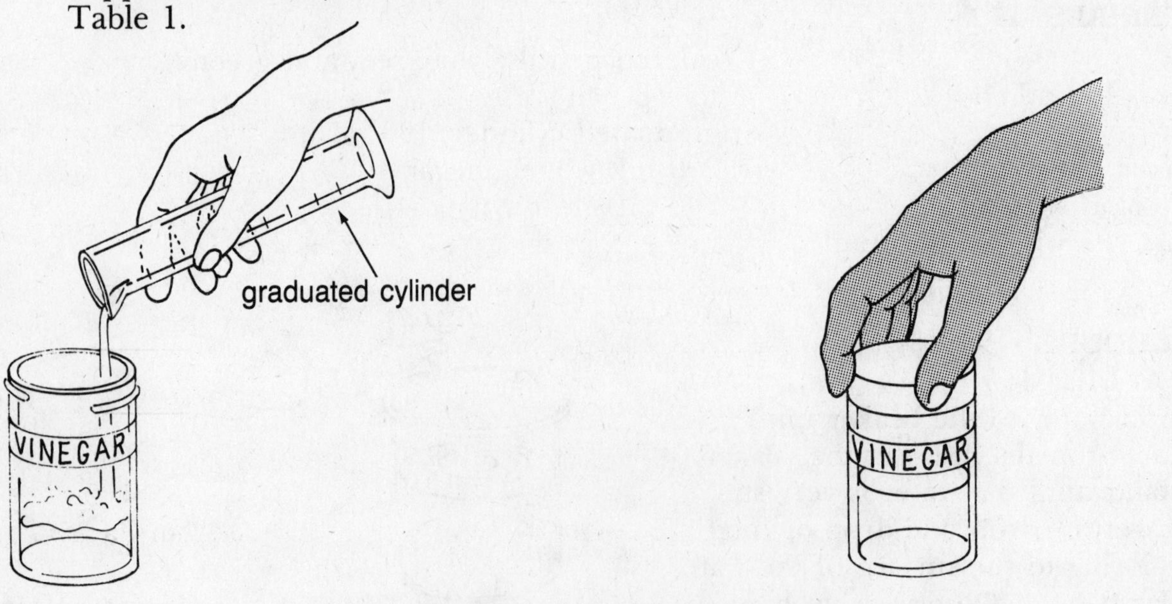

FIGURE 3. Adding vinegar and egg

10

Name _____ Class _____ Period _____

5. Put 200 mL syrup in the syrup jar. Place the egg in the syrup jar. Leave it for one day.
6. Measure the amount of vinegar remaining in the vinegar jar. Write the amount in the table. Then empty the vinegar into the sink.
7. The next day measure 200 mL of water and add it to the water jar.
8. Carefully remove the egg from the syrup jar. Make observations and record them. Remove the egg from the syrup and carefully rinse it with water. Use the red pencil and trace the egg over the second drawing of the egg. Place the egg in the water jar.
9. Measure the amount of syrup that remains in the syrup jar. Write the amount in the table. Empty the syrup into the sink.
10. After one or two days, remove the egg from the water. Use the green pencil and trace the egg over the third drawing of the egg. Also record your observations of the egg. Measure the amount of water that is left in the water jar. Write the amount in the table.

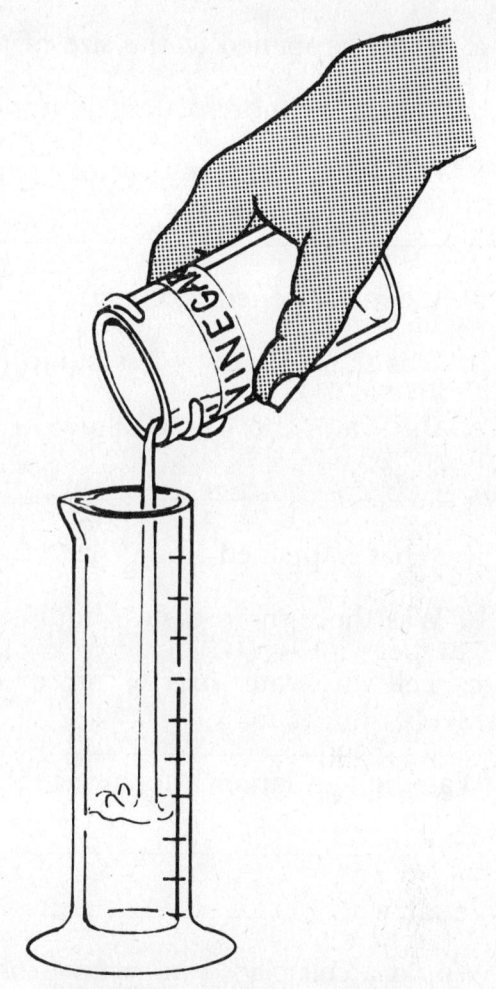

FIGURE 4. Measuring vinegar

Table 1. Results of Osmosis

Jar	Amount present when egg was put in	Amount present when egg was removed	Observations
Vinegar			
Syrup			
Water			

QUESTIONS
1. What did you observe when food coloring was dropped on the water in Part A?
_____

2. What is this process called? _____

3. In Part B, what happened to the shell of the egg? _____

4. Vinegar is made of acetic acid and water. Which part of the vinegar dissolved the shell? _____

5. a. What happened to the size of the egg after remaining in vinegar? _____

   b. Was there more or less liquid left in the jar? _____

   c. Did water move into or out of the egg? How do you know? _____
   _____

6. a. What happened to the size of the egg after remaining in syrup? _____

   b. Was there more or less liquid left in the jar? _____

   c. Did water move into or out of the egg? How do you know? _____
   _____

7. a. What happened to the size of the egg after remaining in water? _____

   b. Was there more or less liquid left in the jar? _____

   c. Tell why water moved into or out of the egg. _____
   _____

8. Was the egg larger after remaining in water or after remaining in vinegar? ____
   _____
   _____

9. Would a cell placed in syrup probably lose or gain water? _____
   Why? _____

10. Why are fresh fruits and vegetables sprinkled with water at a market? _____
    _____

11. Roads are sometimes salted to melt ice. What does this salting do to the plants along the roadside? _____
    Why? _____

12. Why do dried fruits and dried beans swell when they are cooked? _____
    _____
    _____

Name _____ Class _____ Period _____

2-2 What Cell Parts Can You See With the Microscope?

Living things are made of cells. All cells have parts that do certain jobs. Cells have an outer covering called the cell membrane. Cell membranes give cells their shapes and control what enters and leaves the cells. The clear, jellylike material inside the cell is the cytoplasm. The nucleus is the control center of the cell. Plant cells have a thick outer covering called the cell wall. It is on the outside of the cell membrane.

Cell parts can be studied by making wet mount slides. A wet mount slide is a temporary slide. It is not made to last a long time. You can make wet mount slides of living and once-living materials to study cell parts.

EXPLORATION

OBJECTIVES
In this exercise, you will:
  a. make wet mount slides for examination under the microscope.
  b. study four cell parts—the cell wall, cytoplasm, nucleus, and cell membrane.

KEYWORDS
Define the following keywords:

cell wall _____

cytoplasm _____

nucleus _____

wet mount _____

MATERIALS
glass slide   dropper   cork shaving
coverslip   forceps   bamboo shaving
light microscope   iodine stain   onion skin
water   elodea leaf   prepared slide of frog blood

PROCEDURE
Part A. Making a Wet Mount
1. Get a clean glass slide and coverslip. Handle the slide and coverslip by the edges to keep them clean (Figure 1).

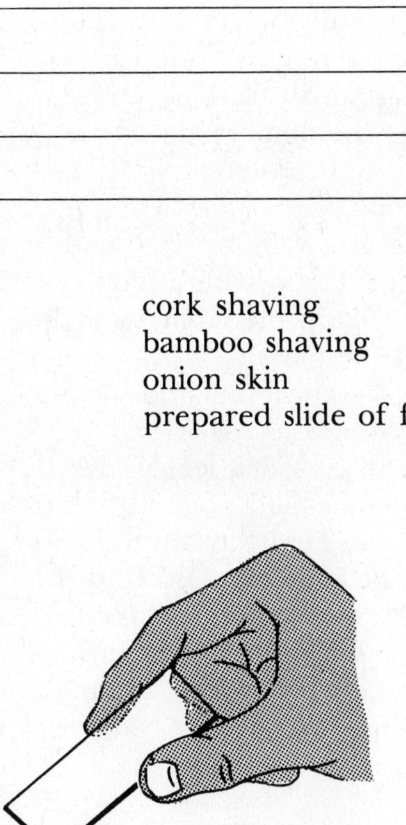

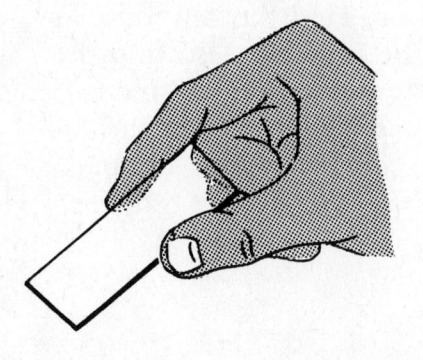

FIGURE 1. Holding slide by edges

13

2. Use a dropper to put a drop of water in the center of the slide.
3. With forceps, place the object to be examined in the drop of water.
4. Hold the coverslip at an angle. Gently lower it onto the drop of water (Figure 2).

Part B. Looking at Cells
1. Prepare a wet mount of the cork shaving. Follow the steps given in Part A.
2. Examine the slide of cork under low power of the microscope. Switch to high power. Draw the cells you see in the circle in Figure 3. Label the cell wall.
3. Prepare a wet mount of a bamboo shaving. Examine the bamboo under low and then high power of your microscope. Draw several dead bamboo cells in the circle in Figure 3. Label the cell wall.
4. Peel the thin layer of cells from the inside of an onion as shown in Figure 4. Make a wet mount of the onion skin cells. Add one drop of stain in place of water.
5. Examine the onion slide under low and high power of your microscope.
6. Find the cell wall, nucleus, and cytoplasm. Draw living onion cells that you see in the circle in Figure 5. Label the parts.
7. Prepare a wet mount of an elodea leaf.
8. Examine the elodea leaf under low and then high power of your microscope. Find the cell wall. Try to find the nucleus. Also find the parts that make the leaf green.
9. Draw some elodea cells in the circle in Figure 5. Label the parts you see.

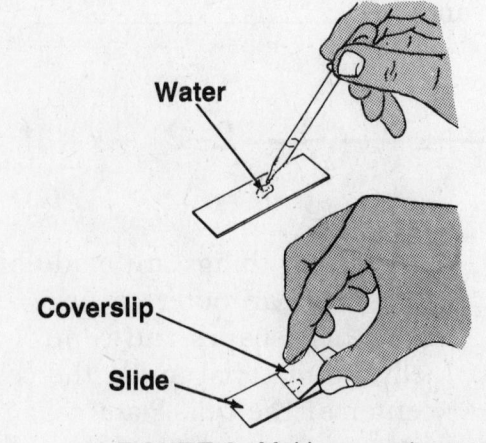

FIGURE 2. Making a wet mount

Cork cells

Bamboo cells

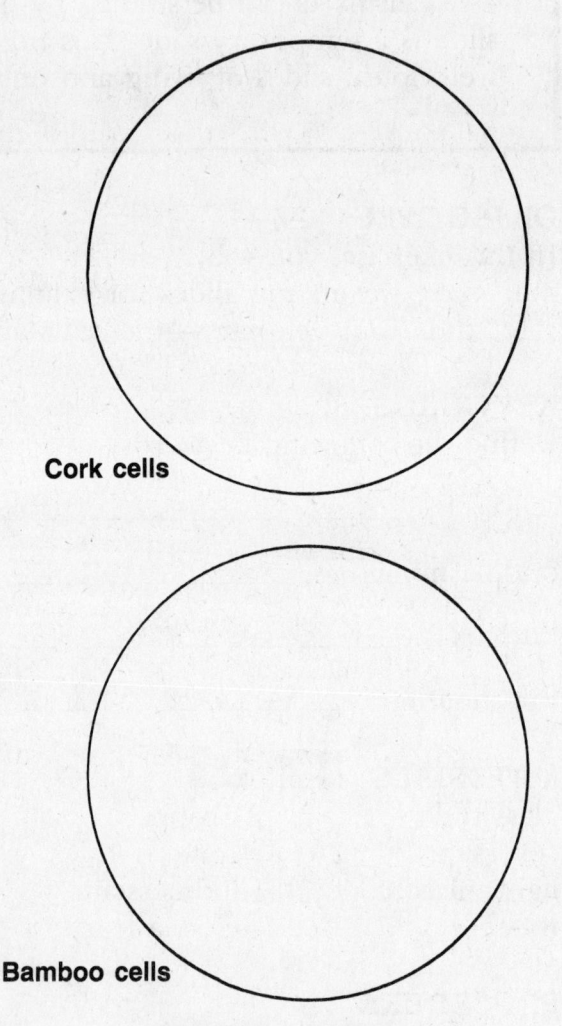

FIGURE 3. Looking at cells

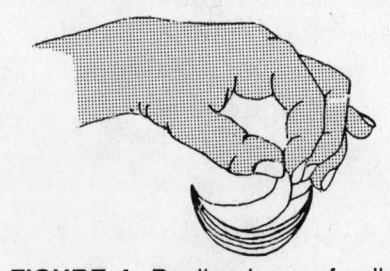

FIGURE 4. Peeling layer of cells

Name _____ Class _____ Period _____

○ ○ ○

Onion skin cells   Elodea cells   Frog blood cells

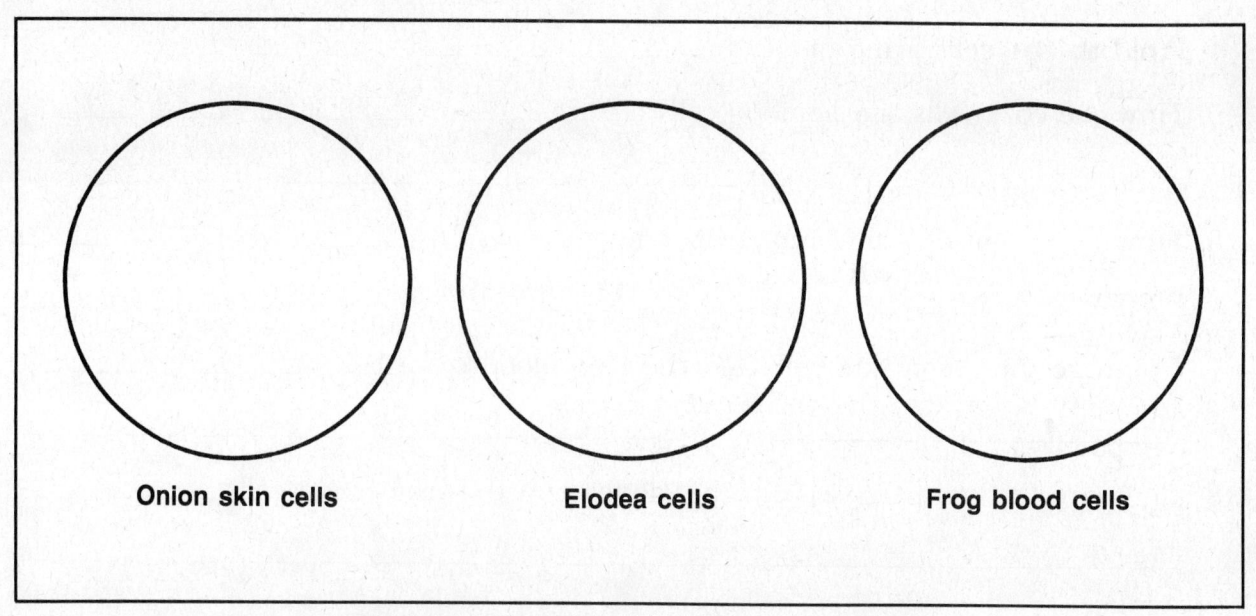

FIGURE 5. Examining cells under the microscope

10. Examine a prepared slide of frog blood with low and then high power. In the circle in Figure 5, draw some frog blood cells. Label the nucleus, cytoplasm, and cell membrane.
11. Complete Table 1.

Table 1. Parts of Cells

Cell type	Cell wall present? (yes or no)	Nucleus present? (yes or no)	Cytoplasm present? (yes or no)	Shape of cell?	Cell living or dead?
Cork					
Bamboo					
Onion					
Frog blood					
Elodea					

QUESTIONS
1. What is the name of the small units that make up cork? _____
2. Are the cork cells filled with living material or are they empty? _____

3. Describe how the small units of cork look. _____
   _____

4. Are bamboo cells living or dead? _____

5. How are cork cells and bamboo cells alike? _____
   _____

6. How are onion cells different from the cork cells? _____
   _____

7. Compare the onion skin cells and the frog blood cells. _____
   _____

8. Compare the cell parts seen in the elodea and the frog blood cells. ___
   _____
   _____

9. a. Are onion cells plant or animal? _____
   b. Are elodea cells plant or animal? _____
   c. Are frog blood cells plant or animal? _____

10. Why do cells have different shapes? _____
    _____

11. Skin cells seem to fit together or overlap. How is this cell arrangement helpful to the organism? _____

12. If blood cells were box-shaped, like onion cells, why would they be unable to do their job as well? _____
    _____

Name _____ Class _____ Period _____

3-1 How Can Paper Objects Be Grouped?

You group, or classify, many things every day. The phone book, the library, and the grocery store are all classified to make finding things easier. In any classification, things with similar traits are grouped together.

Scientists divide living things into groups. The main groups are called kingdoms. Kingdoms are divided into phyla. Phyla are divided into classes. The other groups are order, family, genus, and species. The more similar two living things are, the more groups they share.

INTERPRETATION

OBJECTIVES
In this exercise, you will:
  a. group paper objects.
  b. use the words *kingdom,* *phylum,* and *class* in your classification.
  c. decide what traits were used in the classification.

KEYWORDS
Define the following keywords:

class _____

classify _____

kingdom _____

phylum _____

trait _____

MATERIALS
scissors

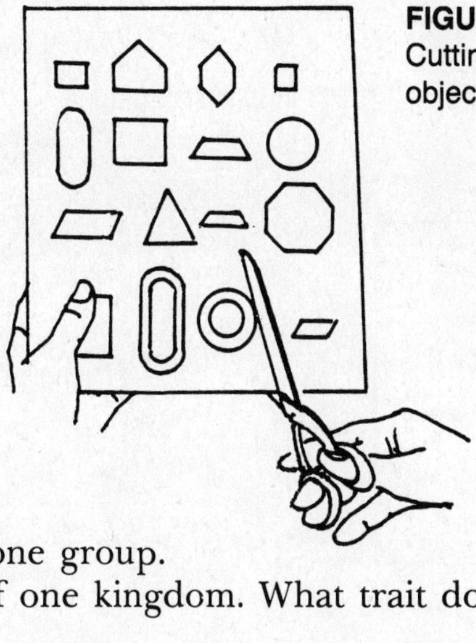

FIGURE 1. Cutting out objects

PROCEDURE
1. Get a copy of the paper objects in Figure 2 on the next page from your teacher.
2. Cut out the objects as shown in Figure 1. **CAUTION:** *Use extreme care with the scissors.*
3. Place the objects on your desk. Divide them into two groups as follows:
   a. Put objects 1, 4, 6, 7, 9, 11, 13 and 16 into one group.
      These will represent the classification level of one kingdom. What trait do all of these objects have in common? _____

17

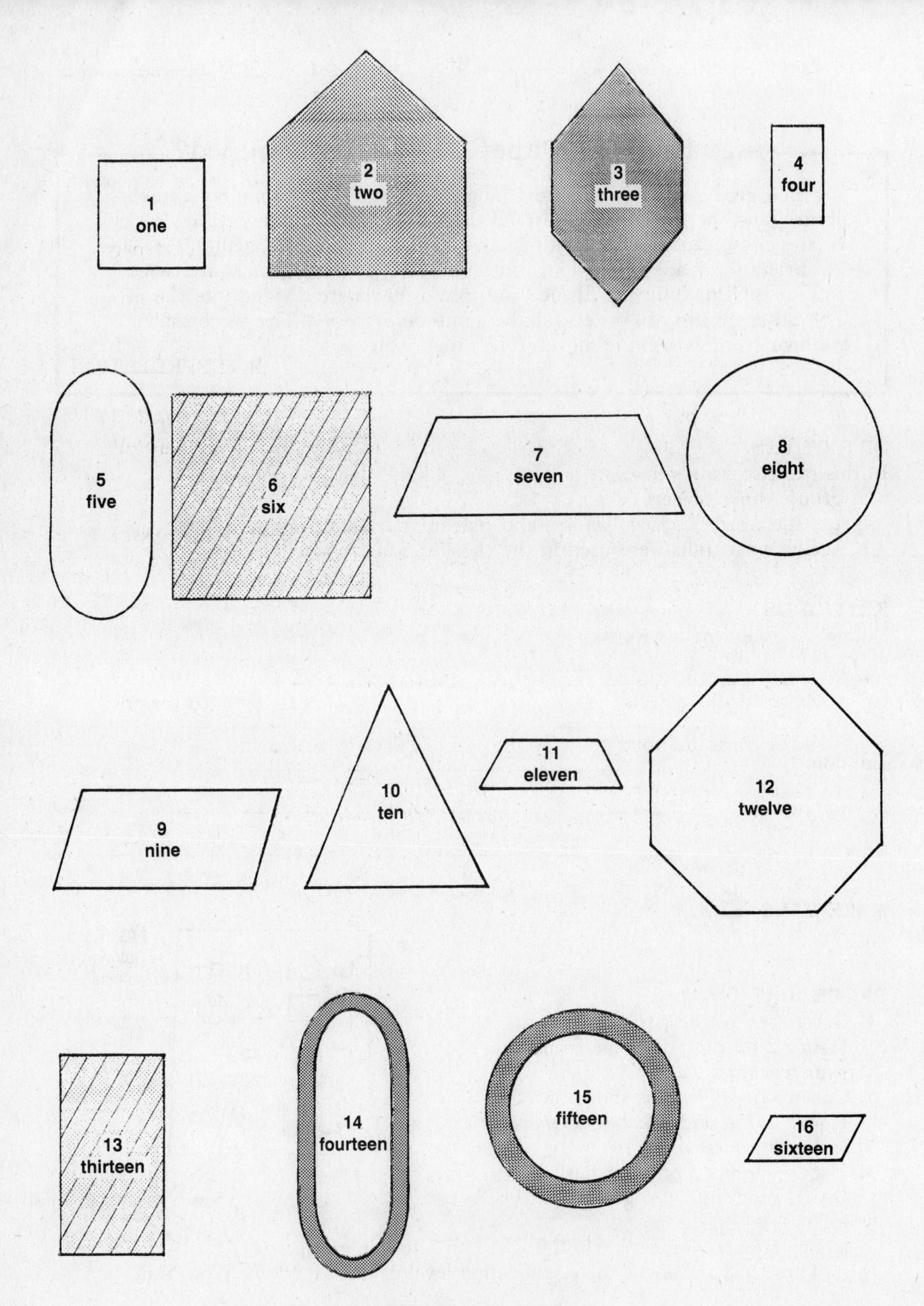

FIGURE 2. Paper objects

Name _____ Class _____ Period _____

    b. Put objects 2, 3, 5, 8, 10, 12, 14, and 15 into a second group.
These will represent a second kingdom. What trait do all these objects have in

common? _____
    c. Write a good kingdom name for each group in Chart 1 on the next page.
4. a. Divide the objects from the first kingdom only in this way. Put objects 1, 4, 11, and 16 into one group. Put objects 6, 7, 9, and 13 into a second group. This represents the next classification level called Phylum.

    What trait was used to separate the two groups? _____
    b. Write a good phylum name for each group in Chart 1.
5. a. Divide the objects from the second kingdom in this way. Put objects 2, 3,, 10, and 12 into one group. Put objects 5, 8, 14, and 15 into a second group.

    What trait did you use to separate the objects into two groups? _____
    b. Write a good phylum name for each group in Chart 1.
6. a. Use objects 1, 4, 11, and 16. Separate them into two groups as follows. Put objects 1 and 4 into one group and objects 11 and 16 into a second group. This represents the next classification level called Class.
What trait did you use to separate the objects into two

groups? _____
    b. Write the class level name for these groups in Chart 1.
7. a. Separate out objects 2, 3, 10, and 12. Then group them into two groups.

    What traits did you use to group them as you did? _____
    b. Write the objects' classification level and group names in Chart 1.
8. a. Repeat Step 7 for objects 6, 7, 9, and 13. Complete Chart 1.
    b. Repeat Step 7 for objects 5, 8, 14, and 15. Complete Chart 1.

QUESTIONS
1. Underline the correct word choice in each of these sentences.

    a. Objects in the same class also belong to the same (phylum, genus, family).

    b. Objects in the same phylum also belong to the same (family, order, kingdom.

2. List the classification levels in order from largest to smallest. _____

_____

3. Classifications are not all alike. Suppose objects 14 and 15 were put in one

group. What trait do they have in common?_____
4. Write a name below for each of these groups.

    Objects 1 and 4 _____

    Objects 2 and 3 _____

    Objects 10 and 12 _____

    Objects 8 and 15 _____

5. What are used to determine if living things belong to a particular group? _____

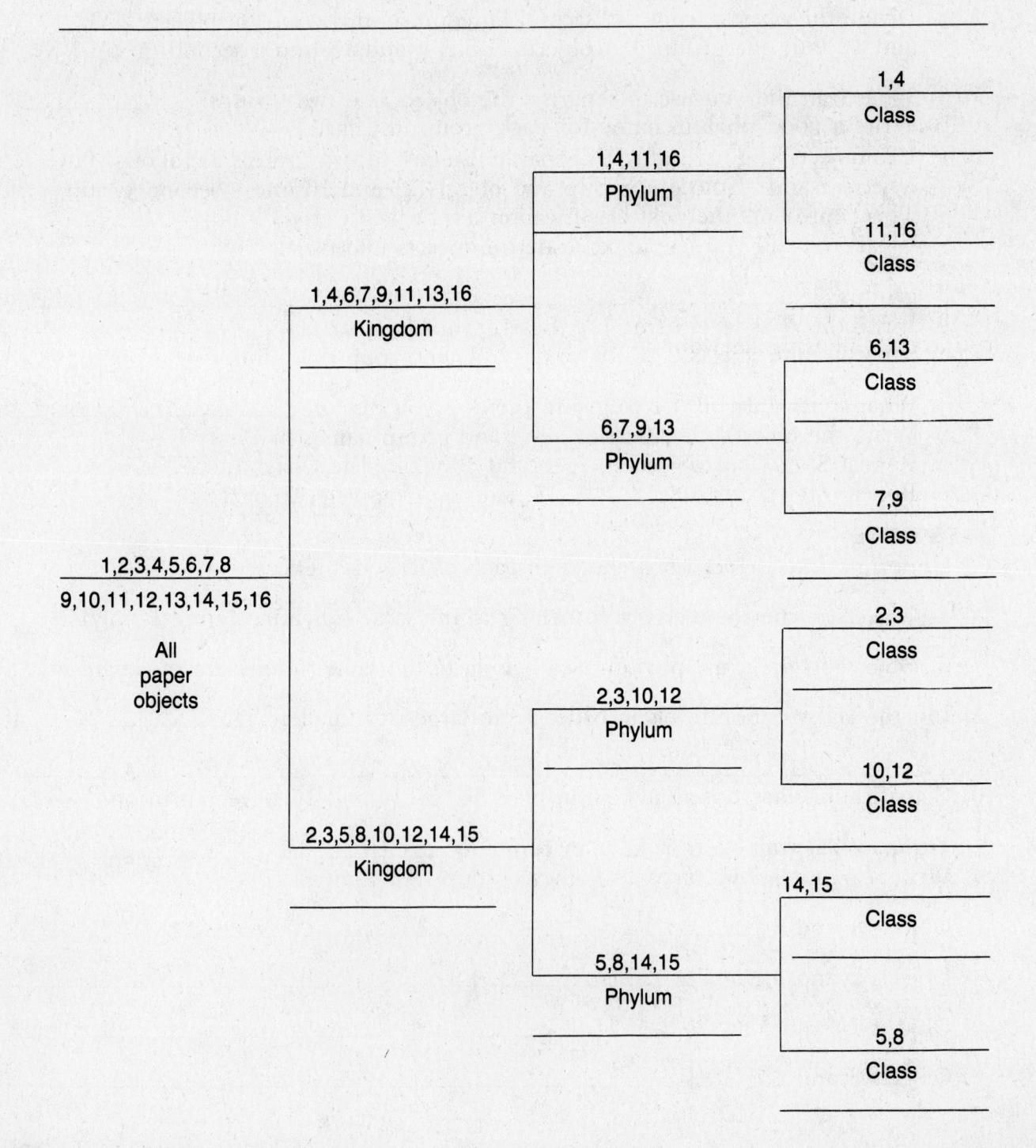

6. How are pieces of clothing classified in a department store? _____

7. How are athletic teams grouped? _____

8. What are two reasons for classifying things? _____

Name _____ Class _____ Period _____

3–2 How Can Living Things Be Grouped?

Living things can be grouped or classified just like nonliving things. Scientists classify living things to put them in order and to show how they are alike. This makes the study of the many different living things easier. One way of classifying is by placing all living things into five main groups. Each group is called a kingdom. The five kingdoms are the Moneran kingdom, the Protist kingdom, the Fungus kingdom, the Plant kingdom, and the Animal kingdom.

INTERPRETATION

OBJECTIVES
In this activity, you will:
  a. observe and record the traits of living things.
  b. use the observed traits to see similarities and differences in living things.
  c. group living things into kingdoms.

KEYWORDS
Define the following keywords:

classify _____

kingdom _____

trait _____

MATERIALS
specimens of living things
microscope
hand lens

PROCEDURE
Part A. Describing Organisms
1. Look at Figures 1 and 2 of living things on the next page. Use Figure 3-10 on page 63 of your text.
2. Fill in the blanks for each organism in the space provided as follows:
   a. For 'number of cells' write *one* or *many*.
   b. For 'cell nucleus' write *yes* or *no*.
   c. Describe the color of the organism.
   d. For 'makes food' write *yes* or *no*.
   e. For 'can move' write *yes* or *no*.
   f. Determine and write the kingdom to which it belongs.

FIGURE 1

Organism	Bread Mold	Moss	Bacteria

Number of Cells _____ _____ _____

Cell Nucleus _____ _____ _____

Color   _____ _____ _____

Makes Food   _____ _____ _____

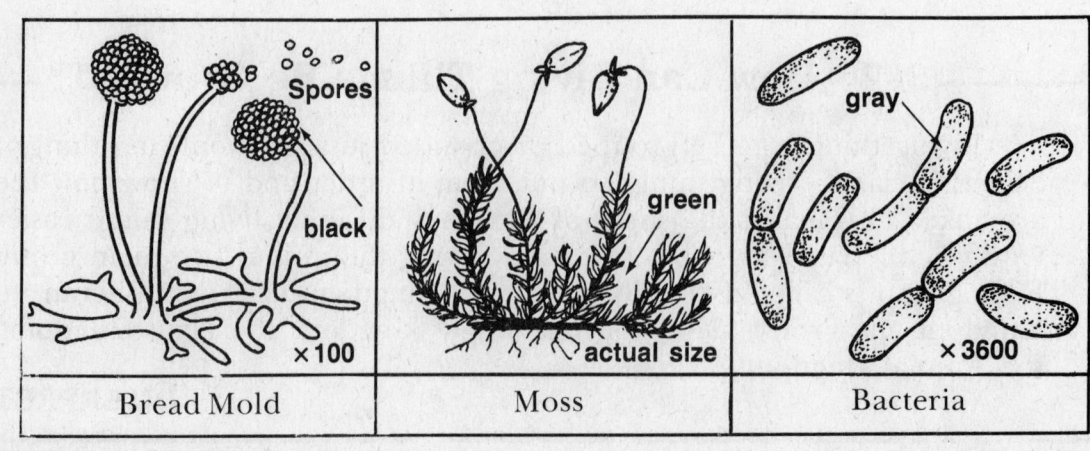

Can Move   _____ _____ _____

Kingdom   _____ _____ _____

FIGURE 2

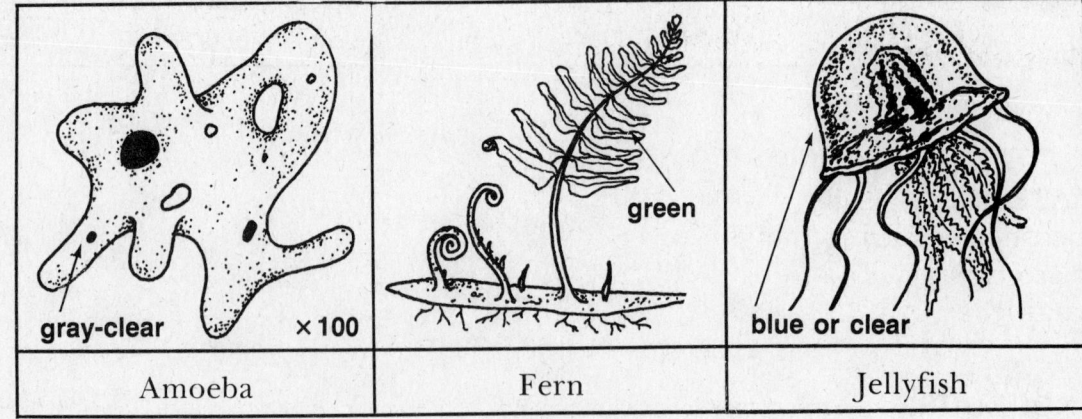

Organism	Amoeba	Fern	Jellyfish

Number of Cells _____ _____ _____

Cell Nucleus _____ _____ _____

Color   _____ _____ _____

Makes Food   _____ _____ _____

Can Move   _____ _____ _____

Kingdom   _____ _____ _____

Name _____ Class _____ Period _____

Part B. Classifying Organisms
1. Look at the six living specimens provided for you to study.
2. Examine each one carefully and make drawings of the organisms in the spaces provided.
3. Fill in the blanks as for Part A.
4. Determine the kingdom to which each organism belongs.

Organism

Number of Cells _____ _____ _____

Cell Nucleus _____ _____ _____

Color _____ _____ _____

Makes Food _____ _____ _____

Can Move _____ _____ _____

Kingdom _____ _____ _____

Organism

Number of Cells _____ _____ _____

Cell Nucleus _____ _____ _____

Color _____ _____ _____

Makes Food _____ _____ _____

Can Move _____ _____ _____

Kingdom _____ _____ _____

QUESTIONS

1. In what kingdoms are one-celled organisms placed? _____
2. How are producers different from consumers? _____
   _____
3. In what kingdoms do you find producers? _____
4. How do fungi get their food? _____
5. What are some examples of fungi? _____
6. Which of the organisms were placed in a kingdom on the basis of one trait?
   Explain. _____
   _____
7. What organisms were classified as plants? _____
8. What traits do these plants have in common? _____
   _____
9. What organisms were classified as animals? _____
10. What traits do these animals have in common? _____
    _____
11. What organisms are classified as protists? _____
12. What traits do these protists have in common? _____
    _____
13. How are these protists different? _____
14. What traits are used to place living things into the following kingdoms?

    a. Monera _____

    b. Protist _____

    c. Fungus _____

    d. Plant _____

    e. Animal _____

Name _____ Class _____ Period _____

4–1 How Can You Test for Bacteria in Foods?

Bacteria are living things too small to be seen without a microscope. Bacteria are said to be microscopic. Often a test can be done to show the presence of microscopic living things. The test can be done with a chemical indicator.

As its name says, a chemical indicator is a chemical that "indicates" something. One type of chemical indicator will tell you if carbon dioxide is present. The chemical turns from blue to yellow or green if carbon dioxide is present.

Respiration in bacteria and other living things produces carbon dioxide. Thus, this chemical can tell if bacteria or other living things are present in a sample being treated.

Bacteria are used to make many dairy products. The samples you will be testing in this exercise are different dairy products.

INVESTIGATION

OBJECTIVES
In this exercise, you will:
  a. learn how to do a chemical test for the presence of carbon dioxide.
  b. test dairy products to determine if they contain bacteria.

KEYWORDS
Define the following keywords:

control _____

microscopic _____

respiration _____

variable _____

MATERIALS
6 test tubes   wax pencil
6 cork stoppers   2 droppers
test tube rack   3 wooden splints
150 mL bromthymol blue solution
dairy products—milk, buttermilk, yogurt, cottage cheese, sour cream

FIGURE 1. Adding dairy products

Adding liquid dairy products
Adding thick dairy products
Wooden splint
Dairy product

PROCEDURE
1. Label 6 test tubes 1 through 6.
2. Fill each test tube with the blue chemical solution.

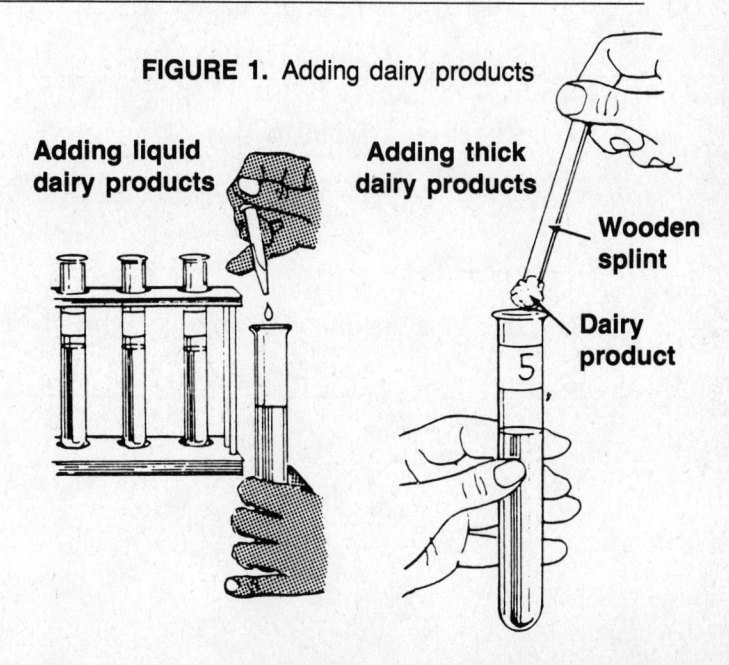

25

3. Use a clean dropper to add one drop of the following to each tube. (Do not let the dairy products go down the sides of the test tubes.) Figure 1 shows you how.
   Tube 1–nothing   Tube 2–milk   Tube 3–buttermilk
4. Using a wooden splint, add an amount about the size of a green pea of the following to each tube.
   Tube 4–plain yogurt   Tube 5–cottage cheese   Tube 6–sour cream
5. Stopper each tube with a cork. Record in Table 1 the color of the tube contents. Leave the tubes undisturbed. Do not shake.
6. Observe and record the color of tube contents at the bottom of each tube, both at the end of the class period and the next day.

Table 1. Test for Bacteria in Food

Tube	Contents	Color at start	Color at class end	Color 1 day later	Carbon dioxide gas present?	Bacteria present?
1	Nothing					
2	Milk					
3	Buttermilk					
4	Plain yogurt					
5	Cottage cheese					
6	Sour cream					

QUESTIONS

1. Was there a color change in the contents of Tube 1 one day later? _____

   What was the purpose of Tube 1? _____

   What name is given to this part of the experiment? _____

2. In which tubes was a color change detected one day later? _____

   What foods were added to these tubes? _____

   Do these foods contain bacteria? _____

   What is your proof? _____

3. What is the variable in this experiment? _____

4. Why are bacteria important to the dairy industry? _____

5. Are bacteria in dairy products helpful or harmful? _____

Name _____ Class _____ Period _____

4–2 Does Soil Contain Bacteria

Bacteria are present almost everywhere. This means that they may be found in soil. How can you find out if bacteria are present in soil? Most living things use oxygen and give off carbon dioxide in cellular respiration. If living bacteria are present in soil, they will use oxygen and give off carbon dioxide. By carrying out an experiment, you can find out if living bacteria are present in soil.

INVESTIGATION

OBJECTIVES
In this exercise, you will:
  a. test soil samples and measure the amount of carbon dioxide given off during 24 hours by the different soil samples.
  b. compare the amount of carbon dioxide given off by different soil samples.
  c. explain why the amount of carbon dioxide may differ in each sample tested.

KEYWORDS
Define the following keywords:

bacteria _____

carbon dioxide _____

cellular respiration _____

distilled water _____

sterile _____

MATERIALS
4 large beakers, 500 mL
4 small beakers, 250 mL
4 plastic stirring sticks
soil^2
sterile soil
Chemical A
8 labels
4 circles of filter paper

distilled water
teaspoon
100 mL graduated cylinder
4 funnels
aluminum foil
glucose or dextrose
2 droppers
Chemical B

PROCEDURE
Part A. Preparing Soil Samples
1. Label 4 large beakers 1 through 4. Add the following as shown in Figure 1 and use the teaspoon to mix the substances in each beaker:

FIGURE 1. Preparation of beakers 1 through 4

100 mL distilled water

100 mL distilled water, 3 teaspoons sterile soil

100 mL distilled water, 3 teaspoons regular soil

100 mL distilled water, 3 teaspoons regular soil, 1 teaspoon glucose

Beaker 1   Beaker 2   Beaker 3   Beaker 4

a. Measure 100 mL of distilled water with the graduated cylinder. Add it to Beaker 1.
b. Add 100 mL of distilled water and 3 teaspoons of sterile soil to Beaker 2.
c. Add 100 mL of distilled water and 3 teaspoons of regular soil to Beaker 3.
d. Add 100 mL of distilled water, 3 teaspoons of regular soil, and 1 teaspoon of glucose to Beaker 4.

2. Cover each beaker tightly with aluminum foil as shown in Figure 2. Allow the beakers to remain overnight.

FIGURE 2. Beaker covers

Part B. Filtering the Soil Samples
1. Label four small beakers 1 through 4.
2. Prepare four circles of filter paper as shown in Figure 3. Fold each one in half. Fold in half again. Place each folded paper into the funnel.
3. Stand each funnel with filter paper into the beakers as shown in Figure 4.
4. Pour the contents of the large beakers into the small beakers.
   a. Pour the liquid from large Beaker 1 into the funnel resting in small Beaker 1. Do not let the liquid overflow onto the sides of the filter paper.
   b. Repeat step a for Beakers 2 through 4.

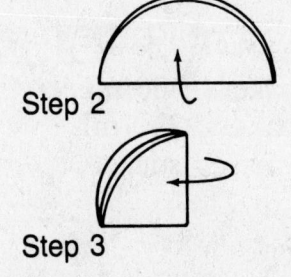

Step 1 — filter paper
Step 2
Step 3
FIGURE 3. Making a funnel

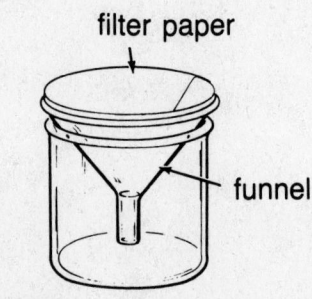

filter paper
funnel
FIGURE 4. Using a funnel

Name _____ Class _____ Period _____

5. Remove the filter paper from each small beaker when filtering is completed.

Part C. Measuring the Amount of Carbon Dioxide Formed by Soil Samples

1. Use a dropper and add three drops of Chemical A to each small beaker.
2. Read all of step 2 before actually doing it.

   a. Use a different dropper for Chemical B. Add Chemical B one drop at a time to the water in beaker 1. **CAUTION:** *Rinse Chemical B with water if it spills on skin or clothes.*
   b. Use a stirring stick to swirl the water in the beaker.
   c. Wait 15 seconds.
   d. Continue to add Chemical B one drop at a time until a light pink color *remains* after swirling and waiting 15 seconds.
   e. Count the number of drops of Chemical B needed to turn the water a *lasting light pink color*.
   f. Record the number of drops needed to turn the water in Beaker 1 a lasting light pink color in Table 1. **NOTE:** The more drops of Chemical B needed to make a lasting light pink color, the more carbon dioxide is present in the water. The fewer drops of Chemical B needed to make a lasting light pink color, the less carbon dioxide is present in the water.

3. Repeat step 2 for Beakers 2 through 4.

Table 1. Drops of Chemical B used in each beaker

Beaker	Contents	Number of drops of Chemical B used
1	water	
2	water and sterile soil	
3	water and regular soil	
4	water, soil, and glucose	

QUESTIONS

1. a. What was the source of bacteria in the experiment? _____

   b. What was used to trap or catch the carbon dioxide given off? _____

   c. What was the purpose of Beaker 1? _____
   _____

2. a. What was being measured by the number of drops of Chemical B added to the water?_____
   b. What do a few drops of Chemical B show about carbon dioxide amount present in the water?_____
   c. What do many drops of Chemical B show about carbon dioxide amount present in the water?_____
3. a. How many drops of Chemical B were used to give a pink color to Beaker 1?_____
   b. Beaker 3?_____
4. a. Which beaker, 1 or 3, had more carbon dioxide present?_____
   b. On the first day, what was put into Beaker 1?_____
   c. On the first day, what was put into Beaker 3?_____
   d. Where did most of the carbon dioxide come from, the bacteria in the soil or the bacteria in the water?_____
5. a. Which beaker, 2 or 3, had more carbon dioxide present?_____
   b. On the first day, what was put into Beaker 2?_____
   c. Does sterile soil contain any bacteria?_____
   d. Where did most of the carbon dioxide come from, the bacteria in the soil, or the soil itself?_____
6. a. Which beaker, 3 or 4, had more carbon dioxide present?_____
   b. What do bacteria do with glucose?_____
   c. On the first day, what was put into Beaker 4?_____
   _____
   d. Which beaker, 3 or 4, carried on more cellular respiration?_____
7. Why was a beaker with only water used in this experiment?_____
   _____
8. Why was sterile soil used in this experiment?_____
   _____
9. What are your conclusions from this experiment?_____
   _____
   _____
   _____

Name _____ Class _____ Period _____

4-3 What Are the Traits of Blue-green Bacteria?

Living things can be classified into five kingdoms. One kingdom, Monera, includes bacteria and blue-green bacteria. Organisms in this kingdom do not have a nucleus or other cell parts, such as chloroplasts.

Blue-green bacteria are different from other bacteria. They have pigments inside their cells and can make their own food. Most are blue-green in color and have the pigment chlorophyll. Blue-green bacteria live as single cells, in colonies or in filaments. The cells are often surrounded by a jelly layer.

EXPLORATION

OBJECTIVES
In this exercise, you will:
  a. observe the traits of two blue-green bacteria.
  b. draw and label the parts observed in each blue-green bacterium.
  c. record the traits of blue-green bacteria.

KEYWORDS
Define the following keywords:

blue-green bacteria _____
_____

colony _____

filament _____

jelly layer _____

MATERIALS
microscope
glass slide
coverslip
pencil, green

methylene blue stain
Gloeocapsa living culture
Lyngbya living culture
dropper

PROCEDURE
1. Place a drop of *Gloeocapsa* on a microscope slide. Add one drop of stain to the slide. Add a coverslip. **CAUTION:** *Methylene blue will stain clothing.*
2. Observe the *Gloeocapsa* under low and high power. The stain will form a dark background. This will allow you to determine if the cells are surrounded by a clear jelly layer.

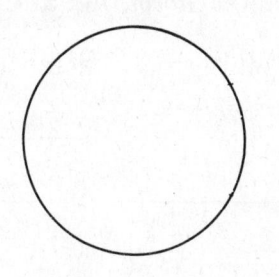

Gloeocapsa X 450   *Lyngbya* X 450

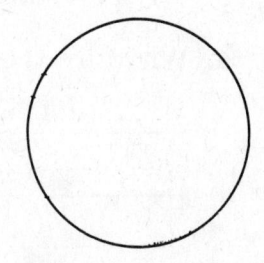

FIGURE 1. Circles for drawing blue-green bacteria

31

3. Make a drawing of *Gloeocapsa* in the circle provided in Figure 1. Note the color and location of the chlorophyll. Color those parts that are blue-green.
4. Complete the first column in Table 1 for *Gloeocapsa*.
5. Repeat steps 1 through 4 for *Lyngbya*.

Table 1. Comparison of blue-green bacteria

Kingdom		
Filament or colony		
Nucleus present?		
Chlorophyll present?		
Chloroplasts present?		
Makes own food?		
Jelly layer present?		

QUESTIONS

1. What is the largest number of cells that you can count within one jelly layer of *Gloeocapsa*? _____
2. a. Is the color of *Gloeocapsa* the same or different from that of *Lyngbya*? _____

   b. What chemical provides the color? _____

   c. What does this chemical allow these organisms to do? _____

3. Compare the jelly layer of *Gloeocapsa* and *Lyngbya*. Which is thicker? _____
4. What are the traits of blue-green bacteria? _____

Name _____ Class _____ Period _____

5–1 Where Are Protists Found?

Almost all protists are single-celled organisms that have a nucleus and other cell parts. Protists must have water in which to live. Most move around in order to search for food, darkness or other, suitable living conditions. There are three kinds of protists: animal-like protists, plantlike protists and fungilike protists. Protists can be found wherever water is found and many protists actually live inside other animals.

EXPLORATION

OBJECTIVES
In this activity, you will:
a. observe and identify protists that can be found in water samples, a termite, diatomaceous earth and scouring powder.
b. determine where animal-like and plant-like protists are likely to live.

KEYWORDS
Define the following keywords:

cellulose _____

cilia _____

false feet _____
_____

flagellum _____

protist _____

MATERIALS
microscope   marking pen   household cleaning powder
5 glass slides   5 droppers   buffer solution
5 coverslips   water samples   1 to 3 living termites
forceps   probe
prepared slide of diatomaceous earth (or materials to
  make a wet mount of diatomaceous earth)

PROCEDURE
Part A. Examination of Water Samples
1. Obtain 3, labeled water samples from your teacher. Label a slide with the letter A as shown in Figure 1.
2. Prepare a wet mount of sample A on the slide you marked with an 'A'.

FIGURE 1. Examining water samples

33

3. Observe the slide under low power and search for protists. Watch for movement. Look around pieces of debris. You may need to reduce the amount of light passing through the slide in order to locate some protists that look like amoebas.
4. When you find some protists that are not moving or are moving only very slowly, switch over to high power.
5. Draw, in the proper column in Figure 3, the protists you observe under high power.
6. Use your text and other sources to help you to identify the protists. Look for flagella, cilia, false feet, pigments and other noticeable features.
7. Repeat steps 1 through 6 for water samples B and C.

Part B. Examination of the Inside of a Termite

Not all protists living inside other animals are parasites. Flagellates live in the intestines of termites. The termite cannot digest the wood that it eats. The flagellates make an enzyme that digests the cellulose in the wood particles.

1. Place a drop of buffer solution on a clean slide.
   CAUTION: *Handle buffer solution with care.*
2. Use forceps to place a termite on the slide near the drop of buffer.
3. Cover the termite and the buffer with a coverslip and a second clean slide.
4. With the end of a probe, gently press on the second slide in order to squash the termite.
5. Remove the top slide. Observe the slide under both low and high power.
6. Make a drawing (or drawings) in Figure 4 of the protists that you observe in the termite. The largest protist you will see is called *Trichonympha*.

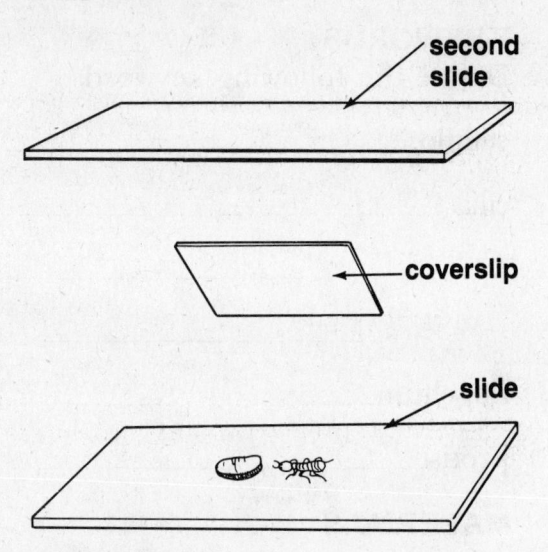

FIGURE 2. Preparing termite slide

Part C. Examination of Diatomaceous Earth and Scouring Powder.

1. Obtain a prepared slide of diatomaceous earth or prepare a wet mount of the powdered diatomaceous earth given to you.
2. Observe the slide under both low and high power.
3. Make drawings of what you observe under high power in Figure 5.
4. Use forceps to place a small amount of scouring powder on a clean microscope slide. Add a drop of water and a coverslip.
5. Observe the slide under both low and high power.
6. Make drawings of what you observe under high power in Figure 5.

Name _____ Class _____ Period _____

FIGURE 3. Protists observed in water samples

A	B	C
Pond	Lake	Stream

FIGURE 4. Termite protists

FIGURE 5. Diatomaceous earth and scouring powder

QUESTIONS

1. Which of the types of protist (animal-like or plant-like) did you find in the greatest numbers in:

   a. pond water _____

   b. lake water _____

   c. stream water _____

2. Which water sample had the least number of protists? _____

3. Which water sample had the greatest number of protists? _____

4. What is the most common type of protist in your water samples? _____
   _____

5. How is *Trichonympha* different from the other protists that you have observed? _____

6. What does the termite use for food? _____

7. In what ways do the termite protists depend upon the termites? _____
   _____

8. Wood is cellulose. Cellulose is made up of simple sugars. Into what compounds do the protists probably digest the wood? _____

9. Why is the termite dependent upon the protists? _____
   _____

10. What are the glasslike parts that you see in the diatomaceous earth. ____
    _____

11. Can you see similar parts in the scouring powder? _____

12. What might be the reasons for adding diatomaceous earth to scouring powder? _____
    _____

Name _____ Class _____ Period _____

5-2 What Do Molds Need in Order to Grow?

Molds are a group of living things in the Kingdom Fungi. Many molds are called saphrophytes because they feed on things that were once living. They also feed on and destroy many foods that humans eat.

When a mold spore that is carried in the air lands on food, it sprouts and sends out rootlike parts into the food. These parts break down the food on which they are growing. The food becomes a source of energy for the molds. In a short time, the mold* can completely cover the food that the spore landed upon.

INVESTIGATION

OBJECTIVES
In this exercise, you will:
   a. observe the growth of molds on different kinds of food.
   b. determine what conditions promote the growth of molds.
   c. determine how mold growth can be slowed or stopped.

KEYWORDS
Define the following keywords:

dehydrate _____

mold _____

saphrophyte _____

spore _____

MATERIALS
bleach
alcohol
instant potatoes
bread, 3 days old
raisins
5 droppers
spoon

marking pen
household cleaner
household ammonia
3 brown paper bags
16 small jars with covers
6 cotton swabs

PROCEDURE
Part A. Growth of Mold on Foods
1. Label 6 small jars 1 through 6. Place your name on each label.
2. Add the following materials to the jars.
   Jars 1 and 2–1 piece of bread, cut to fit into the jar.
   Jars 3 and 4–1 spoonful of instant potatoes.
   Jars 5 and 6–1 spoonful of raisins.
3. Add 20 drops of water to jars 2, 4 and 6.

4. Complete the first three columns in Table 1.
5. Get a cotton swab with mold spores on it from your teacher. Rub the swab across the surface of the food in all 6 jars. Figure 1 shows you how.
6. Place the covers on the jars and store them on a shelf for 3 days.
7. After 3 days, observe the jars for any growth of mold.
8. Complete the last column in Table 1.

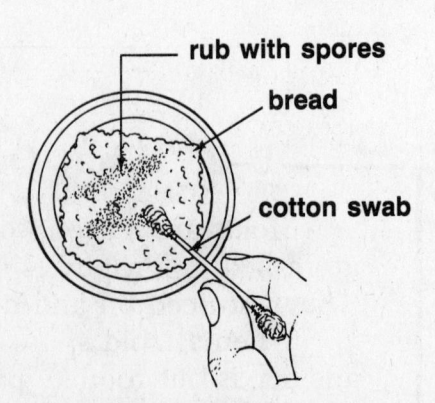

FIGURE 1. Top view of jar

Table 1. Growth of Molds

Jar number	Type of food	Water added?	Is mold growing on food?
1			
2			
3			
4			
5			
6			

Part B. Examination of Growing Conditions for Molds
1. Label 6 jars with numbers from 7 through 12.
2. Cut 6 pieces of bread and place one into each jar.
3. Add 20 drops of water to jars 7, 8 and 9.
4. Get a cotton swab with mold spores on it from your teacher. Rub the swab across the surface of the bread in each of the 6 jars.
5. Write your name on 3 brown paper bags. Put jars 7 and 10 into one paper bag. Place it in a refrigerator.
6. Put jars 8 and 11 into a second paper bag. Place this bag in an incubator at 30°C or leave it in a warm place, perhaps near a radiator.
7. Put jars 9 and 12 into the third paper bag. Place this bag on a shelf at room temperature.
8. Record the conditions for growth (use the words moist or dry, cold, warm or room temperature and light or dark) in the proper column in Table 2.
9. Begin checking for mold growth on the third day. Continue to check the jars each day for 6 more days.
10. Record in Table 2 the amount of mold you find each day in each jar. Use the following terms: none, little, some and much.

Name _____ Class _____ Period _____

Table 2. Growing Conditions for Molds

Jar number	Conditions	Day 3	Day 4	Day 5	Day 6	Day 7	Day 8	Day 9
7	moist, cold, dark							
8	moist, warm, dark							
9	moist, room temp, dark							
10	dry, cold, dark							
11	dry, warm, dark							
12	dry, room temp, dark							

Part C. Slowing or Stopping Mold Growth

1. Label 4 jars with numbers 13 through 16.
2. Cut 4 pieces of bread and place one into each jar.
3. Add 20 drops of water to each jar.
4. Get a cotton swab with mold spores on it from your teacher. Rub the swab across the surface of the bread in all 4 jars.
5. Add nothing to jar 13.
6. Place 10 drops of bleach across the surface of the bread in jar 14.
   CAUTION: *Avoid spilling or inhaling bleach.*
7. Place 10 drops of household cleaner across the surface of the bread in jar 15.
8. Place 10 drops of household ammonia across the surface of the bread in jar 16.
   CAUTION: *Avoid spilling or inhaling ammonia.*
9. Record in the table below the substances you have added to each jar.
10. Place the jars in a warm, dark area.
11. Observe the jars for mold growth after 3 days. Record your observations in the proper column in Table 3.
12. Continue to check on the jars each day for 6 more days.
13. Record in Table 3 your observations of mold growth for each of these days.

Table 3. Slowing or Stopping Mold Growth

Jar number	Substance added	Day 3	Day 4	Day 5	Day 6	Day 7	Day 8	Day 9
13								
14								
15								
16								

QUESTIONS

1. Consider the observations you made in Part A of this exercise.

    a. Does mold grow on moist food? _____ dry food? _____

    b. Which jars show mold growth under these conditions?

    dry _____ moist _____

    c. What types of food does mold grow on? _____

2. Consider the observations you made in Part B of this exercise.

    a. What conditions caused the most growth of mold?_____
    _____

    b. In order to prevent mold growth, how should food be stored?_____
    _____

    c. Why does bread become moldy faster than crackers?_____
    _____

3. Consider the observations you made in Part C of this exercise.

    a. What was the purpose of jar 13?_____

    b. What substances seem to slow mold growth?_____
    _____

4. What three things are needed for mold growth?_____
    _____

5. Foods are often packaged several days or weeks before they are sold. Food companies add preservatives to keep molds from growing on many of these foods. How could you tell if the preservatives that are added are able to keep the molds from growing?_____
    _____

6. Many dehydrated foods do not contain chemical preservatives. What prevents molds from growing on these foods?_____
    _____

Name _____ Class _____ Period _____

6-1 What Are the Traits of Vascular and Nonvascular Plants?

Roots, stems and leaves are special plant structures. Roots anchor plants in the soil and take in water and minerals. Stems carry water to all parts of the plant, support the plant and hold the leaves up to the sunlight. Leaves make food. These structures are traits of plants such as ferns, conifers and flowering plants. These are vascular plants. Many simple plants do not have these parts. Vascular plants have special tubelike structures in the roots, stems and leaves to carry food and water. Nonvascular plants do not have the tubelike structures.

EXPLORATION

OBJECTIVES
In this exercise you will:
  a. use models to determine the differences in vascular and nonvascular plants.
  b. examine and compare vascular and nonvascular plants under the microscope.

KEYWORDS
Define the following keywords:

nonvascular plant _____

_____

tubelike cells _____

vascular plant _____

_____

MATERIALS
green pencil   2 drinking straws   liverwort
scissors   lettuce   microscope
razor blade   celery   2 glass slides

PROCEDURE
Part A. Constructing Plant Models.
 1. The diagrams in Figure 2 on the last page of this lab are outline models of two plants. One is marked vascular plant and the other is marked nonvascular plant. Color parts A and B green. These are the parts that contain chlorophyll.
 2. Tape properly cut straws along the dashed lines *on only one model*. The straws represent tubelike cells. Review the definitions of vascular and nonvascular plants in order to determine on which model to place the straws. **CAUTION:** *Use care with scissors.*

3. Plant parts that appear to be leaves but lack the tubelike cells are said to be leaflike. Label parts A and B as either leaf or leaflike.
4. Plant parts that appear to be stems but lack the tubelike cells are said to be stemlike. Label parts C and D as either stem or stemlike.
5. Plant parts that appear to be roots but lack the tubelike cells are said to be rootlike. Label parts E and F as either a root or rootlike.
   These leaflike, stemlike and rootlike parts are features of nonvascular plants.

Part B.   Examination of vascular and nonvascular plants.
1. Obtain a piece of lettuce. Use a razor blade to cut out a small section (Figure 1). **CAUTION:** *Use extreme care with razor blades.*
2. Place the section on a glass slide. Add two or three drops of water.
3. Place a second slide over the first slide. Squash the slides together (Figure 1).
4. Examine the lettuce under low power magnification only.
5. Look for long, tubelike cells. They may appear as spirals or train tracks. Draw a diagram of what you see in the appropriate place in Table 1.
6. Obtain a piece of liverwort from your teacher. Repeat steps 1 through 5. Draw a diagram in Table 1 of what you see.
7. Obtain a piece of celery from your teacher. Prepare the celery as shown in Figure 1. Cut off a 1 cm piece of the string. Repeat steps 2 through 5. Draw a diagram in Table 1 of what you see.
8. Complete the last two columns in Table 1.

Table 1. Comparison of Vascular and Nonvascular plants

Plant	Diagram	Tubelike cells present or absent	Vascular or nonvascular
Lettuce			
Liverwort			
Celery			

Name _____ Class _____ Period _____

Preparing lettuce

Remove a small section as shown here.

Lettuce

Second slide

Squashing slides together

FIGURE 1. Slide preparation

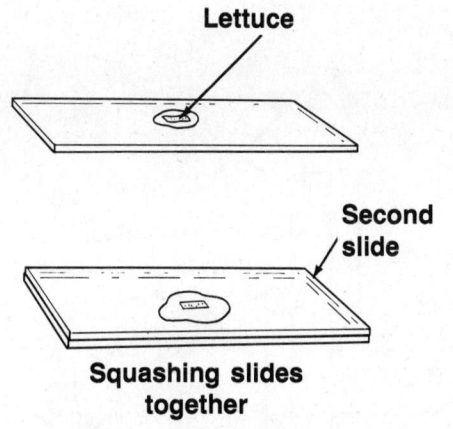

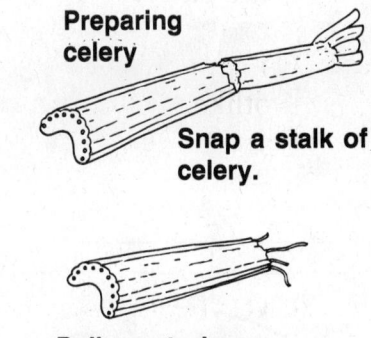

Preparing celery

Snap a stalk of celery.

Pull apart pieces. This will expose tubelike structures (strings).

QUESTIONS

1. What does the green color used in coloring the plant models represent? _____
_____

2. What do the straws used in the model represent? _____

3. List one way in which vascular and nonvascular plants are alike. _____
_____

4. List three ways in which vascular and nonvascular plants are different. Use your model labels for help. _____
_____

5. Explain how tubelike cells in vascular plants are used. _____
_____

6. Explain the difference between stems and stemlike parts. _____
_____

7. Explain the difference between roots and rootlike parts. _____
_____

8. Consult the models. Are tubelike cells continuous throughout the plant if the plant is vascular? _____

   How is this helpful to the plant? _____

9. Would you expect to find roots, stems, and leaves in lettuce, liverwort, and celery? _____

   Why? _____
_____

43

A _____

B _____

C {

D {

E

F

Vascular Plant

Nonvascular Plant

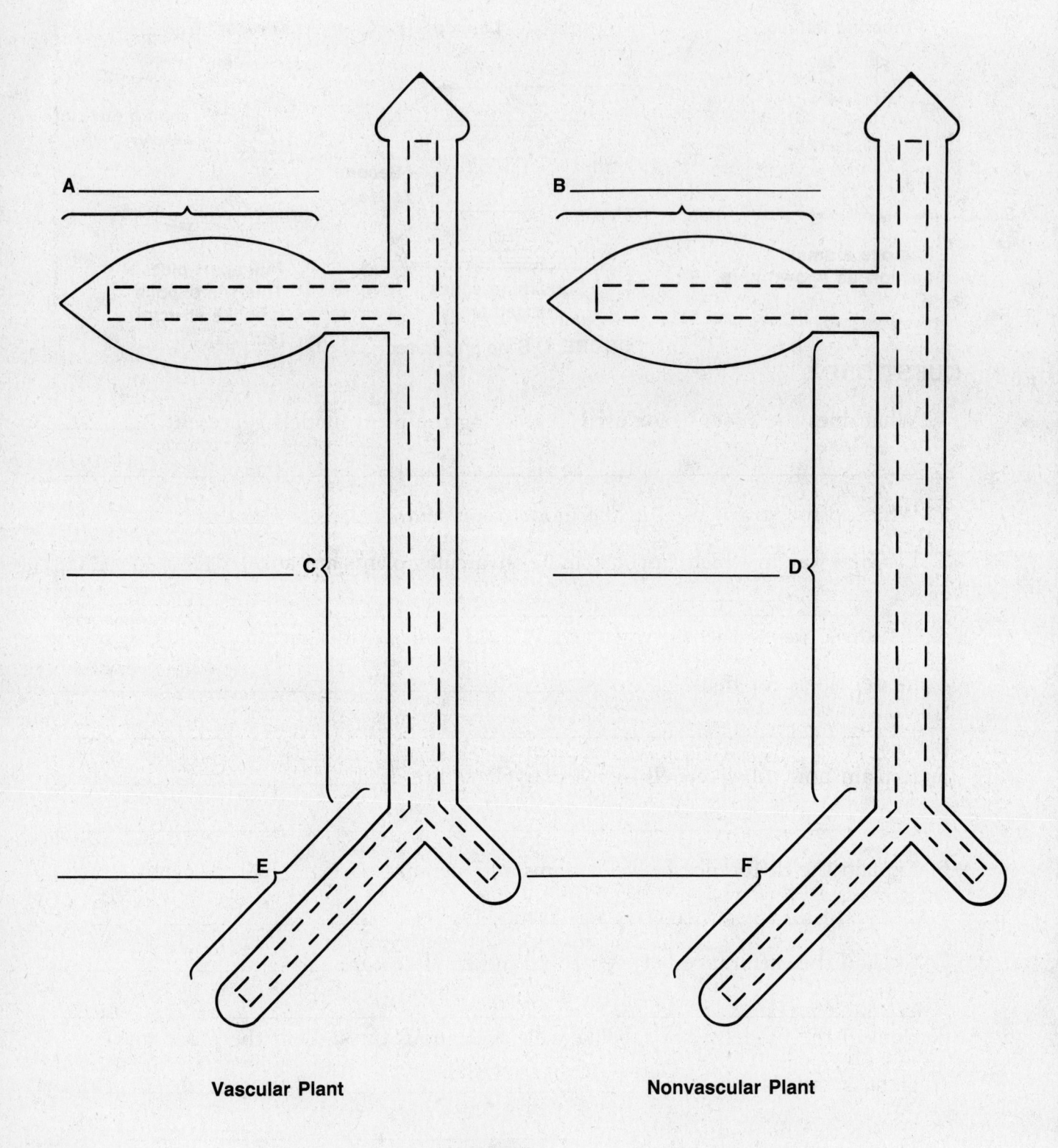

FIGURE 2.

44

Name _____ Class _____ Period _____

6–2 How Can Conifers Be Identified?

Perhaps you have followed a set of directions to get to some place that you wanted to go. Biologists use a set of directions to identify the many kinds of living things. The set of directions used to identify something is called a key. If you follow directions and make correct choices with a key, you can identify living things.

A group of plants called conifers can be identified by their leaves. Conifers are cone-bearing plants. Leaf shape and how the leaves are arranged on stems are features used to identify the different conifers.

EXPLORATION

OBJECTIVES
In this exercise, you will:
a. examine the leaves of some conifers.
b. use a key to identify the conifers from which the leaves came.

KEYWORDS
Define the following keywords:

conifer _____

key _____

MATERIALS
twigs of various conifers
hand lens   metric ruler

PROCEDURE
1. Examine the numbered twigs of the conifers carefully with the hand lens.
2. Record the features of the conifers in Table 1. Use the diagrams shown in Figure 1 to help you.
3. Pick one twig to key. Read question 1 of the key in Table 2. Answer the question, yes or no, using the twig you selected.

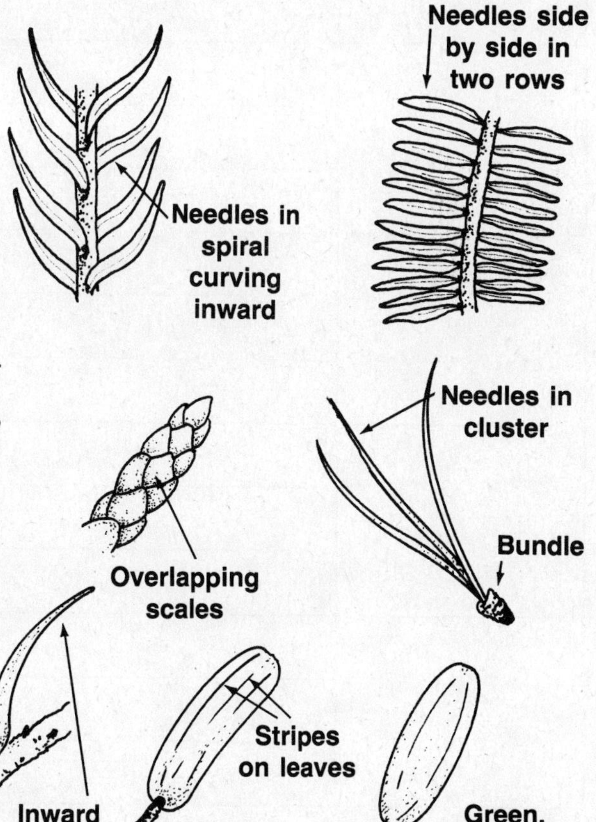

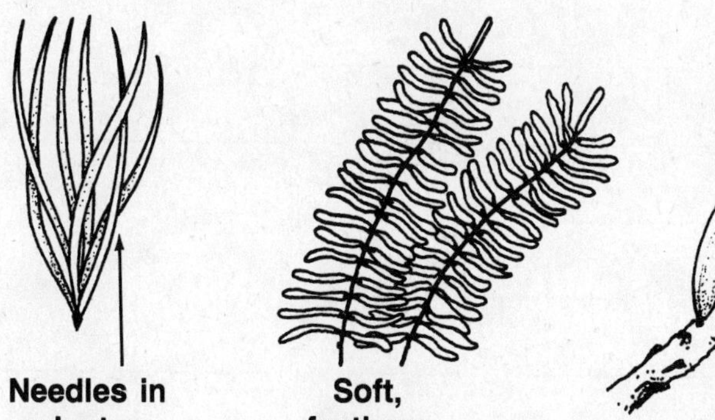

FIGURE 1. Characteristics of conifers

45

Table 1. Identifying Conifers

Plant	Observations	Kind of Conifer
1		
2		
3		
4		
5		
6		
7		
8		

Name _____ Class _____ Period _____

Table 2. Conifer Key

Questions	If you answer Yes	No
1. Does the twig have leaves that look like needles?	Go to 2.	Go to 12.
2. Are the needles in groups of two or more?	Go to 3.	Go to 5.
3. Are the needles at least 5 cm long and are there 2 to 5 enclosed in a bundle?	This is a pine.	Go to 4.
4. Are the needles at least 5 cm long and are there 6 or more present but not enclosed in a bundle?	This is a larch.	Go to 5.
5. Do the needles curve inward and grow in a spiral pattern?	Go to 6.	Go to 8.
6. Are the needles soft and flat?	This is a Douglas fir.	Go to 7.
7. Do the needles have four sides? Are the needles sharp?	This is a spruce.	Go to 8.
8. Do the needles grow straight and side by side in two rows on the twig?	Go to 9.	Go to 10.
9. Are the leaves light colored, soft, and featherlike?	This is a bald cypress.	Go to 10.
10. Do the needles have gray stripes on the bottom and are they attached to the stem by a short woody stalk?	This is a hemlock.	Go to 11.
11. Do the needles have two silver stripes on the bottom and are they attached to the stem by a short green stalk?	This is a fir.	Go to 12.
12. Do the twigs have only one kind of leaf that looks like overlapping fish scales?	This is a cedar.	Go to 13.
13. Do the twigs have two kinds of leaves—one that looks like overlapping scales and the other that looks needlelike?	This is a juniper.	This is the end. If you haven't named the plant, go to 1 again.

4. Based on your answer, follow what the key says to do. Move to the next question where the key says to go. Answer this question, yes or no.
5. Based on your answer, do what the key says to do. Follow the key until it tells you the identity of your conifer twig. Record your answer in the last column of Table 1.
6. Repeat steps 3 through 5 with the other twigs. Record your results.

QUESTIONS

1. List two shapes of conifer leaves. _____
2. Which conifers have leaves that are flat? _____
3. Using the traits in the key, list two traits of the bald cypress. _____
_____
4. List the kinds of conifers that have leaves that look like overlapping scales. _____
_____
5. Which conifers have leaves in clusters? _____
6. List two ways in which hemlock and fir leaves are alike? _____
_____
7. Describe a spruce leaf. _____
8. A tree has leaves that have two stripes. What else do you need to know before you can identify the tree? _____
9. Below are some descriptions of different conifer leaves. If you can identify the tree from the description, write the name on the blank. If you cannot identify the conifer, write "cannot identify" on the blank.

   a. needles present _____

   b. leaves like overlapping fish scales _____

   c. needles flat _____

   d. needles in bundles of 2 to 5 _____

   e. leaves light colored and featherlike _____

   f. needles with four sides _____

   g. leaves grow in spiral on twig _____

Name _____ Class _____ Period _____

7-1 What Traits Does a Sponge Have for Living in Water?

Sponges are simple animals that live in water. They do not move about. Sponges do not have the same traits that other animals in the animal kingdom have. They are divided into classes based upon structures that support the sponge body.

Sponges do not have real skeletons. Many have tiny glasslike parts inside their bodies that give support. These glasslike parts are called spicules. Sometimes people refer to the spicules of a sponge as a skeleton.

EXPLORATION

OBJECTIVES
In this exercise you will:
  a. observe and compare two different sponge types, *Grantia* and *Spongia*.
  b. observe the spicules that support each sponge.

KEYWORDS
Define the following keywords:

body _____

osculum _____

pores _____

spicules _____

sponge _____

MATERIALS
probe
scissors
10 mL bleach
  in bottle with dropper

hand lens
microscope
water
  in bottle with dropper

2 glass slides
2 coverslips
petri dish
Grantia sponge
Spongia sponge

PROCEDURE
Part A. *Grantia* Sponge
1. Place the sponge in a small glass dish. Examine the sponge with a hand lens. Observe the many tiny holes. These bring water into the sponge. Food is present in the water that enters the sponge. The food is trapped by cells which line the hollow inside of the sponge. Observe the larger hole. The larger hole allows the water to leave the sponge. It is called the osculum.
2. Label the body, pores and osculum of the sponge in Figure 1.

49

3. Carefully place the probe into one of the small holes. Determine if the probe enters the center of the animal.
4. Place the probe in the larger hole called the osculum. Determine if the probe enters the center of the animal.
5. Insert the point of the scissors into the osculum of the sponge. Carefully cut the sponge lengthwise. Use the hand lens to examine the inside of the sponge. Sponges have only two cell layers. **CAUTION:** *Use care with scissors.*
6. Locate several canals passing through the body wall. The canals are lined with flagella. The flagella set up currents of water that cause the water to move through the sponge.
7. In Figure 2 draw arrows to show how water enters and leaves the sponge. Refer to steps 1 and 2 if necessary.

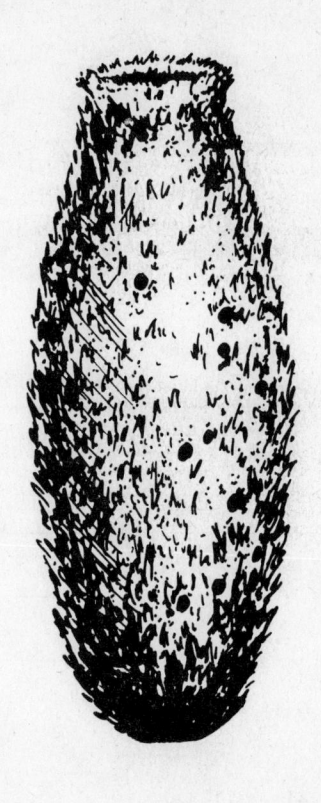

FIGURE 1. Sponge body parts

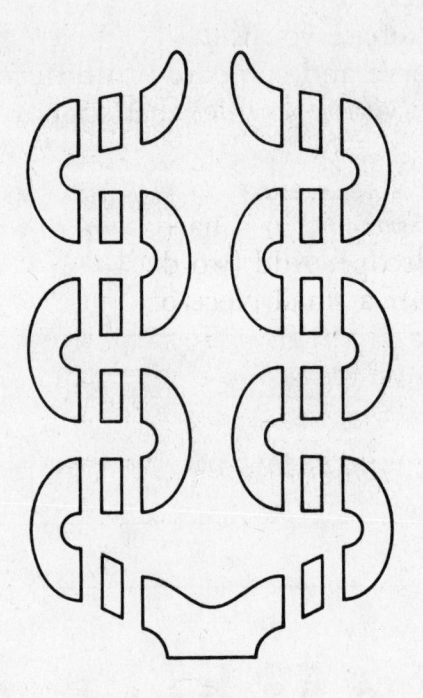

FIGURE 2. Water flow in a sponge

8. Add two drops of bleach to a microscope slide. **CAUTION:** *Be careful not to spill the bleach on your hands or clothing. Avoid inhaling.*
9. Put a small piece of *Grantia* in the bleach. When a sponge is placed in bleach the animal's cells begin to break apart while the spicules remain unchanged.
10. Add a coverslip. Observe *Grantia* using low power. The structures that you see are called spicules. They should look similar to the ones shown in Figure 3. *Grantia* spicules are made of calcium carbonate.

Name _____ Class _____ Period _____

11. Draw some spicules in the space provided.

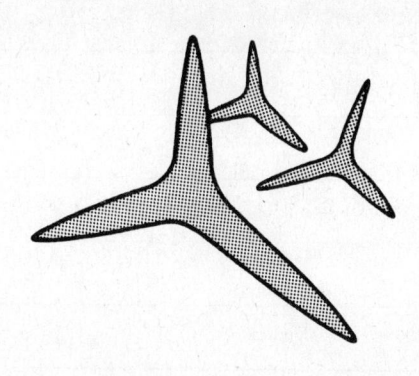

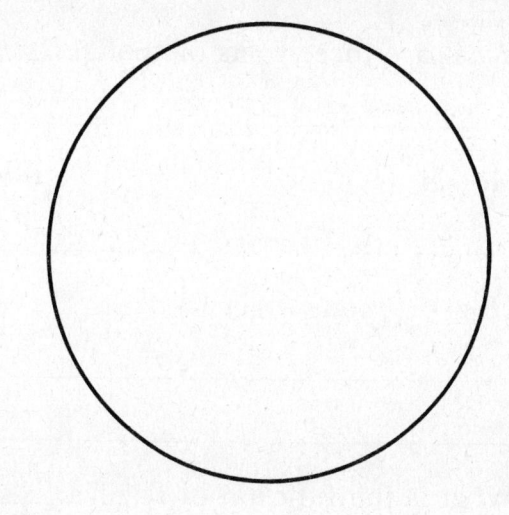

FIGURE 3. *Grantia* spicules

Part B. *Spongia* Sponge
1. Examine the *Spongia*. Note the many pores present that allow food and water to enter the animal.
2. The *Spongia* has a hard support system that can be observed with the microscope. Add two drops of water to a microscope slide.
3. Pull off a small piece of *Spongia* and place it in the water.
4. Add a coverslip. Using low power, observe the *Spongia*.
5. Draw what you see in the space provided. The spicules that you see are composed of spongin. Spongin is made of protein fibers.

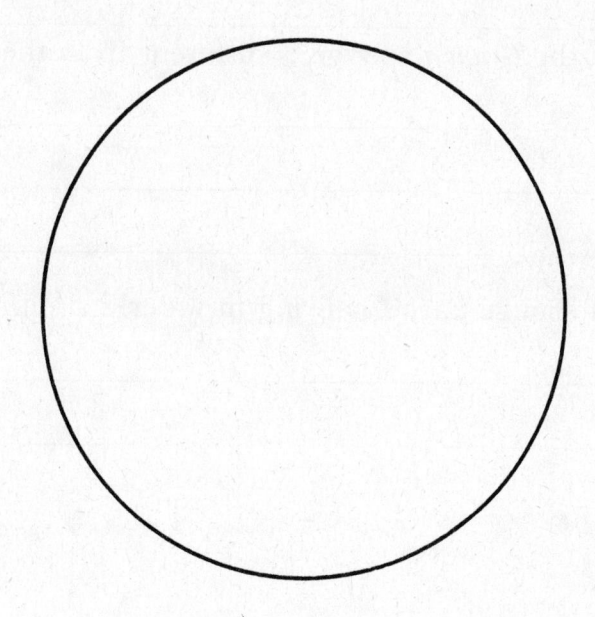

QUESTIONS

1. Where do sponges live? _____
2. From what you have observed do sponges have any parts that might help them move about? _____
3. What are three traits of sponges? _____
_____
4. How does water enter and leave the sponge? _____
_____
5. How do sponges get food? _____
_____
_____
6. What is the function of flagella? _____
_____
7. Are spicules made of the same material as the rest of the sponge? _____
Explain your answer. _____
8. What do spicules do? _____
9. Describe the shape of the spicules from the *Grantia* sponge.
_____
_____
10. Describe the shape of the spicules from the *Spongin* sponge.
_____
11. Why do you think the *Grantia* sponge is different from the kinds used to wash cars or walls? _____
_____
_____
12. What traits does a sponge have for living in water? _____
_____
_____

Name _____ Class _____ Period _____

7-2 What Are the Parts of the Squid?

Squid are soft-bodied animals that live in the ocean. They are among the quickest members of the invertebrates. Some can swim as fast as 20 km per hour and leap as much as 3 m from the surface of the water. They vary in size from 5 cm to 6 m in length. The giant squid is the largest of all invertebrates. It has tentacles that range from 9 to 12 m in length.

Squid feed on smaller invertebrates. They use two methods to avoid being eaten, themselves. They release an inky fluid that hides them and allows them to swim away unseen, and they can disappear into their surroundings by changing color to match the background.

EXPLORATION

OBJECTIVES
In this exercise, you will:
 a. observe the external parts of the squid.
 b. dissect and locate the internal parts of the squid.

KEYWORDS
Define the following keywords:

external _____

gonad _____

internal _____

mantle _____

pen _____

MATERIALS
squid   forceps
scissors   hand lens
dissecting pan   colored pencils: red, green, gray, blue, yellow, and purple

PROCEDURE
Part A.   External Parts of Squid
1. Place the squid right side up in the dissecting pan. Stretch out the two tentacles and eight arms.
2. Examine one tentacle with the hand lens. Note the small suction cup parts on the tentacle.
3. The body is the largest part of the animal. Find the head. It appears loosely attached to the body. The head has eyes on each side.

53

4. Behind the head the body is covered by the mantle. The two small flaps at the tail end of the body are fins.
5. The opening to the body just behind the head is the collar. Locate the water jet on the underside where the collar opening and head meet. Squid swim by squirting water away from their water jet. The direction the water jet faces allows the animal to swim backward or forward.
6. Separate the arms and find the mouth. Notice the dark spot in the center of the mouth. These are the jaws. With the forceps, carefully pull on the upper jaw and remove it. Use the same procedure to remove the lower jaw. Place it with the upper jaw and observe how they work together.
7. Label the tentacle, suction-cup part of the tentacle, arm, body, mantle, head, eye, fin, collar, water jet, and mouth on Figure 1.

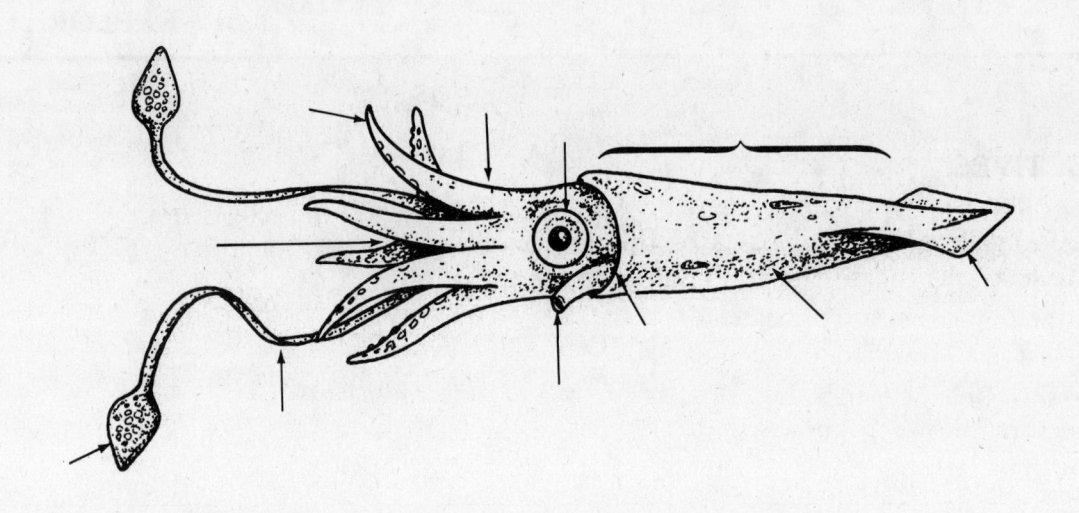

FIGURE 1. Squid external anatomy

Part B. Internal Parts of the squid
1. Turn the squid over so that it is on its back. Use forceps to hold the mantle up away from the internal organs. With scissors, carefully cut through the mantle. **CAUTION:** *Use care with scissors.* Place the blade of the scissors under the collar and cut toward the fins all the way to the end.
2. Lay the mantle open. Use care in finding the internal organs. Find the silvery black ink sac.
3. Locate the esophagous, the tube that leads from the mouth. Food passes through the esophagous.
4. Locate the stomach, the white organ attached to the esophagous at the end opposite the mouth.
5. Locate the liver. It is located beneath the place where the ink sac was found.
6. The gonad is the reproductive organ. It is a mass near the far end of the mantle. The gonad is white in the male and orange in the female.
7. Locate the gills, two long organs that look like feathers. The gill hearts are located at the base of the gills.

Name _____ Class _____ Period _____

8. Probe the mantle near the fins with the scissors. Find a hard object. Grasp this hard object with the forceps and pull straight out from the body. This tough, transparent part is called the pen. It is the shell of the squid.
9. Remove an eye. Make a small cut in the center and find a round object. This is the lens.
10. In Figure 2 label the water jet, ink sac, stomach, liver, gonad, gill and gill heart.

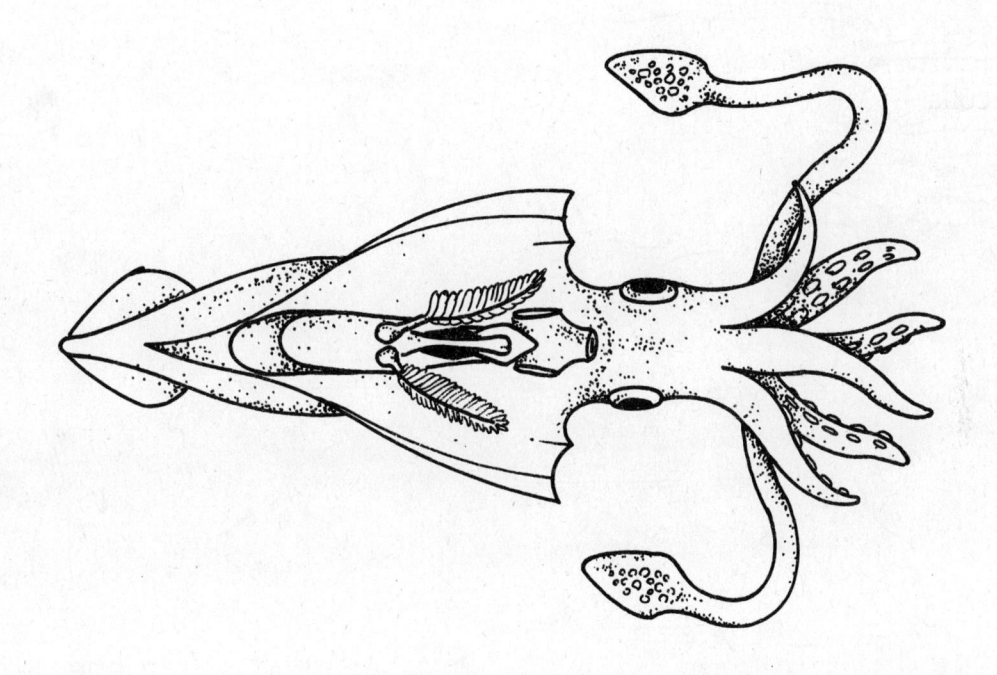

FIGURE 2. Internal parts of the squid

QUESTIONS

1. Describe the shape of the squid's body. _____
   _____

2. How might the body shape help it as it swims? _____
   _____

3. What do squid feed upon? _____

4. How might the eyes, tentacles and arms help the squid to capture food? ____
   _____
   _____

5. How does the squid use its ink sac? _____
   _____

55

6. What is the function of the squid shell? _____
7. Figure 3 shows several squid parts.
    Color red the part used for finding food.
    Color green the part used for holding food.
    Color gray the part used for tearing food.
    Color blue the part used to support the body.
    Color yellow the part used to help the squid swim forward or backward.
    Color purple the part that forms the body covering.

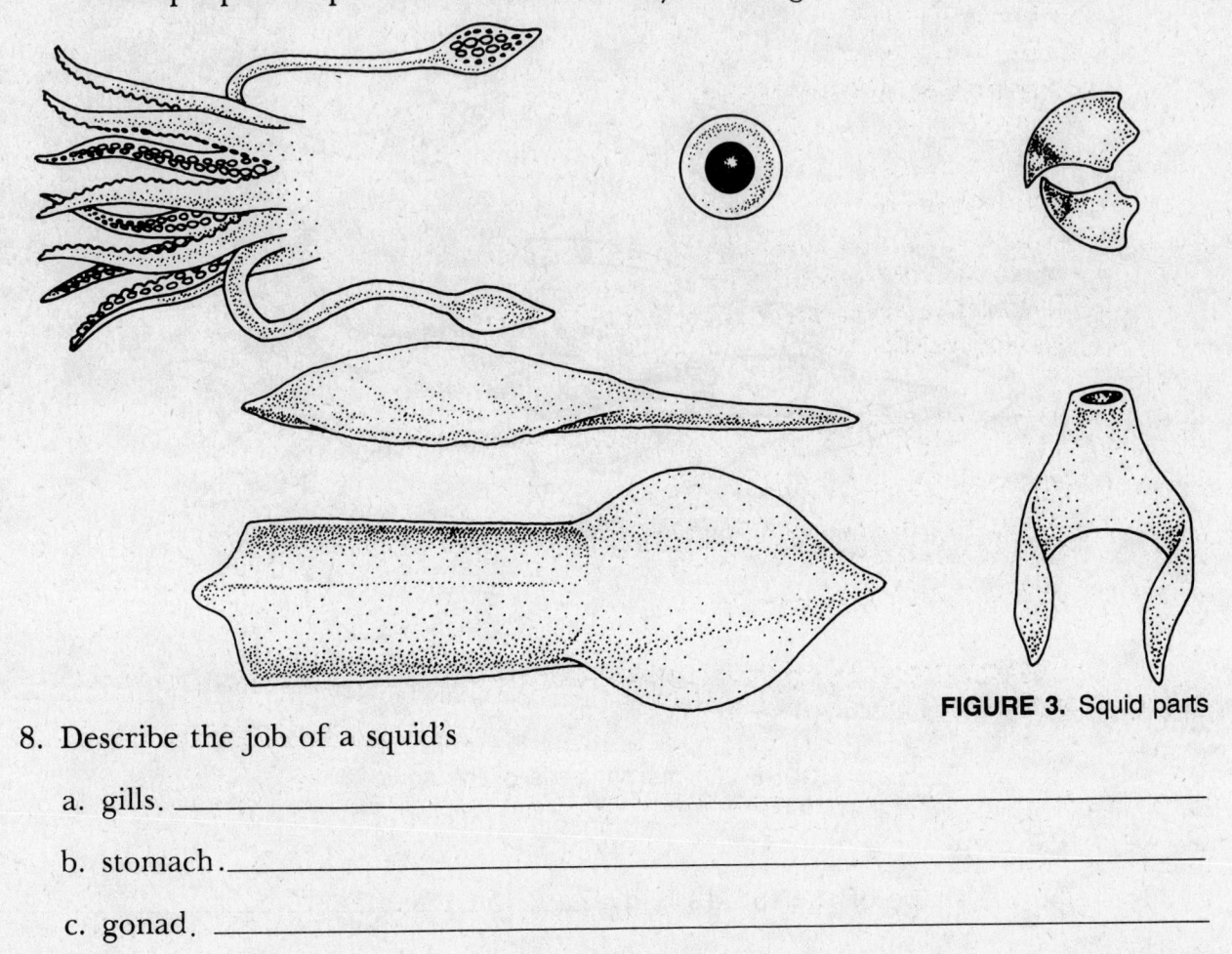

FIGURE 3. Squid parts

8. Describe the job of a squid's

    a. gills. _____

    b. stomach. _____

    c. gonad. _____

9. Shade in the correct arrow ends to show the direction water will squirt out from each jet. Shade in the correct arrow ends to show the direction the animal moves as water squirts from its jet.

FIGURE 4. Squid and water directions

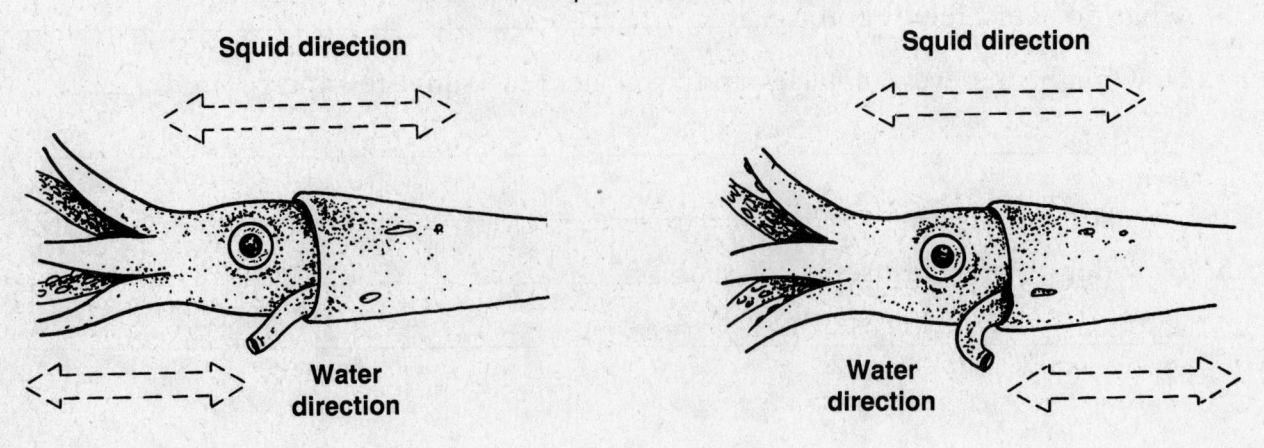

Squid direction

Squid direction

Water direction

Water direction

Name _____ Class _____ Period _____

7-3 What Are the Parts of a Stinging-cell Animal?

Stinging-cell animals have two basic body forms. One is free-swimming and the other remains attached to one place. The basic difference between the two is that the free-swimming form has a mouth and tentacles facing downward. The form that remains attached has a mouth and tentacles that face upward. The sea anemone is an example of the attached type. The sea anemone is so named because it resembles a flower called the anemone.

EXPLORATION

OBJECTIVES
In this exercise, you will:
   a. observe the external parts of a sea anemone.
   b. dissect and locate the internal parts of a anemone.

KEYWORDS
Define the following keywords:

mouth _____

pharynx _____

tentacle _____

MATERIALS
forceps
scalpel
scissors

hand lens
dissecting pan
preserved sea anemone

PROCEDURE
1. Place the sea anemone in the dissecting pan.
2. Examine the sea anemone and find each of the parts labeled in Figure 1. Write the functions of the parts in the table as you study them.
3. Look at the top of the sea anemone and find the mouth and tentacles. The bottom of the sea anemone is the disc by which the animal can attach itself to a rock or other surface.
4. Use the scalpel to cut through the center of the sea anemone from mouth to disc, to make two halves.
   CAUTION: *Always be careful when using a scalpel.*

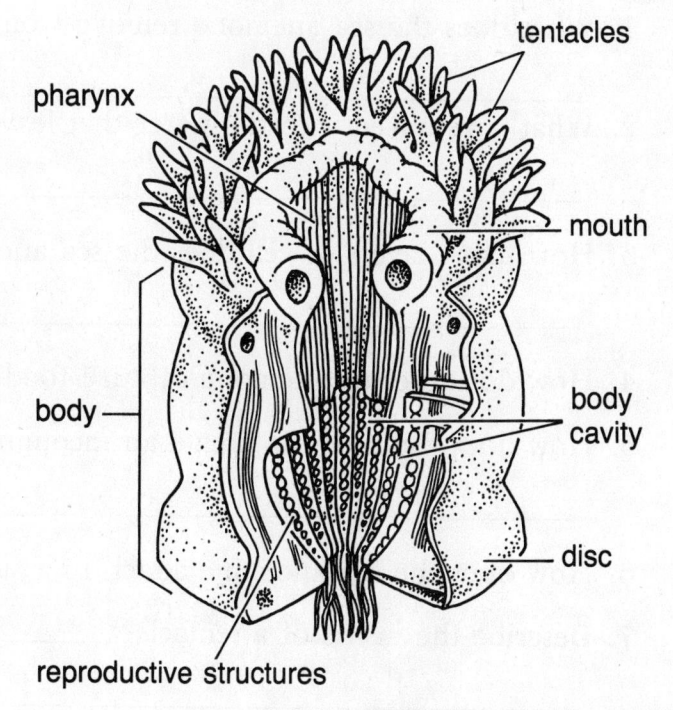

FIGURE 1. Sea Anemone

5. Use the hand lens to examine the outer and inner layers. Find the jellylike layer between them.
6. Find the mouth with its ringlike canal. The mouth leads into the pharynx. The pharynx extends from on-half to two-thirds the way down into the body cavity. Use the hand lens to find the grooves lined with cilia that run lengthwise in the pharynx. These cilia beat downward to create an incoming current of water that brings in oxygen and small particles of food. Other cilia beat upward to remove water from the anemone. This water takes with it carbon dioxide and other wastes.
7. Find the reproductive structures located on the sides of the body cavity below the pharynx.
8. Use the scalpel to open a tentacle and look inside it.
9. Dispose of the dissected sea anemone as your teacher directs. Wash your equipment and your hands.

Table 1. Stinging-cell Animal Parts

Mouth	
Tentacles	
Body	
Disc	
Pharynx	
Body cavity	
Reproductive structures	

QUESTIONS

1. What does the sea anemone remove from the water it takes in? _____

2. What is removed with the water that leaves a sea anemone's body? _____

3. How many layers of cell does the sea anemone have? _____

4. How does the sea anemone capture food? _____

5. How does the pharynx create an incoming current of water? _____

6. How does the sea anemone attach to a surface? _____

7. Describe the inside of a tentacle. _____

Name _____ Class _____ Period _____

8–1 What Are The Traits Of Certain Jointed-leg Animals?

Ticks, fleas, bedbugs, and lice belong to the phylum of jointed-leg animals. Jointed-leg animals share certain traits that separate them from other animal groups. Which traits do they share?

Each jointed-leg animal has a life cycle in which it changes from an egg to a young animal to an adult. The life cycle of a louse is typical of other jointed-leg animals.

EXPLORATION

OBJECTIVES
In this exercise, you will:
a. observe four jointed-leg animals and determine the traits of jointed-leg animals.
b. complete the life cycle of a louse.
c. determine what diseases are carried by certain jointed-leg animals.

KEYWORDS
Define the following keywords:

antennae _____

jointed-leg animal _____

life cycle _____
_____

segment _____

MATERIALS
scissors   ruler
tape   preserved tick, flea, bedbug and louse
petri dish   stereoscopic microscope
forceps

PROCEDURE
Part A. Traits of Four Jointed-leg Animals
1. Obtain a glass dish that contains a tick.
2. Examine the tick using a stereoscopic microscope.
3. Use forceps to turn the animal over so that you can observe both sides.
4. Measure the length of the body of the tick by placing a metric ruler under the glass dish alongside the animal. You can read the millimeters on the ruler through the stereoscopic microscope.
5. Complete the column in Table 1 for the tick. You will notice that one feature is already filled in.
6. Repeat steps 1 through 5 using a flea, a bedbug, and a louse.

59

Table 1. Traits of Four Jointed-leg Animals

Trait	Tick	Flea	Bedbug	Louse
Wings present?				
Antennae present?				
Antennae in segments?				
Three body regions easily seen?				
Abdomen in segments?				
Legs in segments?				
Number of legs?				
Claw-like ends on legs?				
Feed on human ___				
Body length in mm				

Part B. Life Cycle of a Louse

Locate Figure 1 of the life cycle of a louse. Complete the life cycle by following the steps below.

1. Note the drawings numbered 1 through 4 below the life cycle (Figure 1). Cut each one out along the solid lines that enclose the drawing. **CAUTION:** *Use care with scissors.*
2. Arrange them on the life cycle figure according to the position of the numbers.
3. Tape them in place.
4. Also beneath the life cycle are boxes containing a series of facts. Cut these boxes of facts out along the solid lines.
5. Place the facts onto the figure so that they best describe the correct diagram or event. The facts should be placed in the spaces outlined by the dashes.
6. Tape the facts in place.

Part C. Diseases Carried by Jointed-leg Animals

Each of the four animals studied in Part A can carry diseases to humans. The animal itself is not the cause of the disease. The animals carry certain disease-causing organisms in or on their bodies. When they bite a person, they may pass the disease-causing organism on to that person.

1. Table 2 lists five diseases. Complete the table by writing in the name of the jointed-leg animal that can carry that disease to a human.
2. The disease-carrying animal is pictured within Table 2.

Name _____ Class _____ Period _____

FIGURE 1. Life cycle of head louse

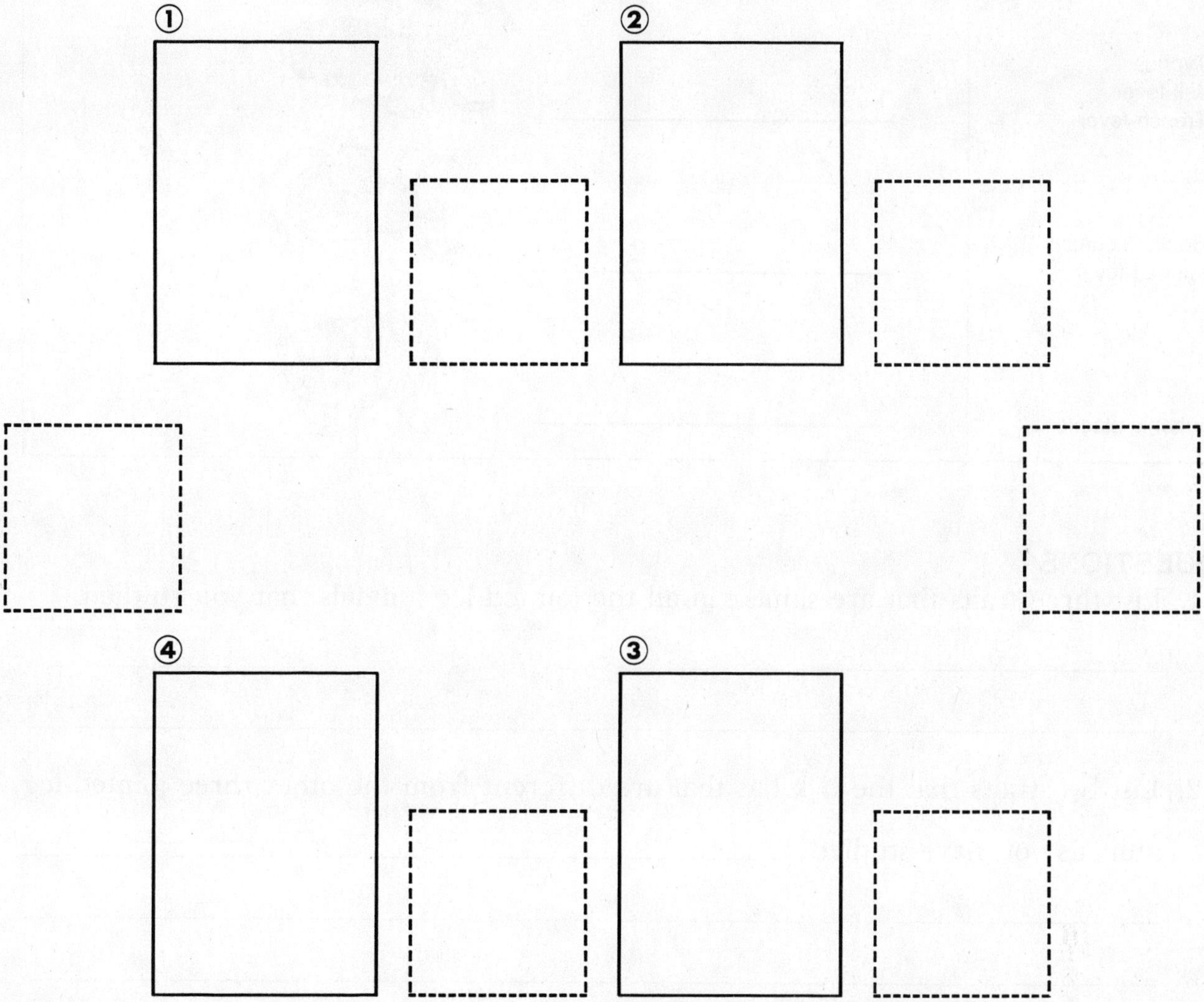

① ②

④ ③

| Adult lives for 35–40 days. | Adult lice crawl about on hair of head and mate. | Young reach adult stage in 10 days. | Young lice bite skin and feed on blood. | 80 to 100 eggs are attached to hairs by each female. | Eggs hatch in 8–10 days. |

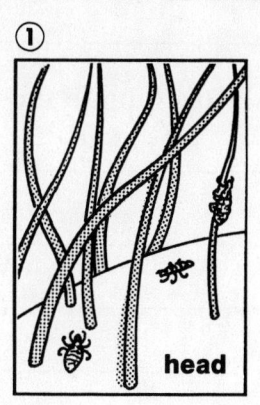

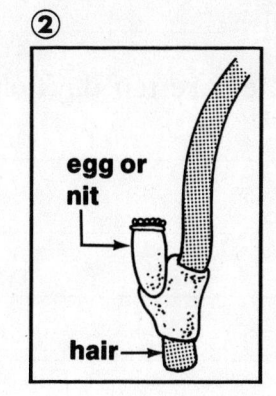

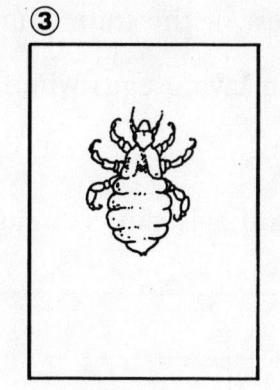

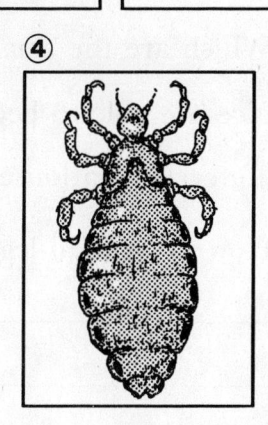

① head ② egg or nit / hair ③ ④

61

Table 2. Diseases Carried by Jointed-leg Animals

Disease	Jointed-leg animal that carries the disease
Typhus Jail fever Trench fever	_____
Rocky Mountain spotted fever	_____
Bubonic plague	_____

QUESTIONS

1. List three traits that are similar in all the jointed-leg animals that you studied.

   _____

   _____

2. List two traits that the tick has that are different from the other three jointed-leg animals you have studied. _____

   _____

   _____

3. Which of the four animals studied could be grouped into a different class? Why? _____

   _____

4. Which is the largest of the four animals? _____

5. Which are the smallest of the four animals? _____

6. Lice are able to begin laying eggs when they are ten days old. How long is the life cycle of a louse? _____

7. How do jointed-leg animals cause disease? _____

   _____

   _____

Name _____ Class _____ Period _____

8–2 How Do Fish Classes Compare?

Fish are vertebrates that live in water and breathe with gills. There are three classes of fish. The most primitive class is jawless fish. Jawless fish have cartilage skeletons and lack jaws. There are fewer than 50 species in this class. The sharks and rays belong to the cartilage-fish class. They also have cartilage skeletons but possess jaws. The largest class is the bony fish. They have bony skeletons and jaws. There are more than 30,000 species of bony fish. Besides the characteristics of the skeleton and the jaws, each fish class has certain traits that separate it from the other two classes.

EXPLORATION

OBJECTIVES
In this exercise, you will:
   a. observe a jawless fish, a cartilage fish, and a bony fish.
   b. record the traits of each fish.

KEYWORDS
Define the following keywords:

cartilage _____

chordate _____

free-living _____

parasite _____

MATERIALS
probe
dissecting pan
preserved shark, preserved lamprey, and a preserved perch

PROCEDURE
1. Examine the lamprey.
2. Complete the column in Table 1 for the lamprey. Select the correct answers from the column marked "choices."
3. Examine the shark.
4. Complete the column in Table 1 for the shark by using the correct answers from the column marked "choices."
5. Examine the perch.
6. Complete Table 1 for the perch by using the correct answers from the column marked "choices."

63

Table 1. Comparison of Jawless, Cartilage, and Bony Fish

Trait	Choices	Jawless fish	Cartilage fish	Bony fish
Skin	Scales Smooth, no scales Rough, no scales			
Mouth location	Under chin, jaws At front, jaws At front, no jaws			
Mouth shape	Round Horizontal			
Teeth	Yes No			
Fins present	Yes No			
Number of body fin pairs (do not include fins along back)	0 2 3			
Location of fins	Draw on body			
Nostrils present	Yes No			
Number of nostrils	0 1 2			
Gills present	Yes No			
Gills covered by a flap of skin	Yes No			
Number of gill openings	1 5 7			
Eyes present	Yes No			
Eyelids present	Yes No			
Skeleton	Endoskeleton Exoskeleton			
Kind of skeleton	Bone Cartilage			
Way of life	Parasite Free-living			
Backbone	Yes No			

Name _____ Class _____ Period _____

QUESTIONS

1. To what phylum do jawless fish, cartilage fish and bony fish belong? _____
   _____

2. List two traits of all chordates. _____
   _____

3. Are jawless fish, cartilage fish, and bony fish invertebrates or vertebrates? _____
   _____

   Explain your answer. _____
   _____

4. List three ways in which all fish are alike. _____
   _____

5. List three ways that jawless fish and cartilage fish are alike. _____
   _____
   _____

6. List three ways in which jawless fish and cartilage fish are different. _____
   _____
   _____
   _____

8. List three ways that cartilage fish and bony fish are different. _____
   _____
   _____
   _____

9. Explain the meaning of the following terms:

   a. jawless fish _____

   b. cartilage fish _____

   c. bony fish _____

10. Complete the classification for the following three animals. Write the correct name on the lines provided below each fish.

Kingdom _____   _____   _____

Phylum _____   _____   _____

Class   _____   _____   _____

Kingdom _____   _____   _____

Phylum _____   _____   _____

Class   _____   _____   _____

66

Name _____ Class _____ Period _____

8–3 What Are the Parts of a Starfish?

Starfish are perhaps the best known spiny-skin animals. They live all over the world in coastal waters and along rocky shores. They may be seen in various colors such as red, purple, green, blue, and yellow. They range in diameter from about 2 centimeters to almost a meter. These slow moving animals are economically important because they prey on oysters, clams, and other organisms that are used by people for food.

EXPLORATION

OBJECTIVES
In this exercise, you will:
   a. study the internal and external structures of a starfish.
   b. observe how a starfish is adapted for living in water.

KEYWORDS
Define the following keywords:

disc _____

dorsal _____

mouth _____

spines _____

tube feet _____

ventral _____

MATERIALS
preserved starfish
dissecting pan
scissors
forceps
stereomicroscope or hand lens

PROCEDURE
Part A. External Structure of the Starfish
1. Place the starfish in the dissecting pan with the dorsal side up.
2. Use the information in Figure 1 to find the external parts of the starfish.
3. Locate the disc, the center raised portion from which the arms radiate.

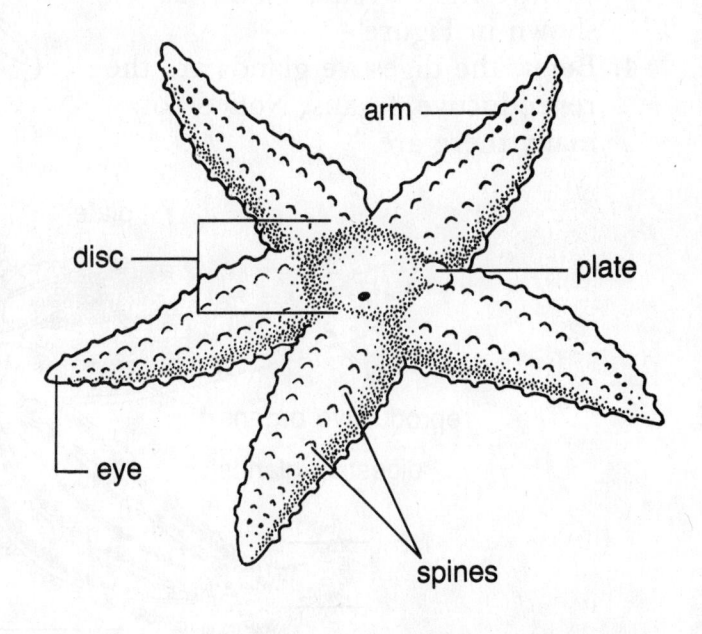

FIGURE 1. Dorsal

67

4. Find the small, round plate on one side of the starfish near its center. The plate takes in water.
5. Find the red spot at the tip of each arm. These are the eyes.
6. Examine the skin of the starfish. Look at the spines under the stereomicroscope or with the hand lens.
7. Turn the starfish so that the ventral side is up. Use Figure 2 to find the mouth in the middle of the starfish. Examine the ring of small spines around the mouth.
8. Find the groove that extends from the mouth to the tip of each arm.
9. Locate the tube feet that line the groove. Observe the tube feet under the stereomicroscope or with the hand lens.

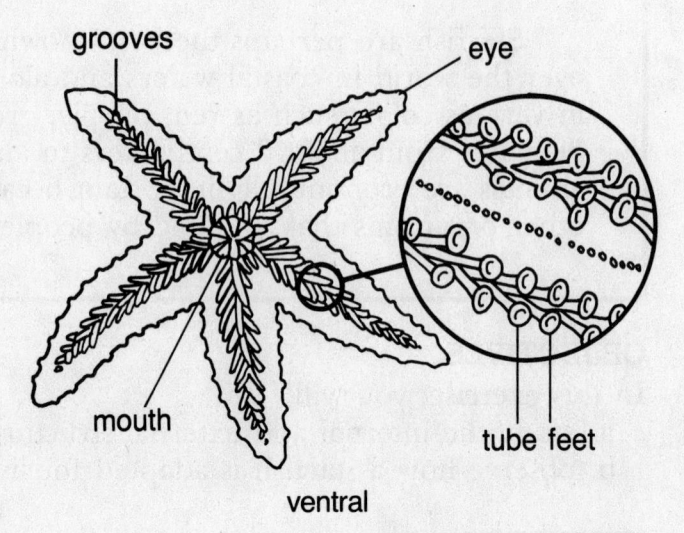

FIGURE 2. Ventral

Part B. Internal Structure of the Starfish

1. Place the starfish dorsal side up in the dissecting pan.
2. Use the scissors to remove the top skin and skeleton carefully from one arm as shown in Figure 3.
   CAUTION: *Always be careful when using scissors.*
3. Find the pair of digestive glands inside the arm. Use forceps to remove the digestive glands as shown in Figure 4.
4. Below the digestive glands are the reproductive organs. Notice how many there are.

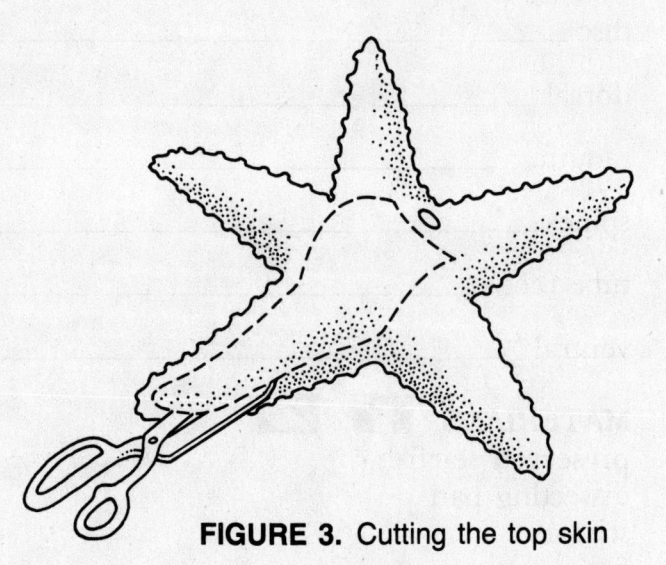

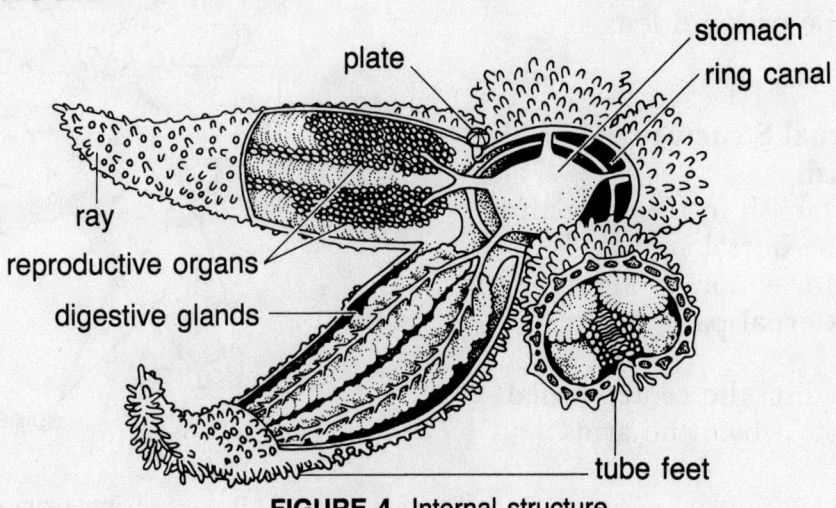

FIGURE 3. Cutting the top skin

FIGURE 4. Internal structure

Name _____ Class _____ Period _____

5. Remove the skin and skeleton from a second arm.
6. Carefully remove the remaining skin and skeleton from the central disc. The part directly below is the stomach.
7. Remove all organs of the digestive system and reproductive system so that you can see the water vascular system as shown in Figure 5.
8. Dispose of the dissected starfish as your teacher directs. Wash your equipment and your hands.

FIGURE 5. Water vascular system

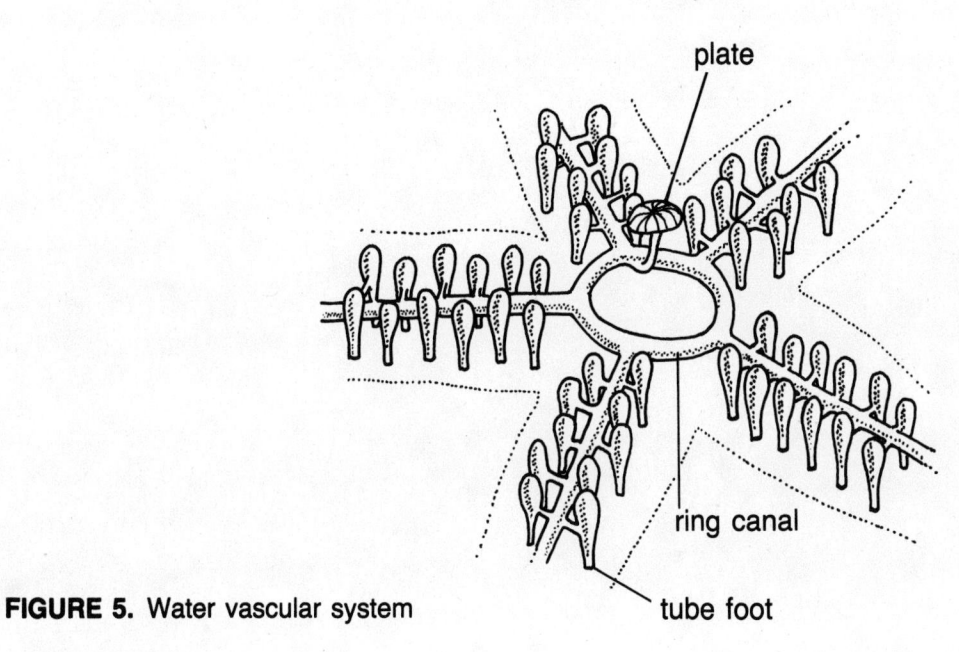

QUESTIONS

1. How many arms does the starfish have? _____

2. What kind of symmetry does the starfish have? _____

3. Why can the starfish move equally well in any direction? _____
   _____

4. Describe the surface of the starfish? _____
   _____

5. How many rows of tube feet does the starfish have? _____
   _____

6. What are the functions of the tube feet? _____
   _____

7. Where is the mouth on the starfish? _____

8. Where is the mouth located in relation to the stomach? _____
   _____

9. How many digestive glands does the starfish have? _____

10. How many reproductive organs does the starfish have? _____

11. How is the starfish adapted for living in shallow marine waters? _____

Name _____ Class _____ Period _____

9–1 What Are the Tests for Fats and Proteins?

All foods contain some of the six nutrients. However, you cannot tell what nutrients are in food by looking at the food. You can tell if certain nutrients are present by doing certain tests. This activity will let you test a number of different foods to find out if fats and proteins are present.

INVESTIGATION

OBJECTIVES
In this exercise, you will:
a. test a food containing fat to observe the test results when fat is present.
b. test a food containing protein to observe the test results when protein is present.
c. test a variety of foods to see if fat and protein are present.

KEYWORDS
Define the following keywords:

control _____

_____

fat _____

nutrient _____

protein _____

translucent _____

MATERIALS
brown paper
cooking oil
8 droppers
test tube rack
8 stoppers
Biuret solution
8 test tubes
labels or wax pencils
food samples
sticks, wood or plastic

FIGURE 1. Rubbing foods on brown paper

PROCEDURE
Part A. Testing a Fat and a Not-fat
1. Use a dropper to place one small drop of cooking oil (fat) onto a piece of brown paper. Label the drop *fat*.

71

2. Rub the drop around on the paper, making a circular area (Figure 1).
3. Use a different dropper to place one small drop of water onto the piece of brown paper away from the oil spot. Label the spot *Non-fat*.
4. Rub the drop around on the paper, making a circular area. Don't use the same finger that you used for the oil.
5. Wait 5 to 10 minutes for the liquids to dry. You can speed up the drying time by waving the paper in the air.
6. Hold the paper up toward window light.
7. Light passes through the oil spot. This spot is said to be translucent. *Fat forms a translucent spot on brown paper*.
8. Record your results in Table 1.

Part B. Testing for Fat

1. Test the following foods for fat: butter, egg white (solid), peanut butter, bacon, mayonnaise, cottage cheese, lard, and apple. Apply a small amount of each food to the large piece of brown paper. Make sure you label the food used at each circular area on the paper. **CAUTION:** *Do not taste food samples. Chemical spills may have made them unsafe.*
2. Wait 5 to 10 minutes and then check to see if a translucent spot appears.
3. Record the results of your tests in Table 1.

Table 1. Testing for Fat

Food	Translucent spot present?	Fat present?
Cooking oil (fat)		
Water (non-fat)		
Butter		
Egg white (solid)		
Peanut butter		
Bacon		
Mayonnaise		
Cottage cheese		
Lard		
Apple		

Name _____ Class _____ Period _____

Part C. Testing a Protein and a Non-protein

1. Add 10 drops of egg white liquid (a protein) to a test tube. Label this tube *protein*.
2. Add 10 drops of water to a second test tube. Label this tube *water*.
3. Label a third test tube *Biuret*.
4. Add 10 drops of Biuret solution to each of the three test tubes. **CAUTION:** *If the Biuret solution is spilled, wash it off your hands or clothing immediately.*
5. Place a stopper into the first two tubes. Shake both tubes gently for one minute.
6. Compare the color of all three tubes. Biuret solution forms a purple color when mixed with protein.
7. Record your results in Table 2.

Table 2. Testing for Protein

Tube number and/or contents	Color with Biuret solution	Protein present?
Egg white liquid (protein) + Biuret		
Water (non-protein) + Biuret		
Biuret solution		
1. Biuret		
2. Biuret + cottage cheese		
3. Biuret + lard		
4. Biuret + veal		
5. Biuret + tuna		

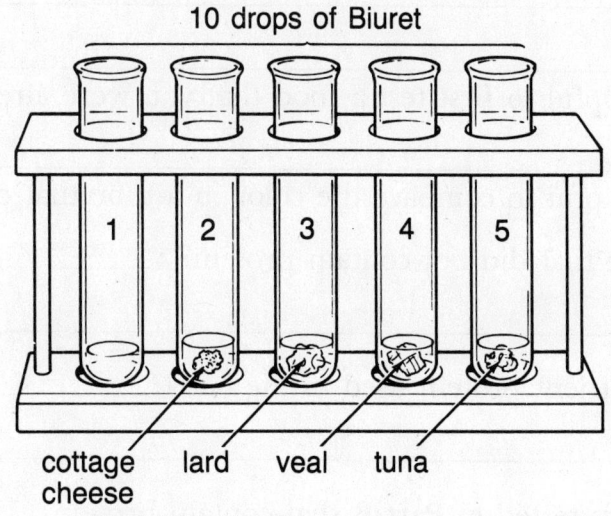

FIGURE 2. Testing for protein

Part D. Testing Foods for Protein

1. Number five test tubes 1 through 5 and place the tubes in a test tube rack.
2. Add the following to each tube using Figure 2 as a guide. Sticks may be used to add very small amounts of each food (about the size of 1/2 a pea) to each tube.

73

Tube 1: 10 drops of Biuret solution.
Tube 2: 10 drops of Biuret solution and cottage cheese.
Tube 3: 10 drops of Biuret solution and lard.
Tube 4: 10 drops of Biuret solution and veal.
Tube 5: 10 drops of Biuret solution and tuna.

3. Place a stopper in tubes 2 through 5. Shake tubes 2 through 5 for 1 minute.
4. Compare the color of the solution in tubes 2 through 5 with tube 1.
5. Record your results in Table 2.

QUESTIONS

1. Describe how you can test for the nutrient fat in foods. _____

2. Why was it helpful to first test a food that you were told already had fat in it? _____

3. Why was it helpful to compare the spot left by fat to the spot left by water? See your definition of *control* in the keywords section. _____

4. How is the nutrient fat used in the body? _____

5. List those foods tested in Part B that contained fat _____

6. Your friend's brown lunch bag has a large translucent spot on the bottom. How would you explain why the spot is there? _____

7. Describe how you can test for the nutrient protein in foods. _____

8. Why was it helpful to first test a food that you were already told had protein in it? _____

9. Why was it helpful to compare the color in a tube that contained protein to the color in a tube that did not contain protein? _____

10. How is the nutrient protein used in the body? _____

11. List those foods tested in Part B that contain protein. _____

Name _____ Class _____ Period _____

9-2 How Do You Test for Vitamin C in Foods?

Vitamin C is also known as ascorbic acid. It is a water soluble vitamin. This means that it will dissolve in water. As a vitamin, it is present in many different foods. Tomatoes and salad greens are examples of foods that are good sources of vitamin C. Citrus fruits, such as lemons, limes, and oranges are excellent sources of vitamin C.

Vitamin C can be detected in foods by using a special blue testing chemical. You will use this chemical to test for the vitamin in a variety of different foods. If a food contains vitamin C, the blue color of the testing chemical will turn colorless. If no vitamin C is present, the color will remain blue.

INTERPRETATION

OBJECTIVES
In this exercise, you will:
a. learn how to test for vitamin C.
b. test citrus fruits to find out which ones do or do not have vitamin C.
c. find out if cooking removes vitamin C from food.
d. find out if canning food changes the amount of vitamin C present in the food.

KEYWORDS
Define the following keywords:

citrus fruit _____

vitamin _____

vitamin C _____

water soluble vitamin _____

MATERIALS

9 test tubes
blue testing chemical
10 droppers
water
vitamin C
raw cabbage juice

cooked cabbage juice
fresh lemon juice
fresh orange juice
fresh grapefruit juice
fresh tomato juice
canned tomato juice

NOTE: all liquids to be tested may be placed in small, labeled plastic dropping bottles for student use.

PROCEDURE

Part A. Testing for Vitamin C

1. Label one test tube *Vitamin C* and a second test tube *Water*.
2. Use a dropper to add 10 drops of blue testing chemical to each test tube. **NOTE:**
3. Using a different dropper, add one drop at a time of vitamin C to the test tube labeled *Vitamin C* until the blue testing chemical turns from blue to clear (see Figure 1a). Shake the tube after adding each drop.
4. Count the number of drops of vitamin C needed to turn the blue testing chemical from blue to clear.
5. Record this number in Table 1.
6. Repeat steps 3 through 5 for the test tube labeled *water*, only this time, use water instead of vitamin C in step 3 (see Figure 1b).
7. If the blue testing chemical has not turned clear after 50 drops of water, stop and record the number *50* in Table 1. **NOTE:** *If a liquid has much vitamin C in it, it will take only a few drops to turn the blue testing chemical clear. If a liquid has very little vitamin C in it, it will take many drops to turn the blue testing chemical clear.*

Table 1. Testing for Vitamin C

Liquid being tested	Number of drops of liquid used
Vitamin C	
Water	

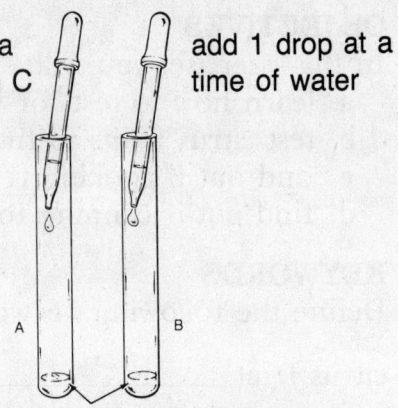

FIGURE 1. Testing for vitamin C

Table 2. Testing Food for Vitamin C

Food Being Tested	Number of Drops of Food Used	Vitamin C Present?
1. Lemon juice		
2. Orange juice		
3. Grapefruit juice		
4. Raw cabbage juice		
5. Cooked cabbage juice		
6. Fresh tomato juice		
7. Canned tomato juice		

Name _____ Class _____ Period _____

Part B. Testing Citrus Fruits for Vitamin C
1. Label three test tubes from 1 through 3.
2. Use a dropper to add 10 drops of blue testing chemical to each test tube.
3. Test for vitamin C as follows:
   a. Using a different dropper, add to Tube 1 one drop of lemon juice at a time until the blue testing chemical turns from blue to clear.
   b. Shake the tube after adding each drop of lemon juice.
   c. Count the number of drops of lemon juice needed to turn the blue testing chemical from blue to clear.
4. Record this number in Table 2.
5. Repeat steps 3 through 4 for Tube 2 using orange juice instead of lemon juice.
6. Repeat steps 3 through 4 for Tube 3 using grapefruit juice instead of lemon juice.

Part C. Is Vitamin C Lost During Cooking?
1. Label two test tubes from 4 through 5.
2. Use a dropper to add 10 drops of blue testing chemical to each test tube.
3. Test for vitamin C as follows:
   a. Using a different dropper, add to Tube 4 one drop of raw cabbage juice at a time until the blue testing chemical turns from blue to clear.
   b. Shake the tube after adding each drop of cabbage juice.
   c. Count the number of drops of raw cabbage juice needed to turn the blue testing chemical from blue to clear.
4. Record this number in Table 2.
5. Repeat steps 3 through 4 for Tube 5 using cooked cabbage juice instead of raw cabbage juice.

Part D. Is Vitamin C Lost When Food is Canned?
1. Label two test tubes from 6 through 7.
2. Use a dropper to add 10 drops of blue testing chemical to each test tube.
3. Test for vitamin C as follows:
   a. Using a different dropper, add to Tube 6 one drop of fresh tomato juice at a time until the blue testing chemical turns from blue to clear.
   b. Shake the tube after adding each drop of tomato juice.
   c. Count the number of drops of fresh tomato juice needed to turn the blue testing chemical from blue to clear.
4. Record this number in Table 2.
5. Repeat steps 3 through 4 for Tube 7 using canned tomato juice instead of fresh tomato juice.

QUESTIONS
1. What was the color of the testing chemical

   a. before adding anything to it? _____

   b. after adding vitamin C? _____

   c. after adding water? _____
2. In Part A, which liquid
   a. took the fewest drops to turn the testing chemical from blue to clear?

   _____

b. took the most drops to turn the testing chemical from blue to clear?_____

   c. contains little or no vitamin C?_____

   d. was your control?_____

3. a. In Part B, which food contained vitamin C?_____
   b. Recheck the meaning of the word *citrus fruit* in the Keywords section.

      Were all the foods tested in Part C citrus fruits?_____
   c. If you wanted to increase the amount of vitamin C in your diet, what food

      type would be helpful to you?_____
4. In Part C, which liquid
   a. took the fewest drops to turn the testing chemical from blue to clear?

      _____

   b. took the most drops to turn the testing chemical from blue to clear?

      _____

   c. contains the most vitamin C?_____

   d. contains the least vitamin C?_____
5. a. Based on your results from Part C, does cooking raise or lower the amount of

      vitamin C in food?_____
   b. Does cooking raise or lower the amount of vitamin C in the cooking

      liquid?_____
   c. If you wanted to increase the amount of vitamin C in your diet, should you

      throw away the liquid in which foods like cabbage have been cooked?_____
   d. Recheck the meaning of *water soluble vitamin* listed in the Keywords section.
      Explain why cooking lowers the amount of vitamin C in foods.

      _____

6. In part D, which food
   a. took the fewest drops to turn the testing chemical from blue to clear?

      _____

   b. took the most drops to turn the testing chemical from blue to clear?

      _____

   c. contains the most vitamin C?_____

   d. contains the least vitamin C?_____
7. a. Based on your results from part D, tell what happens to the amount of

      vitamin C in foods when they are canned._____
   b. If you wanted to increase the amount of vitamin C in your diet, should you

      eat more canned or raw foods?_____

Name _____ Class _____ Period _____

10-1 How Do Digestive System Lengths Compare?

You know that the diet of different animals may vary. You can buy cat food, dog food, and bird food in most supermarkets.

The length of the digestive system may also vary. Animals that eat plants usually have longer digestive systems than animals that eat meat.

EXPLORATION

OBJECTIVES
In this exercise, you will:
a. measure the length of the digestive system in three animals.
b. compare these lengths with the type of food eaten.

KEYWORDS
Define the following keywords:

caecum _____

carnivore _____

digestive system _____
_____

herbivore _____

MATERIALS
string   metric ruler   scissors   tape

PROCEDURE
1. Place a piece of string down on the outline drawing of the rabbit digestive system in Figure 2 on the next page. Figure 1 shows you how.
2. Tape the end of the string in place at the label marked "start" on the stomach of the rabbit.
3. Position the string only over the entire length of the *unshaded* organs. It must match, exactly, the many twists and turns of the stomach, the small intestine and the large intestine (the unshaded organs).
4. When you reach the anus, cut the string, remove it from the drawing, and stretch it out its full length.
   CAUTION: *Use care with scissors.*
5. Measure the length of the string in centimeters and record this number in Table 1.

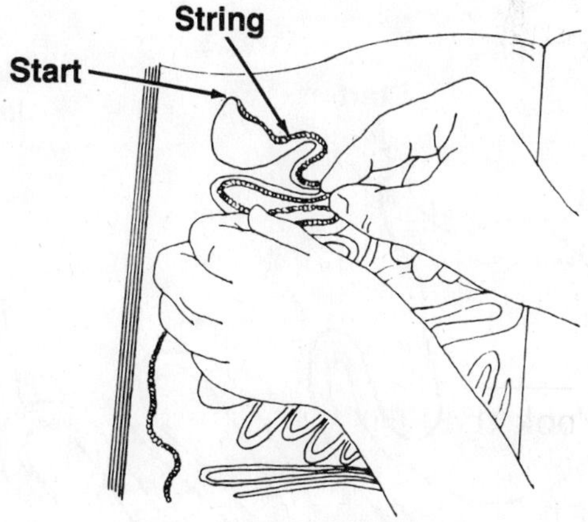

Measuring intestine with string

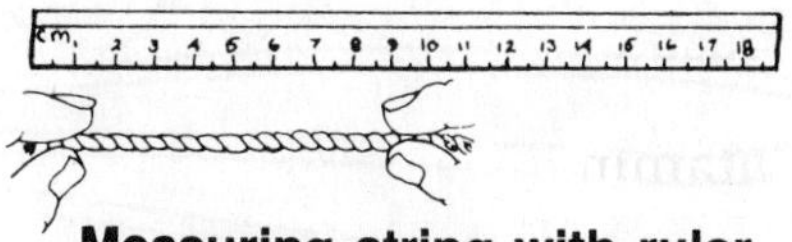

Measuring string with ruler

FIGURE 1. Measuring the digestive system

79

FIGURE 2. Digestive systems

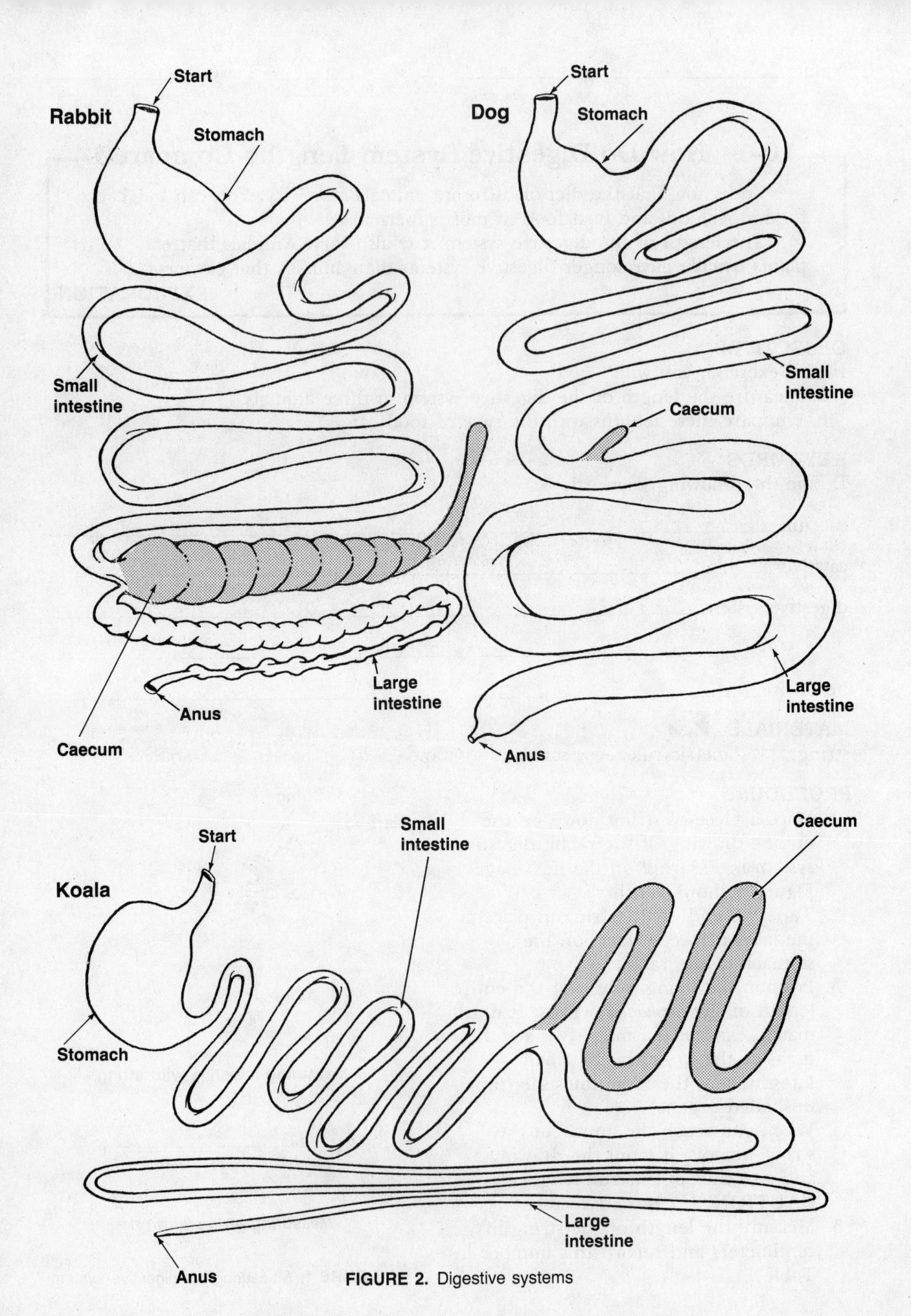

Name _____ Class _____ Period _____

6. Position the string over the shaded portion of the rabbit digestive system and measure the length of the caecum. Record this measurement in centimeters in Table 1.
7. Add together the two numbers that you have now recorded in the table in order to get the total length of the digestive system. Record this number in Table 1.
8. The diagram of the rabbit digestive system is drawn ⅓ smaller than actual size. Multiply the total digestive system length by 3 to complete the first row of Table 1. This number is the actual length of the rabbit digestive system.
9. Repeat steps 1 through 8 for the digestive system of the koala and the dog.

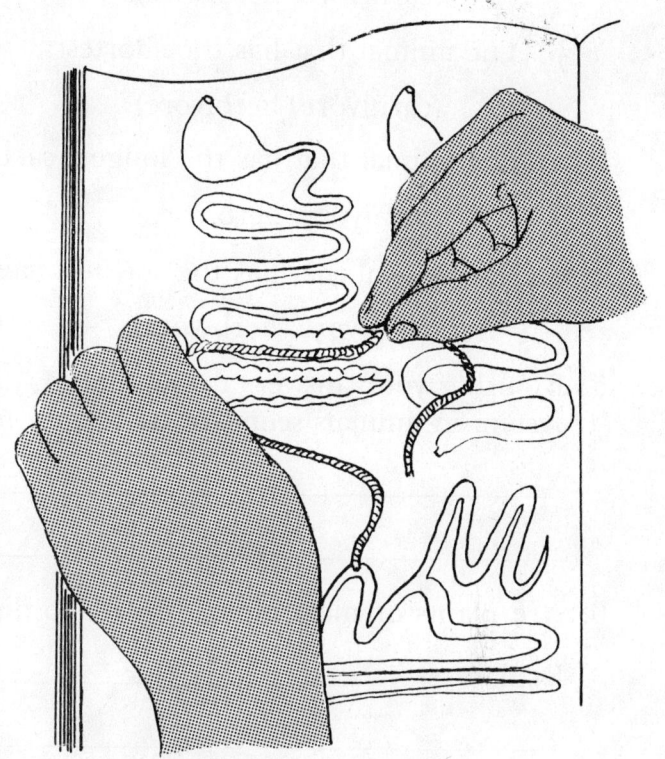

FIGURE 3. Measuring caeca of animals

Table 1. Digestive System Measurements

Animal	Length of stomach, small intestine, large intestine	Caecum length	Total digestive system length	Multiply by 3	Actual length of digestive system
Rabbit	+	=	×	=	
Koala	+	=	×	=	
Dog	+	=	×	=	

QUESTIONS

1. Which animal has the longest actual digestive system? _____

2. Which animal has the shortest actual digestive system? _____

3. Based on what you already know, tell whether the animals used in this experiment are carnivores or herbivores. NOTE: The koala is an Australian animal that feeds only on the leaves and buds of the eucalyptus tree.

   rabbit _____ koala _____ dog _____

4. Circle the correct answer to the following questions:
   a. The animal that has the longest actual digestive system is a

   (carnivore, herbivore).

   b. The animal that has the shortest actual digestive system is a

   (carnivore, herbivore).

   c. The animal that has the longest caecum is a

   (carnivore, herbivore).

   d. The animal that has the shortest caecum is a

   (carnivore, herbivore).

5. By using your answers to question 4, describe how the length of the digestive system in animals seems to be related to the type of food the animals eat.

   _____

   _____

6. Are plants or meat more difficult to digest?_____

   Explain._____

   _____

7. Use the word *long* or *short* to describe what you think the length of the digestive system might be in the

   lion _____ cat _____ horse _____ deer _____

   panther _____ donkey _____ cow _____ wolf _____

8. Two different animals of almost the same size have digestive systems that are of the following lengths:

   Animal A—410 cm   Animal B—145 cm

   a. Which one of these animals is most likely to be a carnivore?_____

   b. Explain your answer._____

   _____

   c. Which one of these animals is most likely to be a herbivore?_____

   d. Explain your answer._____

   _____

9. Describe the function of the:

   a. stomach_____

   b. small intestine_____

   c. large intestine_____

82

Name _____ Class _____ Period _____

10–2 How Does the Fish Digestive System Work?

Biologists study many different animals in order to learn more about animals. These studies often help us to understand more about ourselves.

By studying the organ systems of many animals, we have learned that the organ systems of different animals are often very similar. For example, the digestive system of a frog or a fish is similar to the digestive system of a human.

By studying the digestive system of a fish, you will learn how the different parts work together to break down food. You will also see how the fish digestive system can be compared to the human digestive system.

EXPLORATION

OBJECTIVES
In this exercise, you will:
- a. observe the outside and inside of a fish.
- b. study the fish digestive system.
- c. compare the fish digestive system to that of a human.

KEYWORDS
Define the following keywords:

chemical change _____

digestive system _____

_____

physical change _____

MATERIALS
scissors   metric ruler   preserved perch
scalpel   dissecting pan   colored pencils:
forceps   red and blue

PROCEDURE
1. Place the fish in the dissecting pan.
2. Locate the parts on the outside of the fish as shown in Figure 1.
3. Use the forceps to lift the gill cover and to open the mouth.

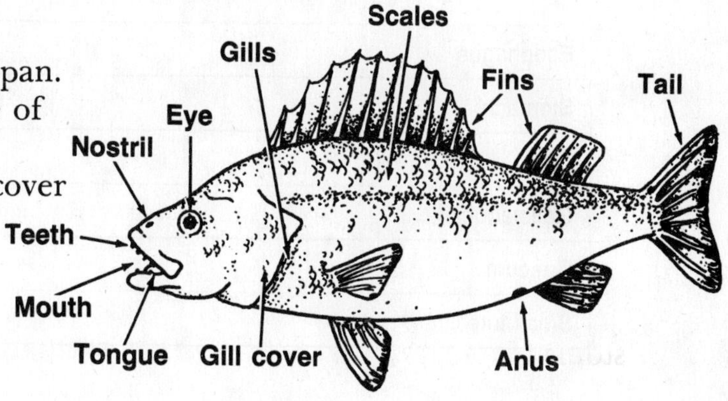

FIGURE 1. Outside of a fish

83

4. Complete Table 1 at this time. It may be necessary to use your textbook and the glossary at the back of this guide as reference.

Table 1. Outside Fish Parts and Their Functions

Fish part	Function	Part of what body system?
Gills		
Eyes		
Scales		
Fins		
Nostrils		
Mouth		
Tail		
Anus		
Teeth		
Tongue		

5. By using a scissors, remove a section of skin and muscle from the side of your fish. **CAUTION:** *Use care with scissors.* Use the directions in Figure 2 as a guide to where to make the first, second, third, and fourth cuts. They are marked cuts 1, 2, 3, and 4. *The body parts shown in Figure 2 should be seen easily now.*
6. Locate and identify each of the body parts shown in Figure 2.
7. Complete Table 2 at this time. It may be necessary to refer to your textbook and the glossary in this guide.

Table 2. Parts of the Fish Digestive System and Their Functions

Digestive system organ	Function
Esophagus	
Stomach	
Liver	
Gallbladder	
Caecum	
Small intestine	

Name _____ Class _____ Period _____

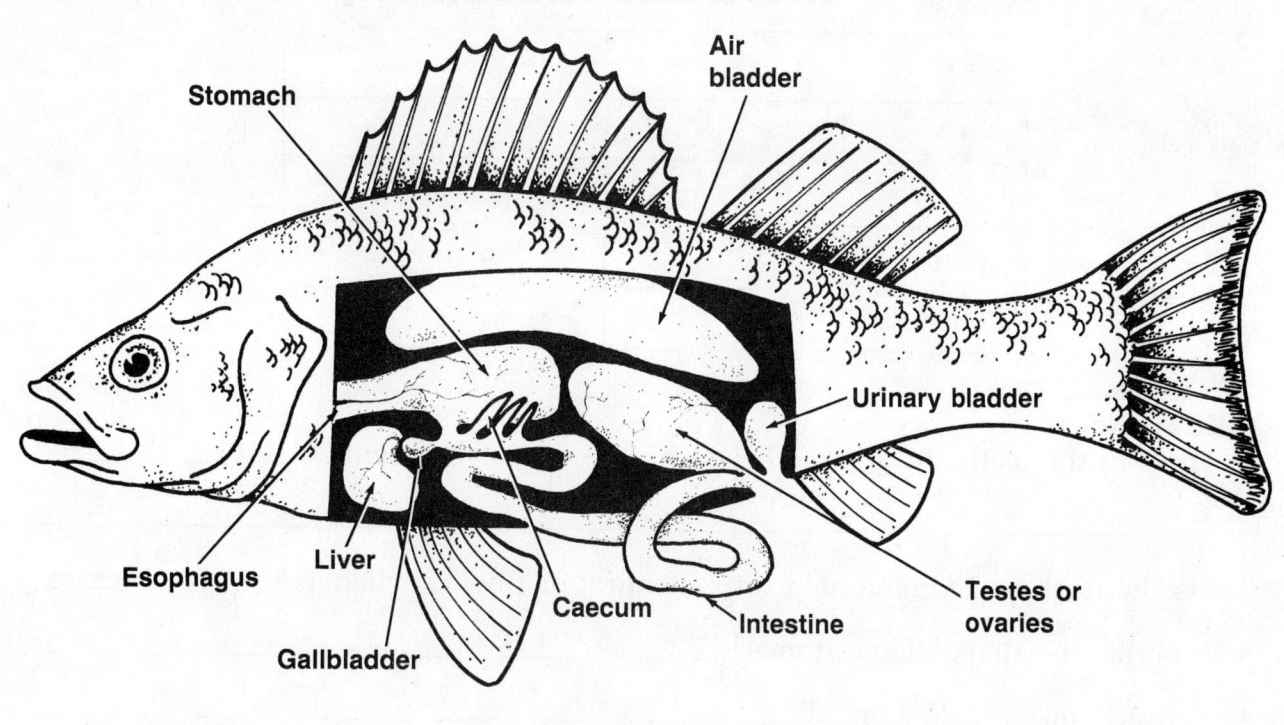

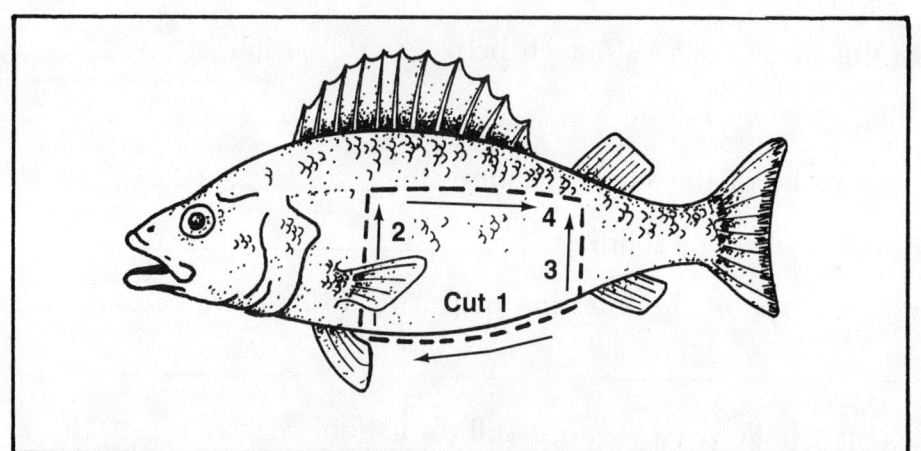

FIGURE 2. Fish digestive system

8. Color red those organs in Figure 2 that are part of the digestive system.
9. Color blue all organs that are not part of the digestive system.
10. Use a scissors to cut loose and remove the entire digestive system.
11. Straighten out the digestive system as best you can and measure its length in centimeters. Record this number after step 12.
12. Measure the entire length of the fish from mouth to tail in centimeters. Record this number in the space provided here.

    Length of digestive system = _____

    Length of fish = _____
13. Use a scalpel to cut through the organs listed in Table 3 to determine if they are hollow or solid. If they are hollow, food will pass through them. If solid, food will not pass through them. **CAUTION:** *Use extreme care with a scalpel.*
14. Complete Table 3. The gallbladder has already been done for you.

Table 3.

Organ	Hollow or solid?	Food passes through?
Stomach		
Liver		
Gallbladder	hollow	no
Caecum		
Small intestine		

QUESTIONS

1. How do the teeth and tongue help with the digestive system? _____

2. Do the teeth and tongue help with chemical or physical changes? _____

3. Describe the shape of the stomach. _____

4. Is the stomach hollow or solid? _____

5. How does the shape of the stomach help it to do its job? _____

6. Describe the shape of the small intestine. _____

7. Is the intestine hollow or solid? _____

8. How does the shape of the intestine help it to do its job? _____

9. The liver is not hollow. How can it still do its job? _____

10. How does the length of the digestive system compare to the length of the fish? _____

11. How does the great length of the digestive system help it to do its job? _____

12. By using the picture of the human digestive system appearing on page 209 in your text, explain how the fish and human digestive systems

    a. are alike _____

    b. are different _____

Name _____ Class _____ Period _____

11-1 How Does the Heart Work?

The heart is a muscular organ which pumps blood. It is divided into four chambers. The two upper chambers take in blood. The two lower chambers pump blood out of the heart. An upper chamber is called an atrium. A lower chamber is called a ventricle.

Blood moves only in one direction in the heart. Between each atrium and each ventricle there is a valve. The valve acts like a door that opens in only one direction.

Blood first moves into the two upper chambers. The top chambers then pump blood through the valves into the lower chambers. As the lower chambers fill with blood, the valves close. When the lower chambers squeeze together, the blood is forced out of the heart. Blood does not move back into the top chambers.

EXPLORATION

OBJECTIVES
In this exercise, you will:
  a. examine the outside and inside parts of a heart.
  b. trace the pathway of blood through the heart.
  c. follow the events within the heart as it pumps blood.

KEYWORDS
Define the following keywords:

atrium _____

contract _____

coronary artery _____

heart valves _____

_____

ventricle _____

MATERIALS
sheep heart on paper towel or
colored pencils: red and blue

PROCEDURE
Part A. Parts of the Heart
1. Obtain a sheep heart from your teacher. Do not turn it over. The right side of the sheep heart is on your left side. The left side of the heart is on your right side.

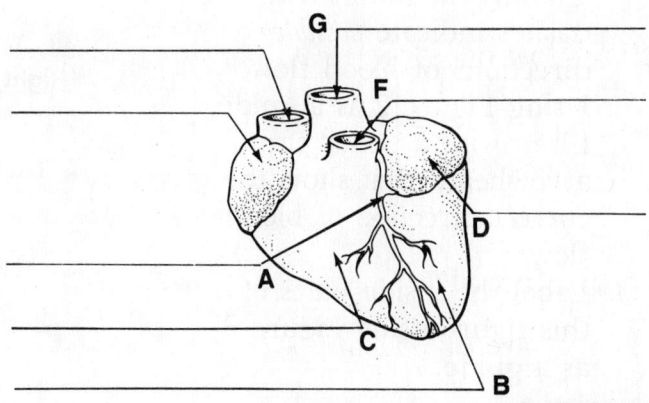

FIGURE 1.

87

2. On your sheep heart, find the parts listed in Table 1. Use the information in the table to help you.
3. Label the eight parts of the heart correctly on Figure 1. To help with the labels use the letters provided in the table and on the figure.

Table 1. Front Parts of the Heart

Part	Location	Traits	Name
A	across front of heart center	small blood vessel	coronary artery
B	bottom right chamber	large muscle section or chamber	left ventricle
C	bottom left chamber	large muscle section or chamber	right ventricle
D	top right chamber	small muscle section or chamber	left atrium
E	top left chamber	small muscle section or chamber	right atrium
F	top center	large blood vessel* from right ventricle	pulmonary artery
G	top center behind F	large blood vessel* from left ventricle; largest artery in body	aorta
H	top left	large blood vessel* from right atrium	vena cava

*All you will see is a hole where the blood vessel was attached to the heart.

Part B. Direction of Blood Flow Through the Heart

1. Examine Figure 2. It is a diagram of the inside of a heart. Arrows show the direction of blood flow.
2. Examine Figure 3 on the next page, which shows the inside of a sheep's heart. The arrows outlined in dashes indicate *possible* directions of blood flow. Using Figure 2 as a guide, fill in with a pencil the arrowheads that show the correct direction of blood flow.
3. Label the inside parts of this figure using Figure 2 as a guide.

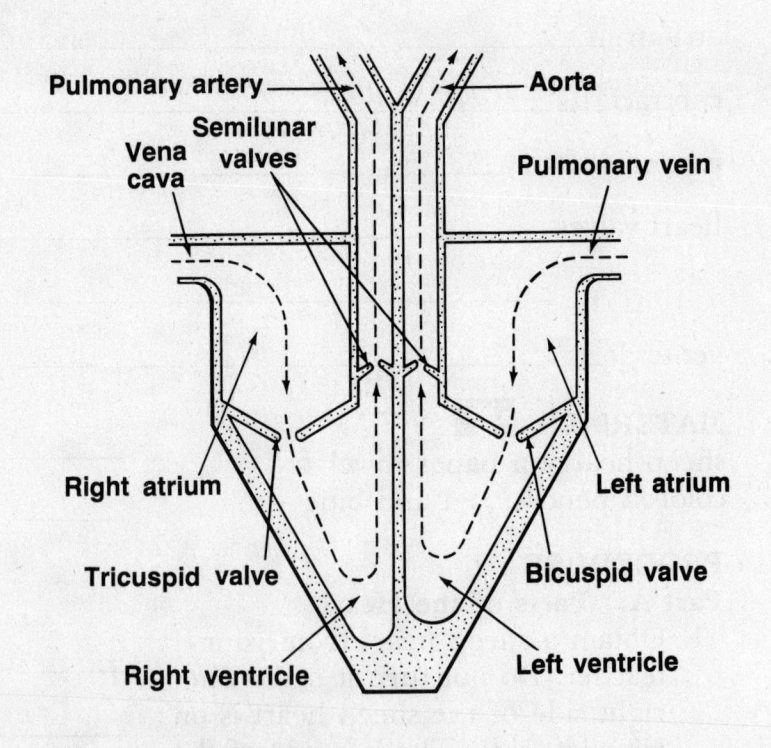

FIGURE 2. Blood flow through the heart

Name _____ Class _____ Period _____

Part C. Condition of Blood in the Heart

All blood on the heart's right side has little oxygen and much carbon dioxide. Blood on the left side has much oxygen and little carbon dioxide.

1. Using colored pencils, fill in the arrows in Figure 3 to show these differences in gas content:
   a. all arrows that indicate blood with much oxygen should be colored red.
   b. all arrows that indicate blood with much carbon dioxide should be colored blue.

Part D. Pumping Action of the Heart

Blood enters the two top chambers of the heart. Because they are made of muscle, they are able to squeeze together or contract. When this happens, blood is pumped to the two bottom chambers which are relaxed. These events are shown in Figure 4.

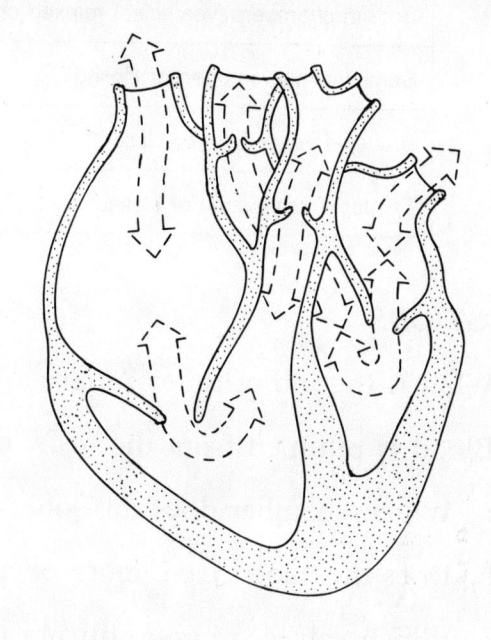

FIGURE 3. Inside of a sheep's heart

1. Note that certain valves in Figure 4 are open while other valves are closed. Complete the first column of Table 2.

Once blood fills the two bottom chambers they contract. Blood is then pumped out of the heart into the rest of the body. These events are shown in Figure 5.

2. Note which valves are open or closed in Figure 5. Complete the second column of Table 2.

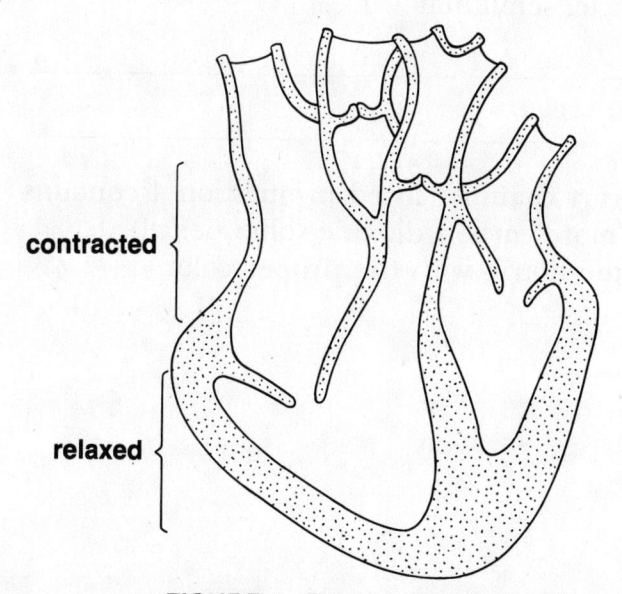

FIGURE 4. Blood entering ventricles

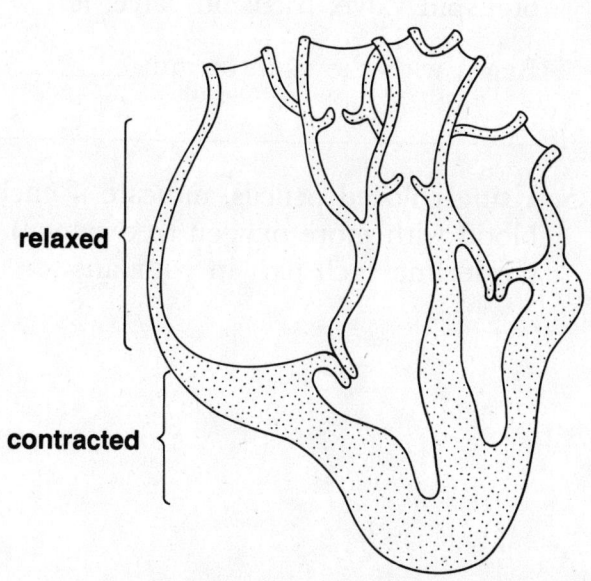

FIGURE 5. Blood leaving ventricles

89

Table 2. The Opening and Closing of Parts of the Heart

	Blood entering ventricles	Blood leaving ventricles
Top chambers (atria) relaxed or contracted?		
Bottom chambers (ventricles) relaxed or contracted?		
Semilunar valves open or closed?		
Bicuspid valve open or closed?		
Tricuspid valve open or closed?		

QUESTIONS

1. What is the job of the coronary artery? _____

2. Blood is pumped from the heart to the body through the aorta.

   a. Which chamber does this job? _____

   b. Does this blood have more oxygen or more carbon dioxide? _____

   c. Which valves are open during this process? _____

3. Blood is pumped from the heart to the lungs through the pulmonary artery.

   a. Which heart chamber does this job? _____

   b. Does this blood have more oxygen or more carbon dioxide? _____

   c. Which valves are open during this process? _____

4. Trace a drop of blood through the heart by putting these heart chambers and valves in proper order: left atrium, semilunar valve, right atrium, right ventricle, bicuspid valve, tricuspid valve, left ventricle, semilunar valve.

   Begin with the right atrium. _____

   _____

5. Using colored pencils, indicate if each heart chamber listed in question 4 contains blood with more oxygen (red pencil) or more carbon dioxide (blue pencil). Underline each part in your answer to question 4 with the proper color.

Name _____ Class _____ Period _____

11-2 How Do Hearts of Different Animals Compare?

The heart is a muscle that pumps blood. Even though they all pump blood, hearts of different animals are different in several ways. Hearts can be different sizes and shapes. Also, they differ in the number of chambers they have. You know that your heart has four chambers—two top chambers and two bottom chambers. Other mammals have four-chambered hearts. Birds also have four-chambered hearts. Amphibians have hearts with three chambers—two top chambers and one bottom chamber. Fish have hearts with two chambers—one top chamber and one bottom chamber. Making models of three different animal hearts will show you how they differ.

EXPLORATION

OBJECTIVES
In this exercise, you will:
   a. build a two-chambered heart model such as the type found in fish.
   b. build a three-chambered heart model such as the type found in amphibians.
   c. build a four-chambered heart model such as the type found in mammals.

KEYWORDS
Define the following keywords:

artery _____

atrium _____

chamber _____

vein _____

ventricle _____

MATERIALS
9 straws   9 paper cups
labels   glue
heavy paper   tape
scissors   ruler
page of model parts on page 92
colored pencils: red, blue
  and purple

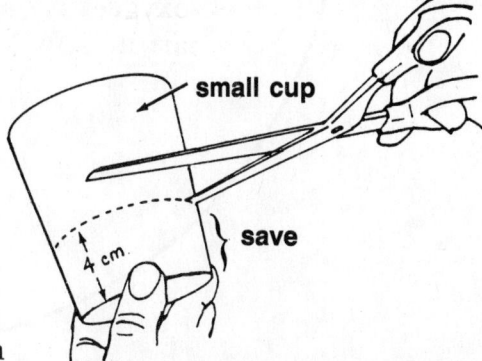

FIGURE 1. Cutting off top of cup

PROCEDURE
Part A. Building a Two-Chambered Heart Model
1. Remove the bottom of a paper cup as shown in Figure 1. **CAUTION:** *Use care with scissors.*

91

Trace three small circles and one large circle.

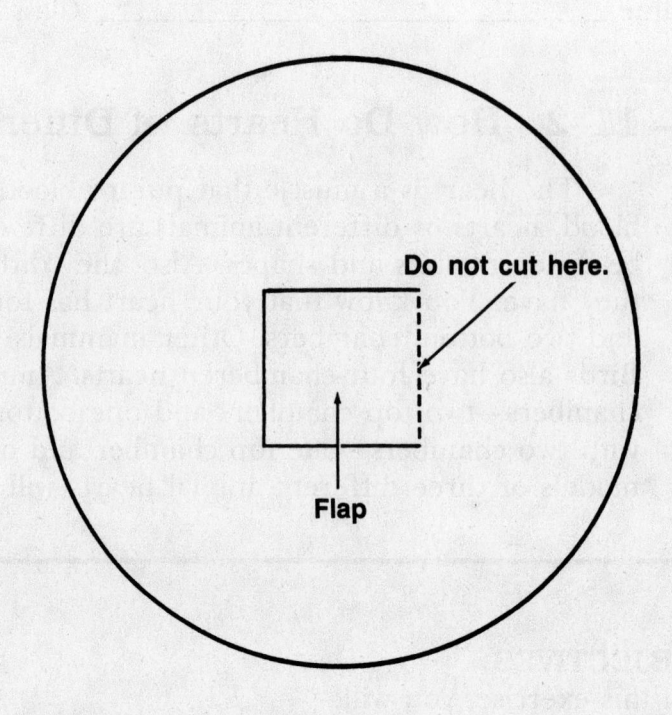

Do not cut here.

Flap

Circles to trace

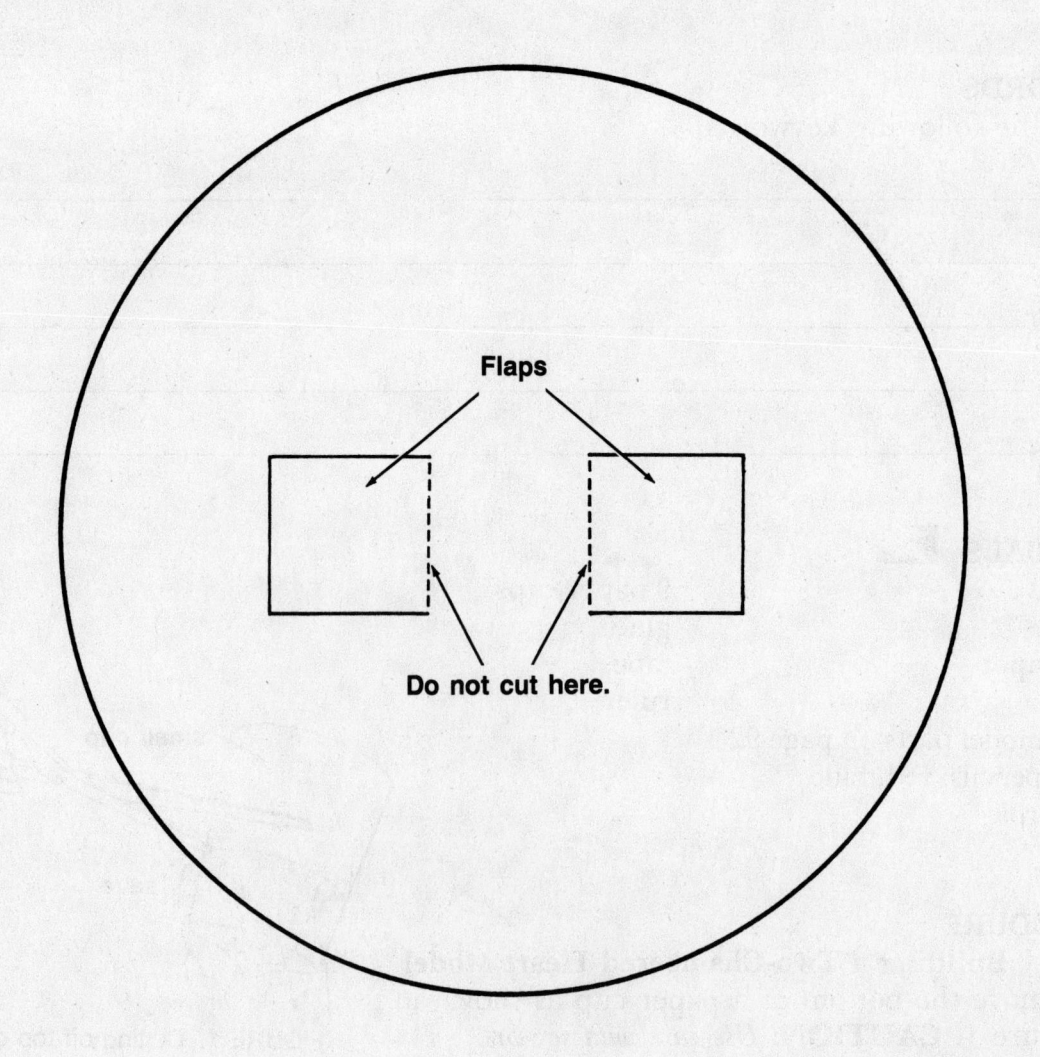

Flaps

Do not cut here.

2. Cut out one small circle from the page of circles given to you.
3. Trace the circle onto heavy paper. Then, cut the circle out from the heavy paper and cut through the flap in the center.
4. Poke holes in the cups as shown in Figure 2. Use the tip end of the scissors. **CAUTION:** *Always be careful with scissors.* Push straws into these holes.
5. Glue the cup pieces and paper circle together.
6. Color and print onto four labels the information shown in Figure 2.
7. Position the labels using Figure 2 as a guide.

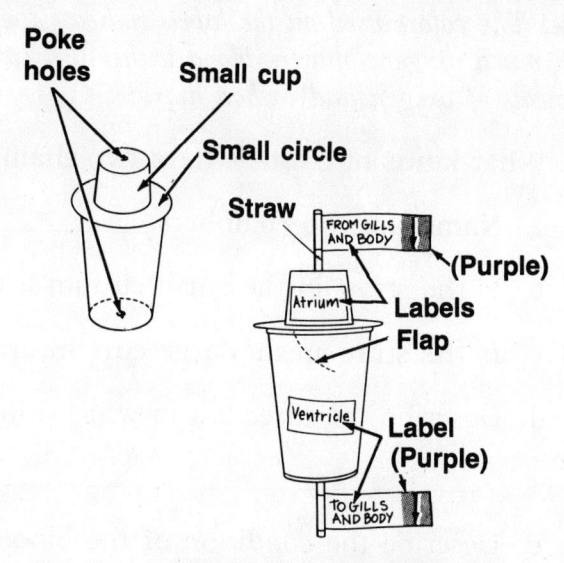

FIGURE 2. Two-chambered heart model

Part B. Building a Three-Chambered Heart Model

1. Prepare this model in the same way as in Part A. Make the following changes: cut off the bottom of *two* paper cups, use a large circle on *heavy* paper rather than a small one, and use *three* straws.
2. Glue the parts together and position the straws as shown in Figure 3.
3. Color and print the information onto six labels using Figure 3 as a guide.
4. Position the labels as shown.

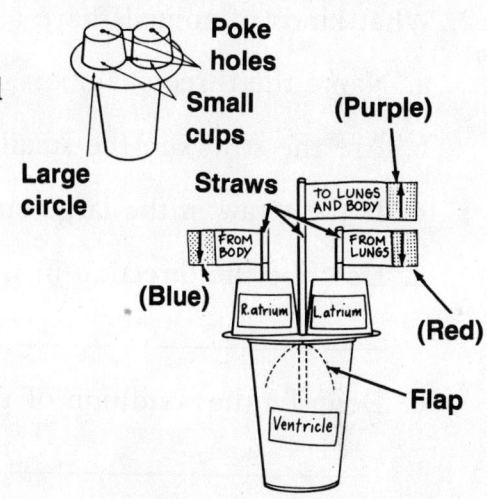

FIGURE 3. Three-chambered heart model

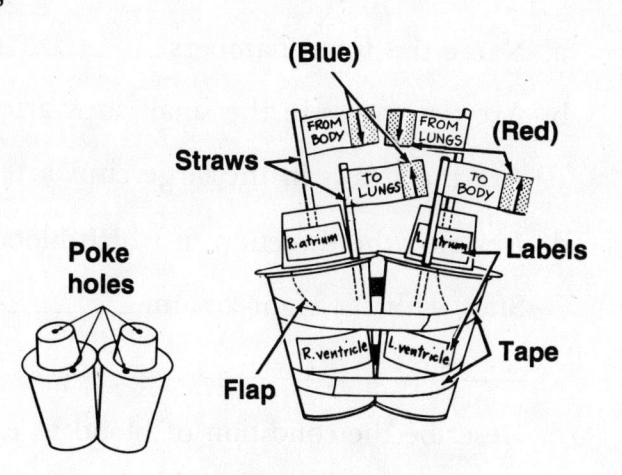

Part C. Building a Four-Chambered Heart Model

1. Prepare this model in the same way as in Part A. Make the following changes: cut off the bottom of *two* paper cups, use *two small circles* on heavy paper, and use *four* straws.
2. Glue the parts together and position the straws as shown in Figure 4.
3. Color and print the information shown onto eight labels.
4. Position the labels using Figure 4 as a guide.
5. You will have to tape the two cups together.

FIGURE 4. Four-chambered heart model

93

QUESTIONS

Note: The colors used on the labels indicate the condition of the blood present. Red = blood with much oxygen, blue = blood with much carbon dioxide, purple = blood with equal amounts of oxygen and carbon dioxide.

1. What kinds of animals have two-chambered hearts?_____

   a. Name the two chambers._____

   b. Is the straw in the small cup an artery or a vein?_____

   c. Is the straw in the large cup an artery or a vein?_____

   d. Describe the direction in which blood flows through a two-chambered heart._____
   _____

   e. Describe the condition of the blood in each chamber._____
   _____

2. What kinds of animals have a three-chambered heart?_____

   a. Name the three chambers._____

   b. Are the straws in the small cups arteries or veins?_____

   c. Is the straw in the large cup an artery or a vein?_____

   d. Describe the direction in which blood flows through a three-chambered heart.
   _____

   e. Describe the condition of the blood in each chamber._____
   _____

3. What kinds of animals have a four-chambered heart?_____

   a. Name the four chambers._____

   b. Are the straws in the small cups arteries or veins?_____

   c. Are the straws in the large cups arteries or veins?_____

   d. Describe the direction in which blood flows through a four-chambered heart. Start with the right atrium._____
   _____

   e. Describe the condition of blood in each chamber._____
   _____

Name _____ Class _____ Period _____

12-1 How Can Blood Diseases Be Identified?

Blood is a tissue. It has many different cells with many different jobs. If you look at blood under the microscope, you will find three different cell types—red cells, white cells, and platelets. In a normal person the numbers of types of blood cells are fairly constant. Sometimes, however, the number of cells will change due to a certain disease. Noticing this change in number can help a physician in the diagnosis of a person's disease.

INTERPRETATION

OBJECTIVES
In this exercise, you will:
 a. learn how to recognize three blood cell types.
 b. examine diagrams of blood samples from six hospital patients.
 c. match the blood samples with certain diseases.

KEYWORDS
Define the following keywords:

diagnosis_____

platelet_____

red blood cell_____

white blood cell_____

PROCEDURE
Part A. Normal Blood Cells
1. Examine Figure 1, which shows human blood cells magnified 1000 times.
2. Count each cell type present.
 HINT: To help avoid counting cells twice place a checkmark on each cell as you count.
   a. red blood cells—round, very numerous, no nucleus.
   b. white blood cells—round, few in number, larger than red blood cells, nucleus present.
   c. platelets—dotlike, many but less than red cells, very small.

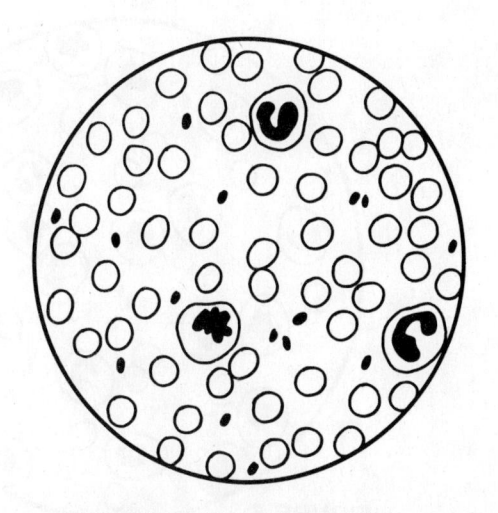

FIGURE 1. Normal blood sample

95

3. Record the number of each cell type for Figure 1 in Table 1. These numbers are for normal blood.
4. Using the numbers 1, 2, or 3, rank the cells in order from the most common (1) to the least common (3). Enter these rankings in the next column in Table 1 marked *Rank*.

Part B. Examining Abnormal Blood Cells
1. Examine Figures 2 to 6. These represent human blood samples from people with certain diseases.
2. Count each cell type and record the number for each sample in Table 1 under the right column.
3. Complete the rank columns using the numbers 1 to 3 as with the normal blood sample.

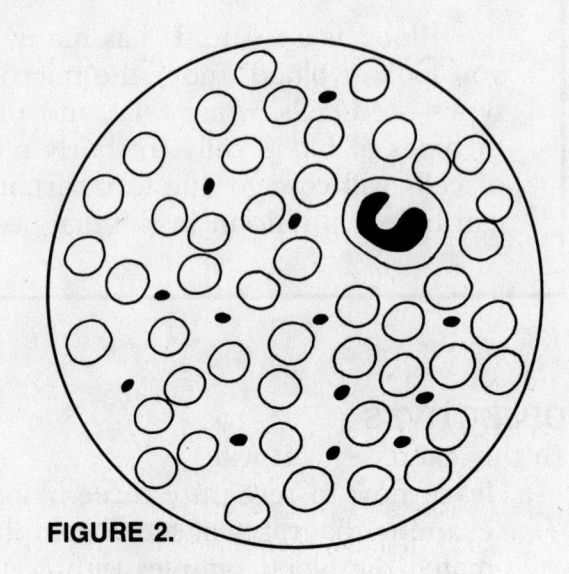

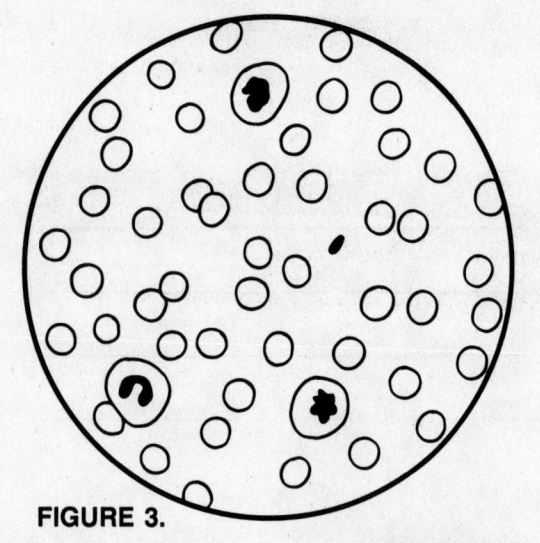

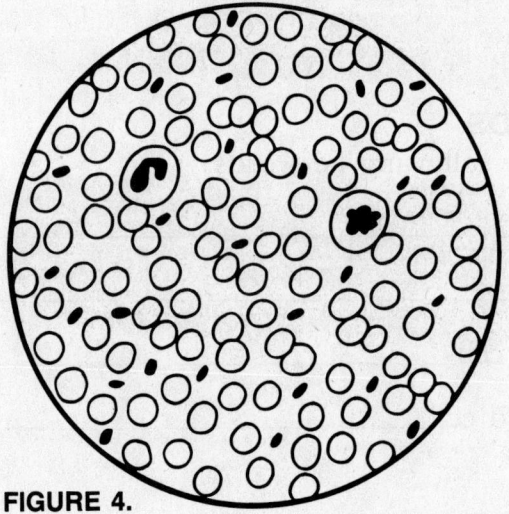

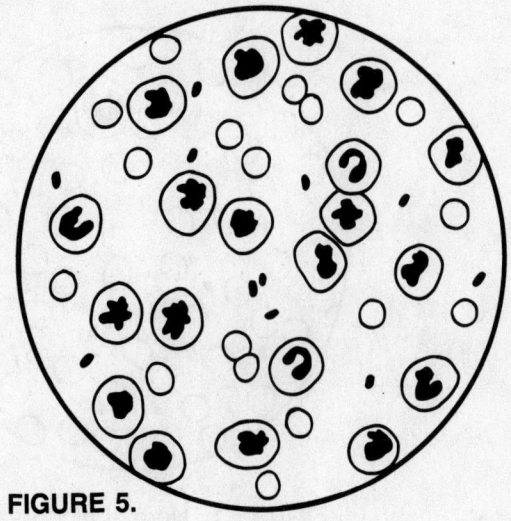

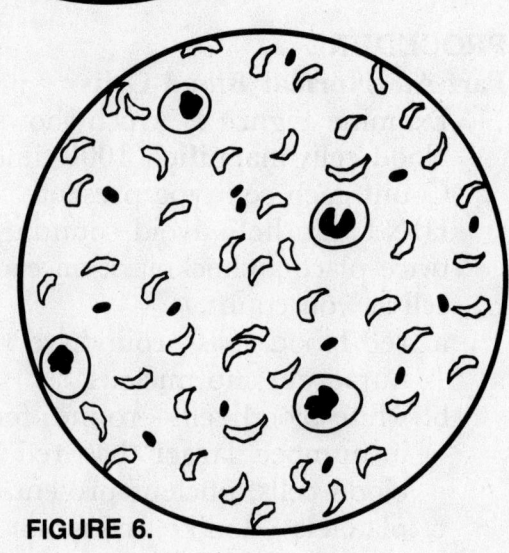

FIGURE 2.

FIGURE 3.

FIGURE 4.

FIGURE 5.

FIGURE 6.

Name _____ Class _____ Period _____

Table 1. Blood Cell Counts

Cell type	Fig. 1 No.	Fig. 1 Rank	Fig. 2 No.	Fig. 2 Rank	Fig. 3 No.	Fig. 3 Rank	Fig. 4 No.	Fig. 4 Rank	Fig. 5 No.	Fig. 5 Rank	Fig. 6 No.	Fig. 6 Rank
Red												
White												
Platelet												
Disease diagnosis	Normal blood											

Part C. Diagnosing Blood Diseases
1. Read over the following case histories for five hospital patients.
2. Match each case history with the appropriate blood sample.
3. Record the name of the disease below each sample in Table 1 in the space provided for disease diagnosis.

Case History:	Male, white, age 28; has admitted to injecting drugs for the past 6 years, has pneumonia and skin cancer
Blood analysis:	Few white cells present
Disease Diagnosis:	AIDS (acquired immunodeficiency syndrome)
Case History:	Male, black, age 15; is always tired and short of breath
Blood Analysis:	Red cells—shaped like crescent moons
Disease Diagnosis:	Sickle-cell anemia
Case History:	Female, oriental, age 14; has a fever, sore throat, and frequent nosebleeds
Blood Analysis:	Red cells—low in number; White cells—high in number Blood cell rank—white = 1, red = 2, platelets = 3
Disease Diagnosis:	Leukemia (leuk = white, emia = blood)
Case History:	Male, white, age 68; has frequent headaches, nosebleeds, shows high blood pressure, a very red complexion
Blood Analysis:	Red cells—a very high number
Disease Diagnosis:	Polycythemia (poly = many, cyth = cell, emia = blood)
Case History:	Female, white, age 22; has sudden appearances of purple marks under the skin, bruises easily, blood does not clot easily after a cut
Blood Analysis:	Platelets—very few in number Blood cell rank—red = 1, white = 2, platelets = 3
Disease Diagnosis:	Thrombocytopenia purpurea (thrombo = platelet, cyto = cell, penia = shortage, purpurea = purple)

QUESTIONS
1. What is the function of
   a. red blood cells?_____
   b. white blood cells?_____
   c. platelets?_____
2. How many
   a. red blood cells are in a drop of normal blood?_____
   b. white blood cells are in a drop of normal blood?_____
   c. platelets are in a drop of normal blood?_____
3. Rank your answers given to question 2 as to the most common (1) to the least common (3)._____
4. Do your rankings for normal blood in Table 1 agree with your answer to question 3?_____
5. Explain why a person with AIDS may also have pneumonia. (Keep in mind the main job of white blood cells)._____
_____

6. The rank of blood cells in a normal person and one with polycythemia is the same. How can you conclude that the person has polycythemia?_____
_____

7. The rank of blood cells in a normal person and one with sickle-cell anemia is the same. How can you conclude that the person has sickle-cell anemia?_____
_____

8. Name a blood disease that shows
   a. too many white blood cells_____
   b. too few platelets_____
   c. too few red blood cells_____
   d. too many red blood cells_____
   e. two few white blood cells._____
9. Explain why a person with thrombocytopenia purpurea shows many bruises or purple marks._____

10. Explain how the counting and appearance of blood cells can help in the diagnosis of blood diseases._____
_____

Name _____ Class _____ Period _____

12-2 What Blood Types Can Be Mixed?

Sometimes patients may lose a lot of blood. In these cases blood from another person can be given to the patient. This giving of someone else's blood to a person is called a transfusion.

There are four main blood types: A, B, AB, and O. Only certain blood types can be mixed when a transfusion is made. Mixing blood types incorrectly during a transfusion can lead to serious illness or the death of a patient.

INVESTIGATION

OBJECTIVES
In this exercise, you will:
a. set up plastic cups filled with water and food coloring to represent the four blood types.
b. mix "blood" to see if color changes take place.
c. judge which blood types can be mixed safely.

KEYWORDS
Define the following keywords:

blood type _____

donor _____

recipient _____

MATERIALS
colored pencils: red, green, and black
food coloring: red and green
graduated cylinder
20 small clear plastic cups
6 droppers

PROCEDURE
Part A. Set Up
1. Turn over the page and examine the grid in Figure 2. Note the columns marked *Recipient* and the rows marked *Donor*.
2. Place one of the small plastic cups onto each of the 20 squares as shown here in Figure 1.
3. Fill each cup with 10 mL of water.
4. Using a dropper, add 4 drops of red food coloring to each of the four cups in the column marked *Recipient A* (red), and to the cup marked *Donor A*.

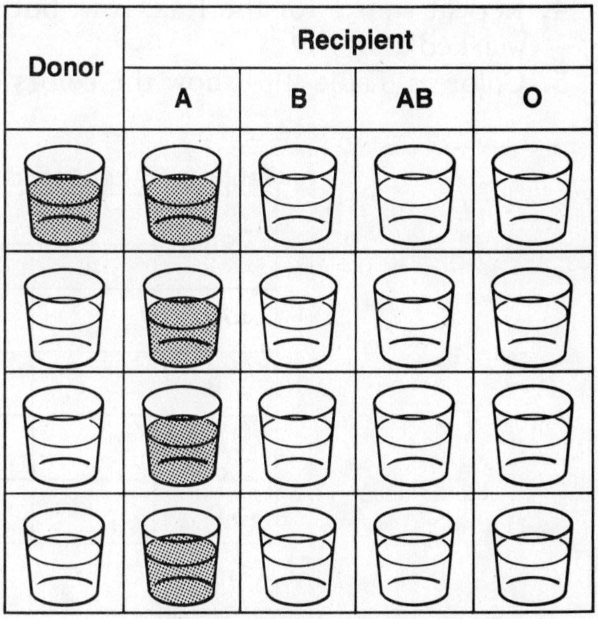

FIGURE 1. Placing cups on grid

99

5. Using a different dropper, add 2 drops of green food coloring to the four cups in the column marked *Recipient B* (green), and to the cup marked *Donor B*.
6. Add 3 drops of red food coloring and 3 drops of green food coloring to each of the four cups in the column marked *Recipient AB* (red and green), and to the cup marked *Donor AB*.
7. Note that the four cups in the column marked *Recipient O*, and the one cup marked *Donor O* have no food coloring added to them.
8. Using colored pencils, color in Table 1 to show the colors of all 16 cups marked *Recipient*.

Table 1. Before Blood Is Mixed

Donor	Recipient			
	A	B	AB	O
A				
B				
AB				
O				

Part B. Mixing Blood Types
1. Using a clean dropper, remove "blood" from the cup marked *Donor A*. Moving across the grid, add 2 droppers full of Type A "blood" to each of the four cups in the same row. This step shows what happens when a donor gives his or her blood to a recipient.
2. Repeat step 1 for the next row, but this time use "blood" from the cup marked *Donor B*.
3. Repeat step 1 for the next row, but this time use "blood" from the cup marked *Donor AB*.
4. Repeat step 1 for the final row, but this time use "blood" from the cup marked *Donor O*.
5. Color in Table 2 to show the colors of all 16 recipient cups.

Table 2. After Blood Is Mixed

Donor	Recipient			
	A	B	AB	O
A				
B				
AB				
O				

Name _____ Class _____ Period _____

FIGURE 2. Grid for mixing food colors

Donor	Recipient			
	A	B	AB	O
A (red)	(red)	(green)	(red + green)	(clear)
B (green)	(red)	(green)	(red + green)	(clear)
AB (red + green)	(red)	(green)	(red + green)	(clear)
O (clear)	(red)	(green)	(red + green)	(clear)

101

Part C. Judging If Blood Is Safe to Mix
1. Compare Tables 1 and 2. Blood is *safe* to mix between donor and recipient if there is *no change in color* in the same cup from Table 1 to Table 2. Blood is *not safe* to mix between donor and recipient if there is *a change in color* in the same cup from Table 1 to Table 2.
2. Complete Table 3. Write the word *safe* or *unsafe* in each of the 16 squares.

Table 3. Is Blood Safe To Mix?

Donor	Recipient			
	A	B	AB	O
A				
B				
AB				
O				

QUESTIONS

1. List the blood types of people to which a Type A donor can safely donate blood._____
2. List the blood types of people to which a Type B donor can safely donate blood._____
3. List the blood types of people to which a Type AB donor can safely donate blood._____
4. List the blood types of people to which a Type O donor can safely donate blood._____
5. List the blood types of people from which a Type AB recipient can receive blood._____
6. A person with Type O blood is often called a "universal donor." Why might this be a good term to use to describe such a person?_____

7. A person with Type AB blood is often called a "universal recipient." Why might this be a good term to use to describe such a person?_____

Name _____ Class _____ Period _____

13-1 How Does Breathing Occur?

If you have ever tried to hold your breath, you know that breathing is automatic. Breathing is moving air into and out of the lungs. Taking in air is called inhalation. Letting out air is called exhalation.

Your ribs and chest help with breathing. A muscle called the diaphragm also helps by contracting as you inhale and relaxing as you exhale. Let's see how these parts work together during the process of breathing.

EXPLORATION

OBJECTIVES
In this exercise, you will:
   a. compare a model to the human chest.
   b. use the model to show how the diaphragm helps inhalation and exhalation.
   c. use the model to show how the chest wall helps inhalation and exhalation.

KEYWORDS
Define the following keywords:

contracted _____

diaphragm _____

exhalation _____

inhalation _____

relaxed _____

MATERIALS
model of the human chest

PROCEDURE
Part A. Model Parts and How They Work
1. Obtain a model of the human chest from your teacher.
2. Note the following parts on your model and in Figure 1:
   a. rubber sheet along the bottom
   b. glass tube that has a short and a long side; tube bottom contains colored water
   c. outer plastic dome
   d. two balloons attached to an upside-down Y-shaped tube
   e. air trapped inside the plastic dome (air cannot escape from inside because of the balloons and the water inside the tube)
3. Push up gently on the rubber sheet and note the water level change in the tube.
4. Record the water level changes for both sides of the tube in Table 1.

103

FIGURE 1. Model of the human chest

FIGURE 2. The human chest

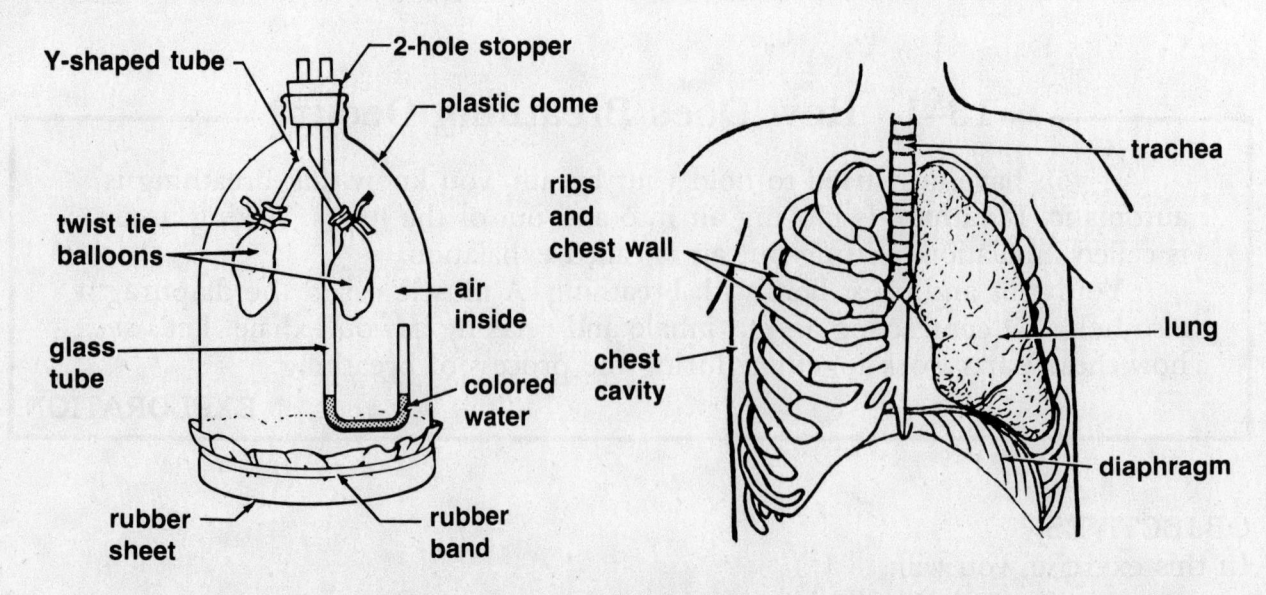

5. Pull down gently on the rubber sheet and note the water level change in the tube.

Air trapped inside the model will cause changes in air pressure when the rubber sheet is pushed up or pulled down. When the rubber sheet is pulled down, air pressure inside the chamber is decreased or low, and water rises on the short side of the tube. When the rubber sheet is pushed up, air pressure inside the chamber is increased or high, and water rises on the long side of the tube.

6. Complete Table 1.

Table 1. Water Levels in the Chest Model

Rubber sheet	Water level on long side	Water level on short side	Change in inside air pressure	Air pressure in model

Part B. Comparing Your Model With the Human Chest
1. Compare Figures 1 and 2. The model you have been working with represents the human chest.
2. Match the parts of the model (Figure 1) with the parts of the human chest (Figure 2) that are listed below.

Model Parts	Parts of the Human Chest
balloons _____	A. trachea and bronchi
rubber sheet _____	B. ribs
Y-shaped tube _____	C. chest cavity
air inside dome _____	D. diaphragm
plastic sides of dome _____	E. lungs (air sacs)

104

Name _____ Class _____ Period _____

Part C. The Role of the Diaphragm in Breathing
1. Gently push up on the rubber sheet (diaphragm) of the model.
2. Note the following: a. what happens to the balloons.
   b. in which side of the tube the water rises.
3. As pressure inside the chest increases, it squeezes the air sacs. This pushes the air within them out.
4. Complete the top row of Table 2. Note that the diaphragm is in a relaxed condition when it pushes up in your body.
5. Gently pull down on the rubber sheet (diaphragm) of the model.
6. Note the following: a. what happens to the balloons.
   b. in which side of the tube the water rises.
7. As pressure inside the chest decreases, the air sacs return to their original shape. They fill with air as you breathe in.
8. Complete the bottom row of Table 2. Note that the diaphragm is in a contracted condition when it pulls down in your body.

Table 2. Using the Chest Model

Rubber sheet	Diaphragm condition (relaxed/ contracted)	Diaphragm position (up/down)	Tube side in which water rises (long/short)	Inside air pressure (high/low)	Balloons (air sacs) (empty/fill)	Person breathing (exhale/ inhale)
Pushed up						
Pulled down						

Part D. The Role of the Ribs and the Chest Wall in Breathing
1. Gently squeeze in the sides at the bottom of the plastic dome (chest wall).
2. Note the following:
   a. what happens to the balloons.
   b. in which side of the tube the water rises.
3. Complete the top row of Table 3. Note that the chest wall and the ribs in Figure 3A move down slightly when the human chest wall moves in.
4. Gently squeeze in the sides at the bottom of the plastic dome, then let go and note:
   a. what happens to the balloons.
   b. in which side of the tube the water rises.
5. Complete the bottom row of Table 3. Note that the chest wall and the ribs in Figure 3B move slightly up when the human chest wall moves out and that the size of the chest cavity gets larger.

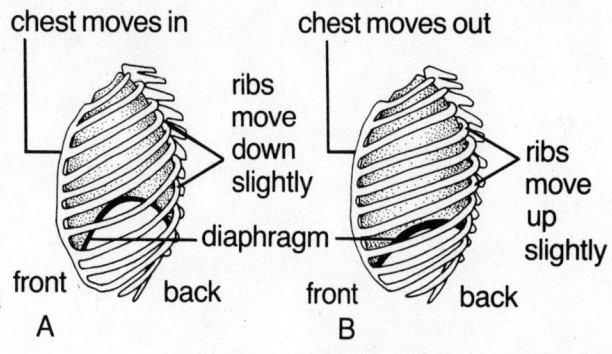

FIGURE 3. Side views of the rib cage

Table 3. The Movement of the Chest During Breathing

	Chest wall pushed in	Chest wall back to original shape
Tube side in which water rises (long/short)		
Inside air pressure (falls/rises)		
Air pressure (high/low)		
Rib cage movement (up/down)		
Chest cavity size (large/small)		
Balloons or air sacs (empty/fill)		
Person breathing (exhale/inhale)		

QUESTIONS

1. Complete this summary table by writing in the correct word or words.

	Inhalation	Exhalation
Diaphragm pulled up or down?		
Diaphragm relaxed or contracted?		
Chest wall pushed in or out?		
Ribs pulled up or down?		
Air pressure in chest high or low?		
Pressure does or does not squeeze air sacs?		
Chest cavity size increases or decreases?		
Lungs filling or emptying?		
Breathing in or out?		

Name _____ Class _____ Period _____

13–2 What Chemical Wastes Are Removed by the Lungs, Skin and Kidneys?

When foods are broken down in the body, chemical wastes are formed. Carbon dioxide and water are two such waste chemicals. Other waste chemicals given off by the body include urea and salts. Which body organs help to remove certain waste chemicals?

INVESTIGATION

OBJECTIVES
In this exercise, you will:
   a. learn how to test for water using cobalt chloride paper.
   b. learn how to test for salt using silver nitrate.
   c. test to see if the lungs, skin and/or kidneys give off water.
   d. test to see if the lungs, skin and/or kidneys give off salt.

KEYWORDS
Define the following keywords:

distilled water _____

perspiration _____

urine _____

waste chemical _____

MATERIALS
silver nitrate
plastic bag
cobalt chloride paper
5 test tubes
distilled water
test-tube rack

salt water
urine
water
dropper
cotton

PROCEDURE
Part A. Testing for Water
1. Examine a dry piece of cobalt chloride paper and record the color in Table 1.
2. Place a drop of water on the paper (by using a dropper). Record the color of the cobalt chloride paper in Table 2.

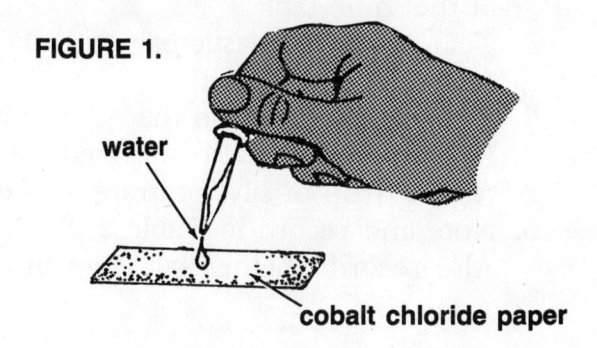

FIGURE 1.

How could you test for the presence of water by using dry cobalt chloride paper? _____

_____

Part B. Testing for Salt
1. Fill a clean test tube ¼ full of distilled water.
2. Fill a second test tube ¼ full of salt water.
3. Add 1 drop of silver nitrate to each tube. **CAUTION:** *Wash with water immediately if silver nitrate is spilled on the skin.*
4. Record the color of both tubes in Table 1. The white cloud that forms in the tube with salt indicates that salt is present. How can you test for the presence of salt using silver nitrate?

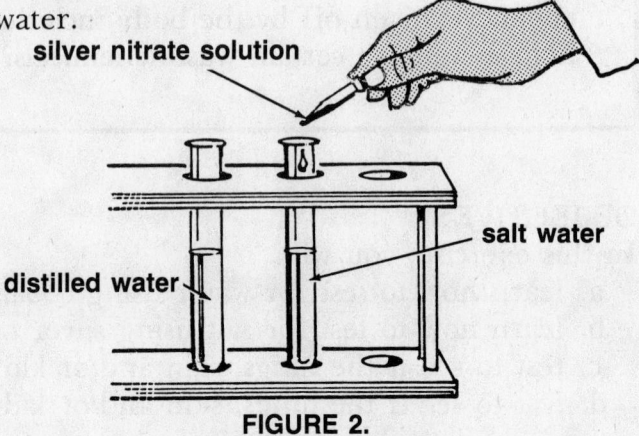

FIGURE 2.

_____

Table 1.

Substance Tested	
Dry cobalt chloride	
Wet cobalt chloride	
Distilled water (no salt)	
Salt water	

Part C. Testing Breath from the Lungs for Water and Salt
1. Place a dry piece of cobalt chloride paper in a plastic bag.
2. Hold the plastic bag to your mouth and breathe into the bag five times.
3. Hold the bag closed for about two minutes.
4. Note and record in Table 2 the color of the cobalt chloride paper. Also record whether water is given off by the lungs in your breath in the same table.
5. Tear open the plastic bag and soak up any liquid inside the bag with a small piece of cotton.
6. Roll up the cotton so that it will fit into a test tube.
7. Fill the tube ¼ full of distilled water.
8. Add 1 drop of silver nitrate to the tube.
9. Note and record in Table 2 if a white cloud appears in the tube. Also record whether salt is given off by the lungs.

Name _____ Class _____ Period _____

Part D. Testing Skin Perspiration for Water and Salt
1. Hold a piece of dry cobalt chloride paper in your hand with your fist closed for about 5 minutes.
2. Record the color of the cobalt chloride paper in Table 2.
3. Rub the palm of your hand at least 20 times with a clean piece of cotton, as if you were trying to dry your palm.
4. Roll up the cotton and place in the bottom of a test tube.
5. Fill the tube ¼ full with distilled water.
6. Add 1 drop of silver nitrate to this tube.
7. Note and record in Table 2 if a white cloud appears in the tube. Also record whether salt is given off by perspiration.

FIGURE 3.

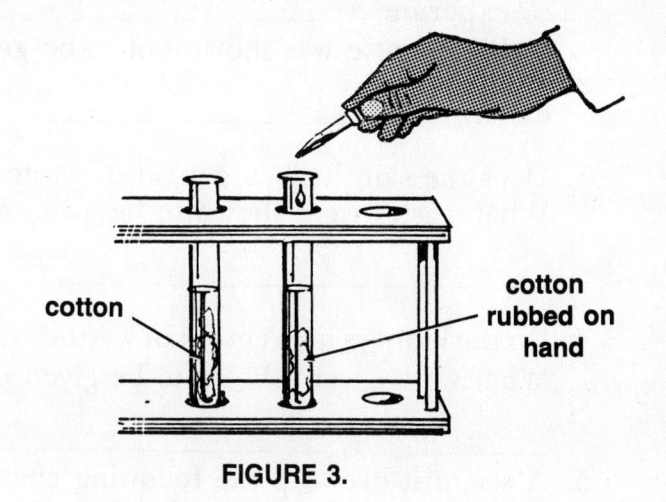

Table 2.

	Cobalt chloride color	Water given off?	Silver nitrate color	Salt given off?
Breath from lungs				
Skin perspiration				
Urine from kidneys				

Part E. Testing Urine for Water and Salt
1. Fill a test tube ¼ full of "urine."
2. Tip the tube and insert a small piece of dry cobalt chloride paper so that it just touches the urine.
3. Record any color change in Table 2. Also record whether water is given off by the kidneys in urine.
4. Add one drop of silver nitrate to the contents of the test tube.
5. Note and record in Table 2 if a white cloud appears in the tube. Also record whether salt is given off by the kidneys in urine.

QUESTIONS

1. Name three organs that give off waste chemicals. _____

2. Do the lungs help to get rid of wastes? _____
    a. What waste was shown to be given off by the lungs in this experiment? _____
    b. What waste was shown not to be given off by the lungs in this experiment? _____

3. Does the skin help to get rid of wastes? _____
   What wastes were shown to be given off by the skin in this experiment?
   _____

4. Do the kidneys help get rid of wastes? _____
   What wastes were shown to be given off by the kidneys in this experiment?
   _____

5. A scientist drew up the following chart based on her results from a similar experiment. The chart appears as follows:

    (a + indicates "present", a − indicates "absent")

Organ	Water	Salt	Carbon dioxide	Urea
Lungs	+	−	+	−
Skin	+	+	−	−
Kidneys	+	+	−	+

    a. Do your results agree with the scientist's for tests on water? _____

    b. Do your results agree with the scientist's for tests on salt? _____

6. Examine the chart in question 5.
    a. List the organ that gives off more *different* wastes. _____

    b. List the only organ that gives off urea. _____

    c. List the only organ that gives off carbon dioxide. _____

    d. List those organs that can give off water. _____

    e. List those organs that can give off salt. _____

Name _____ Class _____ Period _____

14–1 What Causes Sports Injuries?

A number of different kinds of injuries can take place that involve the skeletal system or the muscular system. Many of these injuries result from everyday accidents while others may occur while participating in certain sports.

INTERPRETATION

OBJECTIVES
In this exercise, you will:
   a. learn what the difference is between ligaments and tendons.
   b. relate sprains, torn tendons, and tendonitis to certain injuries.
   c. learn the names of certain body muscles, bones, and tendons.

KEYWORDS
Define the following keywords:

ankle _____

ligament _____

muscular system _____

skeletal system _____

sprain _____

FIGURE 1. Bones of the ankle

MATERIALS
#2 pencil
colored pencils: blue and red

PROCEDURE
Part A. Sprains
1. Examine Figure 1. This is a drawing of the bones that are a part of the human ankle.
2. Examine Figure 2. This is a similar drawing of the ankle except that three ligaments have been added. They are marked 1, 2, and 3.

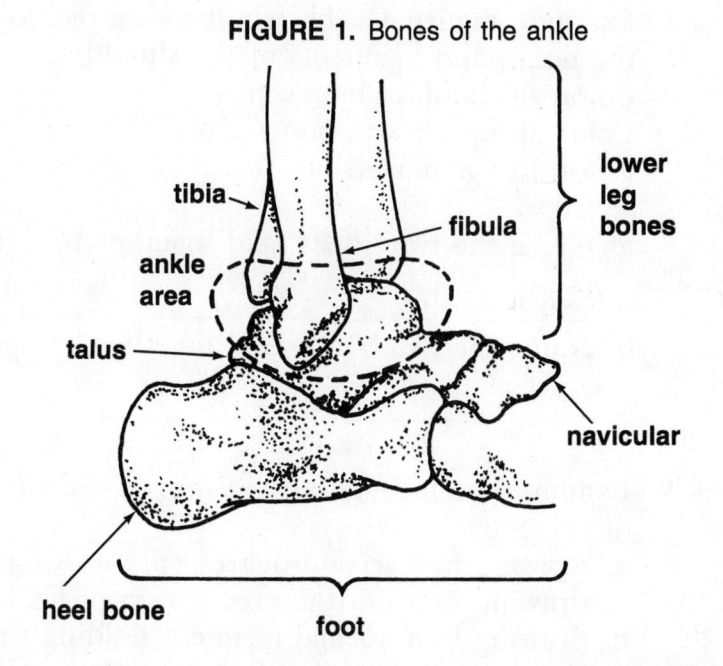

111

3. a. Color all leg bones in Figure 2 grey (use #2 pencil).
   b. Color all foot bones in Figure 2 blue.
   c. Color all ligaments in Figure 2 red.

4. Answer the following questions:
   a. Name the two bones held together by ligament 1.

   _____

   b. Name the two bones held together by ligament 2.

   _____

   c. Name the two bones held together by ligament 3.

   _____

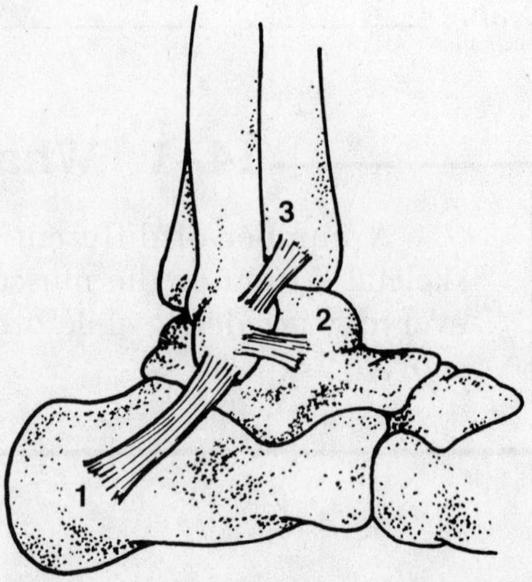

FIGURE 2. Ligaments of the ankle

5. Examine Figure 3 showing the three types of sprains. They are:
   First-degree sprain—ligaments are only stretched.
   Second-degree sprain—ligaments are only partly torn.
   Third-degree sprain—ligaments are torn completely.

6. a. Which ligament (1, 2, 3) shows a

      first-degree sprain? _____
   b. Which ligament (1, 2, 3) shows a

      second-degree sprain? _____
   c. Which ligament (1, 2, 3) shows a

      third-degree sprain? _____

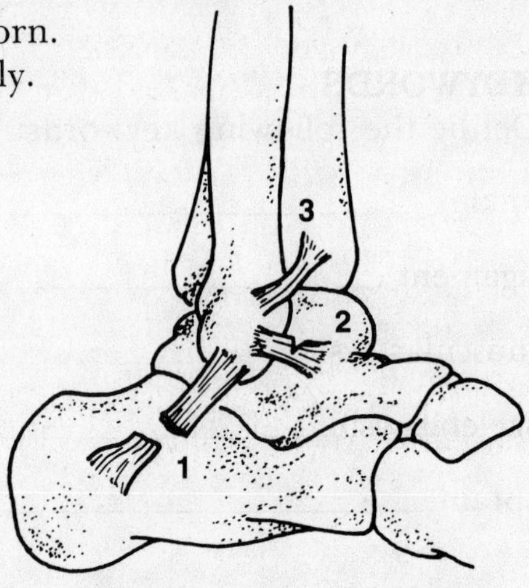

7. Examine Figure 4. This is a drawing of the bones and ligaments of the shoulder.
   Color all shoulder bones grey.
   Color all upper arm bones blue.
   Color all ligaments red.

FIGURE 3. Sprained ligaments

8. a. Name the two bones held together by ligament 1.

   _____

   b. Name the two bones held together by ligament 2.

   _____

9. Examine the incomplete drawing of the shoulder in Figure 5. Finish the drawing by:
   a. drawing in a second-degree sprain of ligament 1.
   b. drawing in a third-degree sprain of ligament 2.
   c. drawing in a normal ligament holding the humerus to the scapula.

112

Name _____ Class _____ Period _____

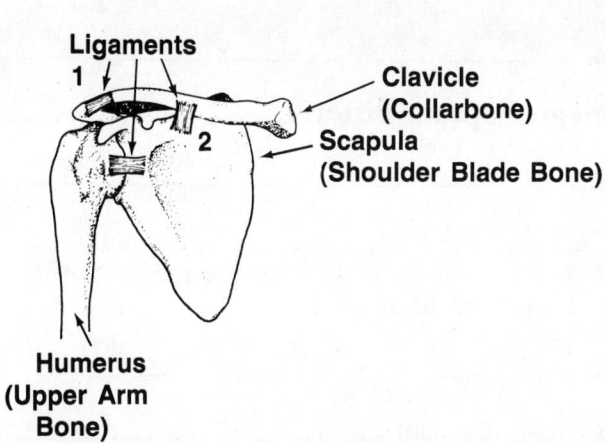

FIGURE 4. Ligaments of the shoulder

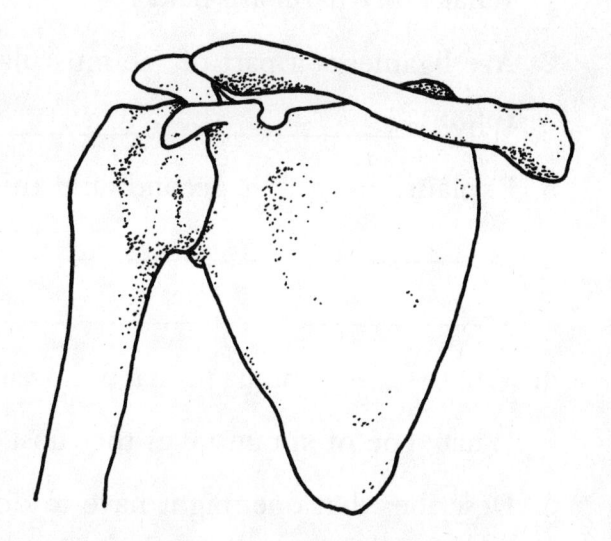

FIGURE 5. Sprains of the shoulder

Part B. Totally Torn Tendons—Tendonitis

1. Locate your calf muscle (your Gastrocnemius muscle). Run your hand down your calf until you nearly reach the back of your heel. You should now be able to feel a thick cord at the back of your heel. This cord is a tendon (your Achilles tendon.)
2. Examine Figure 6a. This drawing shows an actual view of the back of a person's leg. The skin has been removed.
3. Finish Figure 6b by showing what a totally torn Achilles tendon would look like. Draw an arrow pointing to the torn area and label it.
4. Finish Figure 6c by showing what tendonitis of the Achilles tendon would look like. Tendonitis is a soreness of the tendon. It is caused by small tears which occur along the tendon. Draw an arrow pointing to the tears and label them.

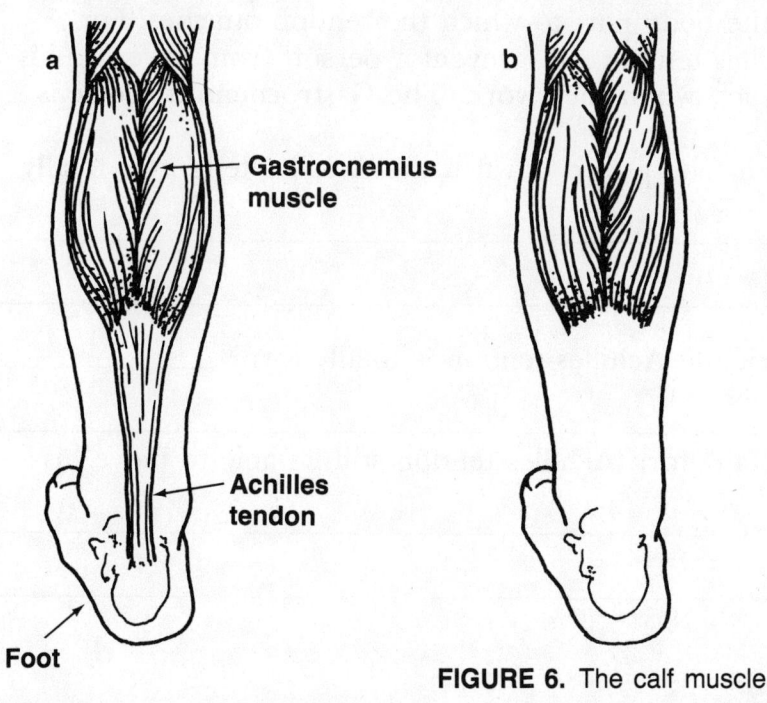

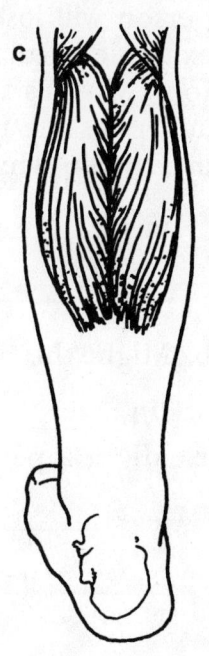

FIGURE 6. The calf muscle

113

QUESTIONS

1. What body parts are held together by ligaments? _____
2. Are ligaments a part of the muscular system or the skeletal system? _____
   Why? _____
3. Explain how a first, second, and third degree sprain differ? _____
   _____
   _____
4. What type of sprain probably takes the least time to heal? _____
5. What type of sprain takes the most time to heal? _____
6. Describe what one might have to do to cause a sprain. _____
   _____
7. What body parts are connected by tendons? _____
8. Are tendons a part of the muscular system or the skeletal system? _____
   Why? _____
9. Explain how tendons differ from ligaments. _____
   _____
10. Describe what one might have to do to cause a tendon to totally tear or develop
    tendonitis. _____
    _____
11. A totally torn tendon is a serious problem for an athlete or anyone else. A person will lose the use of the body part to which the tendon attaches. For example, a totally torn Achilles tendon will prevent a person from lowering his foot. Muscles shorten (contract) when they work. The Gastrocnemius shortens and pulls the foot down.
    a. Explain why the foot cannot be pulled down if the Achilles tendon is totally
       torn. _____
       _____
    b. Might the foot be raised if the Achilles tendon is totally torn? _____
       Why? _____
    c. Might a person with a totally torn Achilles tendon still be able to move his
       leg? _____ Why? _____
       _____

114

Name _____ Class _____ Period _____

14-2 How Do Male and Female Skeletons Differ?

A skeleton is found. A doctor reports to the police that it is an adult male skeleton. How could the doctor determine if the skeleton were from a male or a female?

Several differences exist between the skeleton of a male and that of a female. The main difference is in the shape of the pelvis. The female usually has a wider pelvis. Let's see how some measurements compare.

EXPLORATION

OBJECTIVES
In this exercise, you will:
1. examine and measure diagrams of a male and female pelvis.
2. determine how these measurements differ in male and female pelvises.
3. use your data to determine if a third pelvis is male or female.

KEYWORDS
Define the following keywords:

femur _____

pelvis _____

sacrum _____

skeleton _____

MATERIALS
metric ruler

PROCEDURE
1. Examine Figure 1. Figure 1a is the pelvis from an adult male. Figure 1b is the pelvis from an adult female.
2. Measure the length (in millimeters) of the following dashed lines on Figures 1a and 1b: lines *a*, *b*, *c*, and *d*.
3. Record these numbers in Table 1. Note that lines b and c are part of the sacrum bone (shaded). This bone is found at the back of the pelvis and does not block the pelvis opening. It does, however, appear to block the opening in the figures.

115

a

pelvis width (a)

sacrum length (b)

This line is vertical—caution students.

sacrum width (c)

pelvis opening (d)

(e)
(f)
(g)

b

(a)
(b)
(c)
(d)

(e)
(f)
(g)

c

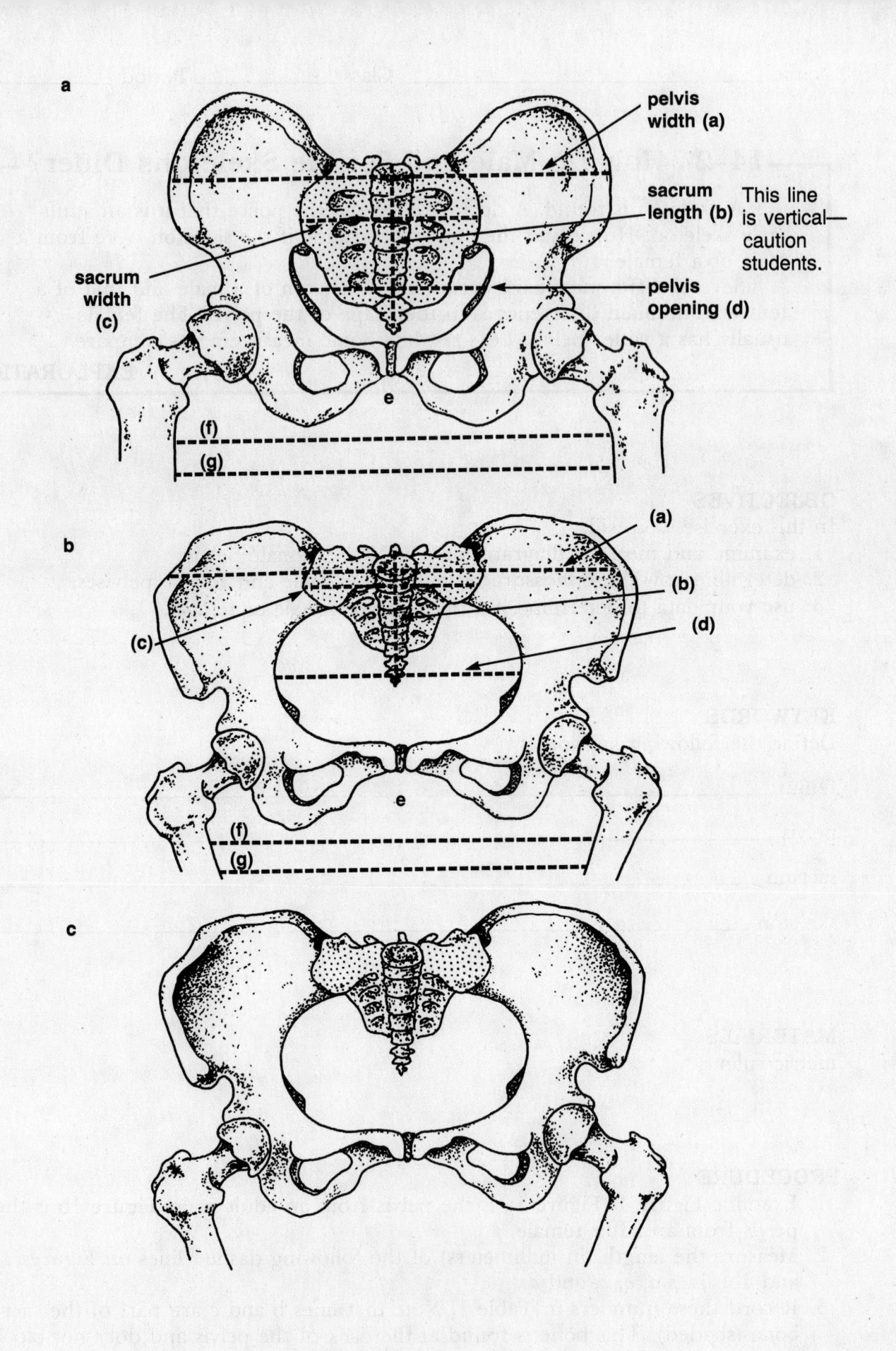

FIGURE 1. Human pelvises

116

Name _____ Class _____ Period _____

4. Locate letter *e* on Figures 1a and 1b. Note that the bottom of each pelvis is either round or pointed at this location. Record in Table 1 if the bottom is pointed or round.
5. Measure and record in Table 1 the lengths of dotted lines *f* and *g* on Figures 1a and 1b. If the femur bones hang *straight down*, the lengths of lines *f* and *g* will be the same. If the femur bones *slant inward*, the lengths of lines *f* and *g* will differ. This position of femur bones (thigh bones) provide a clue as to whether the skeleton is male or female.
6. Record your measurements and position of the femurs in Table 1.
   Note that now you have a way of telling male from female skeletons by using all the data in Table 1.
7. Measure and record all the lengths of the pelvis and femur parts in Figure 1c just as you did for Figures 1a and 1b. The dashed lines are not included in the figure. Record your data in Table 1.
8. Indicate in Table 1 if Figure 1c represents a male or female skeleton.

Table 1. Pelvic Bone Measurements

Figure	Sex	Pelvis width line a	Sacrum length line b	Sacrum width line c	Pelvis opening line d	Bottom shape	Line f	Line g	Position of femurs
1a	Male								
1b	Female								
1c									

QUESTIONS

1. Explain how each of the following differs in adult male and adult female skeletons:

   a. pelvis width _____

   b. sacrum _____

   c. pelvis opening width _____

   d. bottom shape of pelvis _____

   e. position of femur bones _____

2. Figure 1c is from a _____ (male or female.) List three things that helped you with your answer. _____

_____

117

3. The approximate age of a skeleton can be told by measuring the length of the femur bone. The graph shown in Figure 4 gives you these measurements. By using this graph, determine the approximate age of a skeleton whose femur measures

   a. 200 millimeters. _____

   b. 300 millimeters. _____

   c. 350 millimeters. _____

4. Explain why the graph in Figure 2 cannot be used to determine the age beyond 18 years. _____

FIGURE 2. Age of human skeletons

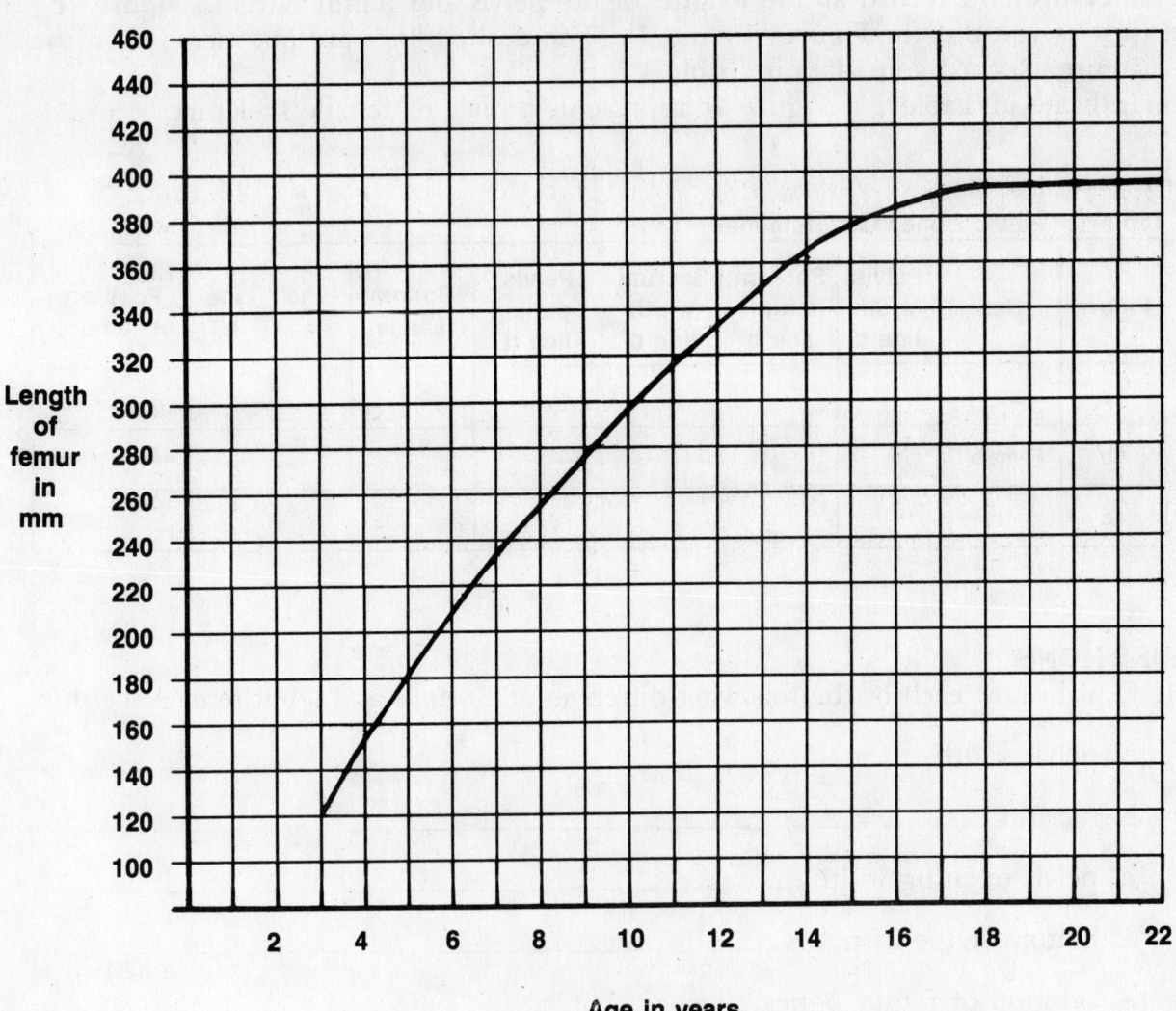

118

Name _____ Class _____ Period _____

15-1 Which Brain Side Is Dominant?

The human brain is divided into a left and a right side. Many things that you do with the right side of your body are controlled by your brain's left side. Many things that you do with the left side of your body are controlled by your brain's right side. If much of what you do is done by your body's right side, your dominant brain side is the left side. If much of what you do is done by your body's left side, your dominant brain side is the right side.

INVESTIGATION

OBJECTIVES
In this exercise, you will:
 a. check to see how many activities you do using your left hand or your right hand.
 b. check how many activities you do using your left foot or your right foot.
 c. find out if you draw or see objects more to the right side or the left side.
 d. find out if the left side or the right side of your brain is dominant.

KEYWORDS
Define the following keywords:

cerebrum _____

dominant _____

left cerebrum side _____

right cerebrum side _____

MATERIALS
paper   red pencil

PROCEDURE
1. Place a check mark in the proper column in Table 1 to show which hand you usually use to do the following tasks. Note: If you use either hand just as often, then check both columns.
   Tell which hand you use to
     a. write your name.
     b. wave "hello."
     c. bat while playing baseball.
     d. which thumb is on top when folding your hands.
     e. hold your spoon or fork while eating.

2. Place a check mark in the proper column in Table 1 to show which foot you usually use to do the following tasks. Note: If you use either foot just as often, check both columns.
   Tell which foot you use to
   a. start down a flight of stairs.
   b. start up a flight of stairs.
   c. catch yourself from falling as you lean forward.
   d. start skipping.
   e. place most weight on when you are standing.
   f. start to run.
   g. kick a ball.

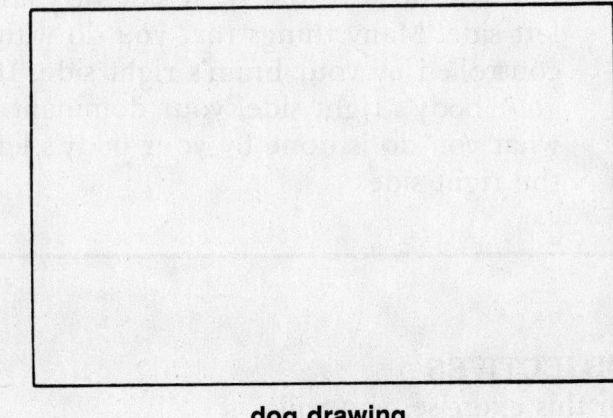

   dog drawing

3. Draw, in the space provided, a simple side view of a dog. Place a check mark in the column in Table 1 that shows the direction the nose faces.

4. Draw a circle in the space provided with your *right* hand. Note the direction in which you made this circle. Now draw a circle with your left hand. Note the direction in which you made this circle. If both circles were drawn clockwise, mark the right column in Table 1. If both circles were drawn counterclockwise, mark the left column in Table 1. If you drew one circle in each direction, check both columns.

   left hand   **right hand**

5. Roll a sheet of paper into a tube. Look through the tube at some distant object with both eyes open as shown in Figure 1. Then while looking through the tube at that distant object, close one eye and then the other. The eye that sees the object through the tube is your dominant eye. Place a check mark in the proper column in Table 1.

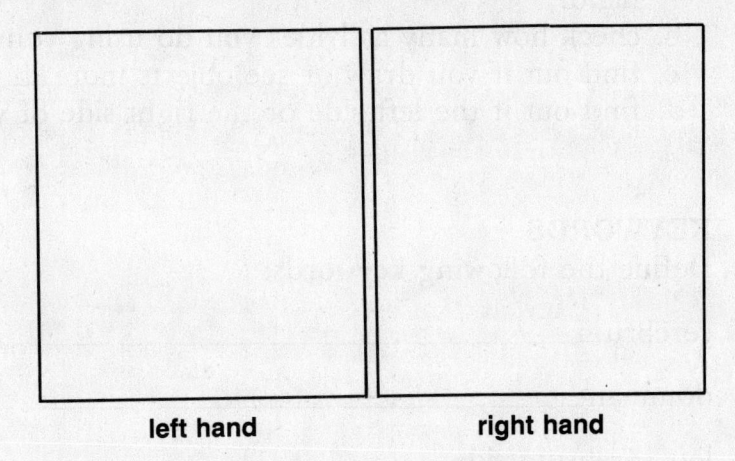

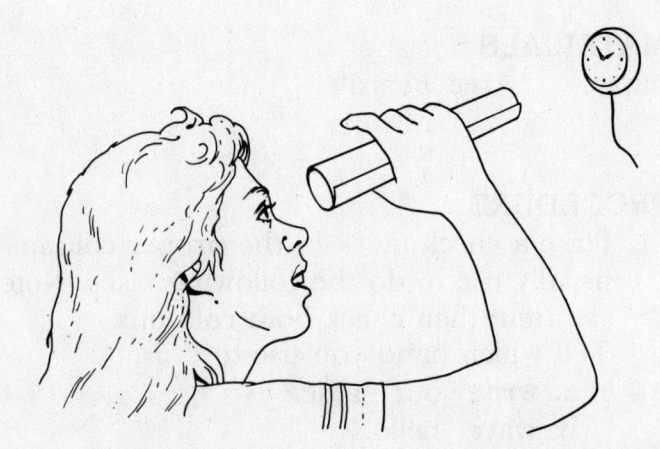

   FIGURE 1. Finding your dominant eye

Name _____ Class _____ Period _____

6. Total up the check marks for each column of Table 1 and place the total at the bottom of the columns.

Table 1. Finding Your Dominant Side

Task	Left side	Right side
Write name		
Wave "hello"		
Bat		
Thumb position		
Hold spoon		
Walk down stairs		
Walk up stairs		
Catch from falling		
Skipping		
Standing		
Start to run		
Take off shoe		
Leg on top		
Kick		
Dog drawing		
Circle drawing		
Dominant eye		
Totals =		

QUESTIONS

1. Which column in Table 1 has the most check marks?_____

2. Which column in Table 1 has the fewest check marks?_____

3. Which body side seems to be your dominant side?_____

4. The human cerebrum is divided into left and right sides.

    a. Which brain side controls the left side of your body?_____

    b. Which brain side controls the right side of your body?_____

5. The brain side that you use the most is said to be your dominant brain side.

    Which is your dominant brain side?_____
    (HINT: The answer will be the opposite side from your answer to question 3.)

6. Look at Figure 2. It shows a top view of the brain. Label the following parts: left cerebrum side, right cerebrum side. Use a red pencil to shade in your dominant brain side.

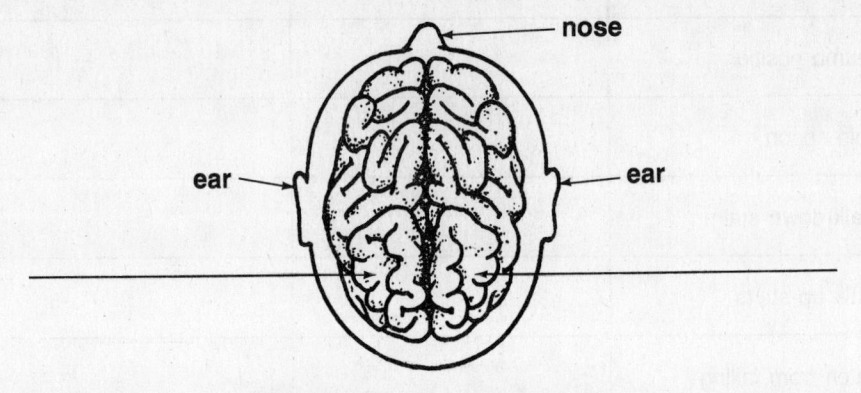

FIGURE 2. Top view of brain

7. Your teacher will ask for a class survey of certain results. Complete the following data for your class:

    a. number of students who are right-handed and show the right body side as dominant._____

    b. number of students who are right-handed and show the left body side as dominant._____

    c. number of students who are left-handed and show the right body side as dominant._____

    d. number of students who are left-handed and show the left body side as dominant._____

8. Using your results from question number 7,

    a. does a person who uses his or her right hand for writing always show a dominant right body side?_____

    b. does a person who uses his or her left hand for writing always show a dominant left body side?_____

Name _____ Class _____ Period _____

15–2 The Brain and Its Functions

The human brain is divided into three different parts. Each part is specialized. Each part has a job to perform that is different from the other parts. The brain is even more specialized in that specific brain sides control only specific body sides.

INTERPRETATION

OBJECTIVES
In this exercise, you will:
  a. identify and label the three brain areas.
  b. determine the jobs of certain brain areas.
  c. match brain areas with their corresponding areas of body control.

KEYWORDS
Define the following keywords:

cerebellum _____

cerebrum _____

involuntary _____

medulla _____

voluntary _____

MATERIALS
#2 pencil   colored pencils: red, green, blue, gray, and yellow

PROCEDURE
Part A. Control Areas of the Brain
1. Examine Figure 1. This shows a side view of the human brain. Label the brackets correctly to show the brain's three parts or areas. Use the following labels: medulla, cerebrum, and cerebellum.
2. Label the functions of certain brain parts by using the following labels:
   A. vision center
   B. speech center
   C. sensation of body pain
   D. muscle control of body
   E. hearing center
   F. heartbeat center
   G. coordination center for body muscle
   H. smell center
   I. personality center
3. Still using Figure 1, color in the voluntary parts of the brain with a red pencil. Color the involuntary parts blue.

FIGURE 1. Side view of the human brain

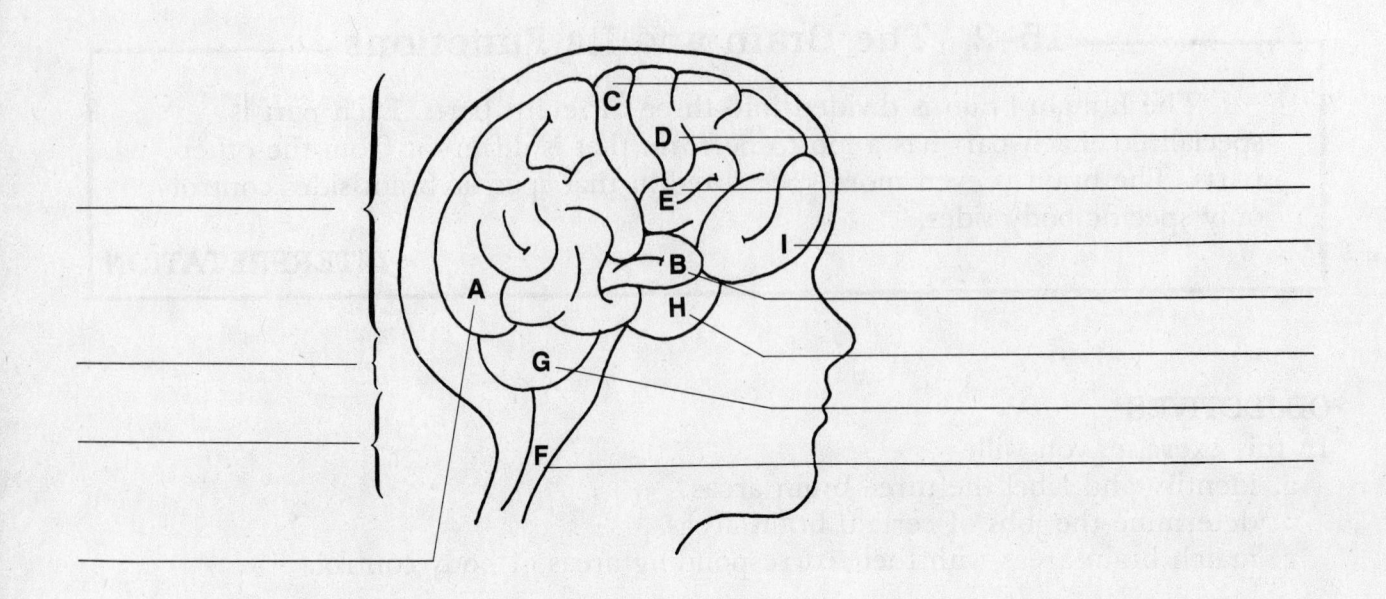

Part B. How the Brain Controls the Body

1. Examine the top view of the cerebrum in Figure 2. Note that the cerebrum is divided into left and right sides. Each side has been marked for you.

2. Locate and examine the two front views of the body in Figure 3. Note that in these views the left and right sides are reversed (this is because they are front views). The body views are marked either "sensation of body pain" or "muscle control of body." Muscle control and body pain are controlled by certain brain areas. The brain area controlling this and the corresponding areas on the body are marked with similar letters.

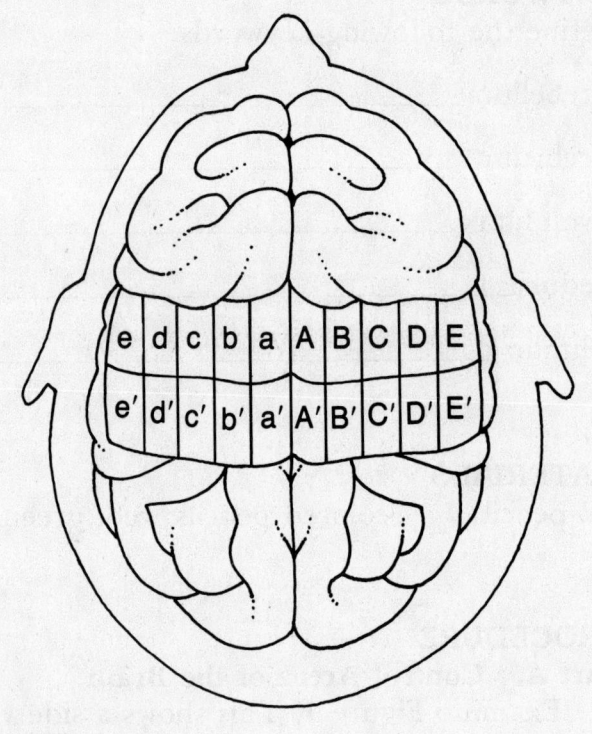

FIGURE 2. The cerebrum

left side of cerebrum   right side of cerebrum

3. Match the brain areas of Figure 2 with their corresponding body areas of Figure 3 by coloring in parts of the figures as follows:

A–solid red	A'–stripe red	a–crisscross red	a'–dot red
B–solid blue	B'–stripe blue	b–crisscross blue	b'–dot blue
C–solid green	C'–stripe green	c–crisscross green	c'–dot green
D–solid yellow	D'–stripe yellow	d–crisscross yellow	d'–dot yellow
E–solid gray	E'–stripe gray	e–crisscross gray	e'–dot gray

124

Name _____ Class _____ Period _____

FIGURE 3. Front views of the human body

sensation of body pain

muscle contol of body

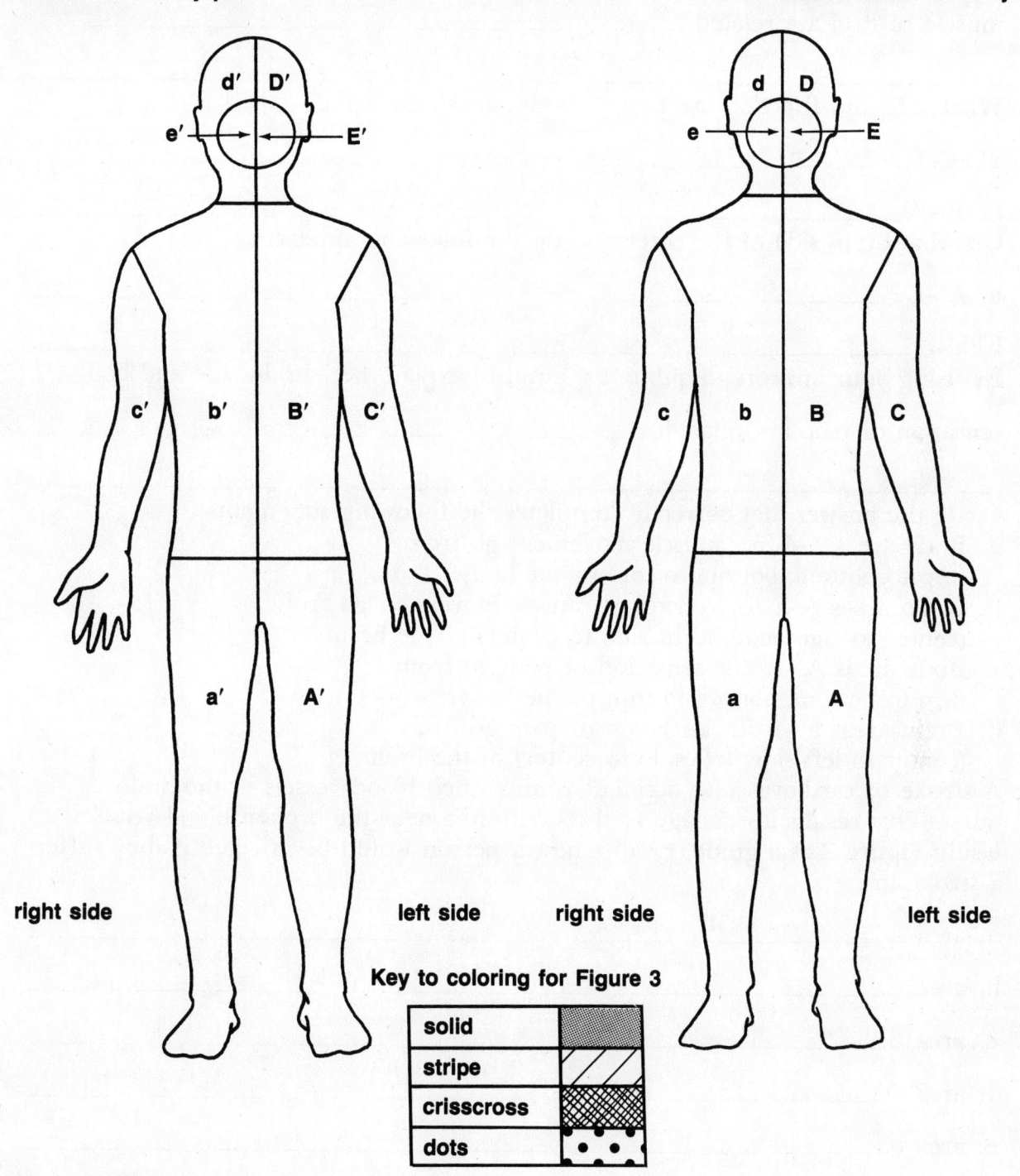

right side   left side   right side   left side

Key to coloring for Figure 3

solid	
stripe	
crisscross	
dots	

QUESTIONS
1. Which brain area is the largest? _____
2. What side and functions are part of body areas
   a. A–E? _____
   b. a–e? _____

125

3. On what brain side of the cerebrum do the following appear:

   a. A–E? _____

   b. a–e? _____

4. By using your answers to questions 2 and 3, explain how brain side and body muscle control are related. _____
   _____

5. What side and function are part of body areas

   a. A'–E'? _____

   b. a'–e'? _____

6. On what brain side of the cerebrum do the following appear:

   a. A'–E'? _____

   b. a'–e'? _____

7. By using your answers to questions 5 and 6, explain how brain side and body sensation of pain are related. _____
   _____

8. Circle the answer that correctly completes the following statements:
   a. Body areas A–E for muscle movement go from
      (top to bottom, bottom to top) of the body.
   b. Brain areas A–E for control of muscle movement go from
      (center to right side, right side to center) of the brain.
   c. Body areas A'–E' for sensation of pain go from
      (top to bottom, bottom to top) of the body.
   d. Brain areas a'–e' for sensation of pain go from
      (center to left side, left side to center) of the brain.

9. A stroke or cardiovascular accident results when blood vessels in the brain burst. This results in damage to the brain area near the broken blood vessel. Using Figure 2 as a guide, predict how a person would be affected if they suffer a stroke in

   a. area A. _____

   b. area C. _____

   c. area E. _____

   d. area D'. _____

   e. area b. _____

   f. area e'. _____

Name _____ Class _____ Period _____

16–1 How Can You Test Your Senses?

Information about your surroundings reaches the brain through sensory neurons. The sensory neurons get information from the receptors. Some of these receptors for pain, pressure, and touch are scattered throughout the skin and other parts of the body. Other receptors for taste, smell, sight, and hearing are contained in special organs. We are dependent, in part, on our sense receptors for our understanding of the world.

INVESTIGATION

OBJECTIVES
In this exercise, you will:
   test your senses of taste, touch, smell, and hearing.

KEYWORDS
Define the following keywords:

receptor_____

sense organ_____

sensory neuron_____

MATERIALS
cotton	10 toothpicks	index card
blindfold	paper plate	apple
clock or timer	plastic spoon	onion
meterstick	paper drinking cup	forceps
2 metal spoons	cinnamon chewing gum	potato
glue	paper towel	

PROCEDURE
Part A. Touch Receptor Test
Complete this test with a blindfolded partner.
1. Look at Figure 1. Use this diagram to place toothpicks on your index card. Measure and mark points on your index card near the edges that are the following distance apart: 2 mm, 4 mm, 8 mm, 10 mm, and 20 mm. Glue the toothpicks to the index card at these points to match those shown in Figure 1.

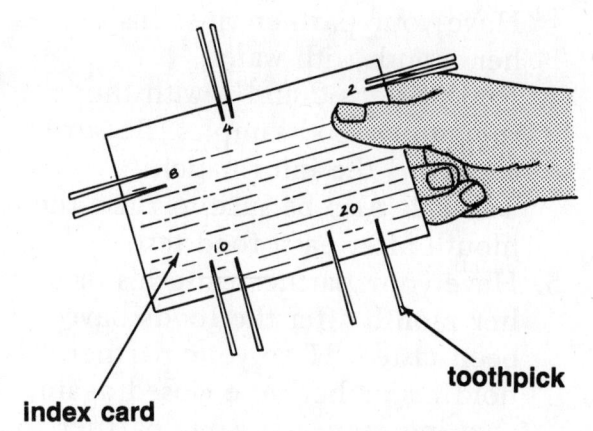

FIGURE 1. Touch test card

127

2. Gently touch your partner's fingertips (palm side) with the points of the toothpicks that are 2 mm apart. Ask your partner to report how many points are touching the skin. If your partner can feel two points, put a check mark in Table 1 under the proper column. If your partner cannot feel two points, put a minus mark in the table. **CAUTION:** *Do not apply heavy pressure when touching your partner's skin.*
3. Repeat step 2 touching your partner's palm, back of the hand, inside of the forearm, back of the neck, and lips. Record your results in Table 1.
4. Repeat steps 2 and 3 using each pair of toothpicks.

Table 1. Testing the Sense of Touch

Body part	Distance between two points of touch				
	2 mm	4 mm	8 mm	10 mm	20 mm
Fingertip (palm side)					
Palm of hand					
Back of hand					
Forearm (inside)					
Back of neck					
Lips					

Part B. Food Test

Complete this test with a blindfolded partner.
1. Using forceps, place a small amount of chopped potato, onion, and apple on a paper plate.
2. Instruct your partner to hold his or her nose with thumb and forefinger so that he or she cannot smell.
3. Using a plastic spoon, give your partner one of the foods to taste. (Do not let your partner swallow the food.) Be sure the nostrils are tightly closed while tasting foods. Ask your partner to tell you what food is being tasted. Record in the proper row in Table 2 what food your partner thought he or she tasted. Dispose of the food in a paper towel. Have your partner rinse his or her mouth with water.
4. Repeat steps 2 and 3 with the remaining food samples. Be sure to record the information in Table 2. Also be sure to rinse the mouth after each food test.
5. Have your partner rinse his or her mouth after the foods have been tasted. Have your partner hold his or her nose closed again. Give your partner a piece of cinnamon chewing gum. Ask your partner to describe the taste. Record the answer in Table 2. Then have your partner release the nostrils and describe the taste.

Table 2. The Sense of Taste

Food sampled	Tasted like
Potato	
Onion	
Apple	
Cinnamon gum	

Name _____ Class _____ Period _____

Part C. Sound Test

1. Hold a ticking clock or timer at the opening of your partner's left ear. Slowly back away until your partner can no longer hear the ticking. Measure and record in Table 3 the distance at which the sound can no longer be heard.
2. Repeat step 1 with your partner's right ear.
3. Have your partner block his or her right ear with cotton (or your partner could use a finger.) Repeat step 1. Record the results in Table 3.
4. Have your partner block his or her left ear and repeat. Record the results in Table 3.
5. Blindfold your partner. Strike two metal spoons together at a distance of 1 meter from the front of the person's head. Have your partner indicate the direction of the sound. Place a check mark in the proper place in Table 4 if your partner is correct and a minus sign if your partner is not correct. Then strike the spoons together on the right side, the left side, in back of, and above the head at a distance of 1 meter. Record the results of each of these tests in Table 4.

Table 3. The Sense of Hearing

Sound source	Distance from	
	Left ear	Right ear
Ticking clock		
Ticking clock with ear blocked		

Table 4. How Distance Affects the Sense of Hearing

Location	Ears open		Left ear blocked		Right ear blocked	
	1 m	3 m	1 m	3 m	1 m	3 m
Front of head						
Left side of head						
Right side of head						
Back of head						
Above the head						

6. Repeat step 5 at a distance of 3 meters. Record the results in Table. 4.
7. Block only your partner's right ear (as before.) Repeat steps 5 and 6. Record the results in Table 4.
8. Block only your partner's left ear (as before.) Repeat steps 5 and 6. Record your results in Table 4.

QUESTIONS

1. Which area of the body as tested in Part A seems to have the greatest number of touch receptors? _____

2. How does this help you? _____

3. Which area seems to have the least number of touch receptors? _____

4. What do the answers to questions 1 and 3 tell you about the distribution of touch receptors in the skin? _____

5. How many basic tastes can your tongue sense? _____
   What are they? _____

6. a. Is it possible to detect correctly the three foods tested in part B with the nose closed? _____
   b. Would it be possible to detect the three foods correctly if the nose were not closed? _____

7. What does the gum taste like with the nostrils closed? _____
   With the nostrils open? _____ Why? _____
   _____

8. How can you tell the difference between two things that taste sweet?
   _____

9. When you eat vanilla ice cream, what senses are you using? _____

10. Why do you lose the ability to taste foods when you have a cold? _____

11. In Part C, do you hear sounds that originate at greater distances better with one ear or two ears? _____

12. Which sense organ was being used in Part A? _____

13. Which sense organ was being used in Part B? _____

14. Which sense organ was being used in Part C? _____

Name _____ Class _____ Period _____

16-2 How Are Your Senses Sometimes Fooled?

What you see and what you think you see are not always the same. Seeing is done with your eyes. The message your eyes pick up is sent to the brain where it is interpreted. The brain may then "tell" you that you see something that is not present. The mistaken idea that you get is an illusion. If the mistaken idea is because of what you see, or your eyes, it is called an optical illusion.

INVESTIGATION

OBJECTIVES
In this exercise, you will:
  a. look at several diagrams and record what you see.
  b. compare your results with what you know is present.

KEYWORDS
Define the following keywords:

illusion _____

optical illusion _____

sense _____

MATERIALS
ruler   red marker
file card   yellow marker
blank paper

FIGURE 1.

PROCEDURE
Part A. Triangle Illusion
1. Examine Figure 1 for a white triangle. Record in Table 1 on page 127 if you see it.
2. Now record in the table if the white triangle is really present. Is there a white triangle drawn on the diagram? _____

Part B. Bent Line Illusion
1. Examine Figure 2. Record in Table 1 if you think the lines across the diagram could meet without the resulting line being bent.

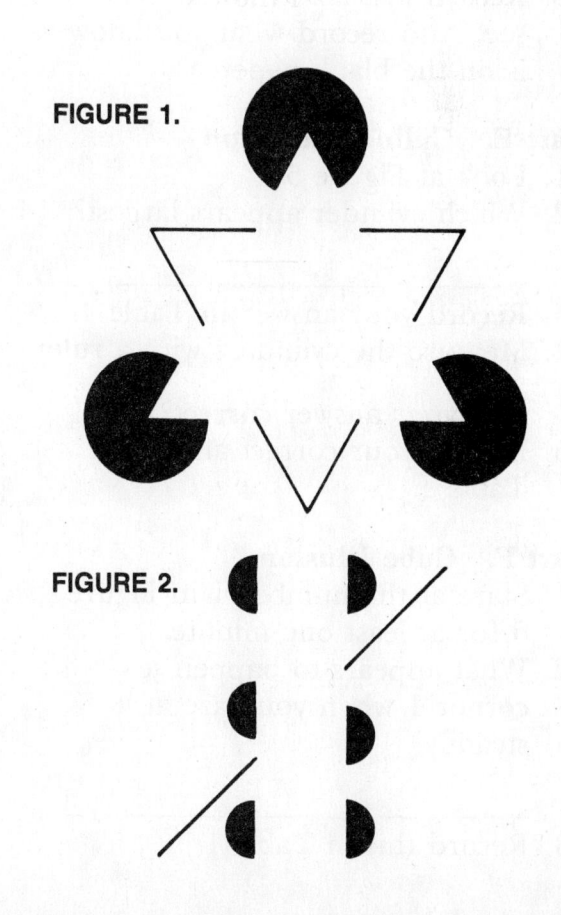

FIGURE 2.

131

2. Lay a ruler on Figure 2 next to both lines. Does it now appear that the lines would meet in the center without the resulting line being bent? _____
3. Record your results in Table 1.

Part C. Paper Fold Illusion
1. Fold a file card as shown here in Figure 3. Lay it on your desk about 20 cm from your eyes.
2. Stare at the figure for about 20 seconds with one eye closed. Does the fold appear to always be in the top of the figure? ____
3. Record your answer in Table 1. Also record where you know the fold really is.

Part D. Flower Illusion
1. Color Figure 4 so that the center is red and the petals are yellow.
2. Stare at the diagram for 30 seconds. Then stare at a blank piece of paper.
3. Record in Table 1 what you see. Also record what you know is on the blank paper.

Part E. Cylinder Illusion
1. Look at Figure 5.
2. Which cylinder appears largest?

   _____
3. Record your answer in Table 1.
4. Measure the cylinders with a ruler.

   Was your answer correct?_____
5. Record your correct answer in Table 1.

Part F. Cube Illusion
1. Stare at the number 1 in Figure 6 for at least one minute.
2. What appears to happen to corner 1 when you gaze at it steadily?

   _____
3. Record this in Table 1.

FIGURE 3.

FIGURE 4.

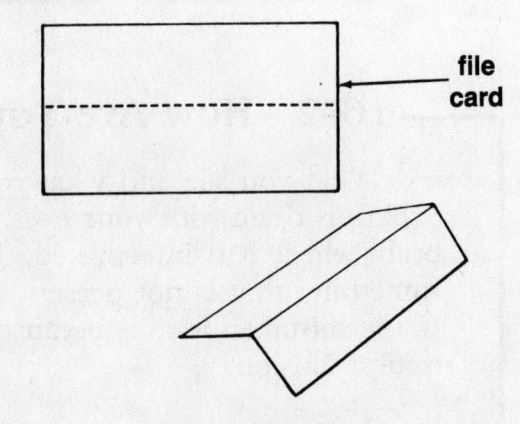

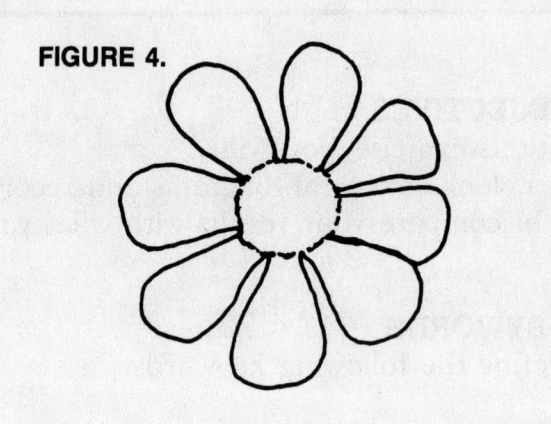

FIGURE 5.

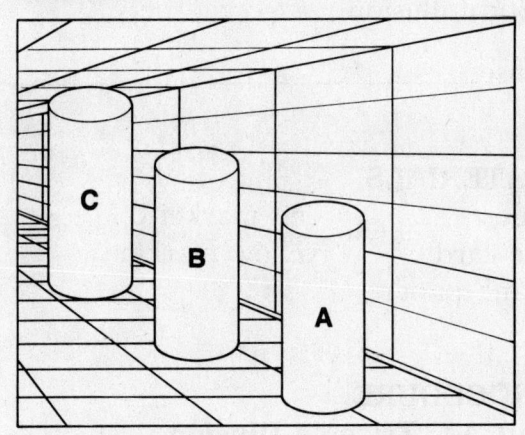

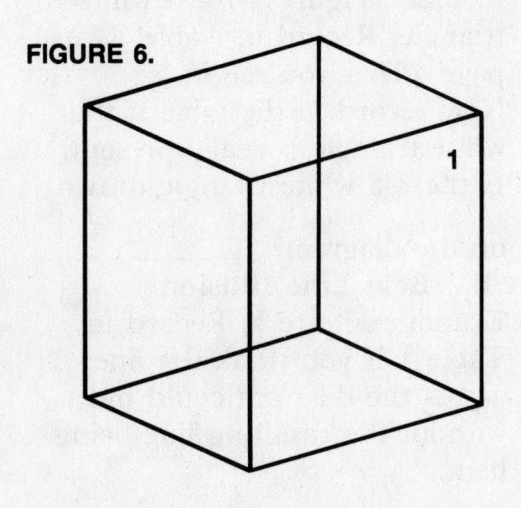

FIGURE 6.

132

Name _____ Class _____ Period _____

4. Also record in Table 1 what Figure 6 really shows.

Table 1. Observing Illusions

Illusion	Appears	Really Is
Triangle		
Bent line		
Paper fold		
Flower		
Cylinders		
Cube		

FIGURE 7.

Part G. Figure Reversals

1. Examine Figure 7. What do you see?
   _____

2. When you look a second time, do you see anything different? _____

3. Do both figures ever appear at the same time? _____

4. Examine Figure 8. This figure was used in 1900 by psychologist Joseph Jastrow. What do you see when the face looks to the left? _____

FIGURE 8.

5. What do you see when the face looks to the right? _____
Can you see both at the same time?
   _____

133

6. Examine Figure 9. What do you see?_____

FIGURE 9.

7. Look at the figure again. Do you see anything different? What do see?_____
Can you see both things at the same time?_____

Part H. Creating Illusions

1. Examine the figure of the two girls below. They are both exactly the same size. Without changing their size, create the illusion that one girl is taller than the other.

FIGURE 10. Make your own illusion

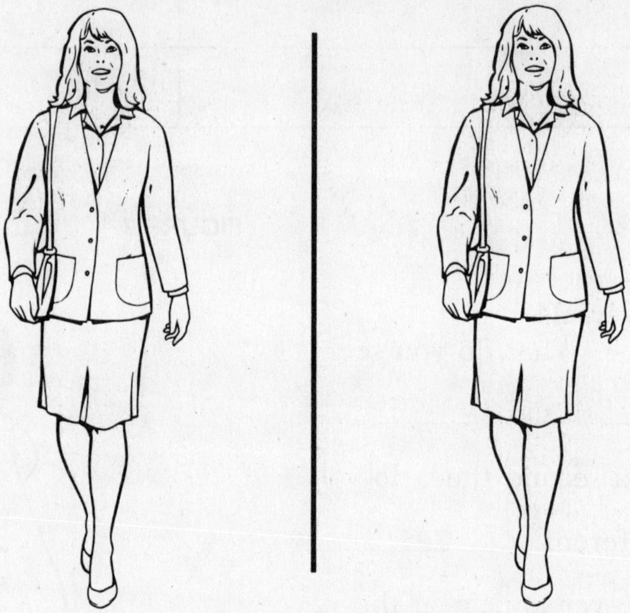

QUESTIONS

1. Which of the illusions in this activity were optical illusions?_____

2. Sometimes a driver sees water on a dry highway. Why is this water an optical illusion?_____

3. Name some jobs people have in which they must be aware of optical illusions.
_____
_____

Name _____ Class _____ Period _____

16–3 What Are the Parts of an Eye?

The eye is a most complex sense organ. It can detect differences in light intensity, color, and motion. It's a perfect example of a body organ where form follows function. This means that each eye part (form) is perfectly designed to carry out its specific job (function).

EXPLORATION

OBJECTIVES
In this exercise, you will:
a. observe and identify the parts on the exterior of a pig eye.
b. build an eye model to study the action of four eye muscles.
c. observe and identify the parts on the interior of a pig eye.

KEYWORDS
Define the following keywords:

exterior _____

interior _____

muscle contraction _____

MATERIALS
preserved pig eye
colored pencils: blue, black
polystyrene ball (7.5 cm diameter)
dissecting pan
large paper clip
4 cloth strips

scissors
ruler
paper
pen
tape

PROCEDURE
Part A. The Exterior Parts of an Eye
1. Examine a preserved pig eye that has been placed in a dissecting pan.
2. Locate the following parts using Figure 1 as a guide: sclera, iris, pupil, eye muscles, optic nerve, cornea.
3. Describe the job or function of these eye parts. Use your text or other references for help in completing the column marked "Function" in Table 1.
4. Complete the column marked "Description." As an example, the iris can be described as "part of the eye that gives it color."

FIGURE 1.

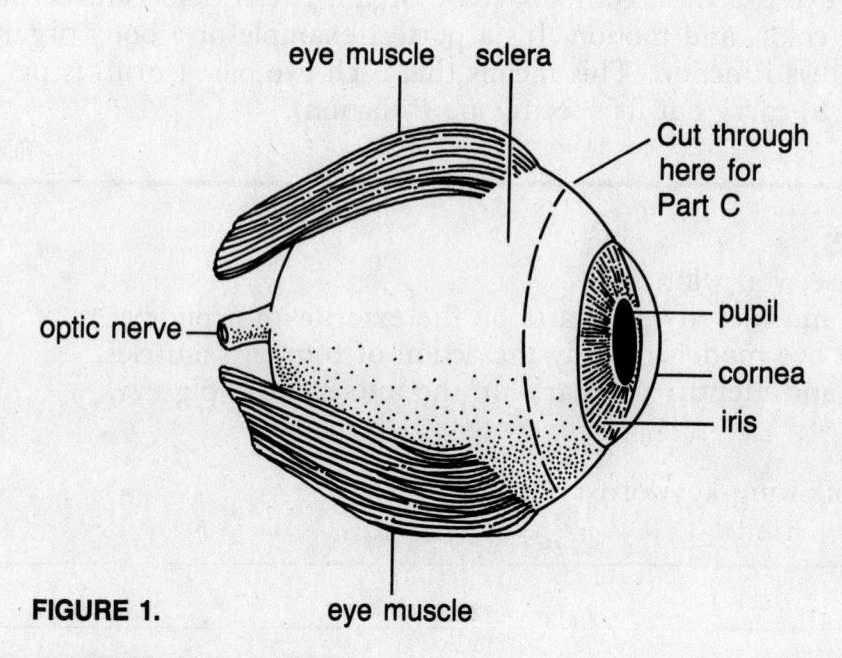

Table 1. Function and Description of Exterior Eye Parts

Part	Function	Description
sclera		
iris		
pupil		
eye muscles		
optic nerve		
cornea		

Part B. Eye Muscles and Movement

1. Trace Figure 2 onto a sheet of paper. Color the pupil black and the iris blue. Cut the traced drawing out and tape it onto a polystyrene ball (eye model) as shown in Figure 3.
2. Use scissors to cut four cloth strips, each measuring 2 x 8 cm.
3. Use a pen to number the strips 1–4.

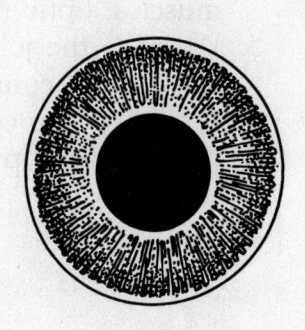

FIGURE 2.

Name _____ Class _____ Period _____

4. Tape the cloth strips onto the eye model as shown in Figure 3. These strips will represent eye muscles.
5. Determine the action of muscles 1 and 2 as follows:
   a. Stick the eye model onto a large paper clip as shown in Figure 4.
   b. Push the paper clip about 1 cm into the ball.
   c. Hold the paper clip between the fingers of one hand and gently pull cloth strip 1 toward the back of the eye with your other hand. The pulling of the cloth shows what happens in the eye as this muscle contracts or shortens.
   d. Note the direction that the eye model turns and record it in Table 2.
   e. Repeat steps c and d, only this time pull on cloth strip 2.
6. Determine the action of muscles 3 and 4 as follows:
   a. Stick the eye model onto the open end of a large paper clip as shown in Figure 5.
   b. Push the paper clip about 2 cm into the ball.
   c. Hold the paper clip between the fingers of one hand and gently pull cloth strip 3 toward the back of the eye with your other hand.
   d. Note the direction that the eye model turns and record it in Table 2.
   e. Repeat steps c and d, only this time pull on cloth strip 4.

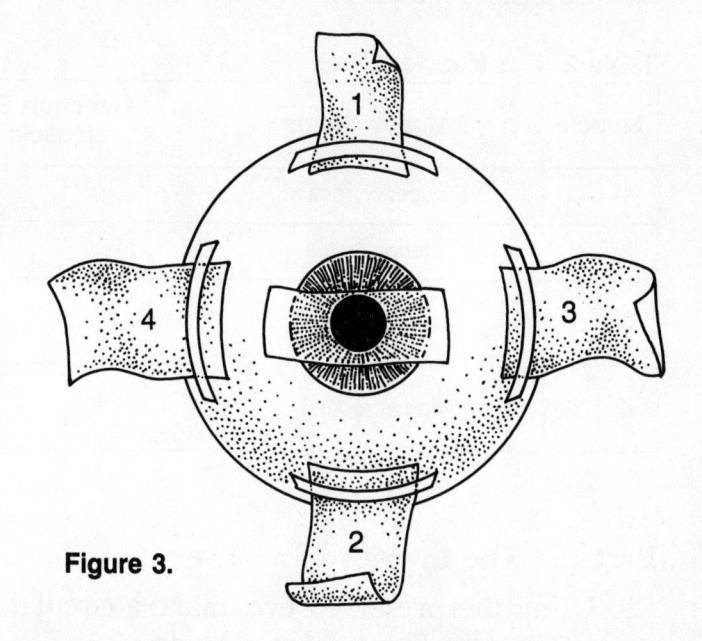

Figure 3.

NOTE: Muscle 3 is on the eye side next to the ear while muscle 4 is on the eye side next to the nose.

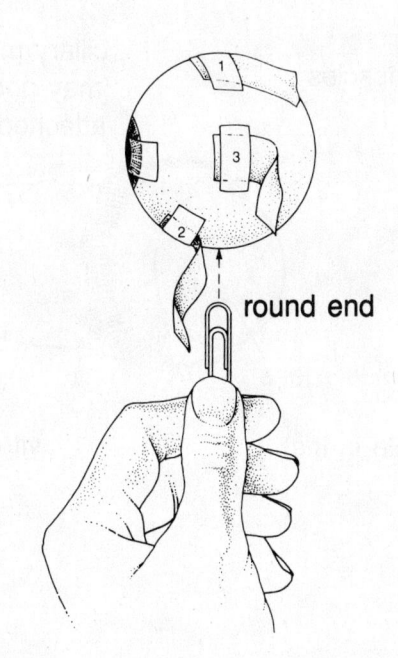

FIGURE 4. round end

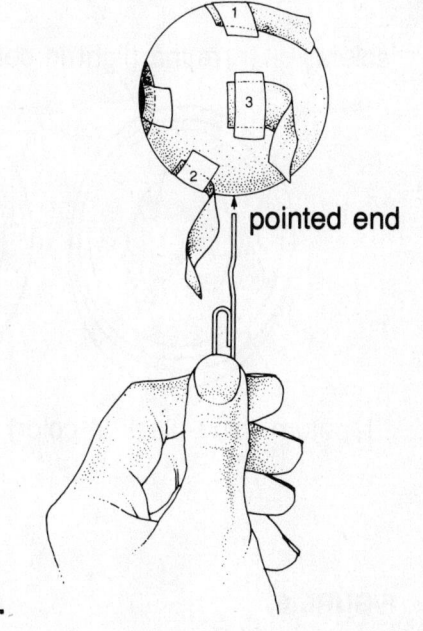

FIGURE 5. pointed end

137

7. Complete Table 2.

Table 2. Eye Muscle Action

Muscle	Muscle Name	Direction Eye Moves as Muscle Contracts	Location on Eye
1	Superior rectus		
2	Inferior rectus		
3	Lateral rectus		
4	Medial rectus		

Part C. The Interior of an Eye

1. Using the preserved eye, make a circular cut with your scissors through the sclera. Use Figure 1 as a guide.
2. Separate the eye into two halves. A jellylike material and marble-shaped part may fall out of the cut eye.
3. Locate the following eye parts using Figure 6 as a guide: retina, tapetum, ciliary muscles, lens, vitreous humor.
4. Describe the job or function of these eye parts. Use your text or other references for help in completing the first column of Table 3.
5. Complete the column marked "Description." As an example, the lens can be described as "clear, marble-shaped."

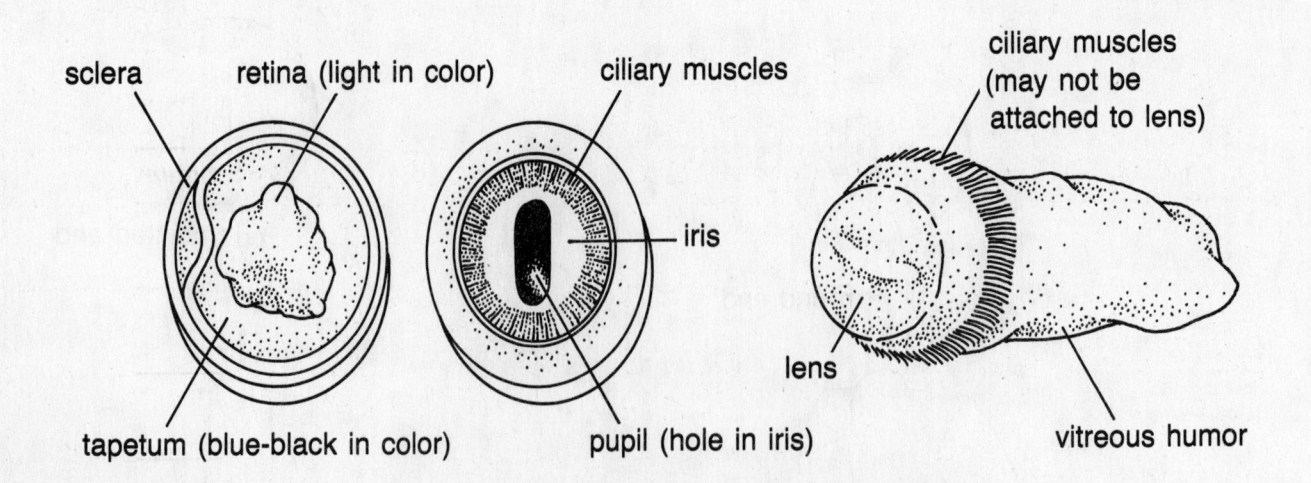

FIGURE 6.

Name _____ Class _____ Period _____

6. To better observe the cornea and iris, do the following:
   a. use a single-edge razor to cut the front half of the eye, as shown in Figure 7.
   b. look at the cut edge and compare it to Figure 8.
7. Label the following parts: cornea, iris, sclera, upper eyelid, lower eyelid, pupil.

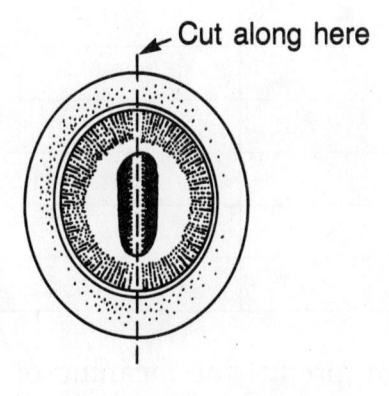

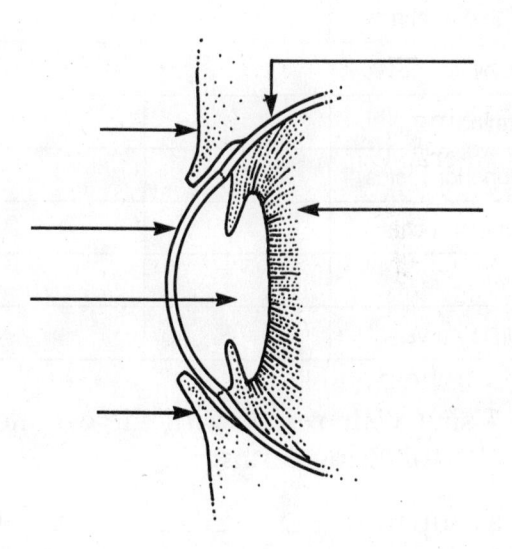

FIGURE 7.

FIGURE 8.

Table 3. Function and Description of Interior Eye Parts

Part	Job or Function	Description
retina		
tapetum		
ciliary muscles		
lens		
vitreous humor		

139

QUESTIONS

1. Complete the following chart. Use check marks. Some eye parts may be marked more than once in a row.

Eye part	Light does pass through	Light does not pass through	Muscle tissue	Can change shape	Made of nerve cells
Sclera					
Cornea					
Lens					
Ciliary muscles					
Vitreous humor					
Pupil					
Retina					
Superior rectus					
Lateral rectus					
Iris					
Optic nerve					

2. Using your results with the eye model and muscle action, predict the meaning of the following terms:

   a. superior _____

   b. inferior _____

   c. lateral _____

   d. medial _____

Name _____ Class _____ Period _____

17-1 What Self-protecting Behaviors Do Isopods Show?

How would an animal behave if given a choice of spending time in a light or a dark place? The answer to this question will be different depending upon the kind of animal. The way that each animal responds to its surroundings is part of its behavior. Usually, most behavior in animals will help rather than harm them.

INVESTIGATION

OBJECTIVES
In this exercise, you will:
  a. experiment with an animal called an isopod.
  b. test the animal's behavior in response to four different conditions.
  c. determine how the isopod behaves under these four conditions.

KEYWORDS
Define the following keywords:

behavior _____

innate behavior _____

isopod _____
_____

learned behavior _____

MATERIALS
isopods   tape
paper towel   scissors
black paper   wall clock
petri dish and cover   stick, plastic

PROCEDURE
Part A. Do Isopods Prefer a Smooth or a Rough Surface?
1. Place the bottom of a round dish (petri dish) onto a paper towel.

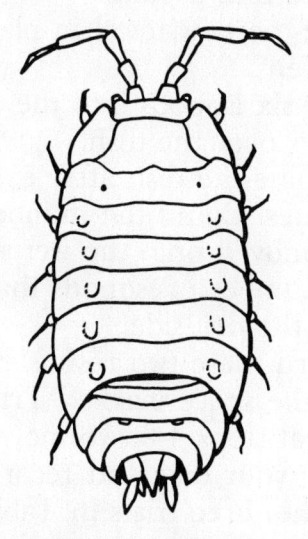

FIGURE 1. Isopod

141

2. Trace the bottom of the dish onto the towel.
3. Use scissors to cut out the paper outline of the dish and fold it in half.
4. Place the folded paper towel in the bottom of the dish. Tape the paper towel in place along all sides as shown in Figure 2.
5. Place six isopods into the center of the dish. Cover the dish.
6. Examine the dish after exactly 5 minutes. Count the number of isopods that moved onto the toweled side. Count the number of isopods that moved onto the untoweled side.
7. Record your data in Table 1. Use the space marked Trial 1.
8. Repeat steps 5–7 two more times. Total your data and figure and record averages for the three trials in Table 1.

Part B. Do Isopods Prefer Wet or Dry Conditions?

1. Repeat steps 1–3 from Part A.
2. Moisten the paper towel and place it into the dish so that it is next to the dry towel used in Part A. Examine Figure 3 as a guide.
3. Tape the new towel in place, as before.
4. Place six isopods into the center of the dish. Cover the dish.
5. Examine the dish after exactly 5 minutes. Count the number of isopods that moved onto the wet side. Count the number of isopods that moved onto the dry side.
6. Record these two results in Table 1. Use the space marked Trial 1.
7. Repeat steps 4-6 two more times. Total your data and record averages for the three trials in Table 1.

FIGURE 2.

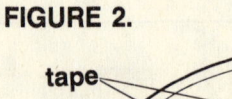

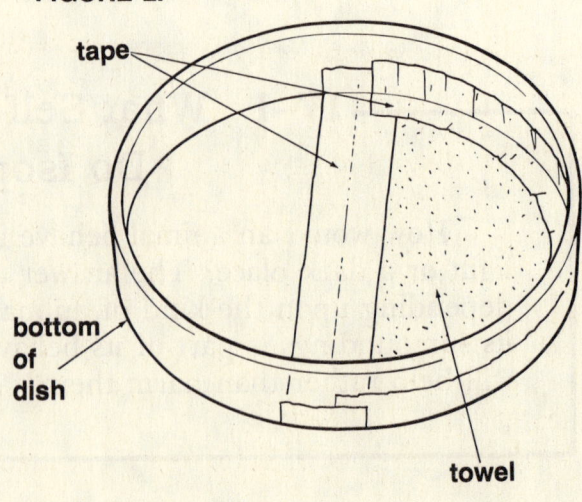

FIGURE 3.

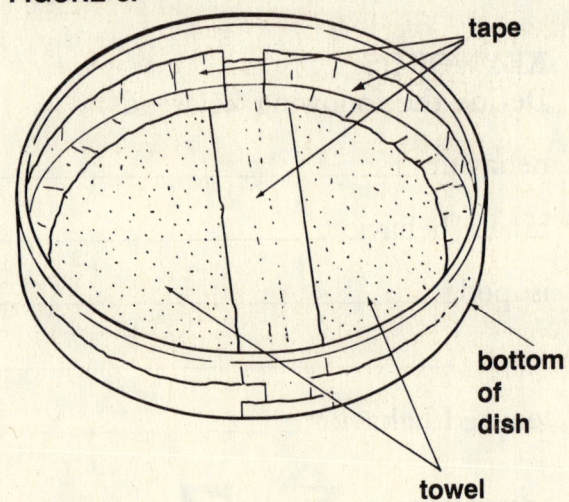

FIGURE 4.

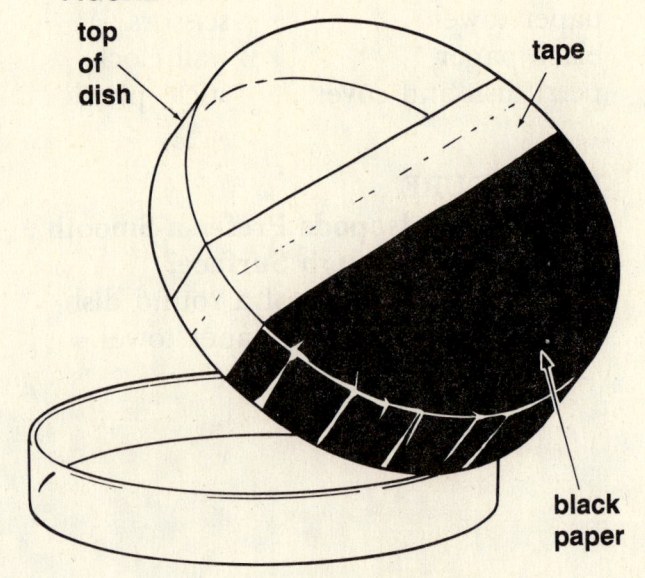

Name _____ Class _____ Period _____

Part C. Do Isopods Prefer Light or Dark?
1. Cover one-half of a new dish with black paper.
2. Tape the paper in place. Use Figure 4 as a guide.
3. Place six isopods into the center of the dish.
4. Cover the dish.
5. Examine the dish after exactly 5 minutes. Count the number of isopods that moved into the dark side of the dish. Count the number of isopods that moved into the lighted side of the dish.
6. Record your data in Table 1. Use the space marked Trial 1.
7. Repeat steps 4–7 two more times. Total your data and record averages for the three trials in Table 1.
8. Return five of the animals to your teacher. Keep one for Part D.

Part D. How Do Isopods Respond to Touch?
1. Place an isopod into a petri dish. Wait until the animal has uncurled itself before continuing on with the next step.
2. Lightly touch the animal with a plastic stick.
3. Record your observations in Table 2. Use the space marked Trial 1.
4. Repeat steps 2 and 3 for three more trials. Be sure to wait between trials until the animal has uncurled itself before touching it again.
5. Return the animal to your teacher when finished.

Table 1.

Trial	Rough Surface	Smooth Surface	Dry Condition	Wet Condition	Dark	Light
1						
2						
3						
Total						
Average						

Table 2.

Trial	Observations
1	
2	
3	
4	

QUESTIONS

1. With your data on averages from Parts A–C of this experiment, do isopods:

    a. prefer a rough or a smooth surface?_____

    b. prefer a wet or a dry condition?_____

    c. prefer a light or a dark place?_____

2. Offer a reason why an isopod would choose the type of surface noted in Part A of the experiment._____

3. Offer a reason why an isopod would choose the type of condition noted in Part B of the experiment._____

4. Offer a reason why an isopod would choose the type of light noted in Part C of the experiment._____

5. Explain how the behavior seen in Part D may be protective to the animal.

_____

_____

6. a. Are the behaviors seen in the isopod learned or innate?_____

    b. How can you tell?_____

    _____

7. a. Might the results to the four experiments have been the same if you had used a different animal?_____

    b. Why?_____

    c. How could you find out if another animal would behave in the same way that the isopod did?_____

    _____

8. Explain why you used three trials rather than just one for each experiment.

_____

_____

Name _____ Class _____ Period _____

17-2 What Happens When You Learn By Trial and Error?

Suppose you were given a ring of keys and you were not told which key would open a door. How could you solve this problem? You could keep trying keys until you found the right one. Each time you made an error, you would try again. The process you used to find the correct key was trial and error. You learned by repeating over and over until you discovered the right key.

The next time you needed to use that same key, however, you would not have to go through the whole process again. You had already learned which key opens the door. This process of finding the key that fits is an example of learned behavior.

INVESTIGATION

OBJECTIVES
In this exercise, you will:
  a. solve the same paper puzzle several times.
  b. run the same maze pattern several times.
  c. see the effect of trial and error on learning how to solve puzzles.

KEYWORDS
Define the following keywords:

innate behavior _____

learned behavior _____

maze _____

repeating or repetition _____

trial and error _____

MATERIALS
paper   ruler
file card   wall clock
scissors   folder
tape

FIGURE 1.

PROCEDURE
Part A. Paper Puzzle Trial and Error
1. Trace the shapes shown in Figure 1 onto a piece of paper.

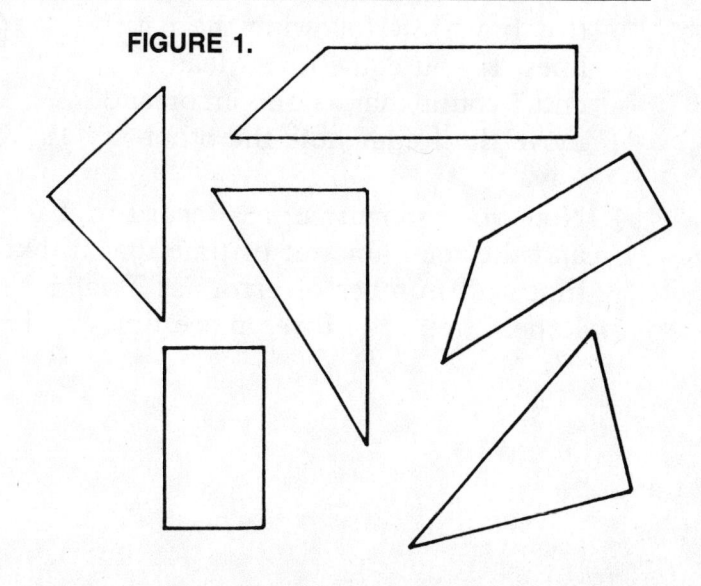

145

2. Tape the paper tracings onto a file card.
3. Cut out the six pieces with the scissors.
4. Put the six pieces together to form a single square. Time how long it takes.
5. Record in Table 1 the amount of time it took you to put the pieces together for trial 1.
6. Study how the pieces fit together to form a square.
7. Mix up the pieces of the square.
8. Repeat steps 4–7 three more times and enter your times in Table 1 as trials 2, 3, and 4.

Table 1.

Trial	Paper Puzzle Time	Maze Time	Number of Errors
1			
2			
3			
4			

Part B. Maze Trial and Error
1. Make a hole 1 cm in diameter in the exact center of an opened folder. Use Figure 2 as a guide.
2. Place the folder hole exactly over the letter "S" (for start) on the maze in Figure 3.
3. Keep the maze covered with the folder during the entire exercise. Move the hole to the letter "F" (for finish) by following the path lines. If you come to a "dead end," count that as one error and move the folder hole the other way.

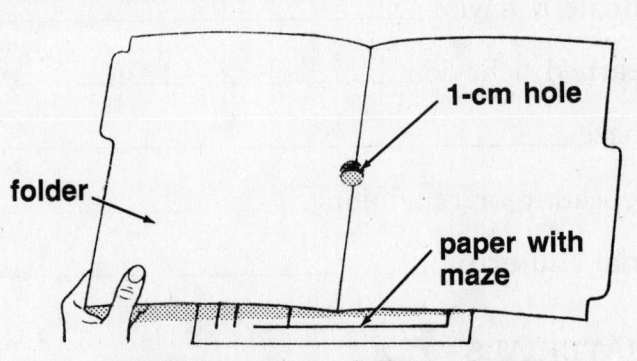

FIGURE 2. Holding folder over maze

4. Note your starting time. Record in Table 1 the number of errors that you make and the total amount of time that it takes you to complete the maze. Record the time and number of errors as Trial 1.
5. Repeat steps 2–4 three more times and record your results as trials 2, 3, and 4.

Name _____ Class _____ Period _____

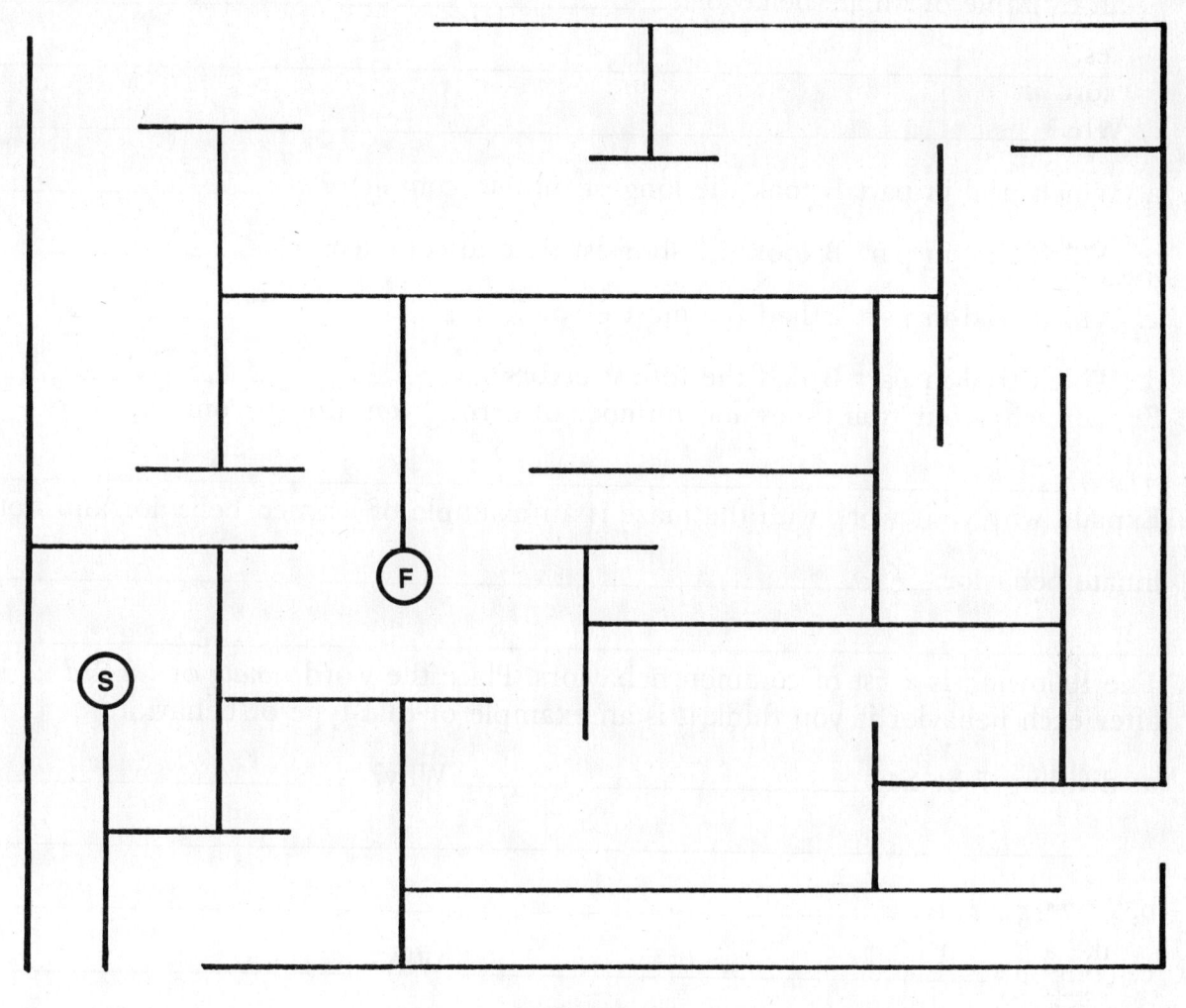

FIGURE 3.

QUESTIONS

1. a. Which trial in part A took the longest time to complete? _____

   b. Which trial in part A took the shortest time to complete? _____

2. Explain why the first trial in part A could be called trial and error learning.

   _____

3. Explain how repeating the puzzle through repetition improved your time at

   solving it. What was happening to your behavior in terms of learning? _____

   _____

   _____

4. a. Explain why part A of the experiment was not an example of innate behavior.

   _____

147

b. How might your data have looked for part A if solving the paper puzzle was an example of innate behavior? _____

c. Why? _____

5. a. Which trial in part B took the longest time to complete? _____
   b. Which trial in part B took the shortest time to complete? _____
   c. Which trial in part B had the most errors? _____
   d. Which trial in part B had the fewest errors? _____

6. Explain why your trial times and number of errors went down from trial 1 to trial 4. _____

7. Explain why your work with the maze is an example of learned behavior and not innate behavior. _____

8. The following is a list of common behaviors. Place the word *innate* or *learned* after each behavior if you think it is an example of that type of behavior.

   a. dividing 7 by 24 _____ Why? _____

   b. driving a car _____

   c. coughing _____ Why? _____

   d. sneezing _____

   e. using a skateboard _____

   f. blinking _____

   g. finding your way out of a building if lost in it _____

   h. memorizing the words to a song _____

Name _____ Class _____ Period _____

18-1 Which Antacid Works Best?

Antacids are a type of drug. The word *antacid* means against. These drugs are sold over-the-counter. They are used by people who have pain from too much acid in their stomach. The acid undergoes a chemical change with the antacid when this drug is taken. The antacid neutralizes the acid by changing it into water, a neutral liquid. The stomach pain disappears because there is no longer too much acid in the stomach.

Not all antacids are alike. Some may neutralize more acid than others. Television commercials are always telling us how much better one brand is than another. You can measure how good an antacid is by measuring how much acid it neutralizes.

INVESTIGATION

OBJECTIVES
In this exercise, you will:
  a. test water to find out if it is a good or a poor antacid.
  b. test three different antacids in order to find out which neutralizes the most acid.
  c. use your data to judge which antacid works best against stomach acid.
  d. find out what chemicals are used as antacids.

KEYWORDS
Define the following keywords:

acid _____

antacid _____

dosage _____

neutralize _____

neutral _____

sodium bicarbonate _____

MATERIALS
labels
wooden sticks
sodium bicarbonate liquid
red indicator chemical
8 100 mL beakers
water
antacids
dropper
acid in dropper bottle

graduated cylinder

149

PROCEDURE
Part A. Water as an Antacid
1. Label two small beakers, Beaker A and Beaker B.
2. Use a graduated cylinder to add 50 mL of water to each beaker.
3. Add 5 drops of red indicator chemical to each beaker as shown in Figure 1. Use a wooden stick to mix the red chemical in each beaker.
4. Record the color of each beaker in Table 1 under the column labeled Color before adding acid.
5. Read over steps 6 through 8 before starting step 6.
6. To *Beaker A only*, add acid one drop at a time. Use a stick to stir the liquid after adding each drop as shown in Figure 2. **CAUTION:** *Acid can burn skin or clothing. If acid is spilled, wash immediately with water and call your teacher.*
7. Count the number of drops of acid needed to change the color of the liquid to a deep red or purple color. Use Beaker B as a control for judging the original color of Beaker A. **NOTE:** *Once the color change has taken place in Beaker A, stop adding acid. At this point, you have neutralized all the acid that you can with the water in Beaker A. The indicator chemical turns color because the water in Beaker A has become acid.*
8. In Table 1, record the number of drops of acid used. Record the color of Beaker A and Beaker B in the column labeled Color after adding acid.

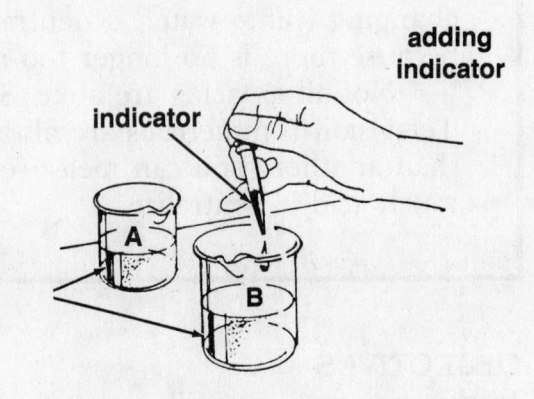

FIGURE 1. Adding red indicator chemical

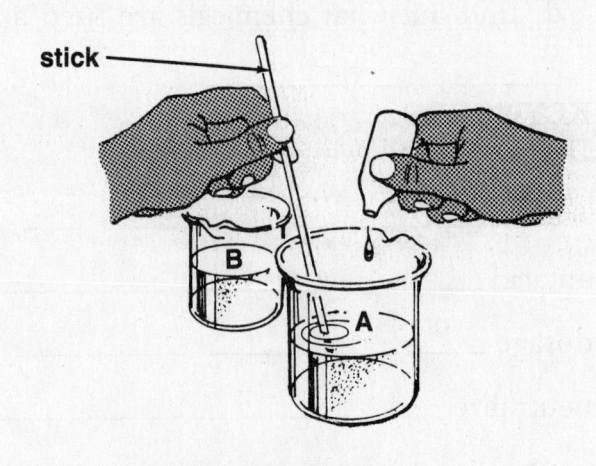

FIGURE 2. Adding acid

Table 1. Amount of Acid Neutralized by Antacid

Beaker	Color before adding acid	Number of drops of acid used	Color after adding acid
A			
B			

Part B. Testing Sodium Bicarbonate as an Antacid
1. Label two small beakers, Beaker C and Beaker D.
2. Use a graduated cylinder to fill each beaker with 50 mL of sodium bicarbonate liquid.

Name _____ Class _____ Period _____

3. Add 5 drops of red indicator chemical to each beaker. Use a clean wooden stick to mix the red chemical in each beaker.
4. To *Beaker C only*, add acid one drop at a time. Use a stick to stir the antacid after adding each drop. **CAUTION:** *Acid can burn skin or clothing. If acid is spilled, wash immediately with water and call your teacher.*
5. Count the number of drops of acid needed to change the color of the antacid from red to a deep red or purple color. Use Beaker D as a control for judging the original color of Beaker C. **NOTE:** *once the color change has taken place in Beaker C, stop adding acid. At this point, you have neutralized all the acid that you can with the antacid in Beaker C. The indicator chemical turns color because now the liquid in the beaker has turned acid.*
6. In Table 2, record the number of drops of acid added to Beaker C.

Table 2. Using an Antacid to Neutralize Acid

Beaker	Antacid name	Number of drops of acid added
C	sodium bicarbonate	
D	sodium bicarbonate	0
E		
F		0
G		
H		0

Part C. Testing Other Antacids
1. Repeat steps 2 through 8 in part B, using a different sample of antacid. This time, label the beakers E and F. Use Beaker F as a control for judging the original starting color, and record your results in Table 2. Complete the column labeled Antacid name. Your teacher will tell you the name of the antacid being tested.
2. Again repeat steps 2 through 8 in part B, using a different sample of another antacid. Your teacher will tell you the name of the antacid being tested. This time, label the beakers G and H. Use Beaker H as a control for judging the original starting color, and record your results in Table 2.

QUESTIONS
1. What color is an:
   a. antacid solution with red indicator? _____

   b. acid solution with red indicator? _____
2. Explain how you were able to tell when the antacid had neutralized as

   much acid as it could? _____

151

3. The fewer drops a liquid takes before changing color, the poorer it is as an antacid. The more drops a liquid takes before changing color, the better it is as an antacid.
   a. How many drops of acid were used to neutralize water (Beaker A)?_____
   b. How many drops of acid were needed to neutralize sodium bicarbonate (Beaker C)?_____
   c. Is water a very good antacid compared to sodium bicarbonate?
      Explain your answer._____
4. a. Which antacid (Beaker C, E, or G) needed the most drops of acid before showing a color change?_____
   b. Which antacid (Beaker C, E, or G) can neutralize the most acid?_____
   c. Which antacid (Beaker C, E, or G) is the most efficient?_____
5. a. Which antacid (Beaker C, E, or G) needed the fewest drops of acid before showing a color change?_____
   b. Which antacid (Beaker C, E, or G) can neutralize the least acid?_____
   c. Which antacid (Beaker C, E, or G) is the least efficient?_____

You will need to look at the packages of the antacids used to answer questions 6, 7, 8 and 9.

6. Check the labels on the antacids used. What is the suggested dosage for:
   a. antacid C?_____
   b. antacid E?_____
   c. antacid G?_____
7. What actual chemical is listed on the label as the antacid for:
   a. antacid C?_____
   b. antacid E?_____
   c. antacid G?_____
8. a. Is antacid C purchased as a liquid, powder, or solid?_____
   b. Is antacid E purchased as a liquid, powder, or solid?_____
   c. Is antacid G purchased as a liquid, powder, or solid?_____
9. a. Compare the cost of a box of sodium bicarbonate with the other antacids used. Which antacid is the least expensive?_____
   b. Does the most expensive antacid seem to neutralize the most acid?_____

Name _____ Class _____ Period _____

18-2 How Do Drugs Affect the Ability of Seeds to Grow into Young Plants?

Drugs change the way in which living things carry out every-day functions, such as growth and repair. It would be very hard to experiment on humans or other animals to show that this is true. It is possible to experiment with drugs and certain living things, such as plants. The living things used in this experiment are seeds.

INVESTIGATION

OBJECTIVES
In this exercise, you will:
  a. test the effect of four drugs on seed growth.
  b. compare the behavior of different seed types when tested with the same four drugs.

KEYWORDS
Define the following keywords:

caffeine_____

drug_____

ethyl alcohol_____

MATERIALS
5 small beakers	scissors	ruler
10 labels	5 petri dishes	paper towels
cheesecloth	2 baggie twist ties	aspirin
50 radish seeds	graduated cylinder	ethyl alcohol
50 pinto bean seeds	4 cigarettes	water
spoon	ground coffee (caffeinated)	

PROCEDURE
Part A. Soaking Seeds
1. a. Label five small beakers as follows: water, alcohol, aspirin, nicotine, caffeine.
   b. Add your name to each beaker.
2. a. Use scissors to cut two pieces of cheesecloth so that each piece measures 12 cm by 12 cm.
   b. Remove the filter (if present) and paper from four cigarettes.
   c. Place the tobacco into the center of one cheesecloth square.
   d. Wrap the tobacco in the cheese cloth. Tie the ends of the bag together with a twist tie.

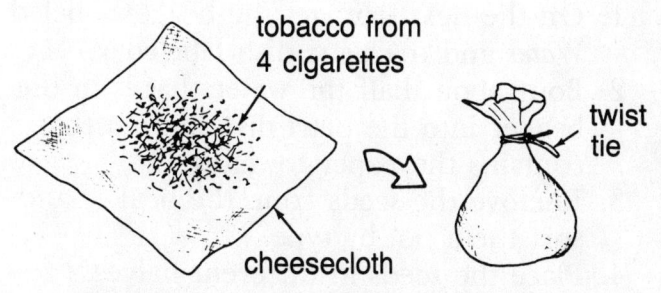

FIGURE 1. Making a cheesecloth bag

153

3. Prepare a second cheesecloth bag. In this bag, place one spoonful of ground coffee instead of tobacco.
4. Using a graduated cylinder to measure the different liquids, add the following to each beaker:
   a. water beaker, add 50 mL of water, 10 radish seeds, and 10 pinto bean seeds
   b. alcohol beaker, add 50 mL of ethyl alcohol, 10 radish seeds, 10 pinto been seeds
   c. aspirin beaker, add 50 mL of water, one aspirin tablet, 10 radish seeds, and 10 pinto bean seeds
   d. nicotine beaker, add 50 mL of water, cheesecloth bag with tobacco, 10 radish seeds, and 10 pinto bean seeds
   e. caffeine beaker, add 50 mL of water, cheesecloth bag with coffee, 10 radish seeds, and 10 pinto bean seeds
5. Allow the seeds to soak overnight in the beakers of liquid.

FIGURE 2. Making growth chambers

Part B. Preparing Chambers for Growing Seeds into Seedlings

1. Label the tops of five petri dishes as follows: water, alcohol, aspirin, nicotine, and caffeine. **NOTE:** *The top of the petri dish is wider than the bottom.*
2. Add your name to each label.
3. Prepare each dish to receive the soaked seeds by doing the following:
   a. Trace the bottom of a petri dish onto several thicknesses of paper toweling.
   b. Use scissors to cut out these circles of paper toweling.
   c. Place the circles of toweling into the bottom of each petri dish. Each dish should have at least five thicknesses of toweling (see Figure 2).

Part C. Adding seeds to Chambers

1. On the next day, get the beaker labeled *Water* and the petri dish labeled *Water*.
2. Pour about half the water that is in the beaker into the petri dish bottom that contains the paper toweling.
3. Remove the seeds from the beaker and sort them out by type.
4. Place the seeds in different halves of the petri dish as shown in Figure 3.

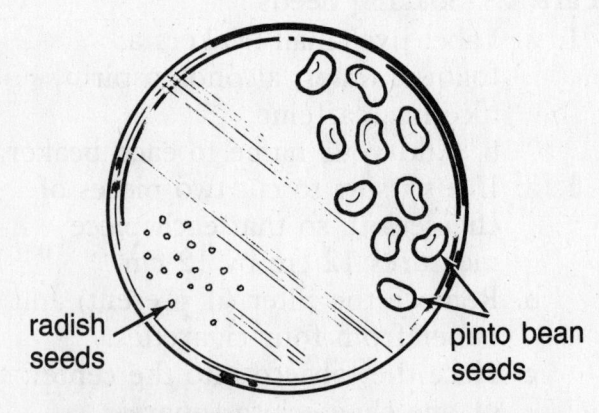

FIGURE 3. Adding seeds to growth chambers

Name _____ Class _____ Period _____

5. Cover the petri dish.
6. Repeat steps 1 through 5 for the remaining four beakers and petri dishes. Make sure that:
   a. the correct beaker liquid is poured into the correct petri dish.
   b. the correct seeds from each beaker are placed into the correct petri dish.
   c. each petri dish is covered with its properly labeled cover.

Part D. Observing Seed Growth
1. Examine all seeds in each petri dish. Look for signs of seed growth. See Figure 4 in order to tell if your seeds are growing. A small root will stick out from the seed if it is alive and growing.

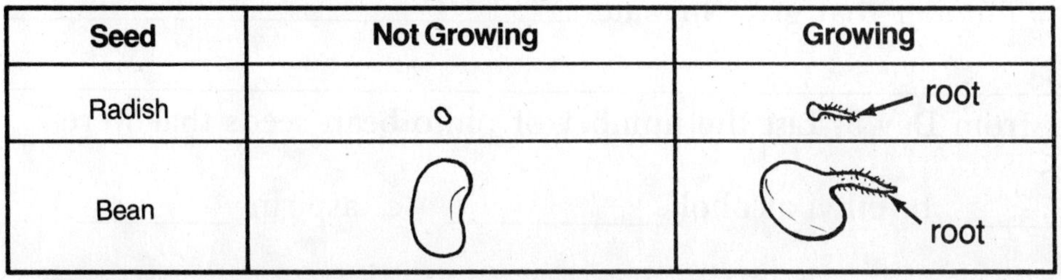

Seed	Not Growing	Growing
Radish		root
Bean		root

FIGURE 4. How to tell if seeds are growing

2. Record in Table 1 how many seeds of each type and treatment are growing. Consider today as Day 1 in the experiment.
3. Place the petri dishes in a place indicated by your teacher.
4. Observe the seeds for four more days and record the number of seeds growing of each type in each petri dish under the correct day.

Table 1. Number of Seeds Growing in Each Petri Dish Under Each Treatment

Day	Water Radish	Water Bean	Aspirin Radish	Aspirin Bean	Alcohol Radish	Alcohol Bean	Nicotine Radish	Nicotine Bean	Caffeine Radish	Caffeine Bean
1										
2										
3										
4										
5										

QUESTIONS
1. a. What different kinds of seeds were used in this experiment?

   _____

   b. What different drugs were used in this experiment? _____

   _____

2. What was the purpose of this experiment? _____

   _____

Name _____ Class _____ Period _____

3. Using your results from Day 5, list the number of radish seeds that were growing in:
   a. water _____   b. ethyl alcohol _____   c. aspirin _____

   d. nicotine _____   e. caffeine _____

4. a. Do the different drugs used seem to affect the number of radish seeds that grew? _____
   b. Support your answer by comparing the number of radish seeds that grew in the drugs to the number that grew in water. _____

5. Using your results from Day 5, List the number of pinto bean seeds that were growing in:
   a. water _____   b. ethyl alcohol _____   c. aspirin _____

   d. nicotine _____   e. caffeine _____

6. a. Do the different drugs used seem to affect the number of bean seeds that grew? _____
   b. Support your answer by comparing the number of pinto bean seeds that grew in the drugs to the number that grew in water. _____

7. Write a short paragraph that sums up your findings in this experiment.
   _____
   _____
   _____
   _____

8. a. Might all living things show the same type of results to these different drugs as did the seeds? _____
   b. How could you find this out? _____
   _____
   c. Why might this type of experiment be difficult to carry out on animals?
   _____
   _____

Name _____ Class _____ Period _____

19-1 What Do the Inside Parts of Leaves Look Like?

A leaf is one of the organs of a plant. The main function of a leaf is to make food for the rest of the plant. The food is carried to other plant parts. What parts of a leaf make and carry food?

Since leaves are organs, they are made of many different kinds of tissues. In leaves these tissues appear as layers of cells. Each tissue has a specific function and the shape of the cells in the tissue is related to its function. Some cells look like small boxes stacked side by side, while others are long and balloon-shaped. Certain cells are round and loosely packed while others look like small tubes stacked together.

EXPLORATION

OBJECTIVES
In this exercise, you will:
  a. build a model of the structure of a leaf.
  b. compare the model with the cells of a leaf as seen through the microscope.

KEYWORDS
Define the following keywords:

epidermis _____

palisade layer _____

spongy layer _____

stoma _____

_____

MATERIALS
scissors
1 sheet white paper
prepared slide of leaf

colored pencils: red, blue, purple, yellow,
microscope   tan, light green, dark green
transparent tape

PROCEDURE
Part A. Making a Model of a Leaf
1. Figure 1 shows the different cell layers of a leaf. Your teacher will provide you with a copy to color as follows:
   a. waxy layer—purple
   b. upper epidermis layer—yellow
   c. lower epidermis layer—tan
   d. spongy layer—light green
   e. palisade layer—dark green
   f. xylem—red
   g. phloem—blue
   h. guard cells—dark green

157

FIGURE 1. Tissue layers of a leaf

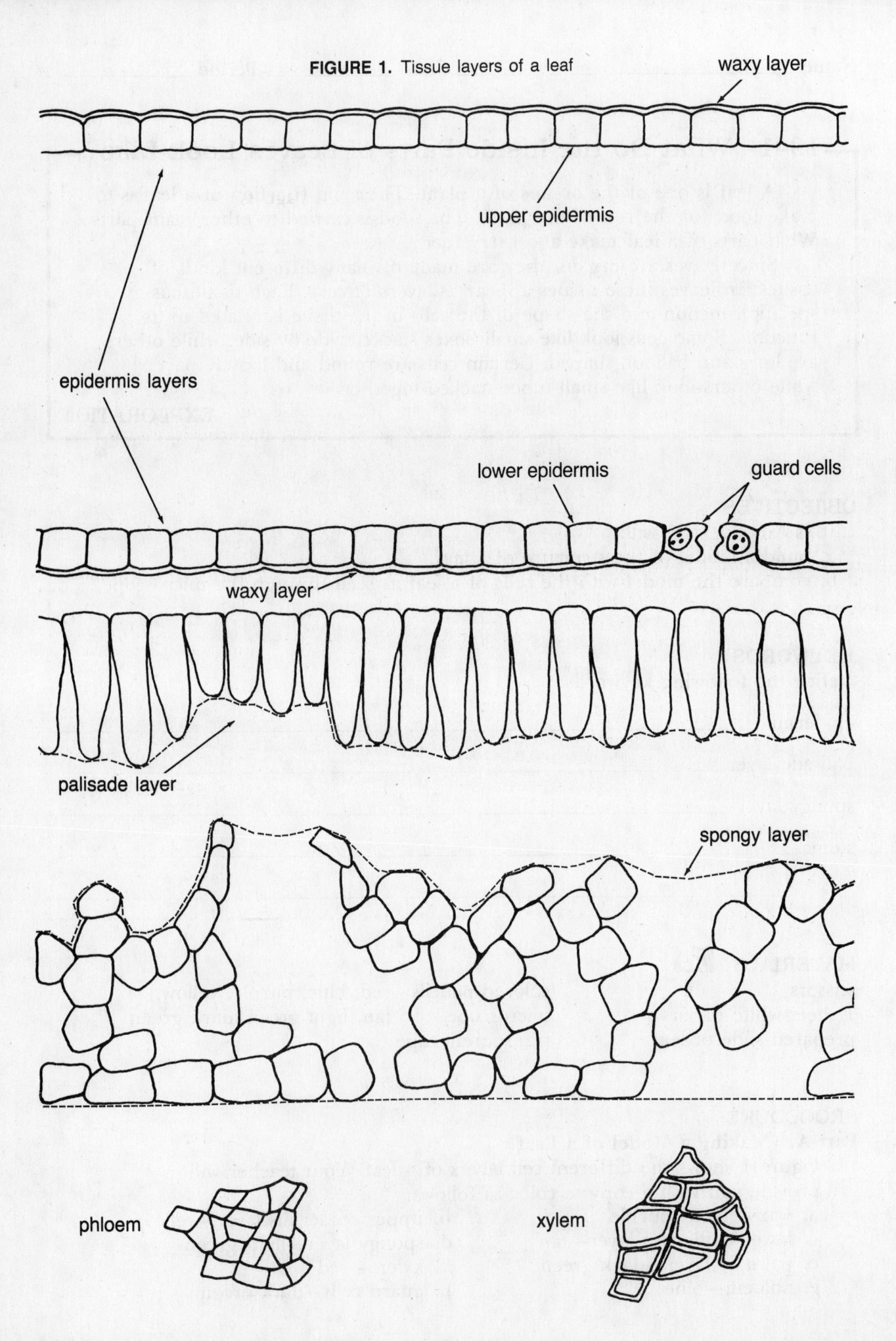

Name _____ Class _____ Period _____

2. Cut out the six layers of your colored copy of Figure 1. Cut along the dotted lines or follow the outlines of the cell layers.
3. Assemble your model of a leaf using Figure 19-5 on page 401 in your text.
4. Starting with the upper epidermis, tape each layer onto the blank sheet of paper. (HINT: The xylem and phloem fit together like parts of a puzzle.)
5. Label your leaf model by using the key in step 1.

Part B. Examining the Tissues of a Leaf
1. Examine the prepared slide of a leaf cross section on the low power of a microscope.
2. Locate and identify the six cell layers in the leaf section.
3. Draw a small section of the leaf that you see through the microscope on low power in the space provided. Label the parts of this leaf drawing by using the key in step 1.

FIGURE 2. Parts of a leaf

QUESTIONS
1. List the job of each of the following leaf parts:

   a. waxy layer_____

   b. upper and lower epidermis layers_____
   _____

   c. guard cell_____

   d. spongy layer_____

    e. palisade layer _____

    f. xylem _____

    g. phloem _____

    h. stoma _____

2. How many cell layers thick are the upper and lower epidermis? _____

3. Is the waxy layer thicker or thinner than the epidemis? _____

4. How is the waxy layer similar to a plastic bag? _____
_____

5. Describe the differences in the shapes of the kinds of leaf cells that make food. _____

6. Where does the phloem get the food that it carries to the stem and roots?
_____

7. Tell which cells in leaves

    a. are box-shaped. _____

    b. are shaped like balloons. _____

    c. have many air spaces around them. _____

    d. are covered with wax. _____

8. What would happen to the gases and water in a leaf if the guard cells were closer together, that is, if the stoma were smaller than it is? _____
_____

9. Many houseplants have very thick, waxy leaves. They do not wilt as quickly as houseplants with thinner leaves. Explain why. _____
_____

10. The celery stalks that you eat are leaf stalks. What kind of cells are the "strings" inside the celery? _____

Name _____ Class _____ Period _____

19–2 Where in a Leaf Does Photosynthesis Take Place?

The entire leaf blade of most plants is green. Some leaves have white, red, and yellow parts as well as green parts. The green color is due to a chemical called chlorophyll. Chlorophyll is the green pigment found inside parts of plant cells called chloroplasts. Chloroplasts are found mainly in the leave of plants.

Chlorophyll is needed by the plant to carry out photosynthesis. Chlorophyll traps the energy from light. This energy is used to change carbon dioxide and water to sugar. Sugar in many plants is then changed to starch. The starch is stored in other parts of the cell until needed by the plant.

Without light, plants cannot make food. If food is not made, cells of the plant can die. Sugar is made in the plant as long as light is present.

INVESTIGATION

OBJECTIVES
In this exercise, you will:
  a. observe the parts of the leaf that make starch during photosynthesis.
  b. decide what effect sunlight and lack of sunlight has on the making of starch in a plant.

KEYWORDS
Define the following keywords:

carbon dioxide_____

chlorophyll_____

chloroplast_____

photosynthesis_____

MATERIALS
1 500 mL beaker
2 250 mL beakers
scissors
50 mL rubbing alcohol
hot plate
bean plant, potted

forceps
30 mL iodine solution
2 petri dishes
2 aluminum foil squares,
  20 cm × 20 cm
Coleus plant, potted

161

PROCEDURE

Part A. Setting Up the Experiment

1. Cut two pieces of foil, each 20 cm square.
2. Fold the pieces of foil in half as shown in Figure 1. Place one-half (bottom half) beneath a leaf of a bean plant as shown in Figure 2.
3. Fold the top half over the bottom half.
4. Crimp three edges of the foil to keep out the light as shown in Figure 3. Be careful not to crush the leaf. Put a label with your name on the foil.
5. Repeat steps 2 through 4 with a *Coleus* plant.
6. Place both plants in the light for two days as shown in Figure 4.

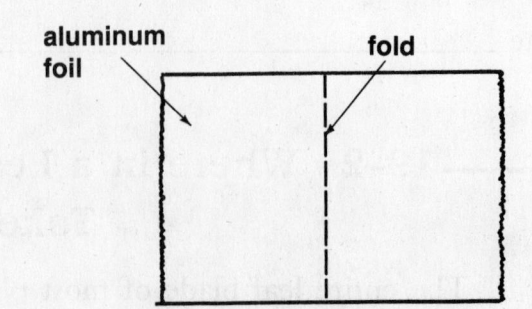

FIGURE 1. Folding aluminum foil

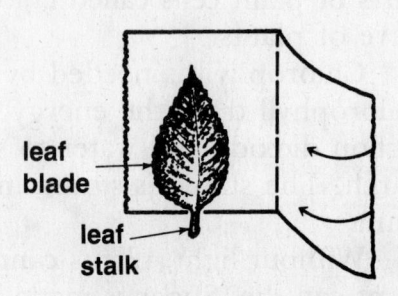

FIGURE 2. Covering a leaf

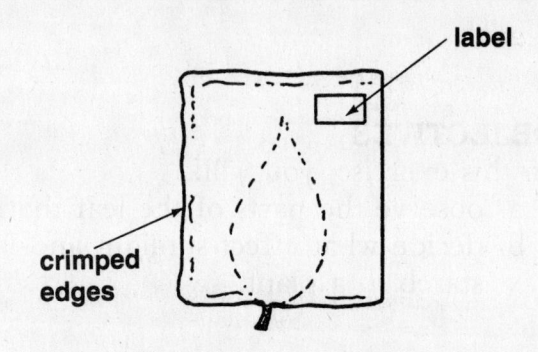

FIGURE 3. A covered leaf

Part B. Testing Leaves for Starch

1. Remove an uncovered leaf from each of the two plants. Cut off the stalks from each of your leaves with scissors.
2. Using forceps, carefully place your leaves into a beaker of boiling water that your teacher has set up on a hot plate as shown in Figure 5. This procedure ensures that the leaf is killed and cannot make more starch by photosynthesis. Remove the leaves with forceps after 3 minutes and place them in a petri dish.

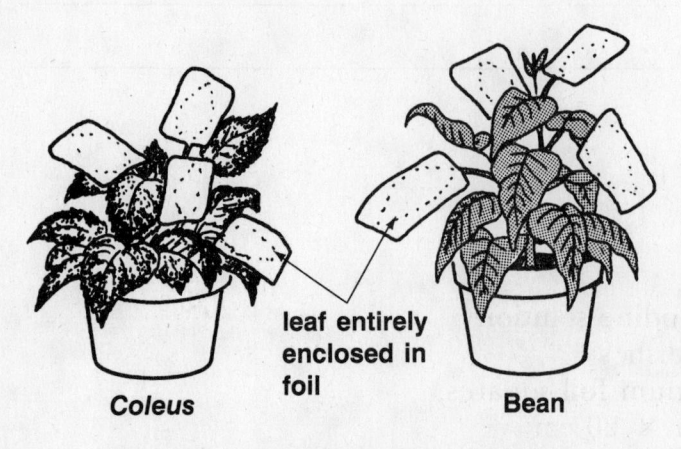

FIGURE 4. Placing plants in the light

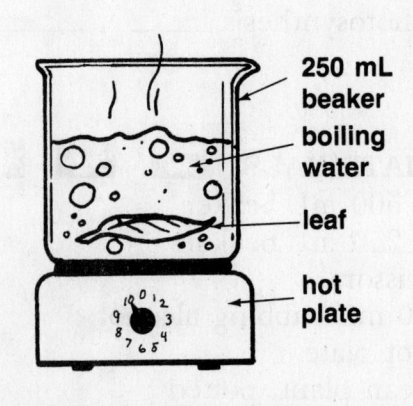

FIGURE 5. Killing leaves

Name _____ Class _____ Period _____

3. Give your leaves to your teacher who will place them into a small beaker of warm alcohol. This is prepared by placing the beaker of alcohol into a beaker of water heated on a hot plate as shown in Figure 6. The warm alcohol will remove the green pigment from the leaves. Your teacher will remove your leaves after 3 minutes. To soften the leaves, use forceps to dip them one more time into the boiling water.

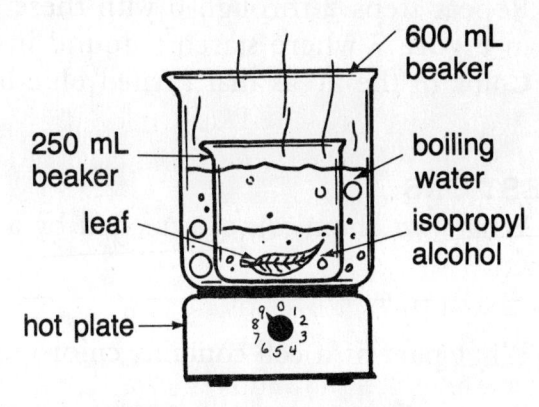

FIGURE 6. Removing chlorophyll

5. Put each leaf in a petri dish. Cover each leaf with iodine solution. Iodine turns starch a blue-black color. This is a positive test for starch.
6. Watch for any color change to appear in the leaves during the next 2 minutes.
7. Make simple outline drawings in Figure 7 to show the areas where starch is found on your uncovered leaves.

Bean leaf with light	***Coleus* leaf with light**
Bean leaf in foil	***Coleus* leaf in foil**

FIGURE 7. Areas of starch in leaves

8. Remove the foil-covered leaves from the two plants. Cut off the stalks from each of your leaves with scissors. Gently remove the foil from the leaves.
9. Repeat steps 2 through 6 with these two leaves and draw the areas of the leaves in Figure 7 where starch is found in these covered leaves.
10. Color in the areas that turned blue-black.

QUESTIONS

1. What two chemicals are needed by a plant for photosynthesis? _____

2. What part of a cell contains chlorophyll? _____

3. What colors are present in the bean leaf and the *Coleus* leaf? _____

4. In what area of the uncovered bean leaf was starch made? _____

5. In what area of the uncovered *Coleus* leaf was starch made? _____

6. In what area of the foil-covered bean leaf was starch made? _____

7. In what area of the foil-covered *Coleus* leaf was starch made? _____

8. How can you explain the differences in the making of starch in the four leaves? _____

9. Explain why photosynthesis takes place only in a certain area of a leaf.

Name _____ Class _____ Period _____

20-1 What Does a Woody Stem Look Like Inside?

Plant parts, like animal parts, are made of layers of cells called tissues. Each tissue is different in size and shape. The size and shape of each tissue is a clue to the job of the cells.

Woody stems have six kinds of tissues. The outer ring of cells, cork, is made of thick box-shaped cells. Cork protects the cells inside the stem. The thinner cells just inside the cork make up the cortex. These cells store food.

Next to the cortex cells are phloem cells. Phloem cells are thin, tubelike cells that carry food. A thin ring of cells just inside the phloem is the cambium. These cells make new phloem and xylem cells. Next to the inside of cambium cells is the xylem. Xylem cells carry water inside the plant. The center of a woody stem is made of pith cells. Pith cells store food.

INTERPRETATION

OBJECTIVES
In this exercise, you will:
  a. build a model of the structure of a woody stem.
  b. compare the model to a diagram of a woody-stem cross section.

KEYWORDS
Define the following keywords:

bark _____

cork _____

pith _____

xylem _____

MATERIALS
scissors   sheet of blank paper
transparent tape
colored pencils: purple, blue, red, green, orange, brown

PROCEDURE
1. Examine the woody-stem model parts on page 168. Obtain a photocopy of those model parts from your teacher. Color the different cell layers as follows:
   pith—purple   phloem—blue
   xylem—red   cambium—green
   cortex—orange cork—brown

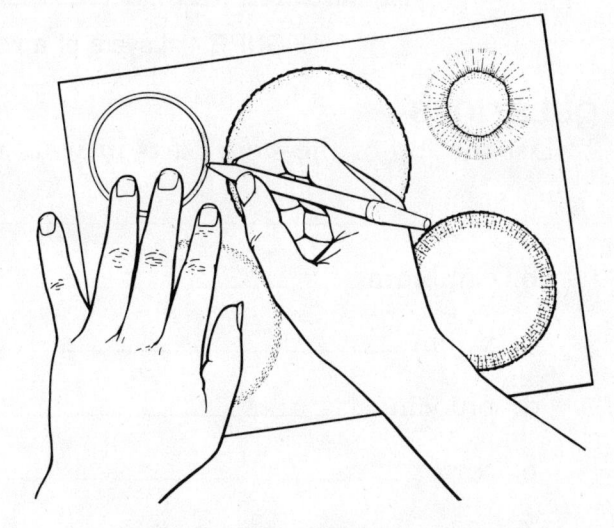

FIGURE 1. Coloring woody-stem layers

165

2. Cut out each layer of cells. Cut along the outer edge only.
3. Start assembling your model by placing one part on another. All parts fit on top of each other.
4. Tape your model on a blank sheet of paper. Label your model using step 1 as a guide.
5. Compare your woody-stem model with Figure 3, a drawing of the cross section of a woody stem.
6. Label the parts of the woody stem in Figures 3 and 4 by using your model as a guide. The diagram in Figure 4 shows what a woody stem looks like as seen through a microscope. Notice that the phloem is not an even layer as it is in your model. Also notice cells that make up each layer.

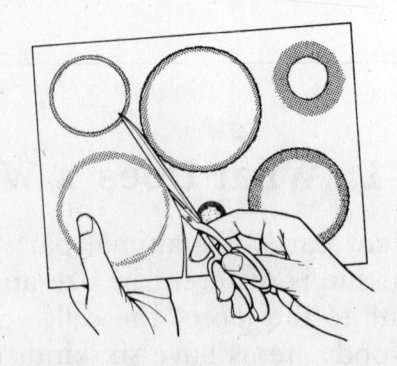

FIGURE 2. Cutting out woody-stem layers

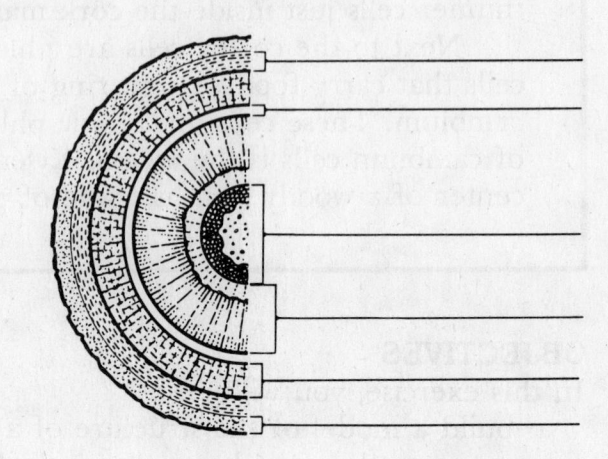

FIGURE 3. Cross section of a woody stem

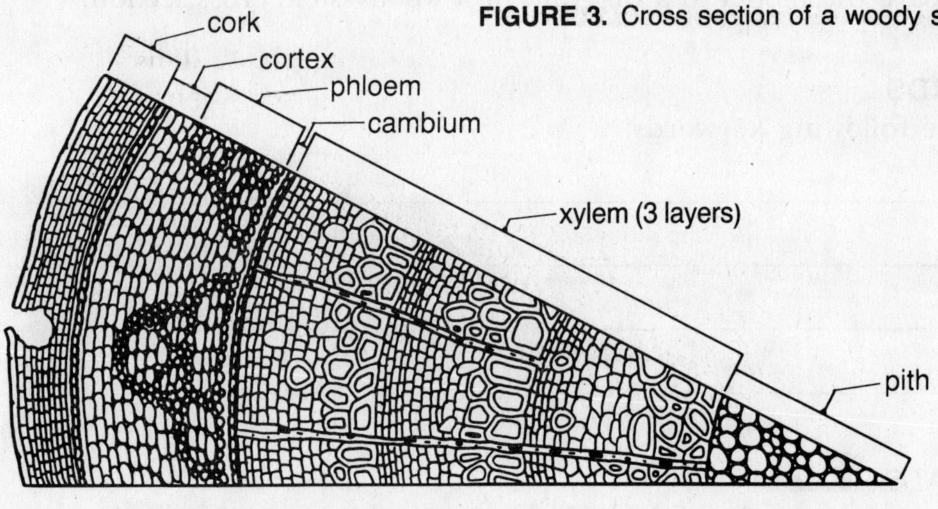

FIGURE 4. Layers of a woody stem as seen through a microscope

QUESTIONS
1. List the job of the stem parts in your model.

   a. cork _____

   b. cambium _____

   c. xylem _____

   d. phloem _____

   e. cortex _____

   f. pith _____

Name _____ Class _____ Period _____

2. Trace the path a beetle would follow as it burrows from the outside to the center of a tree stem. List the parts in correct order going from outside to inside.
   _____

3. The cell layers in your model are called tissues. Why are they called tissues?

   (HINT: Use your text for help.) _____
   _____

4. Bark is the term used to describe the outer layers of a woody stem from cork through phloem. Name the tissue layers that form bark.
   _____

5. Use the drawing of the woody stem cross-section and describe the cell layers. Use a complete sentence.

   a. xylem _____

   b. pith _____

   c. phloem _____

   d. cork _____

6. The thickness of each xylem ring may differ each year. This may be the result of growing conditions during that year. A thick ring means that conditions such as rainfall were good and much growth took place. A thin ring means conditions were poor and little growth took place.

   a. Which ring, A or B, is the thinner ring? _____
   b. Which ring, A or B, may have grown during a year when much rain

      fell? _____ When little rain fell? _____

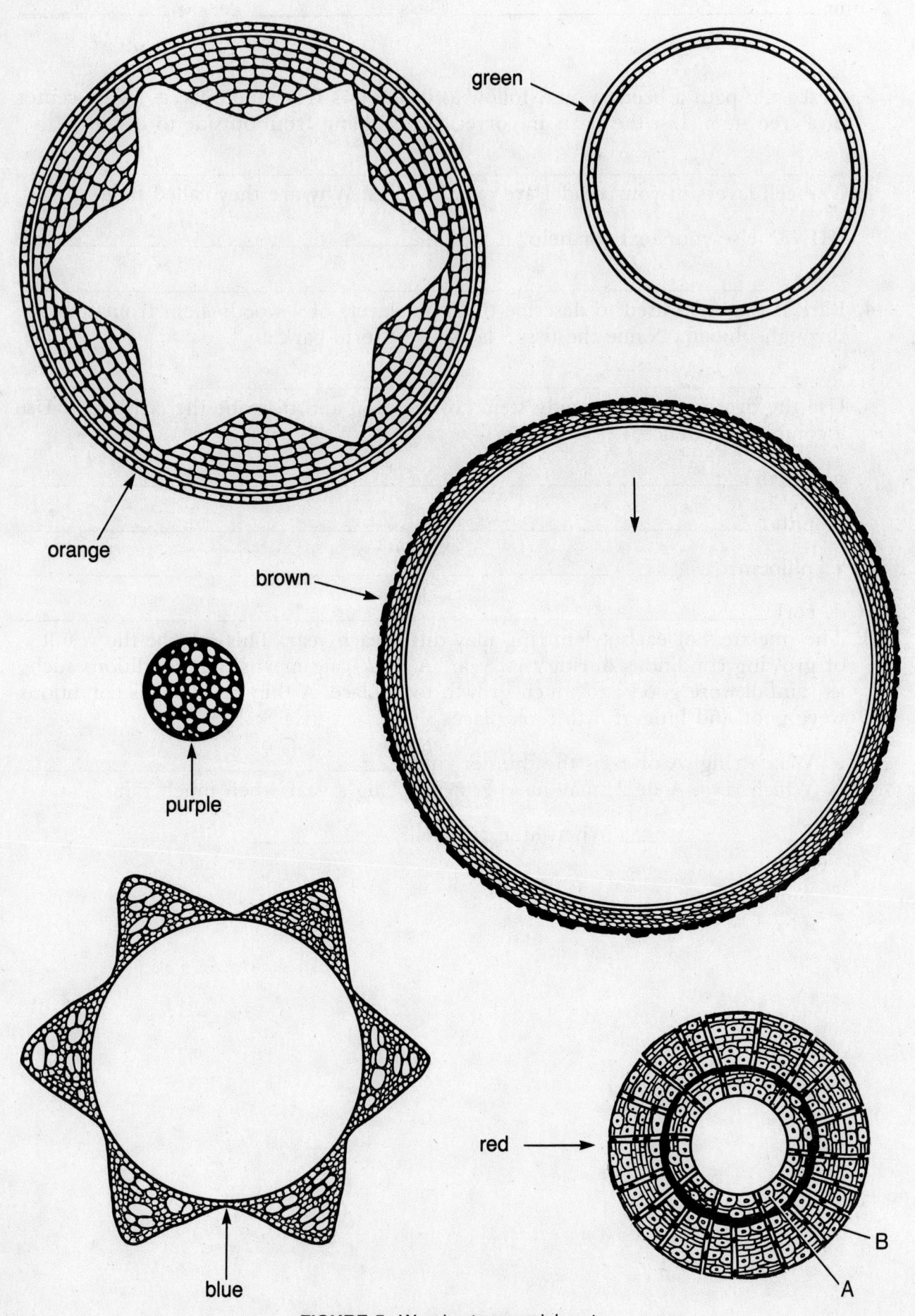

FIGURE 5. Woody-stem model parts

168

Name _____ Class _____ Period _____

20–2 What Do the Inside Parts of a Root Look Like?

Roots have five kinds of cells: epidermis, cortex, endodermis, xylem, and phloem. The cells do not look exactly the same as they do in the stem. The size and shape can be different. The way the cells are packed together can also be different.

The outer layer of cells is the epidermis, a layer of protective cells. These cells form a ring around the outside of the root. Certain long, thin cells of the epidermis that absorb water and nutrients are called root hairs. The layer of cells, or tissue, next to the epidermis is the cortex, a layer of food-storage cells. Within the cortex is a ring of cells called the endodermis. Endodermis cells protect the xylem and the phloem. Phloem cells are found inside and next to the endodermis cells. Phloem cells carry food to other cells. The center of the root is packed with xylem cells. Xylem cells carry water to all cells in the plant.

EXPLORATION

OBJECTIVES
In this exercise, you will:
 a. build a model of the structure of a root.
 b. compare the model with the cells of a root as seen through a microscope.

KEYWORDS
Define the following keywords:

cortex _____

epidermis _____

phloem _____

root hairs _____

xylem _____

MATERIALS
blank sheet of paper
scissors
light microscope
transparent tape
prepared slide of a dicot root
colored pencils: red, blue, green,
            brown, purple

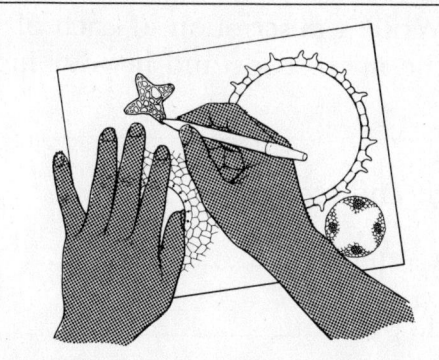

FIGURE 1. Coloring root model parts

PROCEDURE

1. Examine the drawings of the root model parts in Figure 6 on page 160. Obtain a photocopy of the root model parts from your teacher. Color the different layers of cells in the root model parts as follows:
   a. epidermis—green
   b. cortex—purple
   c. endodermis—brown
   d. phloem—blue
   e. xylem—red
2. Cut out each layer of cells as shown in Figure 2. Cut along the outer edge only.
3. Assemble your model by placing one part on another. Start with the thick, outer layer of cells. All parts fit on top of each other.
4. Tape your completed model to a sheet of blank paper.
5. Label your root model parts by using step 1 as a guide.
6. Examine the prepared slide of the dicot root under low power, then high power, of a microscope. Locate each of the cell layers in the slide. Note how each layer looks.
7. Look at Figure 4. This is what the prepared root slide will look like. Use your model and root slide as guides to label the parts in Figure 4.

FIGURE 2. Cutting out root model parts

FIGURE 3. Examining a root slide

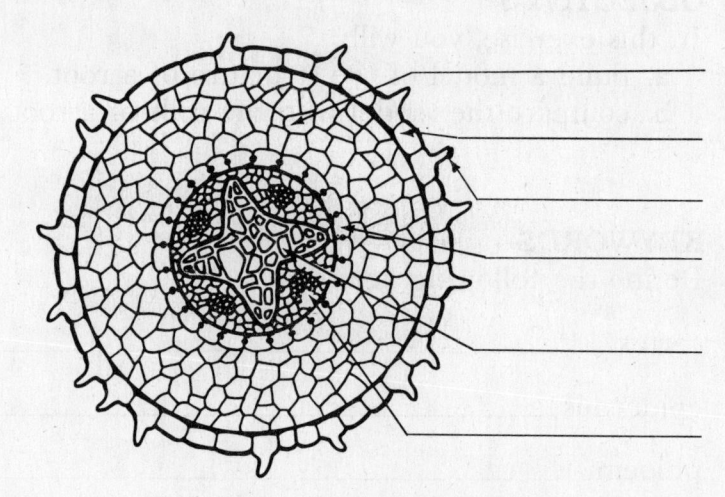

FIGURE 4. Cross section of root

QUESTIONS

1. Write a description of each of the following cell layers as they looked through the microscope and in your model. Use complete sentences.

   a. cortex_____

   b. endodermis_____

   c. epidermis_____

   d. phloem_____

   e. xylem_____

Name _____ Class _____ Period _____

2. Examine Figure 5, the diagram of a carrot cut lengthwise. Label the parts of the carrot by referring to your model.

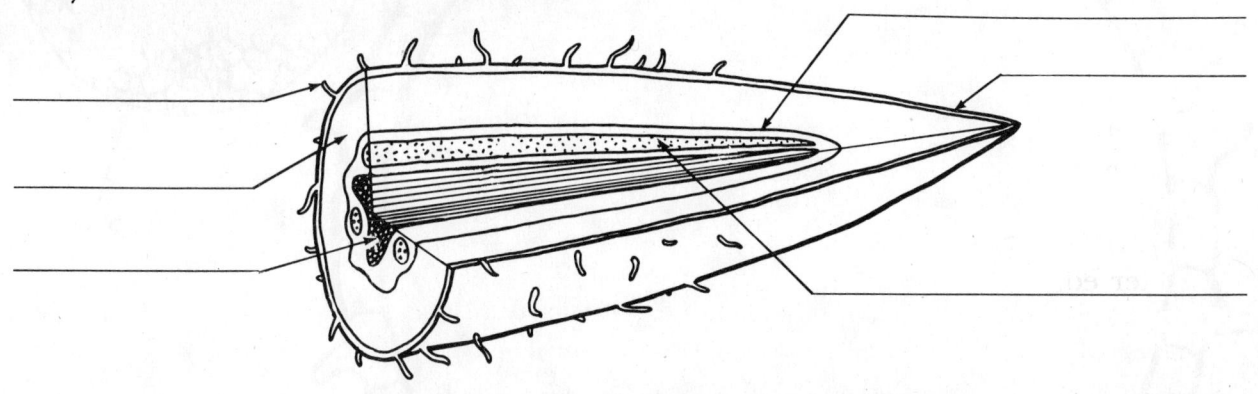

FIGURE 5. Carrot root longitudinal section

3. Which of the three types of cells, xylem, phloem or cortex, has the largest diameter? _____ Why? _____

4. Which of the two types of cells, xylem or phloem, has the largest diameter? _____ Why? _____

_____

5. If the phloem and xylem are tubelike cells, why do they look like circles and boxes in your models? _____

_____

6. What is the advantage of making a model of the cell layers of a root? _____

_____

7. Trace the path that a drop of water in the soil would follow as it moves into a root. List the parts in their correct order, going from outside to inside. Use your root model to help. (HINT: Water does not pass through the phloem.) _____

_____

8. Many of the roots people eat are peeled first.
   a. What cell layer is peeled away? _____
   b. Why do some people peel away the outer covering of carrots, turnips and beets before eating them? _____

_____

9. Some roots grow thicker than others. What group of cells do you think are responsible for the thickness of a root?

_____ Why? _____

171

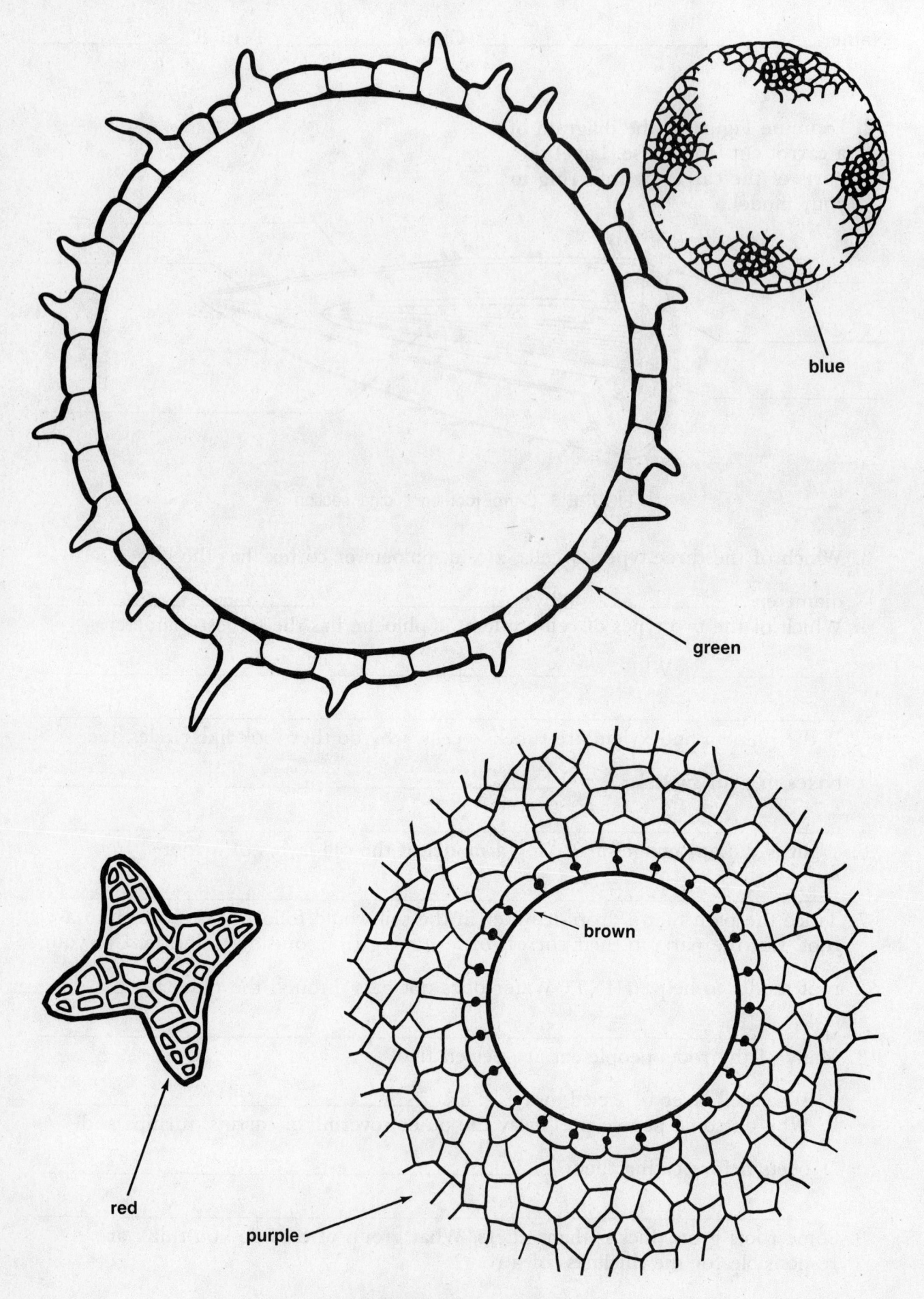

FIGURE 6. Root model parts

172

Name _____ Class _____ Period _____

21-1 What Tropisms Can Be Seen in Growing Plants?

Plants do not have a nervous system as animals do. But, a plant can respond to a stimulus in its environment. A response to a stimulus by a plant is called a tropism.

Most plant tropisms have something to do with growth movements. Unlike animals, plants are attached to the soil and are not able to move about. So, the movements made by a plant are made by a plant's parts, such as the stem, root, leaf, or flower. A growth movement in response to light is called a phototropism while a growth movement in response to touch is called a thigmotropism.

INVESTIGATION

OBJECTIVES
In this exercise, you will:
 a. observe a growing bean plant's response to light.
 b. observe a *Mimosa* plant as it responds to touch.
 c. find out how much time it takes for a Mimosa plant to "recover" after being touched.

KEYWORDS
Define the following keywords:

phototropism _____

stimulus _____

thigmotropism _____

tropism _____

MATERIALS
for Part A
500 mL soil
2 plastic foam cups
4 germinated bean seedlings
metric ruler
cardboard box with slit

for Part B
potted *Mimosa* plant
toothpick
watch or clock

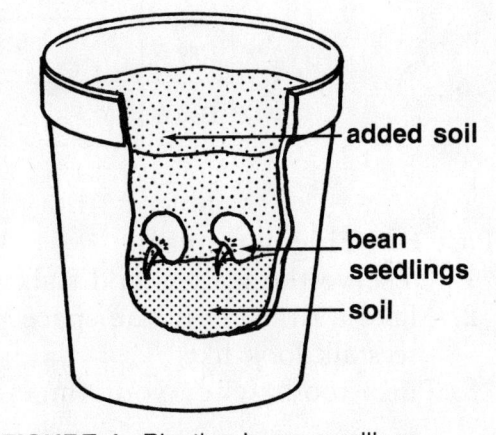

FIGURE 1. Planting bean seedlings

PROCEDURE
Part A. Phototropism
1. Punch four holes in the bottom of each cup. Label the cups A and B and put your name on each cup.
2. Fill the cups nearly full with soil.

173

3. Put two bean seedlings on top of the soil in each cup as in Figure 1.
4. Fill the cups to the top with soil and add water to lightly dampen the soil.
5. Place the cup labeled A in an area of the room that gets even light from all sides.
6. Place the cup labeled B in a box that has been prepared for you. Note that a small strip has been cut out of the side of the box. Place the cup in the box as shown in Figure 2.
7. When a box is filled with cups, it should be closed so that light enters only through the cutout slot.

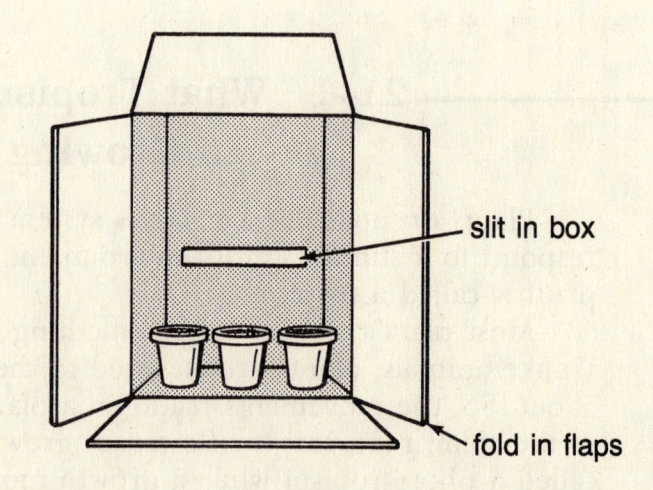

FIGURE 2. Placing bean seedlings in a cardboard box

8. Examine the plants during each class period. Observe changes in the plants as they grow.
9. Measure the height of one of the bean plants in each cup. (Be sure to measure the same plant in each cup throughout this experiment.) Record this measurement for both plant A and plant B in the proper place in Table 1.
10. Make a small sketch of the plant in the proper place in the column marked "Amount of Bending" in Table 1. In the sketch you should show the direction, if any, in which the stem and leaves seem to be turning or bending.

Table 1. Height and Bending Response in Plants

	Plant A (in Light)		Plant B (in Dark)	
	Height in cm	Amount of Bending	Height in cm	Amount of Bending
Day 1	—	—	—	—
Day 2	—	—	—	—
Day 3				
Day 4				
Day 5				
Day 6				

Part B. Thigmotropism.
1. Observe the leaflets and stalk of the *Mimosa* plant without touching the plant.
2. Make a drawing in the space provided in Figure 3 to show what the leaflets and the stalk look like.
3. Put a toothpick in your fingers as shown in Figure 4.

Name _____ Class _____ Period _____

4. Touch a leaflet by letting the toothpick spring forward as shown in Figure 4.
5. Observe the response of the leaflet to touch and note the time at which the response occurred. This is Trial 1. Record this time in the proper column in Table 2 under "Leaf."
6. Make a drawing, in the space provided in Figure 5, of what the leaflet looks like after it has been touched.
7. Record in Table 2 in the column marked "Time Recovered" the time at which the leaflet returns to its normal position.

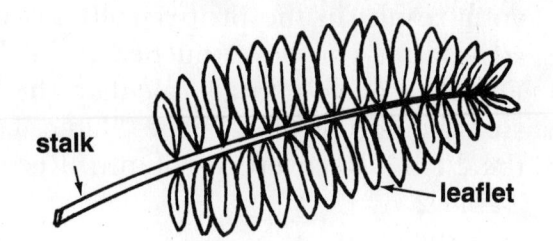

FIGURE 3. *Mimosa* plant

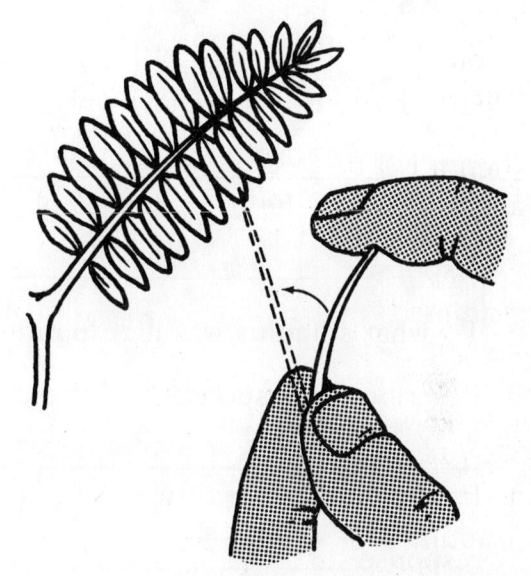

FIGURE 4. Touching *Mimosa* leaflet

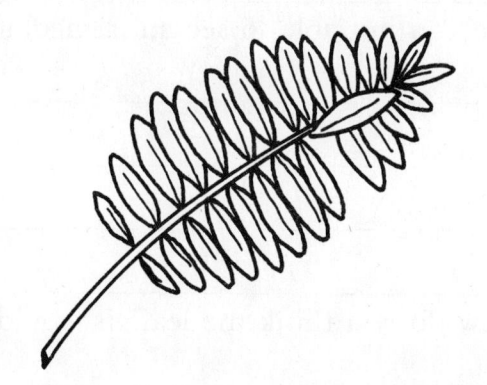

FIGURE 5. *Mimosa* leaflet after touching

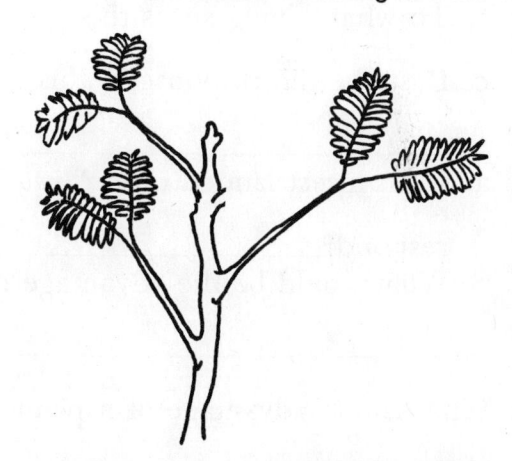

FIGURE 6. *Mimosa* stalk after touching

Table 2. Recovery Time for *Mimosa*

	Leaf			Stalk		
	Time Touched	Time Recovered	Amount of Time for Recovery	Time Touched	Time Recovered	Amount of Time for Recovery
Trial 1						
Trial 2						
Trial 3						
Trial 4						

8. Repeat steps 4–7 at least three more times (Trials 2, 3, and 4) with a leaflet.
9. Repeat steps 4–7 at least four times (Trials 1, 2, 3, and 4) with a stalk. Record your results in the proper columns under "Stalk" and make a drawing of the stalk, after you have touched it, in the space provided in Figure 6.
10. For each of your trials calculate the "Amount of Time for Recovery" by subtracting the time in the "Time Touched" column from the time recorded in the "Time Recovered" column. Record your results in the proper columns in Table 2.

QUESTIONS

1. What is a tropism?_____

2. What are the two types of tropisms being studied in this investigation?
_____

3. a. In which of the two types of plants were you able to see an immediate response?_____

   b. To what stimulus was it responding?_____

   c. Describe the response of this plant._____
   _____

4. a. In which of the two types of plants were you not able to see an immediate response?_____

   b. To what stimulus was this plant responding?_____

   c. Describe the response in this plant._____
   _____

5. a. If an insect landed on a *Mimosa* leaf, how do you think the leaflets would respond?_____

   b. What would be the advantage of this type of movement to the plant?
   _____

6. What is one advantage of a plant's responding to light?_____
   _____

7. What would happen if a plant could not grow toward light?_____
   _____

Name _____ Class _____ Period _____

21-2 Is Light an Important Growth Requirement for Plants?

Plants need a certain number of hours of light each day during their growing season. This light is needed for photosynthesis and normal growth. Some kinds of plants need to get more light each day than other plants.

Did you ever notice that some plants grow best in the shade of other plants? Plants of this kind can grow in a forest where trees provide shade. Most mosses and ferns are plants that grow in shade. Suppose you tried to grow corn or tomato plants in the shade. They probably would not grow well because they grow best in direct sun.

INVESTIGATION

OBJECTIVES
In this exercise, you will:
 a. observe changes in height and color in two growing plants.
 b. determine what effect sunlight has on the color and growth of plants.

KEYWORDS
Define the following keywords:

light _____

photosynthesis _____

shade _____

MATERIALS
2 plastic foam cups   150 mL soil
metric ruler   water
2 bean seedlings

PROCEDURE
1. Punch four holes in the bottom of each cup.
2. Label the cups A and B and place your name on each cup.
3. Fill the cups nearly full with soil. Use Figure 1 as a guide.
4. Place a bean seedling in each cup. Try to choose two, nearly identical seedlings. Fill the cups to the top with soil. Water soil lightly.

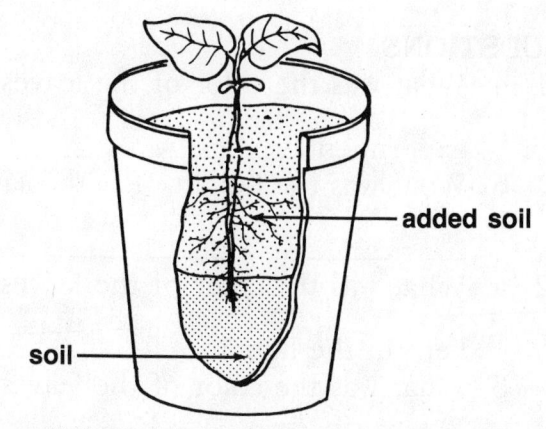

FIGURE 1. Planting bean seedlings

177

5. Measure the length of the stem (in cm) from the soil to the stem tip for both plants. Note the stem's color. Record your observations in the proper columns in Table 1 for Day 0.
6. Put cup A in a lighted area. Put cup B in a dark area.
7. Examine both plants at the same time each day for the next 4 days. Repeat steps 5 and 6 each day. On days 1 through 4, record leaf color. Keep the soil in the cups moist, but not wet, during the entire experiment.

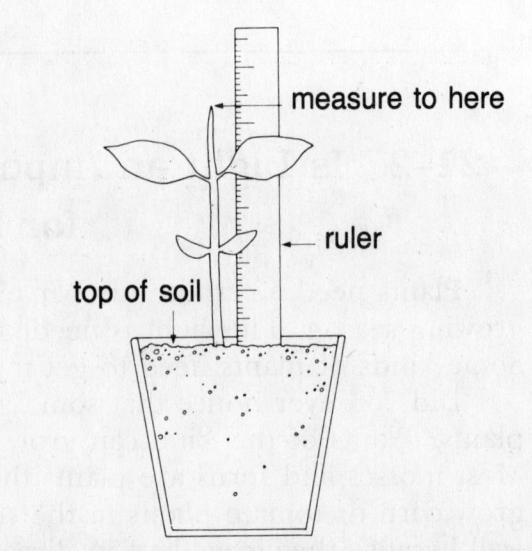

FIGURE 2. Measuring the stem of a bean seedling

8. On day 4, measure the length of one of the first true leaves from the base of the blade to the leaf tip for both plants. Record your observations in the proper columns in Table 1.
9. Make line graphs showing plant height for both plants using your data in Table 1.

Table 1.

Day	Plant A—Light				Plant B—Dark			
	Leaf Color	Leaf Length	Stem Color	Stem Length	Leaf Color	Leaf Length	Stem Color	Stem Length
0								
1								
2								
3								
4								

QUESTIONS

1. a. What was the color of the leaves of both plants at the start of the experiment?_____
   b. What was the height of both plant stems at the start of the experiment?_____

2. a. What was the color of the leaves at the end of the experiment for the plant kept in the light?_____
   b. What was the color of the leaves at the end of the experiment for the plant kept in the dark?_____

Name _____ Class _____ Period _____

3. a. What was the height of the stem at the end of the experiment for the plant kept in the light? _____

   b. What was the height of the stem at the end of the experiment for the plant kept in the dark? _____

4. How were the leaves and stems of the two plants at the end of the experiment different? _____

5. Describe what you think is normal growth in a plant when it gets enough light.

6. Describe what happen to plants that do not get enough light. _____

7. Sometimes potatoes and carrots sprout while being kept in a refrigerator. Explain why they are usually yellow. _____

8. A head of lettuce is mostly leaves. Explain why the leaves near the center of the head are yellow while the outer leaves are green. _____

179

FIGURE 3.

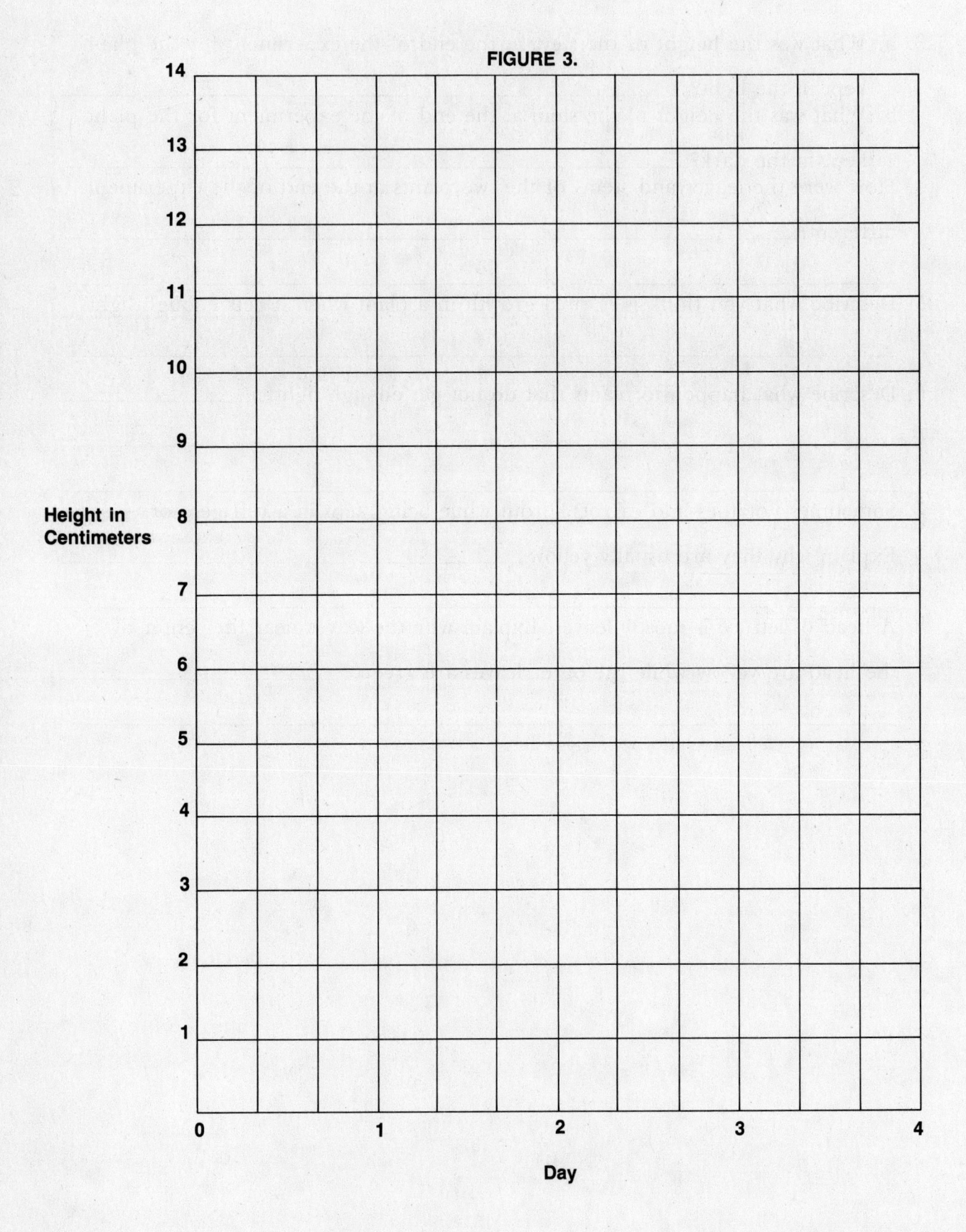

Height in Centimeters

Day

Name _____ Class _____ Period _____

22–1 What Happens When Cells Divide?

Cells form new cells by a process called cell division or mitosis. During mitosis, one cell divides in half to form two new cells. Suppose you could watch a cell divide. You could see that the cell parts called chromosomes move around the cell during mitosis. Because chromosomes move in particular ways, you could arrange the events of mitosis into several steps.

Biologists have been able to arrange the events of mitosis into several steps. They examined many dividing cells in order to learn the steps. What are the steps of cell division? In what order do they occur?

INTERPRETATION

OBJECTIVES
In this exercise, you will:
 a. build models of the steps of mitosis.
 b. compare your models to the steps of animal-cell mitosis.

KEYWORDS
Define the following keywords:

chromosome _____

cytoplasm _____

nucleolus _____

nucleus _____

MATERIALS
4 pieces of different-colored construction paper
scissors   thread
glue   24 toothpicks
yarn   metric ruler

FIGURE 1. Steps of mitosis

PROCEDURE
1. Using Figure 1 and your textbook, review the steps of mitosis.

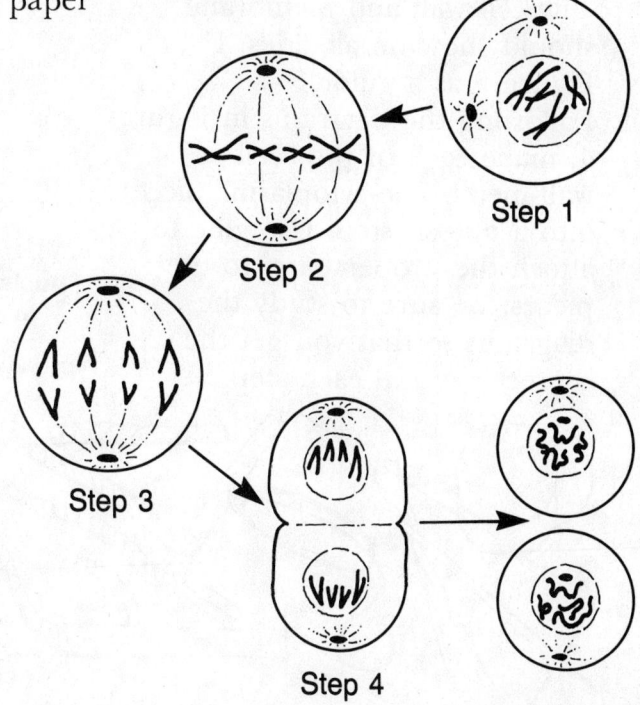

2. Use the materials listed in Table 1 to represent the cell parts. Cut the pieces of paper, yarn, and thread to the sizes given in Table 1.

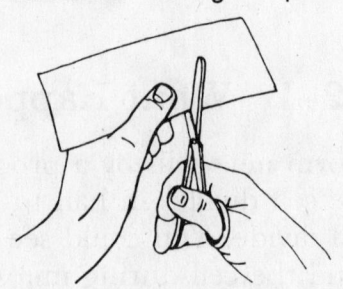

FIGURE 2. Making cell parts

Table 1. Making Cell Parts

Cell Part	Material to Use	Size	Number Needed
Cell wall and membrane	Dark-colored paper	14 × 8 cm	5
Cytoplasm	Light-colored paper	13 × 7 cm	5
Nucleus	Dark-colored paper	5 cm circle	3
Nucleolus	Light-colored paper	1 cm circle	2
Chromosomes	Light-colored yarn	4 cm long 10 cm long	20 2
Fibers	Toothpicks	full size	24
Cell wall between new cells	Dark-colored paper	½ × 8 cm	1
Nuclei in new cells	Thread	½ m	2

3. Begin building the models of the cell division steps by gluing each "cytoplasm" paper to the top of a "cell wall and membrane" paper. The cell wall and membrane should show on all sides. Use Figure 3 as a guide.

4. Following the diagrams in Figure 4, make each of the "cell wall-membrane-cytoplasm" pieces into a mitosis step. Use glue to attach the proper parts to the pieces. Be sure to study the diagrams so that you get the correct parts in each step.

FIGURE 3. Putting models together

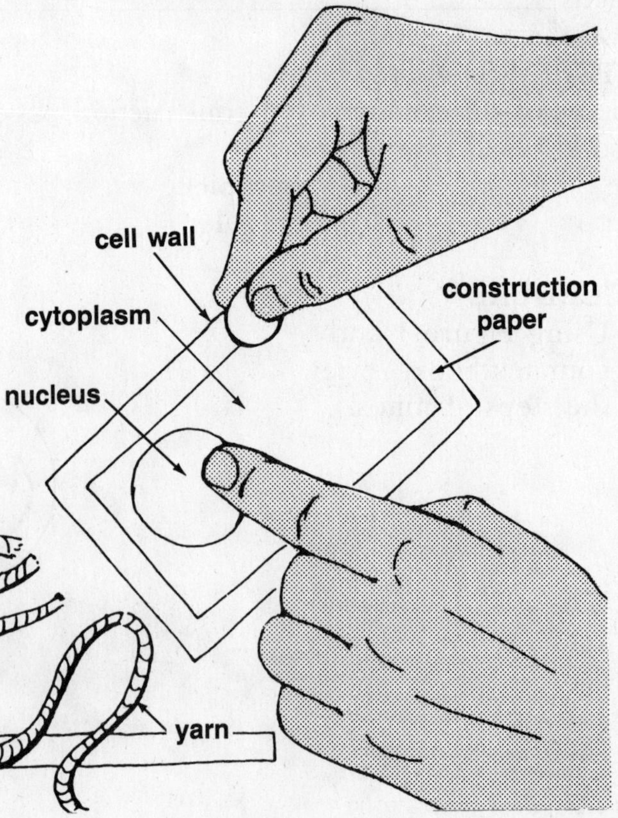

182

Name _____ Class _____ Period _____

FIGURE 4. Steps of plant cell mitosis

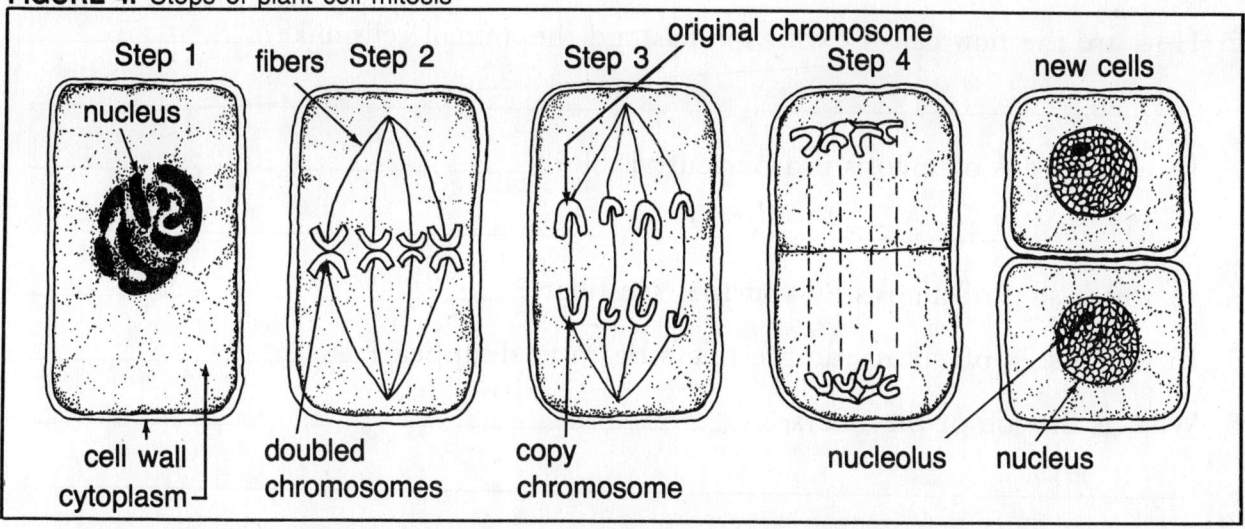

6. Arrange your models in the order in which mitosis occurs. Note how your models differ from those shown in Figure 1. Your models show dividing plant cells.
7. Compare your models with the drawings of the animal cells in Figure 5.
8. Write the number of the step of mitosis below each drawing of the animal cells.

FIGURE 5. Steps of animal cell mitosis

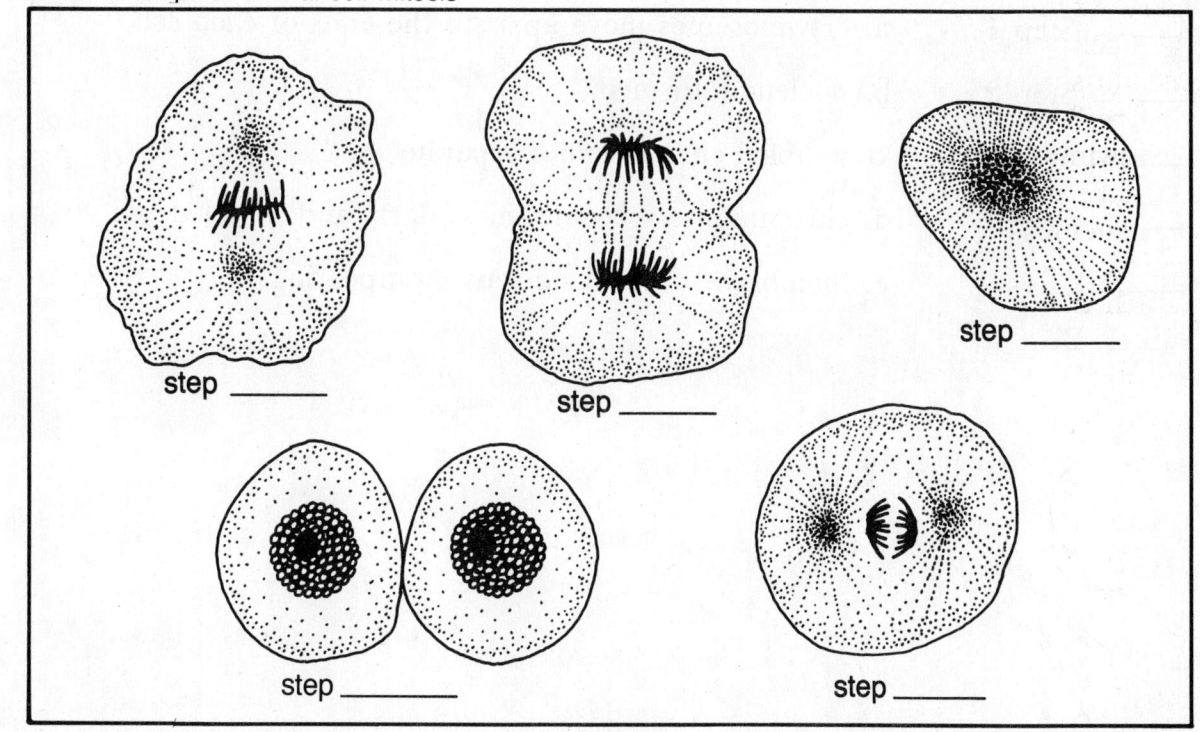

step _____

step _____

step _____

step _____

step _____

183

QUESTIONS

1. What part is present in plant cells but absent in animal cells?_____
   _____

2. How are the new cells of your models and the animal cells alike?_____
   _____

3. In which steps of mitosis is a nucleus visible?_____
   _____

4. In which step of mitosis do you first see fibers?_____

5. In which step of mitosis do the fibers begin to disappear?_____

6. What is the job of the fibers?_____
   _____

7. a. Doubled chromosomes first become visible in which step of mitosis?_____

   b. How many doubled chromosomes are visible in this step?_____

8. What is happening to the doubled chromosomes in steps 3 and 4?
   _____
   _____

9. How many cells does each dividing cell form?_____

10. What forms between the cells after the doubled chromosomes have pulled apart in plant cells?_____

11. Match the following by writing the correct letter in the proper blank.

    _____ Step 1   a. chromosomes move apart to the ends of each cell

    _____ Step 2   b. nucleus reformed

    _____ Step 3   c. doubled chromosomes separate

    _____ Step 4   d. chromosomes become thick, dark, and doubled

    _____ Step 5   e. membrane around nucleus disappearing

Name _____ Class _____ Period _____

22-2 Are There More Dividing Cells or Resting Cells in a Root Tip?

A plant grows in length at the tip of stem and root. In the stem and root tip there is a small group of cells that divide many times; however, not all cells in these parts may be dividing. A dividing cell may be next to several resting cells and a resting cell can be surrounded by several dividing cells.

Cells in mitosis are different from resting cells. Some parts of a cell are seen best only when a cell is dividing. These parts seem to disappear after a cell has divided.

EXPLORATION

OBJECTIVES
In this exercise, you will:
  a. examine dividing and resting cells in an onion root.
  b. compare the number of dividing cells to resting cells.

KEYWORDS
Define the following keywords:

dividing cell _____

resting cell _____

root tip _____

MATERIALS
prepared slide of an onion root tip, L.S.
light microscope

FIGURE 1. Root tip under a microscope

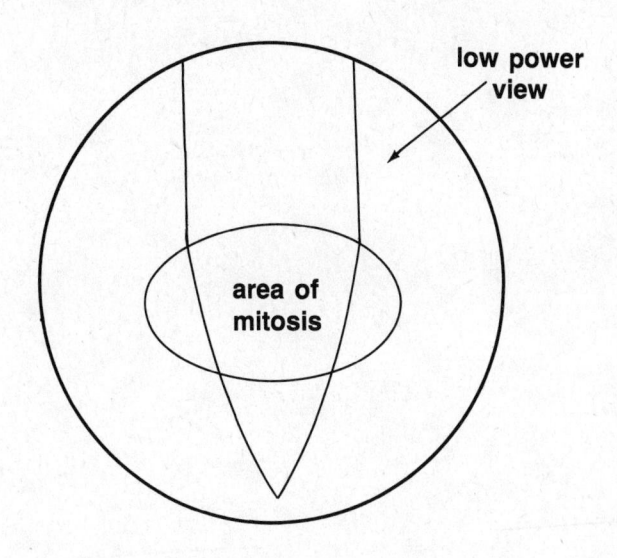

PROCEDURE
1. Obtain a microscope slide of an onion root tip.
2. Place the slide on the stage of the microscope and examine the root tip on low power. The tip of the root should be at the bottom of the field of view as in Figure 1.
3. Change the lens to high power. You will see several columns of cells.
4. Look for a cell that resembles Step 1 of mitosis, as shown in Figure 4 on page 187.

185

5. Draw the cell in the space provided and label the parts.
6. Repeat step 5 for the remaining steps of mitosis shown in Figure 4.
7. Count the cells in your field of view that resemble Step 1 of mitosis. Count the cells a column at a time going from left to right, as in Figure 2.
8. Record in Table 1 the number of cells that resemble Step 1.
9. Repeat steps 7 and 8 for the remaining steps of mitosis shown in Figure 4.
10. Compare your data with those of other students in your class.

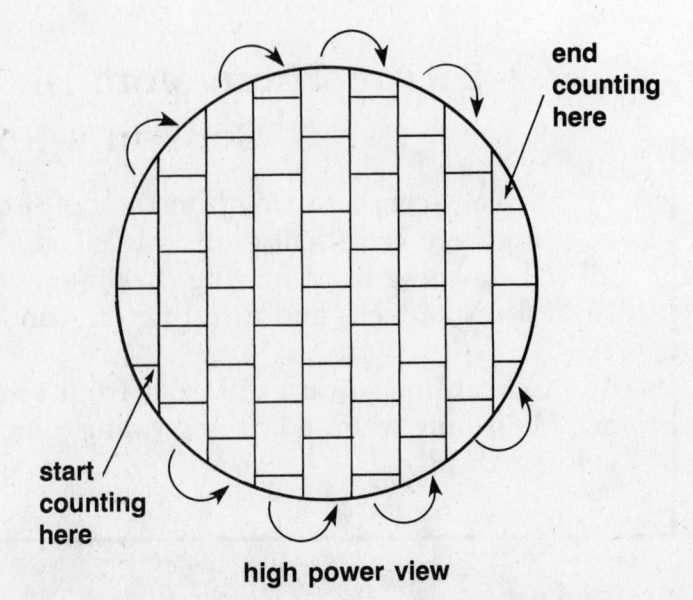

FIGURE 2. How to count cells

FIGURE 3. Student drawings of the steps of mitosis

186

Name _____ Class _____ Period _____

Table 1.

Steps of Mitosis	Number of Cells Seen
1	
2	
3	
4	
Resting Cells	
Total Cells Seen	

FIGURE 4. Steps of mitosis in plant cells

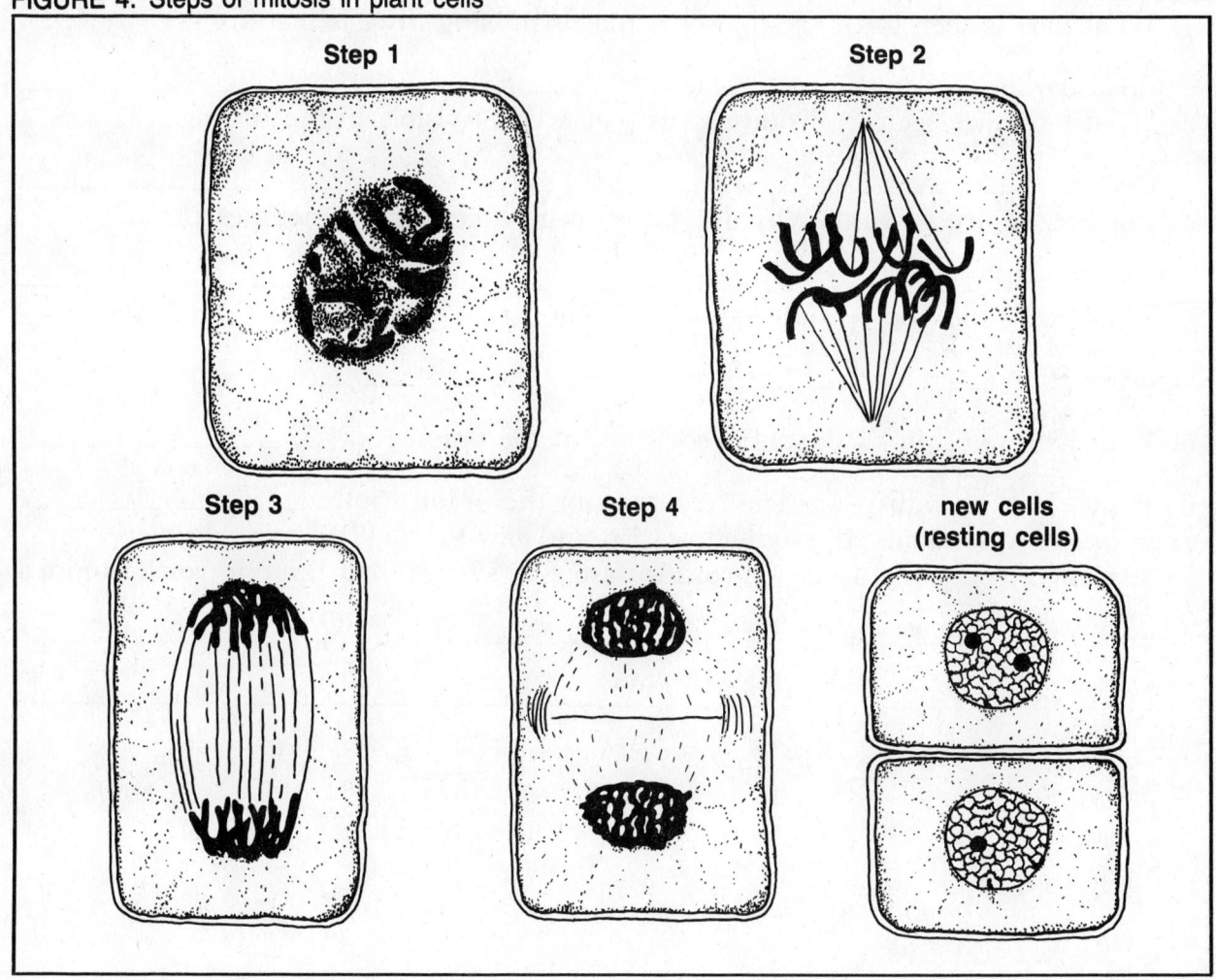

Step 1

Step 2

Step 3

Step 4

new cells (resting cells)

187

11. A biology student was looking at an onion root tip through the microscope and made a drawing of the cells she saw. Record on the chart how many cells you think she saw in each step of mitosis and the number of resting cells.

Table 2. Cells Seen by a Student

Steps of Mitosis	Number of Cells Seen
1	
2	
3	
4	
Resting Cells	
Total Cells Seen	

FIGURE 5. The student's drawing

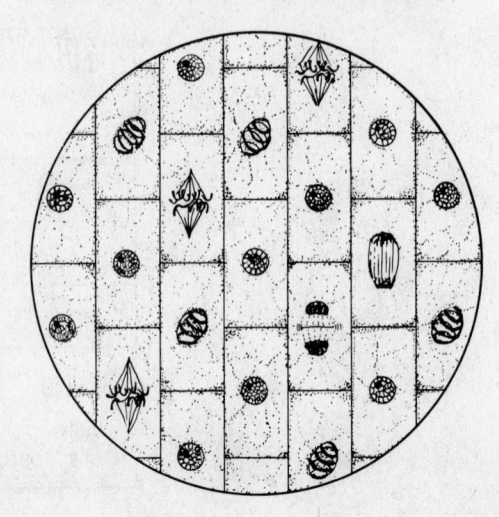

QUESTIONS
1. What part is seen in the resting cells that is missing in cells that are dividing? _____
2. What parts are seen in dividing cells that are not visible in the resting cells? _____
3. Look at Figure 4 again. Why do you think new cells are sometimes called resting cells? _____
4. Which cells did you see more of in the onion root, dividing cells or resting cells? _____
5. Which step of mitosis was most common in the onion root? _____
6. Which step of mitosis was least common in the onion root? _____
7. Suppose you examined another root tip and saw that half of the cells were dividing. Would this root be growing faster or slower than the one you examined in this exercise? Explain your answer. _____

Name _____ Class _____ Period _____

23-1 What Are the Parts of a Flower?

Many common plants you are familiar with form flowers at sometime during the year. The flower is the sexually reproductive part of a flowering plant. Certain flower parts are male, while other parts are female. Certain flower parts are neither male nor female. Which flower parts are male, female, or neither? How exactly do these parts help with sexual reproduction?

EXPLORATION

OBJECTIVES
In this exercise, you will:
  a. dissect and examine the parts of a flower.
  b. learn which parts are male, female, or neither male nor female.
  c. find out how each flower part helps in reproduction.

KEYWORDS
Define the following keywords:

ovule _____

pistil _____

pollen _____

stamen _____

MATERIALS
razor blade
microscope
1 coverslip
glass slide

colored pencils: red, blue, yellow, green, and purple
flower
water

PROCEDURE
Part A. Flower Parts That Are Neither Male nor Female
1. Examine the flower provided by your teacher.
2. Use Figure 1 to help locate the following two flower parts:
   a. sepals—small, green petallike parts forming the outside layer of the flower.

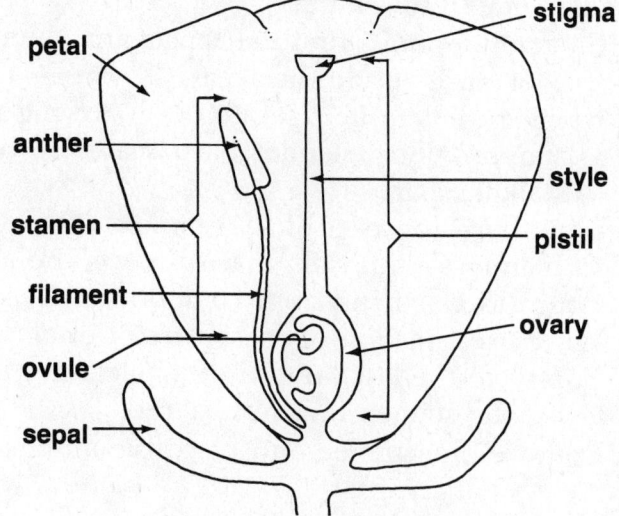

FIGURE 1. Parts of a flower

189

b. petals—large, brightly colored flower parts forming the layer inside the sepals.
3. Count the number of sepals and the number of petals present in your flower. Record these numbers in Table 1.
4. Write the functions of these two flower parts in Table 1.
5. Complete the last column of Table 1 for sepals and petals. (HINT: Read the title of this part of the activity.)
6. Color the sepals green and the petals yellow in Figure 1.

Table 1. The Functions of the Parts of a Flower

Flower Part	Number of Parts	Function	Male, Female, or Neither
Sepals			
Petals			
Stamens			
Anthers			
Filaments			
Pollen Grains	Approximately Thousands		
Pistil			
Stigma			
Style			
Ovary			
Ovules	Approximately 1 to 100		

Part B. Male Flower Parts
1. Carefully remove first the sepals and then the petals. Try not to remove any smaller parts inside the petals.
2. Use Figure 1 to help locate the following male flower parts:
   a. stamen—long, slender parts made up of a top, often yellow, part and a lower stalklike part
   b. anther—top part of the stamen, often yellow in color
   c. filament—thin stalk that supports the anther
3. Count the stamens from your flower. Record this number in Table 1.
4. Write the function of these three flower parts in Table 1.*
5. Complete the last column of Table 1 for the stamen, anther, and filament.
6. Color the stamen in Figure 1 red.
7. Remove one stamen and place the anther on a glass slide.

Name _____ Class _____ Period _____

8. Crush the anther with the eraser end of a pencil. Add two drops of water and a cover slip.
9. Examine the slide under low power, then high power, of your microscope.
10. The small round parts you see are pollen grains. Using high power, draw one or two pollen grains in the circular area provided here.
11. Complete Table 1 for pollen cells.

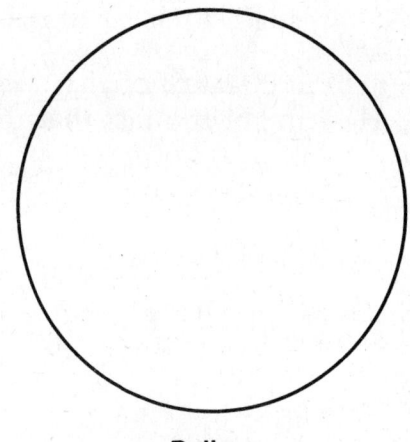

Pollen

Part C. Female Flower Parts
1. Use Figure 1 to help locate the following female flower parts:
   a. pistil—long, slender stalk with a round base in the center of the flower
   b. stigma—tip of the pistil
   c. style—slender stalk part of the pistil
   d. ovary—rounded, swollen bottom part of the pistil.
2. Count the number of each of the flower parts examined in step 1. Record these numbers in Table 1.
3. Write the function of each of these parts in Table 1.
4. Complete the last column of Table 1 for these four parts.
5. Color the pistil blue in Figure 1.
6. Remove the pistil from the flower. Use a razor blade to cut down through the length of the ovary as shown in Figure 2.
   CAUTION: *Always use extreme care with the razor blade.*
7. Note the small, round seedlike parts inside the ovary. These are the ovules.
8. Complete Table 1 for the ovules. Color the ovules in Figure 1 purple.

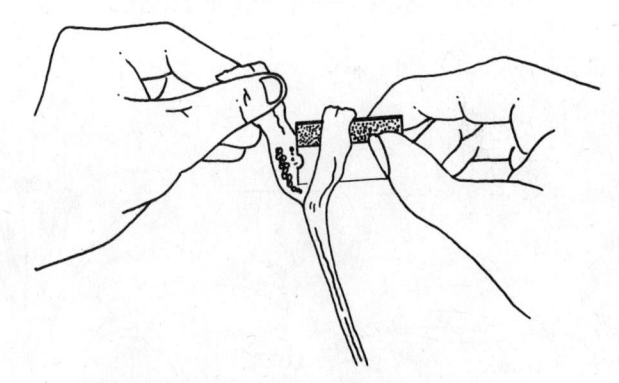

FIGURE 2. Cutting through the ovary

QUESTIONS
1. Group these flower parts—filament, ovary, ovule, petal, pollen grain, anther, stamen, pistil, stigma, sepal, style—under the following three headings:

Male Parts	Female Parts	Neither Male nor Female

191

2. a. What happens to pollen grains during pollination? _____

   b. How might their small size help them for this job? _____

   c. How might the fact that there are so many pollen cells help pollination occur? _____

3. a. What flower part contains egg cells? _____

   b. How do pollen grains on a stigma cause the fertilization of egg cells? _____

4. Look at Figure 3 and label the flower parts that are listed in question 1. Included in this diagram are close up views of the inside of an anther and an ovary. Locate and label a pollen grain and an egg cell.

5. Not all flowers have all the parts you have studied. Examine Figure 4 that shows sections of three different kinds of flowers. Label the parts that are present in each flower. Complete the chart that follows.

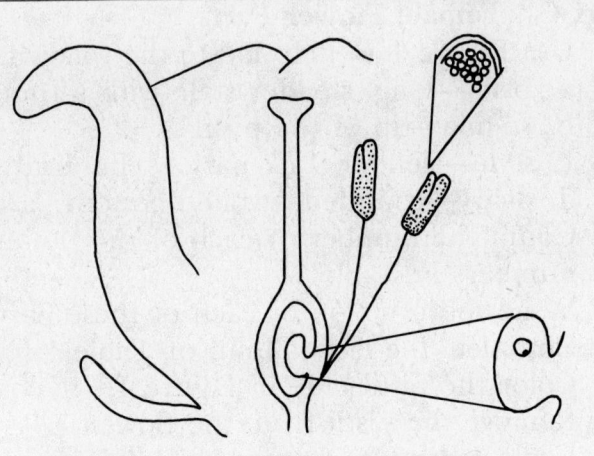

FIGURE 3. Reproductive parts of a flower

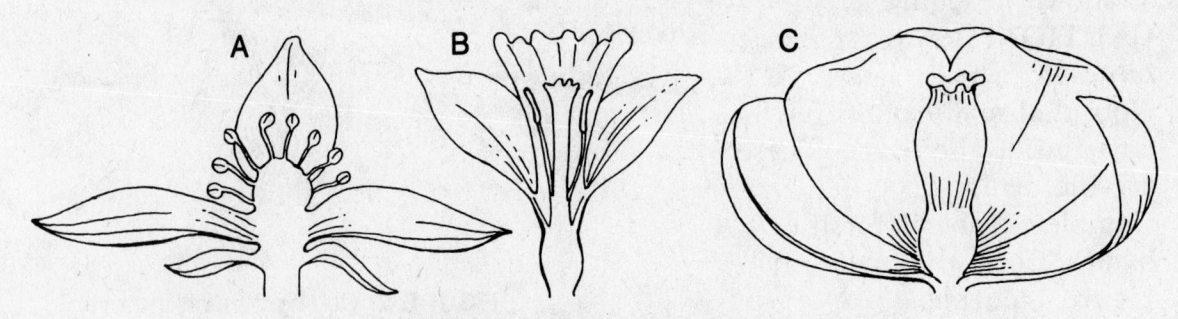

FIGURE 4. Parts of three different flowers

	Flower A	Flower B	Flower C
a. What are the missing flower parts?			
b. Can flower make pollen? Why or why not?			
c. Can flower make eggs? Why or why not?			
d. Can flower self-pollinate? Why or why not?			

Name _____ Class _____ Period _____

23-2 What Plant Part Are You Eating?

Plants store food to provide food for the start of the next year's growth (for example in roots and stems) or for the start of the next generation (for example in seeds or bulbs). Both asexual and sexual reproduction often involve the production of plant parts that store large quantities of food.

We use many of these plant parts for food. Have you ever asked yourself "What part of a plant am I really eating?" You may be eating a plant's root, stem, leaf, flower, fruit, or seed. Sometimes it is difficult to recognize what plant part a particular food may be.

EXPLORATION

OBJECTIVES
In this exercise, you will:
 a. examine identified parts of various food plants.
 b. note the special characteristics of these plant parts.
 c. examine some additional food plants and identify them as either root, stem, leaf, flower, fruit, or seed.

KEYWORDS
Define the following keywords:

asexual reproduction _____
_____

embryo _____

meiosis _____

mitosis _____

sexual reproduction _____

MATERIALS
hand lens or stereomicroscope
glass slide
scalpel
dissecting needle
broccoli
caper

brussel sprout
lettuce
cherry pepper
okra
cucumber

lima bean
pea
sweet potato
carrot
onion
shallot

PROCEDURE
Part A. Examining Known Food Plant Parts
Flower
1. Examine the yellow tip ends of broccoli. These are tiny flowers.
2. Remove two or three of these flowers and place them on a glass slide.
3. Examine the broccoli flowers under a hand lens or stereomicroscope. Use a dissecting needle to spread and separate the flower parts.
4. Using Figure 1, identify the parts of the flower that you can see.

Leaf

5. Examine the flat green part of a piece of lettuce. This is a leaf.
6. Use Figure 2 to identify the veins on the lettuce leaf.

Underground Stem

7. Examine the lower, white part of an onion. This is a bulb.
8. Using a scalpel or a knife, make a section down through the bulb as shown in Figure 3. Notice how the bases of the leaves are thicker than higher up. Also notice how each leaf surrounds and is attached to the small underground stem. **CAUTION:** *Take care when using a scalpel.*
9. Peel away some of the layers of leaf bases that surround the bulb and look for the small buds on the surface of the stem inside each leaf base. Using Figure 3, identify the parts of a bulb.

Root

10. Examine the orange underground part of a carrot. This is a root.
11. Notice how the leaf stalks are attached to the top of the root.
12. Look at the small hairlike roots that come from the sides of the carrot.
13. Notice that the root does not have buds or leaves on its surface as in the underground stem of the bulb.

Fruit

14. Examine the green part of okra. This is a fruit.
15. Using a scalpel, make a cross section through the okra fruit as shown in Figure 4.
16. Notice the small seeds attached to the center part of the fruit.

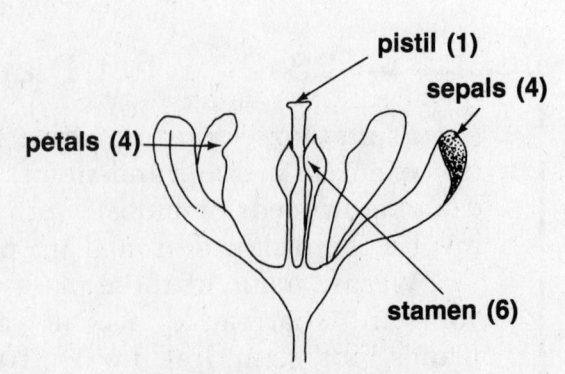

FIGURE 1. Broccoli flower

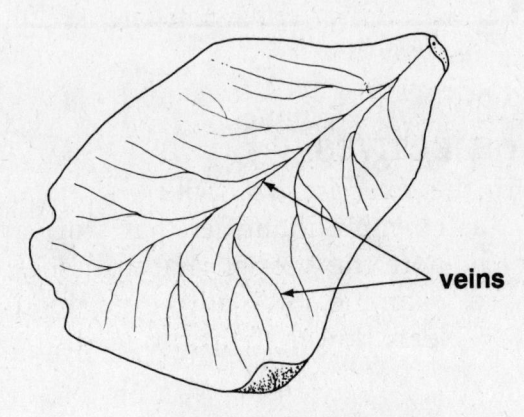

FIGURE 2. Lettuce leaf

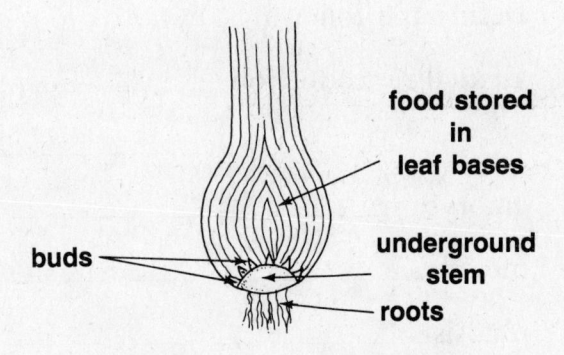

FIGURE 3. Onion underground stem

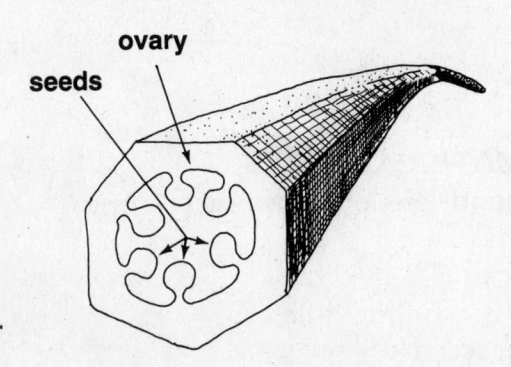

FIGURE 4. Okra fruit

Name _____ Class _____ Period _____

Seed
17. Examine a lima bean. This is a seed.
18. Remove the outer coat and with your fingernails split the bean into its two halves.
19. Using Figure 5, identify the parts of a seed.

Part B. Identifying the Parts of Some Common Food Plants
1. Obtain a brussel sprout, caper, cherry pepper, pea, sweet potato, shallot, and cucumber.
2. Examine each food item.
3. From your observations in Part A, identify the plant parts of each of these foods. Are the parts we eat leaves, fruits, flowers, stems, seeds, or roots?
4. Record your results in Table 1.

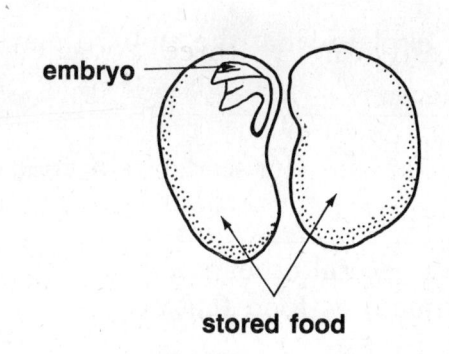

FIGURE 5. Lima bean seed

Table 1. Identifying the Parts of Plants That We Eat

Food	Plant Part We Eat	Evidence That Tells You Which Plant Part We Eat
Pea		
Brussel Sprout		
Shallot		
Caper		
Sweet Potato		
Cherry Pepper		
Cucumber		

QUESTIONS
1. For each of the foods below, describe how you were able to identify what part of the plant it is.

   a. shallot _____

   _____

   b. cherry pepper _____

   c. pea _____

2. List two ways that an underground stem differs from a root. _____

3. Using the information given to you in Figure 5:

   a. explain what the embryo may do with the stored food found in the seed.
   _____

   b. explain why seeds are a good source of nutrients for humans. _____

4. List several other examples (not studied in the experiment) commonly eaten by humans as food that are:

   a. roots _____

   b. fruits _____

   c. seeds _____

5. Plants reproduce sexually and asexually. Identify the type of reproduction involved with the following plant parts:

   a. root _____ b. flower _____ c. bulb _____

6. There are two types of cell reproduction that occur in plants, mitosis and meiosis. Name the type of cell division that occurs in each of the following plant parts:

   a. root _____ b. flower _____ c. bulb _____

7. There are two different types of cells in plants that are involved in reproduction: body cells and sex cells. Name the cell type involved in reproduction in each of the following plant parts:

   a. root _____ b. flower _____ c. bulb _____

8. A classmate tells you that tomatoes, pickles, and squash are fruits. Your classmate is correct. What feature should you look for to identify a plant part as a fruit?
   _____

9. What plant part will form fruit and seeds? _____

10. What is the main function of a flower's stamen and pistil? _____

11. Leaves and stems are plant parts that can also be involved in reproduction.
    a. Are these plant parts involved in sexual or asexual reproduction?
    _____

    b. How? _____

Name _____ Class _____ Period _____

24-1 How Do Some Animals Reproduce Asexually?

Animals, like plants, can reproduce both sexually and asexually. Two parents are needed for sexual reproduction. Only one parent is needed for asexual reproduction. A hydra can reproduce asexually by budding.

Planarians are flatworms that can reproduce asexually by regeneration. Parts that are cut or broken off are replaced by new growth. When an animal is cut in half and the missing parts of each half grow back, the animal has reproduced by regeneration.

Asexual reproduction also occurs in the water flea. This is a small, shrimplike animal that lives in ponds or streams. A water flea produces eggs inside a transparent pouch. This is the brood pouch. Many eggs form at a time within this pouch. These eggs can develop into young water fleas without fertilization by sperm. When the young leave the body of the water flea, new eggs are produced.

EXPLORATION

OBJECTIVES
In this exercise, you will:
   a. observe a planarian as it regenerates.
   b. count the number of young produced by a hydra, planarian, and water flea.
   c. compare asexual reproduction in a hydra, planarian, and a water flea.

KEYWORDS
Define the following keywords:

asexual reproduction _____

brood pouch _____

budding _____

_____

egg _____

regeneration _____

_____

MATERIALS
dropper   light microscope   planarian culture
glass slide   hand lens   2 small jars
paper towel   water flea culture   label
scalpel   *Euglena* culture
metric ruler   1 L aged tap water

PROCEDURE

Part A. Budding in Hydra
1. Look at Figure 1 of the budding hydra.
2. Count how many offspring buds of hydra grow from the parent in six days.

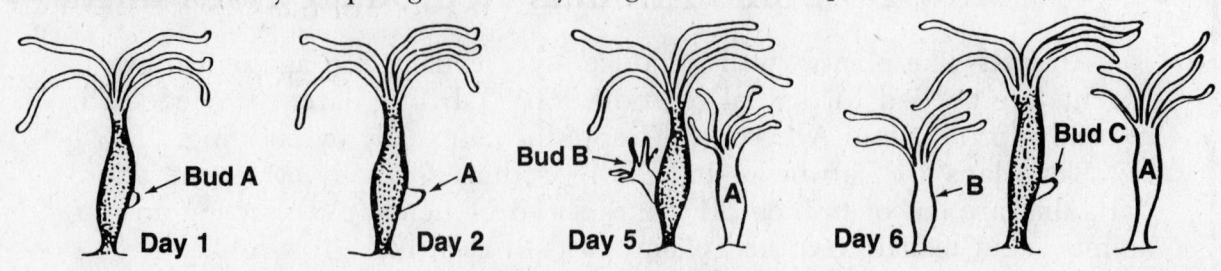

FIGURE 1. Budding hydra

Part B. Measuring Planarians
1. Look at Figure 2 of regenerating planarians.
2. Measure the length of each planarian.
3. Record your results in Table 1.

FIGURE 2. Regenerating planarians

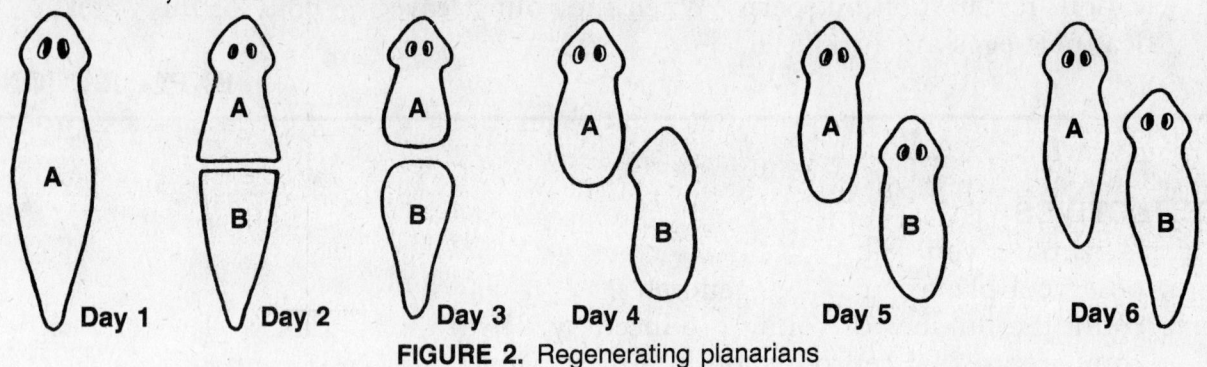

Table 1. Measuring Planarians

Part	Length in millimeters on					
	Day 1	Day 2	Day 3	Day 4	Day 5	Day 6
A						
B	✕					

Part C. How Does a Planarian Reproduce?
1. Remove a planarian from the culture dish and place it on a glass slide with a drop of water.
2. Examine the planarian with a hand lens. Find the two eyespots on the head part.
3. Measure the total length in millimeters of the resting planarian. Figure 3 shows how. Record this length and make a drawing in Table 2.
4. Remove some of the water with a paper towel. Less water will cause the planarian to stretch out.
5. When the planarian is fully extended, cut it in half with a scalpel. **CAUTION:** *Use extreme care with the scalpel.*

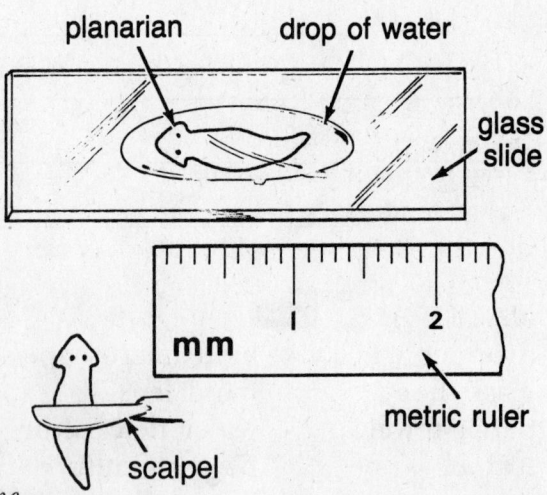

FIGURE 3. Measuring and cutting a planarian

Name _____ Class _____ Period _____

6. Look at each cut area of the head and the body halves of the flatworm. Record the lengths and make drawings of the two halves in Table 2.
7. Use a dropper to move the halves of the planarian to a small jar filled with aged tap water. Put a label with your name on the jar.
8. Place the jar on a shelf in a cool, dark area of the classroom.
9. Add oxygen to the jar each day by squeezing air from a dropper through the water. The bubbles will contain oxygen.
10. Look at the planarian halves each day for the next five class days.
11. Record the changes in size and shape on Table 2.
12. Compare your results with those of your classmates.

Table 2. Regeneration in Planarians

Before Cutting		Day 1	Day 2	Day 3	Day 4	Day 5	Day 6
	Head Part						
	Length in mm						
	Body Part						
	Length in mm						

Part D. How Many Young Can a Water Flea Produce?
1. Study Figure 4 of the water flea. Locate the brood pouch and eggs.
2. Use a dropper to remove a water flea from the culture dish and place it on a glass slide.
3. Remove some of the water with a paper towel as shown, so the water flea cannot move very fast.
4. Examine the slide on low power of the microscope. Look at the brood pouch and eggs.
5. Fill a small jar with aged tap water and place four water fleas into the water.
6. Fill a dropper with *Euglena* culture and add this to the jar.
7. Repeat step 6 three more times. Add a dropper of *Euglena* culture to the jar each day for the next four days.

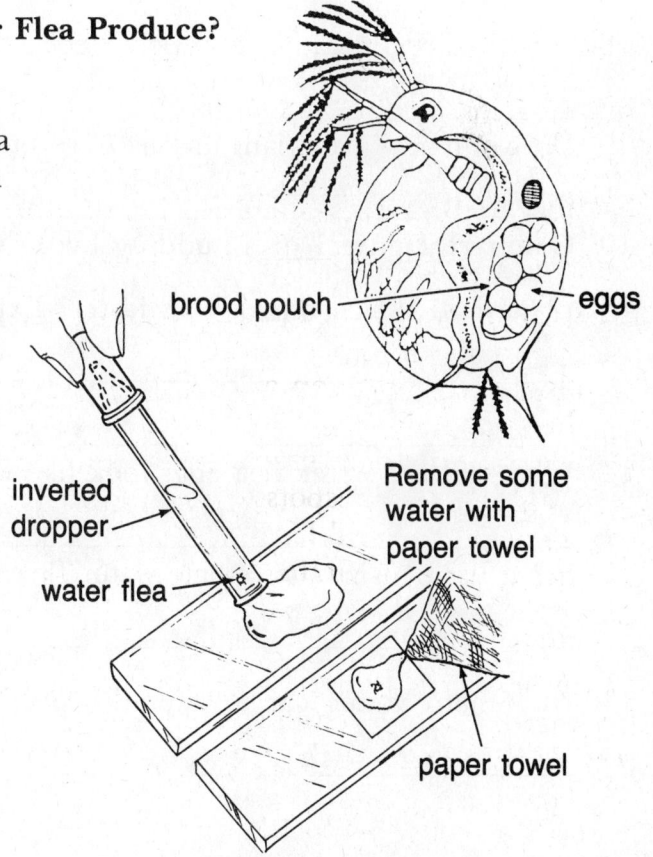

FIGURE 4. Water flea

199

8. Count the number of water fleas in the jar for five days. Record the data in Table 3.

Table 3. Water Flea Offspring

	Day 1	Day 2	Day 3	Day 4	Day 5
Number of water fleas					

QUESTIONS

1. How many new hydra were reproduced in Part A. _____

2. Is reproduction by budding sexual or asexual? Why? _____
   _____

3. How does a planarian reproduce? _____
4. How many new planarians were produced from the planarian that you cut in half? _____
5. What changes did you notice in the area where you cut the planarian?
   _____

6. What changes did you notice in the head part of the planarian over five days?
   _____

7. What changes did you notice in the body part of the planarian over five days?
   _____

8. Why can regeneration also be called asexual reproduction? _____
   _____

9. Did your new planarians in Part C regenerate as fast as those shown in the figure in Part B? _____

10. Compare Figure 1 of a budding hydra with Figure 2 of a regenerating planarian. Which reproduces faster? Explain. _____
    _____
    _____

11. What did the water flea eggs look like when they were in the brood pouch?
    _____

12. What happens to the female water flea after a batch of young leave the brood pouch? _____

13. How many young can a water flea produce? _____
    _____

Name _____ Class _____ Period _____

24-2 How Do Internal and External Reproduction Compare?

Some animals that live in water gather in large groups during their breeding season. The females release their eggs into the water. The males of the species then release their sperm over the eggs. The sperm cells fertilize the eggs in the water. This is called external fertilization because it takes place outside the animal's body.

In other animals, the male releases sperm directly into the female's body where fertilization takes place. This is internal fertilization.

INTERPRETATION

OBJECTIVES
In this exercise, you will:
a. interpret and chart data collected by a student.
b. determine if there is a relationship between the number of young produced, the type of fertilization, and the type of place where an animal lives.
c. determine how parental care helps the young.

KEYWORDS
Define the following keywords:

external fertilization _____

internal fertilization _____

parental care _____

MATERIALS
pencil

PROCEDURE
Part A. Observing Data On Animal Reproduction
Study the data in Table 1 below that was collected by a student.

Table 1.

Animal	Where Animal Lives	Number Of Young Produced	Number Of Young That Survive	Parental Care Yes	Parental Care No
Horse	land	1	1	X	
Spider	land	100	10	X	
Oyster	water	750,000	100		X
Jellyfish	water	1,000	10		X
Tiger Salamander	land	100	8		X
Starfish	water	1,500	5		X
Garter Snake	land	35	3	X	
Falcon	land	5	2	X	
Humpback Whale	water	1	1	X	
Clam	water	1,000,000	30		X

Part B. Using the Table

Use the data from Table 1 in Part A to complete Table 2.
1. Look at Table 1 to find where each animal lives.
2. Look at the number of young produced by each animal in Table 1 and decide if fertilization is internal or external. Hint: If the young produced are greater than 150, then fertilization is usually external.

Table 2.

Lives in Water		Lives on land	
Fertilization		Fertilization	
Internal	External	Internal	External

Part C. Applying The Use of Data

1. Look at Table 3. It gives data on reproduction for another list of animals.
2. Using what you have learned from Parts A and B, complete the information in Table 3 under the columns C and D.

Table 3.

A Animal	B Number Of Young Produced	C Fertilization		D Where Animal Lives	
		Internal	External	Land	Water
Bluegill	15,000				
Chameleon	8				
Leech	60				
Lobster	8,500				
Rainbow Trout	4,000				
Crayfish	40				
Lizard	10				
Cow	1				
Blue Crab	950,000				
Toad	10,000				

Name _____ Class _____ Period _____

Part D. Using Data to Draw Conclusions
Using what you have learned from Parts A, B, and C, fill in Table 4 to show if an animal gives parental care or not.

Table 4.

Animal	Number Of Young Produced	Number Of Young That Survived	Parental Care Yes	Parental Care No
Opossum	10	8		
Grasshopper	300	14		
Carp	12,000	22		
Deer	1	1		
Pig	8	6		

QUESTIONS
1. What kind of fertilization do animals that live in water usually show?_____
2. Where do most of the animals live that show internal fertilization?_____
3. What is an animal that has internal fertilization and lives in water?_____
   _____
4. Do animals that fertilize their eggs internally produce more or less young than animals that fertilize their eggs externally?_____
5. Which eggs do you think have a better change of being fertilized, those that are fertilized internally or those fertilized externally?_____
6. Freshwater clams produce 20 young a breeding season; some freshwater jellyfish produce 500 young. Which animal do you think has internal fertilization? Why?
   _____
7. Do most of the animals that provide parental care have many or few young?
   _____
8. Where do most animals live that give no parental care for their young?_____
9. From Tables 1, 3, and 4 how can you tell which young animals have a better chance of surviving?_____
   _____
10. In Table 3 of Part C, which of the animals do not fit the pattern between numbers of offspring and kind of fertilization as seen in Part A? Explain.
    _____
    _____

203

11. Which of the animals do you think have the best chance of survival? Why?
    _____
    _____

12. The bluegill sunfish cares for its eggs and young while the rainbow trout does not. Which would probably have a greater number of young survive to reproduce? Explain._____

13. Write a short paragraph that explains each of the following points:
    a. The type of fertilization of an animal is different between animals that live on land and those that live in water._____
    _____

    b. The number of eggs produced by an animal is different between animals that have internal fertilization and those that have external fertilization.
    _____
    _____

    c. The number of young that survive differs between animals that have internal fertilization and those that have external fertilization._____
    _____
    _____

    d. The number of young that survive differs between animals that give parental care and those that do not._____
    _____
    _____

Name _____ Class _____ Period _____

24–3 What are the Stages in the Menstrual Cycle?

The human female reproductive system has several different roles to perform. It produces a new egg each month for fertilization. It must prepare the lining of the uterus for a new embryo if fertilization of an egg does occur. This is accomplished by a thickening of the uterus lining.

It must shed the egg and thickened uterus lining if fertilization does not occur. All of these events occur in a cyclic pattern each month in a sexually mature female.

INTERPRETATION

OBJECTIVES
In this exercise, you will:
  a. review the organs that form the human female reproductive system.
  b. prepare a calendar that shows the changes occurring during the human menstrual cycle if no fertilization occurs.
  c. prepare a calendar that shows the changes occurring during the human menstrual cycle if fertilization occurs.

KEYWORDS
Define the following keywords:

fertilization _____

menstrual cycle _____

ovary _____

oviduct _____

uterus _____

MATERIALS
scissors
tape

PROCEDURE
Part A. Review of the Female Reproductive System
  1. Use the following parts and their description for help in properly labeling Figure 1.
     NOTE: The diagram is 1/3 natural size.
     a. ovary—two are present, round in shape
     b. egg—small cells present within ovary
     c. uterus—large muscle, V-shaped, largest part of reproductive system
     d. oviduct—thin tube connecting each ovary to uterus
     e. uterus lining—inner wall or lining of uterus

205

FIGURE 1.

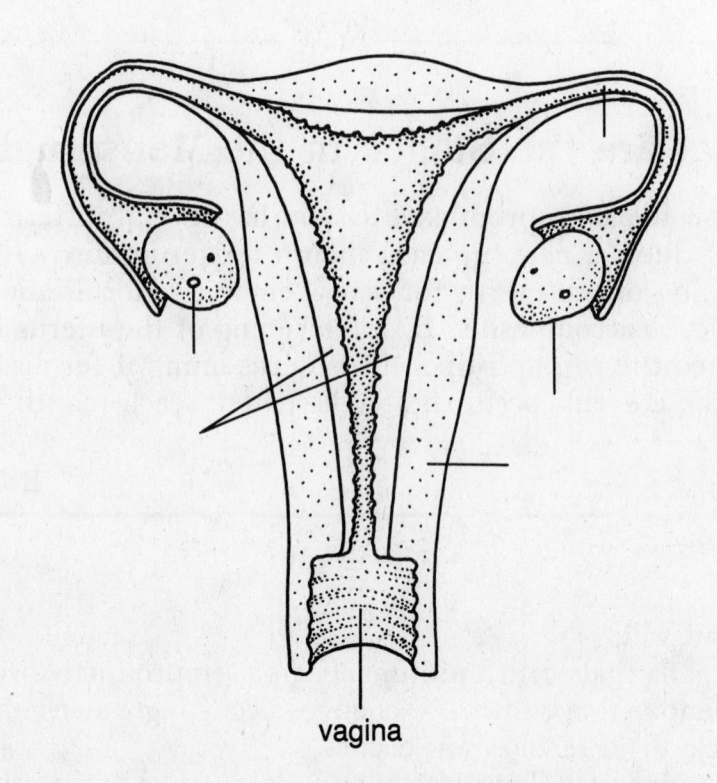

vagina

Part B. Changes in the Menstrual Cycle; No Fertilization of Egg
1. Obtain a copy of Figure 2 from your teacher.
2. Use scissors to cut out the square diagrams in Figure 2. These diagrams show the different stages that occur during the menstrual cycle if fertilization does not occur.
3. Look over the calendar marked Figure 3. It describes a series of events that take place in the female reproductive system if fertilization does not take place.
4. Match the diagrams that you cut out with the events being described in the calendar.
5. When all diagrams have been properly matched, tape them onto the calendar in their proper location to the right of the brackets describing the events.

Part C. Changes in the Menstrual Cycle; Fertilized Does Occur
1. Obtain a copy of Figure 4 from your teacher.
2. Use scissors to cut out the square diagrams in Figure 4. These diagrams show the different stages that occur during the menstrual cycle if fertilization does occur.
3. Look over the calendar marked Figure 5. It describes a series of events that take place in the female reproductive system if fertilization does take place.
4. Match the diagrams that you cut out with the events being described in the calendar.
5. When all diagrams have been properly matched, tape them onto the calendar in their proper location to the right of the brackets describing the events.

Name _____ Class _____ Period _____

FIGURE 2. No fertilization of egg

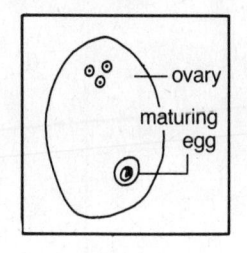

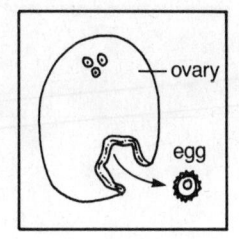

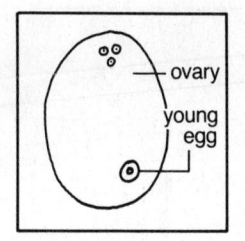

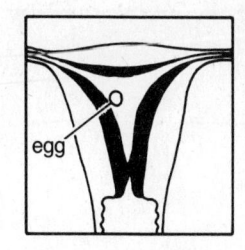

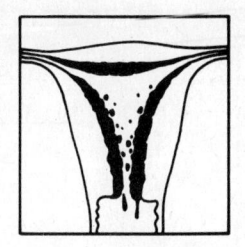

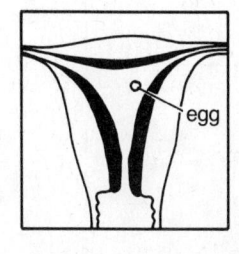

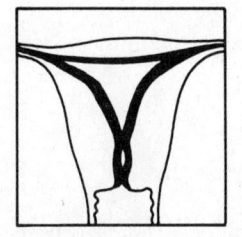

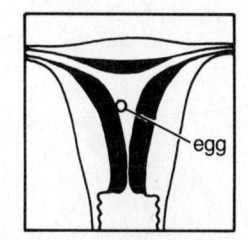

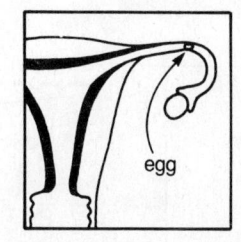

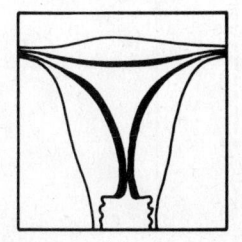

FIGURE 4. Fertilization of egg

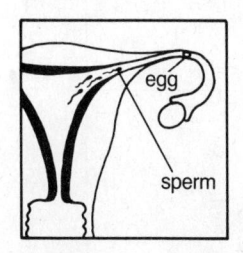

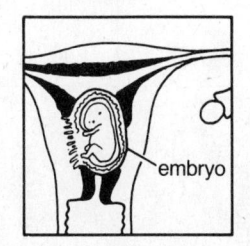

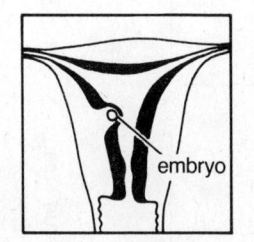

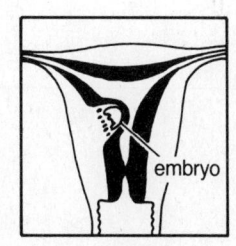

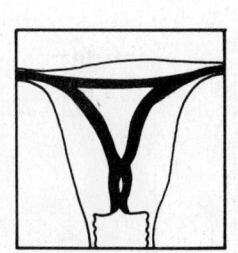

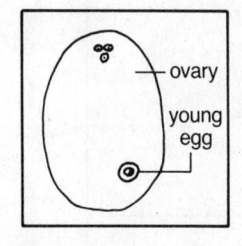

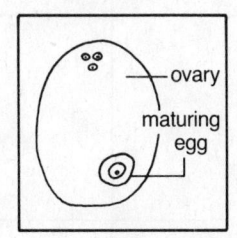

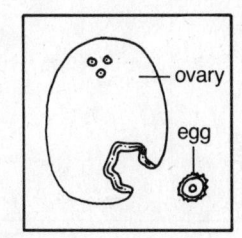

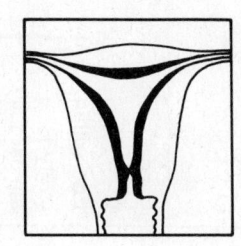

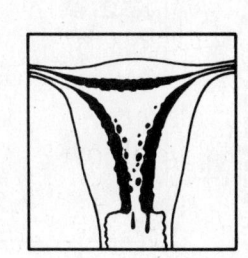

207

Name _____ Class _____ Period _____

Cycle repeats itself.

1 Uterus lining and egg are shed during menstruation.	2	3 New egg is maturing in ovary.	4	5 Uterus lining is thin after blood and tissue have been lost.	6	7
8	9 Egg within ovary is almost fully mature.	10	11 Uterus lining is thickening.	12	13 Mature egg is released from ovary	14
15 Egg is in oviduct. No sperm cells present. Egg is not fertilized.	16	17 Uterus lining continues to thicken.	18	19 Uterus lining is very thick—egg moves lower in uterus.	20	21
22	23	24 Uterus lining is at its thickest.	25	26 Lining of uterus and egg are ready to be shed. They are no longer needed.	27	28 Go back to day 1.

FIGURE 3. Day by day changes in the menstrual cycle—no fertilization of egg

Name _____ Class _____ Period _____

To Figure 3

1 Uterus lining and egg are shed during menstruation.	2	3 New egg is maturing in ovary.	4	5 Uterus lining is thin after blood and tissue have been lost.	6	7
8	9 Egg within ovary is almost fully mature.	10	11 Uterus lining is thickening.	12	13 Mature egg is released from ovary into oviduct.	14
15 Sperm cells fertilize egg.	16	17	18	19	20 6-day-old embryo buries itself in uterus.	21
30 Uterus continues to thicken as embryo grows.	35	90 Embryo is very large —almost a fetus.	91	92	265	266 Last day of pregnancy— Birth occurs. Go back to day 3 of either calendar.

FIGURE 5. Day by day changes in the menstrual cycle—fertilization of egg

209

QUESTIONS

1. Describe the role or function of the following:

    a. ovary _____

    b. uterus lining _____

    _____

    c. uterus muscle _____

    _____

    d. oviduct _____

2. An average menstrual cycle with no fertilization takes how many days?

    _____

3. Describe the changes that take place during the menstrual cycle from day 1–4 to the following:

    a. unfertilized egg _____

    b. uterus lining _____

    c. egg in ovary _____

4. Describe the changes that take place during the menstrual cycle from day 5–13 to the following:

    a. uterus lining _____

    b. egg in ovary _____

5. Describe what happens to the egg during the menstrual cycle on day 14.

    _____

6. Describe the changes that take place to an egg

    a. from day 15–28 if no fertilization occurs. _____

    b. from day 15–21 if fertilization does occur. _____

    _____

    c. from day 21–266 if fertilization does occur. _____

    _____

7. Explain why the female

    a. needs a thick uterus lining if fertilization does occur. _____

    _____

    b. no longer needs a thick lining if fertilization does not occur. _____

    _____

Name _____ Class _____ Period _____

25-1 How Does a Human Fetus Change During Development?

Development in a human takes about 38 weeks. Many changes take place with the fetus during that time. Two changes that do occur are increases in size and mass. How much of a change in mass and size takes place each week?

INTERPRETATION

OBJECTIVES
In this activity, you will:
  a. measure the length of diagrams of the human fetus.
  b. graph the length and mass of a human fetus.
  c. determine when during development most changes in mass and size occur.

KEYWORDS
Define the following keywords:

development _____

embryo _____

fetus _____

mass _____

premature _____

MATERIALS
metric ruler

PROCEDURE
Part A. Development of a Human Fetus
1. Look at Figure 1. It shows six stages of a developing human fetus. They are shown at 40% of their natural size.
2. Follow the steps outlined below to measure the total length of each stage. Use the metric ruler and measure in millimeters. Use the 38-week stage as a guide and record your data in the spaces provided in Table 1.
   a. Measure the body length from the rump to the top of the head.
   b. Measure the thigh length from the rump to the knee.
   c. Measure the length of the leg from the knee to the foot.
4. Add all three measurements together and record the total in the space provided in Table 1.
5. Multiply the total by 2.5 to give a figure that is close to the actual size of the fetus at each stage.
6. Record this actual size in the table.

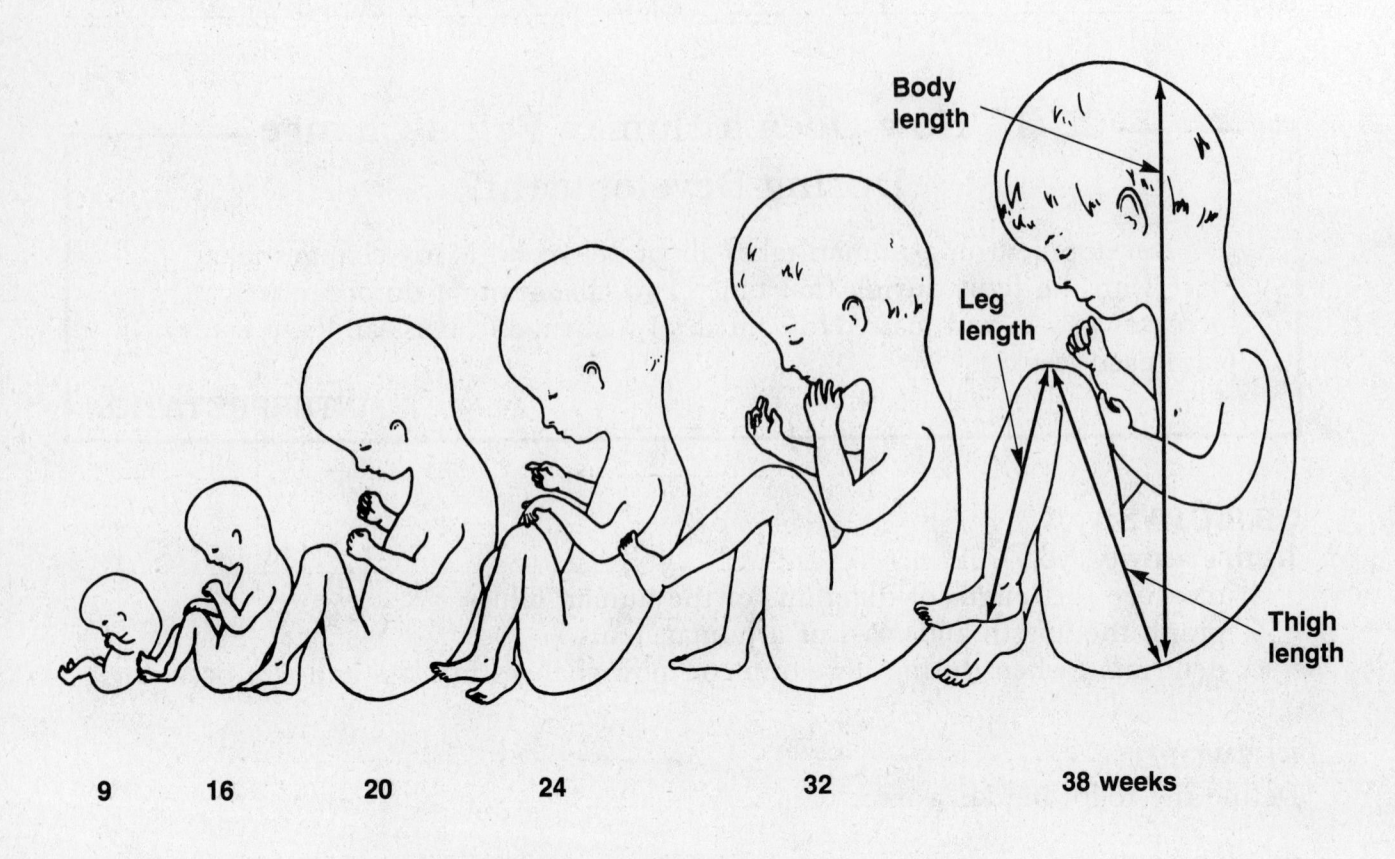

FIGURE 1. Stages in the development of a human fetus

9   16   20   24   32   38 weeks

Body length
Leg length
Thigh length

Table 1. Lengths of a Developing Fetus

Age of fetus in weeks	Body length	+	Thigh length	+	Leg length	=	Total length	× 2.5 =	Actual length
2									
9									
16									
20									
24									
32									
38									

Part B. Plotting Length of a Developing Fetus
1. Plot the data from Table 1 onto the graph in Figure 2.
2. Plot the actual fetal length against the age of the fetus.

Name ______________________ Class ____________ Period ____________

FIGURE 2. Length of a developing fetus

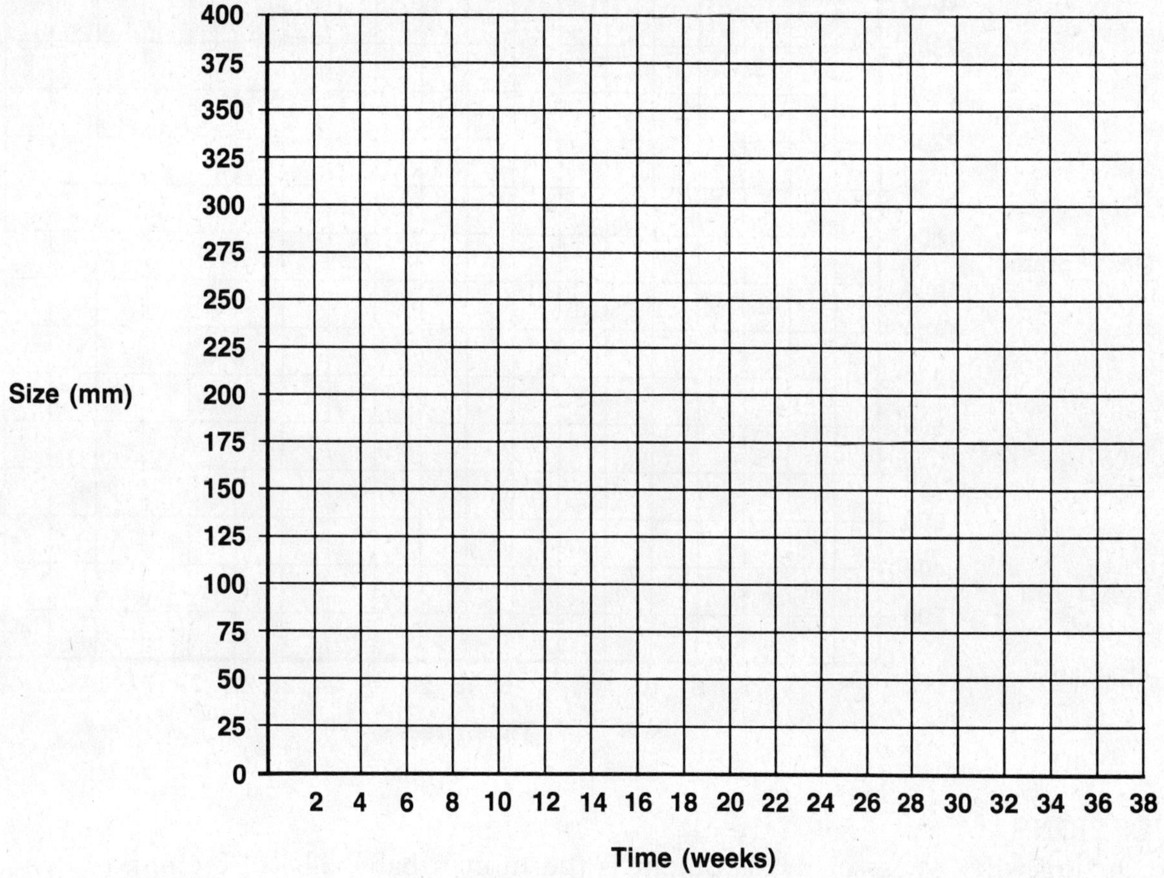

Part C. Plotting Mass of a Developing Fetus

1. Look at the data supplied in Table 2.
2. Plot the data of the developing fetus from Table 2 onto the graph in Figure 3.
3. Plot the mass of the fetus against the age of the fetus.

Table 2. Mass of a Developing Fetus

Time (weeks)	Mass (grams)	Time (weeks)	Mass (grams)
4	0.5	24	650
8	1	28	1100
12	15	32	1700
16	100	36	2400
20	300	38	3300

FIGURE 3. Mass of a developing fetus

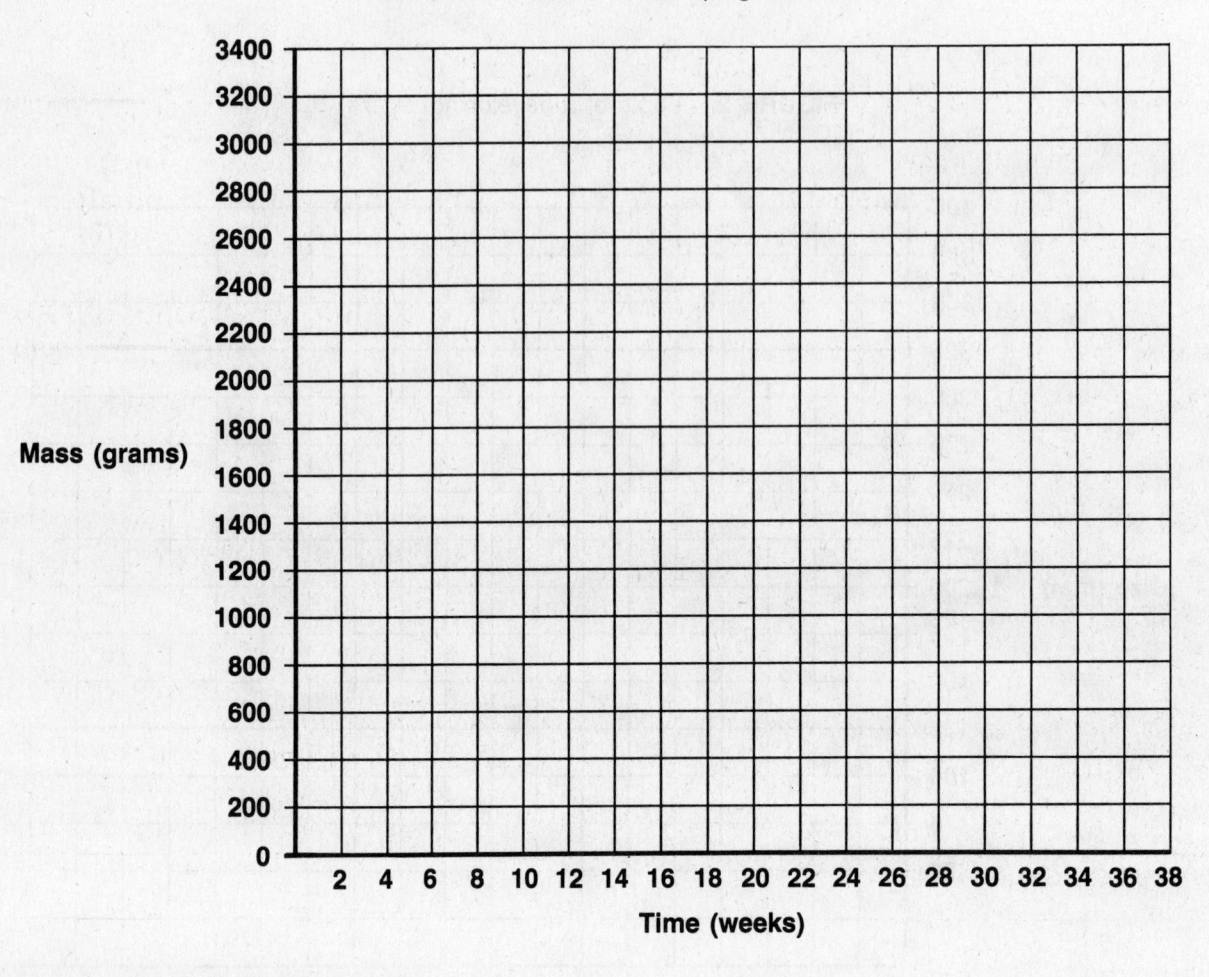

QUESTIONS

1. During what weeks of development is the human baby called an embryo?

   _____

2. What is the length of an embryo during this time? _____

3. How much mass does an embryo gain during this time? _____

4. During what weeks of development is the human baby called a fetus? _____

   _____

5. Look at Figures 2 and 3 for the halfway point in development at week 19.

   a. Is the fetus half of its full length at this time? _____

   b. Is the fetus half of its full mass at this time? _____

6. a. At what week does the fetus reach half its full length? _____

   b. At what week does the fetus reach half its full mass? _____

7. If a premature baby is born with a mass of

   a. 2200 grams, how old is the fetus? _____

   b. 1800 grams, how old is the fetus? _____

214

Name _____ Class _____ Period _____

25-2 What Changes Occur During Birth?

A human baby develops for about 38 weeks inside the mother. Then labor begins and the baby is born. What changes take place during and after birth? Why must a doctor sometimes have to perform a caesarean operation to help in delivery?

INTERPRETATION

OBJECTIVES
In this activity, you will:
   a. compare the changes that occur during birth.
   b. learn why a caesarean delivery may be needed.
   c. compare a delivery through the birth canal with a caesarean delivery.

KEYWORDS
Define the following keywords:

caesarean _____

contractions _____

fetus _____

labor _____

placenta _____

uterus _____

MATERIALS
metric ruler

PROCEDURE
Part A. Stages of Birth
1. Look at the diagrams of four stages of birth shown in Figures 1 and 2.

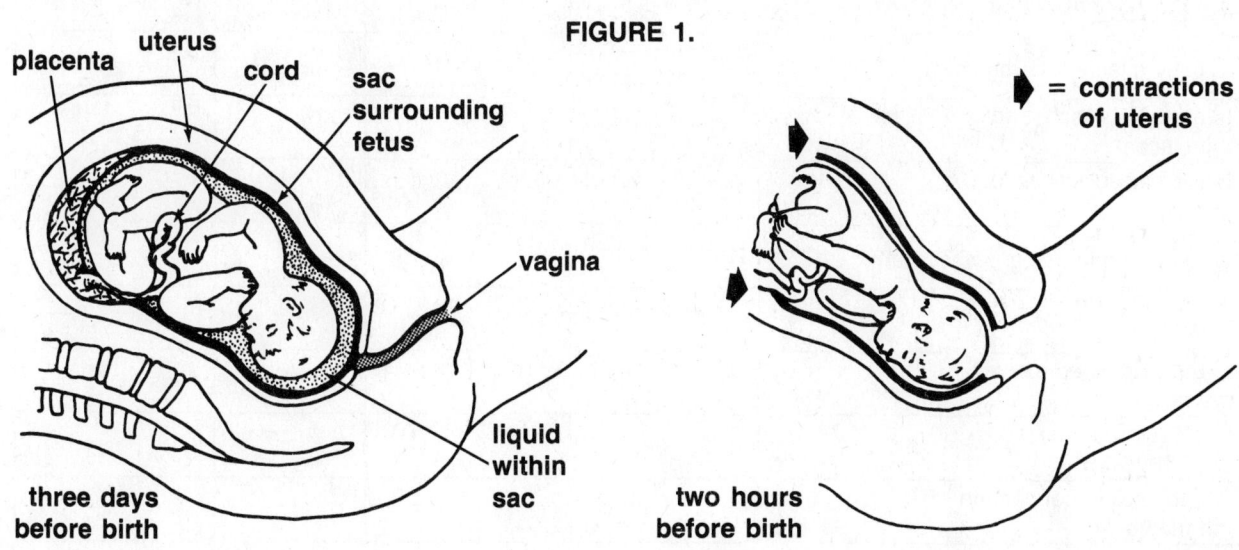

FIGURE 1.

215

FIGURE 2.

during birth

placenta   sac

cord

few minutes after birth

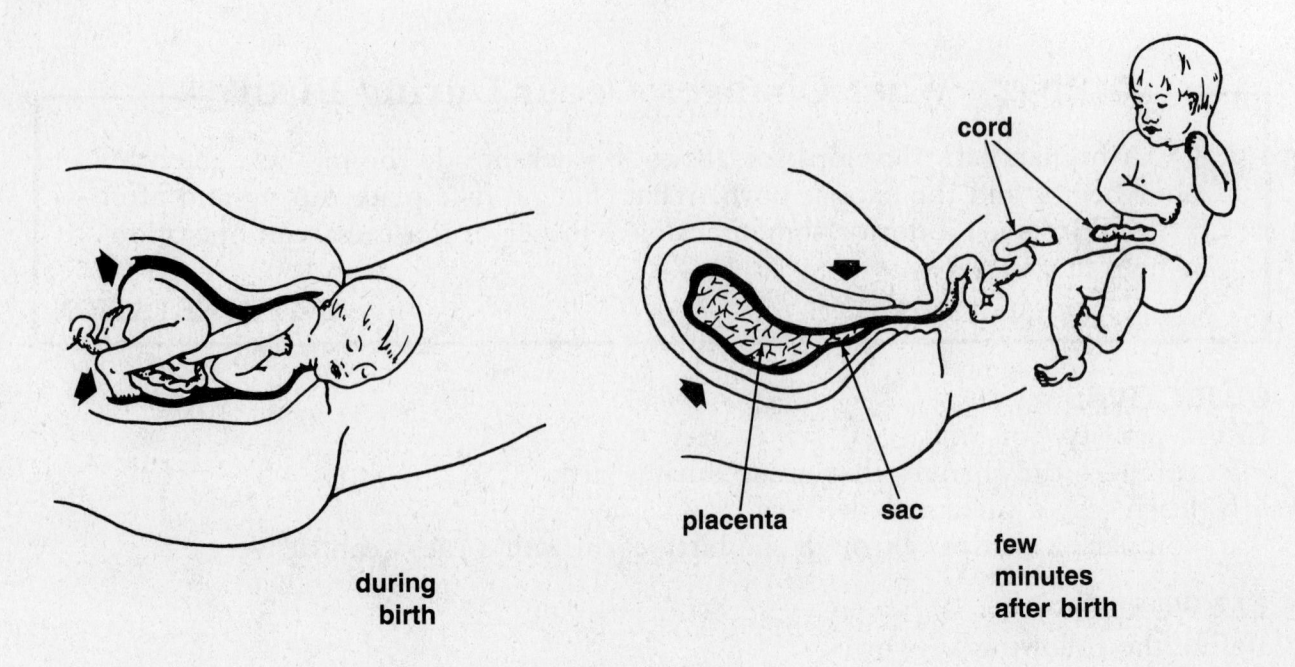

2. Answer *yes* or *no* to each of the following questions in Table 1.

Table 1. Stages During Birth

	Three days before birth	Two hours before birth	During birth	Few minutes after birth
Is baby inside the uterus?				
Is baby inside the vagina?				
Is baby outside the mother's body?				
Is baby inside the sac?				
Has the sac broken?				
Are contractions occurring?				
Is baby attached to the cord?				
Is the cord attached to the placenta?				
Is the placenta attached to the uterus?				
Is the placenta being pushed out?				
Has the vagina opened?				
Is baby attached to the mother?				
Has liquid been lost from the sac?				
Is baby still dependent on the mother?				

Name _____ Class _____ Period _____

Part B. What Is a Caesarean Birth?
1. Look at the diagram in Figure 3 that shows the outline of the pelvis and the head of a fetus just before the time of birth.
2. Note carefully that the head must be able to pass through the opening in the pelvis during birth.
3. Measure line a. This represents the width of the opening in the pelvis.
4. Measure line b. This represents the width of the head of the fetus.

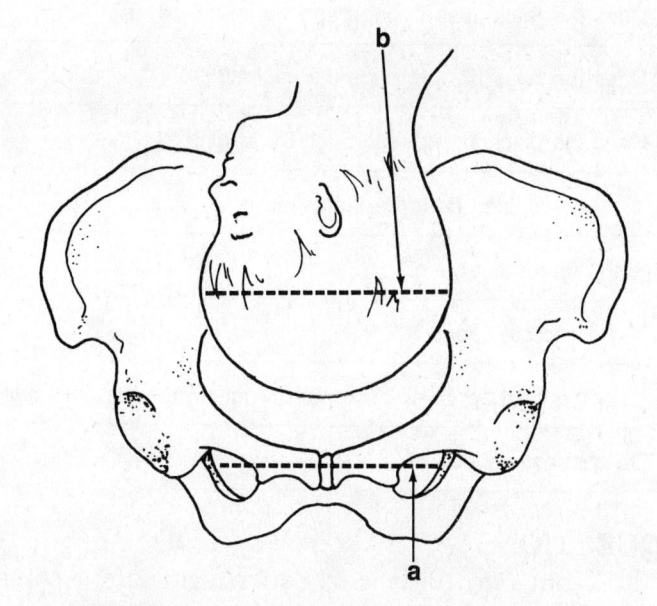

FIGURE 3. Sizes of pelvis and head of fetus

5. Record your data here:

   a. width of pelvis opening _____

   b. width of fetus head _____

6. Notice that this fetus would not be able to pass through this pelvis opening.
7. A caesarean operation must be done to deliver the baby.
8. Look at how a caesarean birth is done in Figure 4. This is usually done before the mother goes into labor.
9. To compare a birth canal delivery with a caesarean delivery, answer the questions in Table 2.

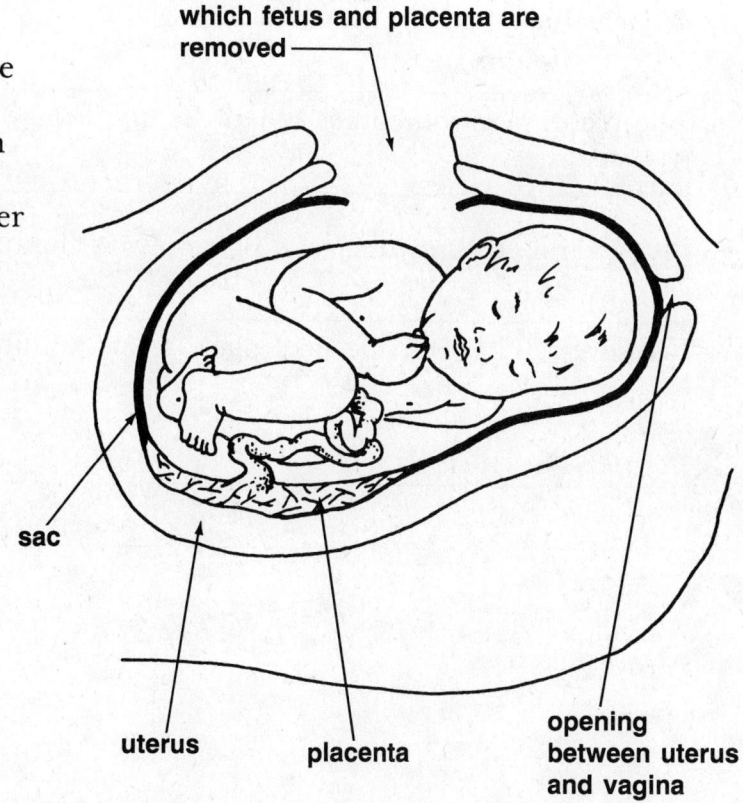

Opening through which fetus and placenta are removed

sac

uterus   placenta   opening between uterus and vagina

FIGURE 4. Caesarean birth

217

Table 2. Comparing a Caesarean Delivery With a Birth Canal Delivery

Trait	Birth canal	Caesarean
Does the fetus pass through opening in the pelvis?		
Does the fetus pass through the vagina?		
Does the placenta move through the vagina?		
Is the fetus lifted from the uterus?		
Is the uterus cut open?		
Is the sac cut open?		
Must the cord be cut to separate the fetus from the placenta?		
Do contractions occur?		

QUESTIONS

1. What two body parts surround and protect the fetus as it develops? _____

2. What is the job of the placenta? _____

3. What is the job of the cord? _____

4. What is meant by the word *labor*? _____

5. The placenta is sometimes called the *afterbirth*. Why is this a good name for this part? _____

6. List several changes that take place several hours before birth. _____

7. List several changes that take place a few minutes after birth. _____

218

Name _____ Class _____ Period _____

26-1 How Can the Genes of Offspring Be Predicted?

The Punnett square can be used to predict expected results from a genetic cross. In mice, black coat color is dominant over white coat color. If two **heterozygous** parents are crossed, we would expect three black to every one white offspring. The Punnett square shows the expected results of the ***Bb* × *Bb*** cross.

You know observed results do not always agree with expected results. Four offspring of the ***Bb* × *Bb*** parents may really all be white. That is, the observed results may all be white. These results are not what would be expected. Of what good are expected results? The expected results help us determine what the observed results will be. When do the expected and observed results agree? When are they the same?

INTERPRETATION

OBJECTIVES
In this exercise, you will:
a. set up a model to compare expected and observed results.
b. find when expected and observed results agree.
c. predict some genes in offspring of mice.

KEYWORDS
Define the following keywords:

expected results _____

observed results _____

Punnett square _____

pure dominant _____

MATERIALS
2 pennies   masking tape

	B	b
B	BB black	Bb black
b	Bb black	bb white

PROCEDURE
Part A. Calculating Expected Results
Assume that a female mouse has several litters of young in one year. She is heterozygous ***(Bb)*** for coat color and mates with a male that is also heterozygous ***(Bb)*** for coat color.

FIGURE 1. Punnett square

219

You can predict what kind of offspring she will have by constructing a Punnett square as shown in Figure 1. Results from mating two mice can be shown by tossing and reading coins.

1. Place two coins in your cupped hands and shake the coins. Drop the coins on a desktop.
2. Examine the coins and determine whether you have two heads, two tails or a head and a tail.
3. Make a mark (/) in Table 1 under the correct combination of genes.
4. Repeat shaking and reading the coins for a total of 40 times. These 40 shakes will represent the combination of genes you might have observed in the offspring of several litters.
5. Count the marks for each gene combination and write the total observed in Table 1.
6. Calculate the expected number for each gene combination by using the Punnett square in Figure 1.
   a. First divide the number of **BB** squares in Figure 1 by 4; multiply that number (a percentage) by 40. Record this number as the expected number for the gene combination of **BB** genes in Table 1.
   b. Repeat these calculations for **Bb** and **bb** squares. Record your values for each gene combination in Table 1.
   c. Record the colors in the last row of Table 1.

Table 1. Results of Coin Tosses (Coat Color in Mice Bb × Bb)

Coin combinations	Head–Head	Head–Tail	Tail–Tail
Gene combinations	BB	Bb	bb
Observed results			
Total observed in 40 tosses			
Total expected in 40 tosses			
Coat color in mice			

Part B. Predicting Mouse Offspring

Suppose you mate a female mouse that is heterozygous (***Bb***) with a male that is pure recessive (***bb***). Predict what kind of offspring she will have by completing a Punnett square like that show in Figure 2.

1. Place tape on both sides of the two coins and mark both sides of one coin with a *b*. Mark one side of the other coin with a *b* and the other side with a *B*.

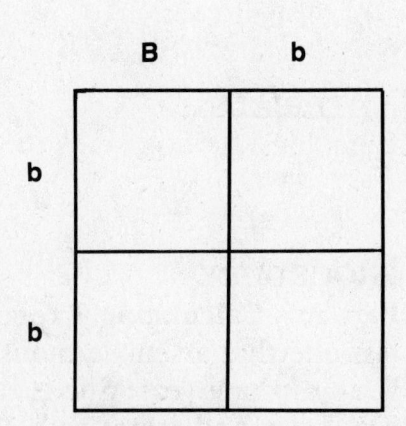

FIGURE 2. Punnett square

Name _____ Class _____ Period _____

2. Place the two coins in your cupped hands and shake the coins. Drop the coins on a desktop.
3. Examine the coins and determine whether the offspring are heterozygous **(Bb)** or pure recessive **(bb)**.
4. Make a mark (/) in Table 2 under the correct combination of genes.
5. Repeat shaking and reading the coins a total of 40 times. The 40 shakes will show the combination of genes you observed in the offspring of several litters.
6. Count the marks for each gene combination and write the total observed in Table 2.
7. Use the method of calculating expected numbers for each gene combination as in Part A step 6 by using the Punnett square in Figure 2.
8. Determine the coat color of the offspring by using the Punnett square in Figure 2. Record the color in the proper part of Table 2.

Table 2. Results of Coin Tosses (Coat Color in Mice Bb × bb)

Coin combinations	B–b	b–b
Gene combinations	**Bb**	**bb**
Observed results		
Total observed in 40 tosses		
Total expected in 40 tosses		
Coat color in mice		

QUESTIONS

1. How often do you expect a tossed coin to land on heads? _____

2. How often do you expect a tossed coin to land on tails? _____

3. When two coins are tossed, how often do you expect to get the following combinations:

    a. heads/heads? _____

    b. heads/tails? _____

    c. tails/tails? _____

4. What gene combinations and features (coat color) do the following coin tosses produce:

                                  gene combinations   features

    a. heads/heads?   _____   _____

    b. heads/tails?   _____   _____

    c. tails/tails?   _____   _____

5. What gene combinations and features (coat color) do the following marked coin tosses produce:

            gene combinations   features

   a. B/b?   _____   _____

   b. b/b?   _____   _____

6. How close were your observed results to what you expected to get?
   _____

7. When will the observed results be close to the expected? _____
   _____

8. What is the expected result of a cross **BB** × **bb**? Use the Punnett square.

	b	b
B		
B		

BB _____

Bb _____

bb _____

9. Suppose you mated two other mice and expected one black mouse for each white mouse. How many of the following mice would you expect to observe out of 100 offspring:

   a. black? _____ b. white? _____

10. A scientist made a cross between two black mice. The cross was repeated between the same two mice several times. The data chart showed the color of all 42 offspring to be black. Use the Punnett squares below to show what you think the gene combinations were of both parents.

	B	__
B	BB	B__
__	B__	

	B	__
B	BB	B__
	B__	B__

222

Name _____ Class _____ Period _____

26-2 What Is a Test Cross?

An organism with a pure dominant trait has two dominant genes for that trait. An organism with a heterozygous trait has only one dominant gene for the trait; the other gene is recessive. Most of the time, both organisms look alike. You cannot tell them apart by looking at them. How can you tell if an organism is pure dominant or heterozygous for a trait? How can you tell what genes the parent with the dominant trait has?

There is a way to tell if an organism is pure dominant or heterozygous for a trait. If the organism is mated to a pure recessive and features of the offspring are examined, you can tell the parent's genes. The cross to a pure organism to determine unknown genes is called a test cross.

Here an example: In dogs, rust red fur color *(R)* is dominant over brown fur color *(r)*. A rust red dog can have either *RR* or *Rr* genes. To determine which genes a rust red dog has, it can be mated to a brown dog *(rr)* and the offspring can be examined. If all the offspring are red *(Rr)*, the parent dog is pure dominant for fur color and has *RR* genes. If about half of the offspring are rust red and about half are brown, the unknown parent is heterozygous for fur color *(Rr)*.

INTERPRETATION

OBJECTIVES
In this exercise, you will:
   a. set up a model to show a test cross.
   b. determine if a parent is pure dominant or heterozygous for a trait.

KEYWORDS
Define the following keywords:

cross _____

heterozygous _____

offspring _____

pure recessive _____

test cross _____

MATERIALS
marking pen   3 prepared bags
                     of beans

PROCEDURE
Part A. Test Cross 1
1. Obtain two bags marked "Parent A" and "Parent B" from your teacher. Do not look inside. These bags represent the parent dogs. The beans inside represent the genes.

223

2. "Parent A" is a red dog. It has one **R** gene and one gene that may be **R** or **r**. Therefore, it has genes **R**___. "___" means that the second gene of the dog is not known.
3. "Parent B" is a brown dog. It has **rr** genes.
4. Remove one bean from each bag without looking into the bags. Place the bean in front of the bag from which it was removed. Record the color of the beans in Table 1. The beans represent the genes of a puppy.
5. Return the beans to the bags from which they came.
6. Shake the bags to mix the beans. Remove another bean from each bag without looking in the bag. Record the colors of the beans in the table.
7. Repeat Steps 5 and 6 until 20 pairs have been recorded. Always return the beans to the bags from which they came.
8. Complete the table by writing in the genes the beans represent and the colors of the offspring. Total the offspring colors and complete the boxes below the table.
9. Make a Punnett square to show what both genes are for each parent. Then fill in the unknown gene of Parent A in the title of Table 1.

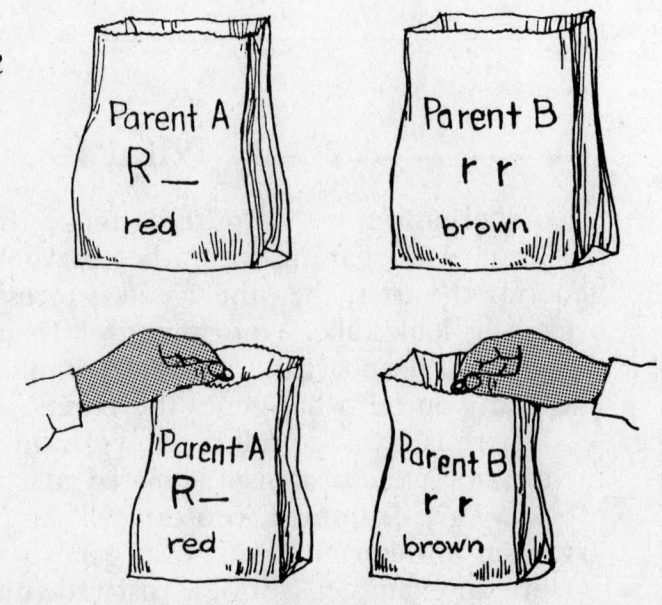

FIGURE 1.

Table 1. Test Cross 1: Parent A (R ___) X Parent B (rr)

Trial	Bean colors	Genes	Offspring color	Trial	Bean colors	Genes	Offspring color
1				11			
2				12			
3				13			
4				14			
5				15			
6				16			
7				17			
8				18			
9				19			
10				20			

Total brown offspring = ☐   Total red offspring = ☐

Name _____ Class _____ Period _____

Part B. Test Cross 2
1. Obtain another bag, "Parent C," from your teacher. C is a red dog. You will determine if Parent C is **RR** or **Rr**. The genes are **R** ___.
2. Repeat Steps 4 to 8 of Part A using Parents C and B. This time you are crossing Parent C with pure recessive Parent B. Recall that using the pure recessive as a parent is a test cross.
3. Record your results in Table 2. Complete the table as before. Total the results.
4. Make a Punnet square to determine what both genes are for each parent. Fill in Parent C's genes at the top of Table 2.

FIGURE 2.

Table 2. Test Cross 2: Parent C (R ___) X Parent B (rr)

Trial	Bean colors	Genes	Offspring color	Trial	Bean colors	Genes	Offspring color
1				11			
2				12			
3				13			
4				14			
5				15			
6				16			
7				17			
8				18			
9				19			
10				20			

Total brown offspring = ☐   Total red offspring = ☐

QUESTIONS
1. Before the test crosses, what possible genes could Parent A have had? _____
2. Before the test crosses, what possible genes could Parent C have had? _____

3. Could Parents A and B have had any brown offspring? _____ Explain.

_____

_____

4. Could Parents C and B have had any brown offspring? _____ Explain.

_____

_____

5. Complete the Punnett squares for the test crosses shown below for dog fur color.

	R	R
r		
r		

	R	r
r		
r		

6. Suppose long fur (*L*) is dominant over short fur (*l*) in guinea pigs. How can you determine whether a particular long-furred guinea pig has *LL* or *Ll* genes? _____

_____

7. Suppose a long-furred guinea pig is crossed to a guinea pig with short fur. They have 4 offspring. Two of the offspring have short fur. The other two have long fur. What genes does the long-furred parent have, *LL* or *Ll*? _____
In the space below draw a Punnett square that shows the parents and the four offspring.

Name _____ Class _____ Period _____

27-1 What Do Normal and Sickled Cells Look Like?

Sickle-cell anemia is a disorder in which red blood cells are sickle-shaped rather than round. Sickle-cell anemia is a genetic disorder. The sickled-cells do not carry as much oxygen as normal red blood cells. A person with this disorder does not get enough oxygen to the body cells.

EXPLORATION

OBJECTIVES
In this exercise, you will:
   a. examine and compare slides of normal and sickled red blood cells.
   b. learn that there are two different conditions to this disease.
   c. solve genetic problems involving sickle-cell anemia.

KEYWORDS
Define the following keywords:

genetic disorder _____

hemoglobin _____

sickle-cell anemia _____

sickle-cell trait _____

MATERIALS
compound light microscope   red pencil
prepared slides of normal and sickled red blood cells

PROCEDURE
Part A. Observation of Normal and Sickled Red Blood Cells
1. Examine a slide of normal blood cells under the microscope. Locate the cells first under low power, then high power.
2. Draw two or three red blood cells in the space provided marked "normal." Draw the cells to scale. (Note: Red cells are round, have no nucleus, and are very pale pink in color. Cells that appear dark blue in color are stained white blood cells.)
3. Label the following cell parts: cell membrane, cytoplasm.
4. Shade, with red pencil, the parts of the cell in which hemoglobin is found.

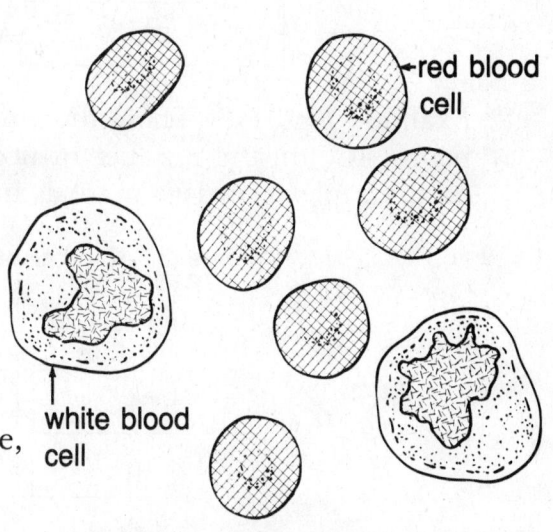

FIGURE 1. Normal blood cells

5. Examine a slide of sickled red blood cells under the microscope. Locate the cells first under low power, then high power.
6. Draw two or three sickled red blood cells in the space provided marked "sickled." (Note: Sickled cells are irregular in shape and may look like crescent moons or teardrops.)
7. Label the following cell parts: cell membrane, cytoplasm.
8. Shade, with red pencil, the parts of the cell in which hemoglobin is found.

Normal cells   **Sickled cells**

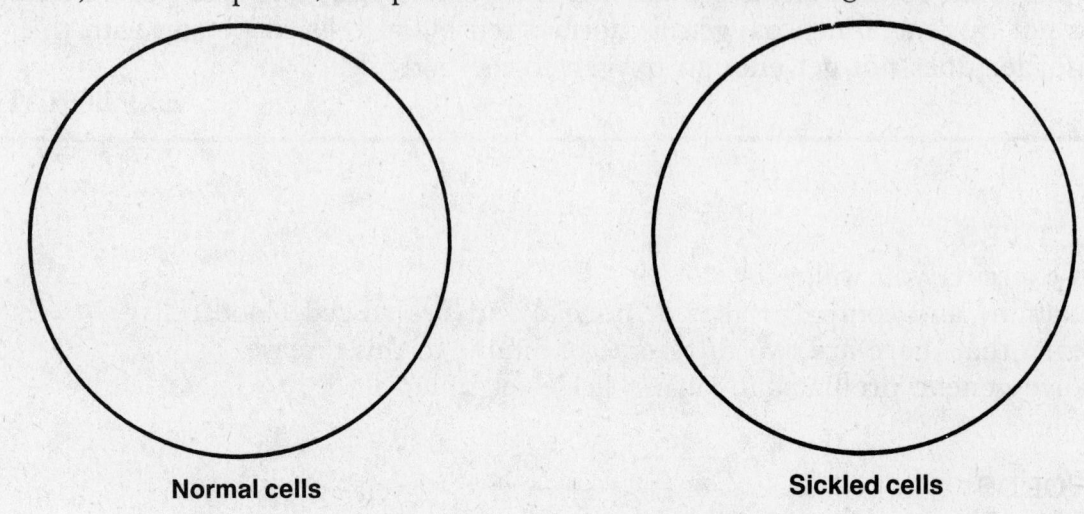

FIGURE 2. Normal and sickled cells

Part B. Comparison of Sickle-cell Trait with Sickle-cell Anemia

1. Examine Table 1. Note that not all people that possess the sickle-cell gene have sickle-cell anemia. There are two different sickle-cell conditions that are determined by the genes that are received from the parents. Sickle-cell trait is less of a problem than sickle-cell anemia. For more information read the material in section 27:5 of your text.

Table 1. Gene Combinations

Gene combination	Blood cell shape	Name of disease
RR	All Round	Normal
RS	Half Round Half Sickled	Sickle-Cell Trait
SS	All Sickled	Sickle-Cell Anemia

2. Examine Figure 3. These drawings represent blood samples from three different people. Count the number of normal and sickled cells in all three samples. For each sample record your totals in Table 2.

Table 2. Number of Normal and Sickled Cells Seen

Blood sample	Number of normal cells	Number of sickled cells
A		
B		
C		

Name _____ Class _____ Period _____

3. Examine your data and Table 1. Determine the blood condition for each of the samples in Figure 3 and write the correct condition ("normal," "sickle-cell trait," or "sickle-cell anemia") below each blood sample.
4. Examine your data and Table 1. Determine the correct gene combination (*RR*, *RS* or *SS*) that produced each sample in Figure 3 and write them below each blood sample also.

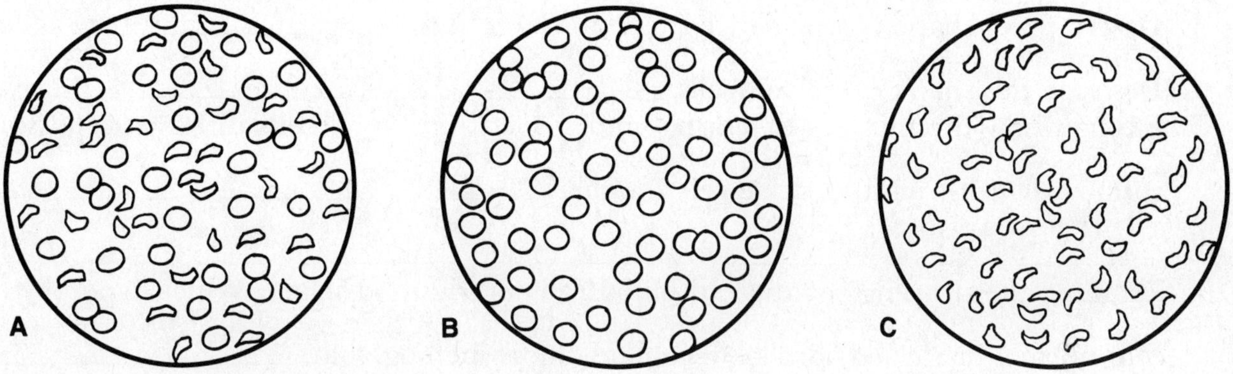

A _____ _____

B _____ _____

C _____ _____

FIGURE 3. Blood samples

Part C. Genetics Problems

Construct and use a Punnett square for each of the following problems. Record your answers in the spaces provided.

1. Two parents have the following genes for blood cell shape: *RS* and *RR*. What kind of blood might their children have?

	number of children
have normal blood	_____
have sickle-cell trait	_____
have sickle-cell anemia	_____

2. Two parents have the following genes for blood cell shape: *RS* and *RS*. What kind of blood might their children have?

	number of children
have normal blood	_____
have sickle-cell trait	_____
have sickle-cell anemia	_____

3. Two parents have the following genes for blood cell shape: **RS** and **SS**. What kind of blood might their children have?

	number of children
have normal blood	_____
have sickle-cell trait	_____
have sickle-cell anemia	_____

QUESTIONS

1. Describe the shape of normal red blood cells._____

2. Describe the shape of sickled cells._____

3. Explain how the number of normal and sickled red blood cells differ in a person with sickle-cell trait and sickle-cell anemia._____

_____

4. Examine the drawings of the red blood cells in Figure 1 again. Which type of cell, normal or sickled, probably contains more hemoglobin?_____

_____

5. The less hemoglobin a person has, the more difficult it is for the cells of their body to get enough oxygen. With too little oxygen, a person will tire easily and cannot do a lot of exercising. Would a person with sickle-cell trait be able to do more or less exercising than a person with sickle-cell anemia?_____
Why?_____

6. Which condition is more of a problem for a person, sickle-cell trait or sickle-cell anemia? _____. Why? _____

_____

7. A scientist gathered the following information while studying people with sickle-cell anemia.
Number of children born in the United States with sickle-cell anemia:

    to white parents—less than 1/100,000 births
    to Afro-American parents — about 200/100,000 births

Which racial group seems to suffer most from this genetic disorder?_____

8. Why do we call sickle-cell anemia a genetic disorder?_____

_____

9. How does this disorder serve as an example of lack of dominance?_____

_____

Name _____ Class _____ Period _____

27-2 How Are Traits on Sex Chromosomes Inherited?

Hemophilia is a disorder in which the person's blood will not clot. The disorder is inherited. If you have the dominant gene *H*, you will have normal blood. If you have only the recessive gene *h*, your blood will not clot.

Color blindness is also a genetic disorder. In this disorder, the person does not see certain colors, such as red and green. This person will see green as a grey color and red as a yellow color. If you have at least one dominant gene *C*, you see all colors. If you have only recessive genes, you cannot see red and green.

INTERPRETATION

OBJECTIVES
In this exercise, you will:
  a. toss coins to show children born in five families.
  b. see how hemophilia and color blindness are inherited in several families.
  c. solve genetic problems involving hemophilia and color blindness in some families.

KEYWORDS
Define the following keywords:

color blindness _____

hemophilia _____

sex chromosomes _____

MATERIALS
8 coins   tape   pen

PROCEDURE
Part A. Hemophilia
Genes for hemophilia are located on the sex chromosomes. Remember, females have two X chromosomes (*XX*) while males have one X and one Y chromosome (*XY*). Only the X chromosomes have the genes for hemophilia. A female can be $X^H X^H$, $X^H X^h$, or $X^h X^h$ for the clotting trait. A male can be $X^H Y$ or $X^h Y$.

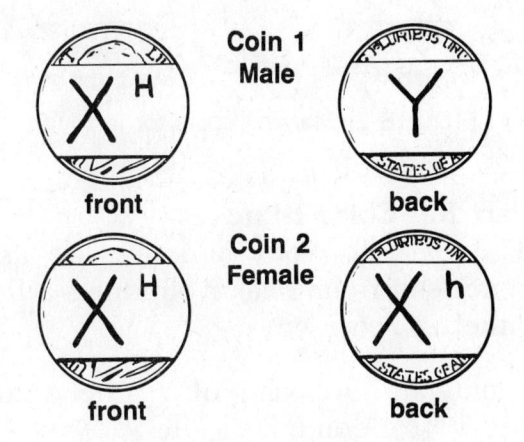

FIGURE 1. Marking coins for family 1

Family 1.   Offspring of parents who are normal; the mother is heterozygous.
  1. Place the tape on both sides of two coins.
  2. Mark the coins as shown in Figure 1. These coins represent the genes of the parents. The coin with the *Y* chromosome is the father. The coin with an *X* on each side is the mother.

231

3. Place both coins in your cupped hands. Shake the coins, and then drop them on your desktop.
4. Read the combination of letters that appears. This combination represents the result that might appear in an offspring of these parents.
5. Make a mark (/) in Table 1 beside the correct gene combination in the column marked "Offspring Observed."
6. Repeat shaking and reading the coins for a total of 40 times.
7. Figure the total marks for each gene combination in Table 1 and write these totals in the proper space in the table.

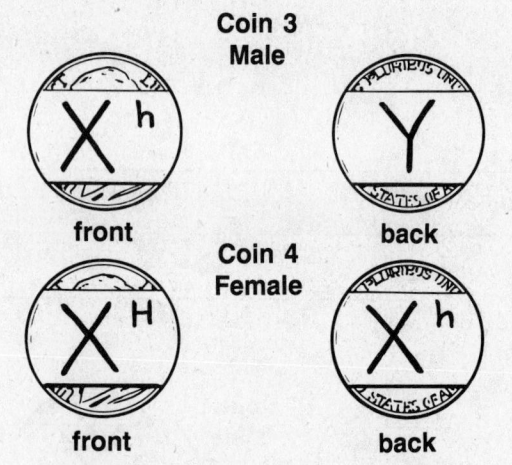

Table 1. Offspring of X^HY Father and X^HX^h Mother

Gene combinations	Offspring observed	Total
X^HX^H		
X^HX^h		
X^hX^h		
X^HY		
X^hY		

Family 2. Offspring of a father who has hemophilia and a heterozygous mother.
1. Place tape on two coins and mark them as shown in Figure 2.
2. Place the coins in your hands and shake. Read the results and make a proper mark in Table 2.
3. Repeat step 2 for a total of 40 times. Total your marks in Table 2.

Coin 3 Male — front: X^h, back: Y
Coin 4 Female — front: X^H, back: X^h

FIGURE 2. Marking coins for family 2

Table 2. Offspring of X^hY Father and X^HX^h Mother

Gene combinations	Offspring observed	Total
X^HX^H		
X^HX^h		
X^hX^h		
X^HY		
X^hY		

Part B. Color Blindness

The genes for color blindness are also located on the sex chromosomes. For the genes controlling color blindness a female can be X^BX^B, X^BX^b, or X^bX^b. A male can be either X^BY or X^bY.

Family 3. Offspring of a father who is color-blind and a mother who has two dominant genes.
1. Place tape on two coins and mark them as shown in Figure 3.
2. Shake the coins and read the results. Place a proper mark in Table 3.
3. Repeat step 2 for a total of 40 times. Total your marks in Table 3.

Family 4. Offspring of parents who are normal but the mother is heterozygous.
1. Place tape on two coins and mark them as shown in Figure 4.
2. Shake the coins and read the results. Place a proper mark in Table 4.

Name _____ Class _____ Period _____

3. Repeat step 2 for a total of 40 times. Total your marks in Table 4.

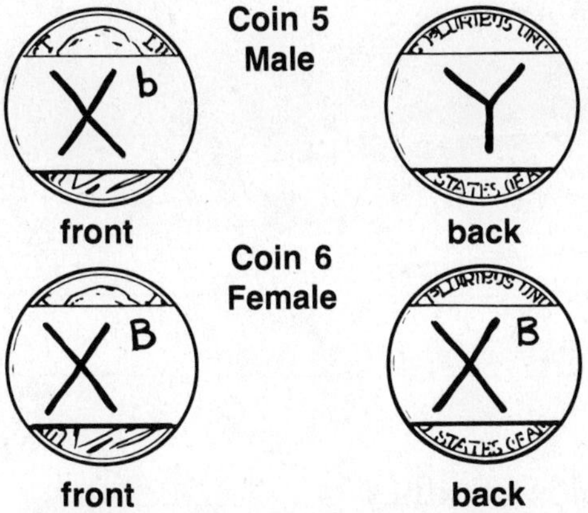

FIGURE 3. Marking coins for family 3

Coin 5 Male: front X^b, back Y
Coin 6 Female: front X^B, back X^B

Table 3. Offspring of X^bY Father and X^BX^B Mother

Gene combinations	Offspring Observed	Total
X^BX^B		
X^BX^b		
X^bX^b		
X^BY		
X^bY		

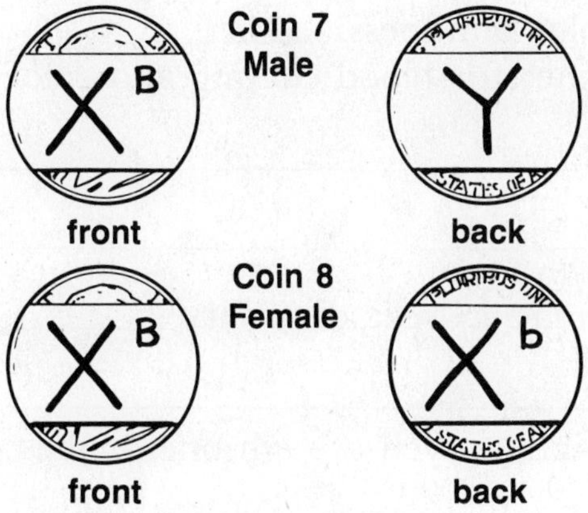

FIGURE 4. Marking coins for family 4

Coin 7 Male: front X^B, back Y
Coin 8 Female: front X^B, back X^b

Table 4. Offspring of X^BY Father and X^BX^b Mother

Gene combinations	Offspring observed	Total
X^BX^B		
X^BX^b		
X^bX^b		
X^BY		
X^bY		

Part C. Problems

For each of the following problems construct and use a Punnett square. Record your answers in the spaces provided.

1. Two parents have the following genes for hemophilia: X^HX^h and X^HY. What type of blood could their children have?

<u>Children</u>
Number of males   Number of females

have normal clotting   _____   _____

have hemophilia   _____   _____

2. Two parents have the following genes for color blindness: X^BX^B and X^bY. What kind of color vision could their children have?

<u>Children</u>
Number of males   Number of females

have normal vision   _____   _____

have color blindness   _____   _____

233

3. Two parents have the following genes for color blindness: X^BX^b and X^bY. What type of color vision could their children have?

Children

	Number of males	Number of females
have normal vision	_____	_____
have color blindness	_____	_____

QUESTIONS

1. What sex chromosomes do female offspring have? _____

2. What sex chromosomes do male offspring have? _____

3. How many genes do females have:

   a. for blood clotting? _____   b. for color blindness? _____

4. How many genes do males have

   a. for blood clotting? _____   b. for color blindness? _____

5. Why is there a difference in the number of genes for blood clotting and color blindness in males and females? _____

6. Which of the two traits studied in this exercise are genetic disorders? _____

7. In Problem 2, why are there no color-blind children even though one of the parents is color blind? _____

8. Which of the parents give the trait of hemophilia to their son? _____

9. Which of the parents give the trait of hemophilia to their daughter? _____

Name _____ Class _____ Period _____

28-1 How Does DNA Make Protein?

DNA directs your cells to make certain proteins. How does DNA make proteins? DNA is a model for making a molecule called messenger RNA (mRNA). Messenger RNA is much like DNA. RNA is made of substances, called nitrogen bases, that must match up with the nitrogen bases in DNA. These nitrogen bases will only match up in certain ways. The production of mRNA occurs in the nucleus.

After it is formed, mRNA leaves the nucleus and attaches to a ribosome in the cytoplasm of the cell. Other RNA molecules, called transfer RNA (tRNA), bring protein parts to the mRNA on the ribosome. The two types of RNA molecules match up, join protein parts together, and make a protein. Figure 1 shows the steps involved in making a protein. DNA determines what proteins are produced.

INTERPRETATION

OBJECTIVES
In this exercise, you will:
  a. use models to show how DNA makes mRNA.
  b. use models to show how mRNA leaves the nucleus and causes tRNA to make proteins.

KEYWORDS
Define the following keywords:

DNA _____

mRNA _____

protein _____

tRNA _____

MATERIALS
scissors
colored pencils: red, blue and green

PROCEDURE
1. Examine Figure 2, a model of a DNA molecule. DNA has two main sides. These sides are often compared with the upright sides of a ladder. The squares in the model represent sugar molecules. The nitrogen bases A, C, G, and T join to connect the two sides.

1. mRNA copies DNA.

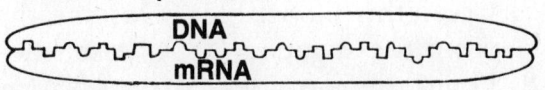

2. mRNA joins tRNA, which has protein parts.

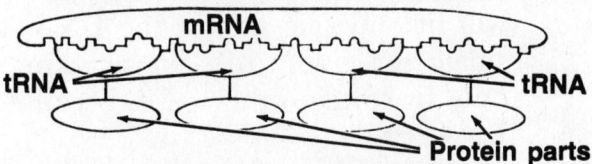

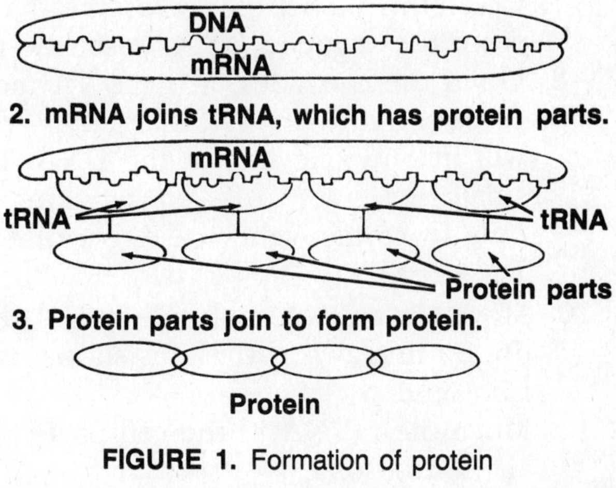

3. Protein parts join to form protein.

FIGURE 1. Formation of protein

235

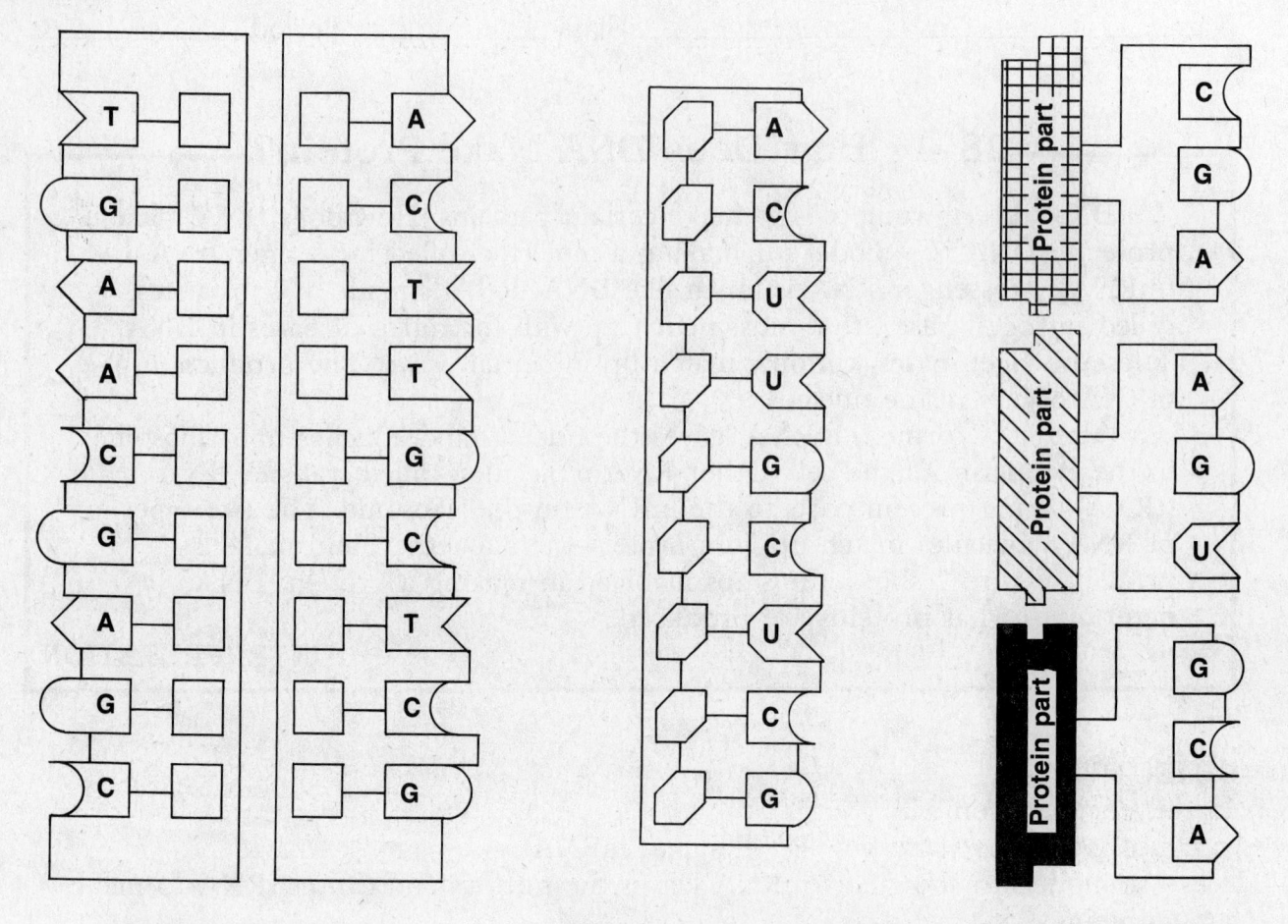

FIGURE 2. DNA molecule   **FIGURE 3.** Messenger RNA   **FIGURE 4.** Transfer RNA

2. Cut out the two sides of the DNA model in Figure 2.
3. Color the two sides red.
4. Put the two sides together so that they fit together like the pieces of a puzzle. Note that nitrogen base A only binds with T and base G only with C.
5. Examine Figure 5, a model of a cell. The nucleus is in the upper left corner. Place the model of DNA in the nucleus. DNA carries the code for making cell proteins. That code is the order in which the nitrogen bases appear.
6. Cut out the model of the mRNA molecule in Figure 3. This molecule has only one side.
7. Color this model blue. Observe that the sugar in the mRNA molecule is different from the sugar in DNA. Also, the nitrogen base U is present instead of T.
8. Open the two sides of the DNA model.
9. Place the mRNA molecule along one side of the DNA model. Note that its bases will fit only one side of the DNA. In an actual cell, the mRNA is assembled from small molecules to fit exactly along one side of the DNA. The nitrogen bases can only fit certain other bases because of their shape. mRNA copies the code of DNA.
10. Move the mRNA molecule out of the nucleus to the cytoplasm by following the dotted line as a path. This shows that mRNA carries the code of the DNA to the ribosomes.
11. Move the mRNA to the cell part called the ribosome. Place it on the dashed lines at the ribosome.

Name _____ Class _____ Period _____

12. Put the DNA model sides back together.
13. Cut out the three tRNA molecules shown in Figure 4. Using a green pencil, color only the lower parts (that contain the letters A, U, C, and G). This type of RNA is different from mRNA in two ways. First, each tRNA molecule has only three nitrogen bases and second, a certain protein part is attached to it. Transfer RNA is found in the cytoplasm of the cell. The top of each tRNA has a specific protein part attached to it.
14. Fit the tRNA molecules to the mRNA molecule again, so the bases fit together tightly. Observe which bases of tRNA bind with which bases of mRNA (A with U, G with C).
15. With the tRNA molecules in place on the mRNA molecule, the protein parts can now join with each other. The linked protein parts carried by the tRNA make a chain. This chain separates from the tRNA molecules and leaves the ribosome to become a protein. The code of the DNA molecule directs certain steps in a cell for the process of forming a certain protein.

QUESTIONS
1. What do the letters DNA stand for? _____
2. In DNA, what nitrogen base always binds with A? _____ G? _____
3. How is mRNA different from DNA? _____
   _____
4. In mRNA, what nitrogen base binds with the DNA base
   A? _____ G? _____ T? _____
5. Where in the cell is mRNA made? _____
6. To what cell part does mRNA attach? _____
7. What carries the protein parts to the ribosome and the mRNA? _____
8. How are mRNA and tRNA alike? _____
   _____
9. What does tRNA have that mRNA does not have? _____
10. Where in the cell are proteins made? _____
11. What determines which proteins are produced? _____

237

FIGURE 5. Model of a cell

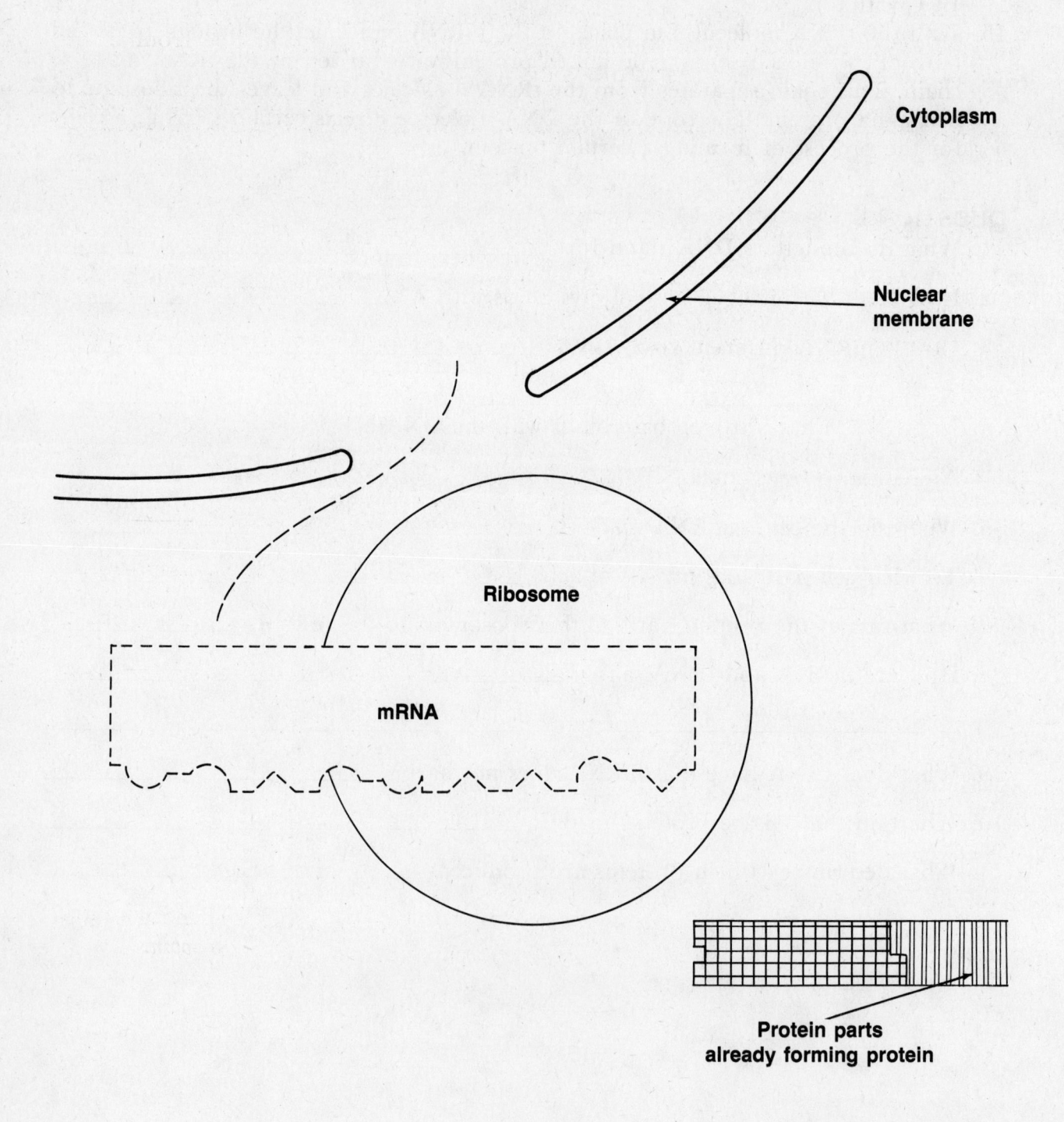

238

Name _____ Class _____ Period _____

28-2 How Can a Mutation in DNA Affect an Organism?

Sometimes the DNA code that makes up a gene has an error in it. This error is called a mutation. When the DNA contains an error, the mRNA it makes will copy that error. When the mRNA contains an error, it will code for incorrect tRNAs and produce an incorrect protein.

Sickle-cell anemia is a disorder that gets its name from the sickle shape of the red blood cells. The sickled red blood cells are caused by a mutation in the hemoglobin of the person with the disorder. Hemoglobin is the main protein in red blood cells. Each hemoglobin molecule carries oxygen from the lungs to all other parts of the body.

INTERPRETATION

OBJECTIVES
In this exercise, you will:
  a. examine the coding errors produced in mRNA and tRNA when there is a mutation in the DNA.
  b. examine the effect of a mutation in the gene that codes for blood hemoglobin.

KEYWORDS
Define the following keywords:

gene _____

hemoglobin _____

mutation _____

sickle-cell anemia _____

MATERIALS
colored pencil

PROCEDURE
1. Examine Table 1. The two columns show a section of normal DNA and a section of DNA that has a mutation in it. The mutation is called *sickle hemoglobin*.

Table 1. Comparing Normal With Sickle Mutation DNA

	This section codes for normal hemoglobin	This section codes for "sickle" hemoglobin
DNA code	G G G C T T C T T T T T	G G G C A A C T T T T T
mRNA code		
tRNA code		
Order of protein parts		
Shape of blood cells		

240

Name _____ Class _____ Period _____

2. In Table 1, in the row marked *mRNA code*, write in the correct letters that will match with the nitrogen base letters of DNA given in the row above. Do this for both columns. Remember that A matches with U, T matches with A, C matches with G, and G matches with C.
3. In the row marked *tRNA code*, write in the correct letters that will match with the nitrogen base letters of mRNA in the row above. Remember that A matches U, U matches with A, C matches with G, and G matches with C.
4. Examine Table 2. This table shows which protein parts are coded for by specific sets of nitrogen bases (three per set) of the mRNA molecule. For example, the mRNA sequence CCC codes for protein part A.

Table 2. Nitrogen Bases of Protein Parts

Protein part	mRNA
A	CCC
B	GAA
C	AAA
X	GUU

5. In Table 1, in the row marked *Order of protein Parts*, write in the correct order of protein parts coded for by the mRNA. Do this for both normal and sickle hemoglobin.
6. In the row marked *Shape of blood cells*, draw in what you think will be the correct shape of blood cells for the kind of protein found in the row above. Use the diagrams in Figure 1 for reference.

Normal red blood cells

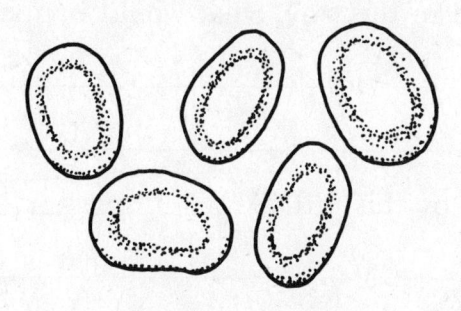

Sickled red blood cells

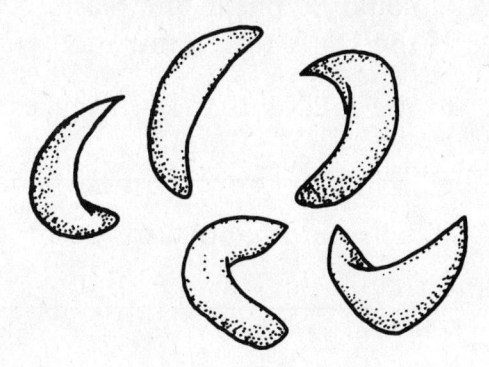

Normal hemoglobin

Sickled hemoglobin

FIGURE 1. Shapes of blood cells

7. In the column marked *This section codes for sickle hemoglobin*, locate the two nitrogen bases that are different in DNA, mRNA, and tRNA from those in the column for normal hemoglobin. Color those bases that are mutations with the colored pencil.

QUESTIONS

1. Look at the two DNA molecules in Table 1. What nitrogen bases in the sickle mutation DNA are different from those of the normal DNA?

   _____

2. If every three nitrogen bases on DNA represent a gene, how many genes are shown on

   a. the section of normal DNA?_____

   b. the section of sickle hemoglobin DNA?_____

3. List the nitrogen bases (examined in Table 1) for

   a. the normal genes of hemoglobin_____

   b. the sickle hemoglobin genes_____

4. How many genes are different in sickle hemoglobin DNA compared with normal hemoglobin DNA?_____

5. How many protein parts are different in sickle hemoglobin compared with normal hemoglobin?_____

6. How many genes are needed to code one protein part into a protein such as hemoglobin?_____

7. Define the word *mutation*

   a. by using the word "gene."_____

   _____

   b. by using the phrase "DNA code."_____

   _____

8. It is possible to move genes from one molecule of DNA to another. A normal gene could be put in the place of a gene with a mutation.
   a. If the DNA with a mutation were corrected in this way, what would happen to the mRNA that DNA makes?_____

   _____

   b. What would happen to the protein formed by this mRNA?_____

   _____

Name _____ Class _____ Period _____

29–1 How Do Some Living Things Vary?

All the living things that are within one species are not alike. The differences in a trait are called variations (ver ee AY shunz.) For example, we see variations in height, eye color, and ear shape.

Variations in traits may be helpful. They may help survival. A living thing with a variation may have an advantage (ud VANT ihj) over a living thing without that variation. We say the variation is an adaptation.

EXPLORATION

OBJECTIVES
In this activity, you will:
  a. observe variations in leaves, bean pods, bean seeds, and humans.
  b. determine how variations may be helpful.

KEYWORDS
Define the following keywords:

adaptation _____

variation _____

MATERIALS
conifer twig
metric ruler
10 pinto bean seeds
10 opened bean pods

PROCEDURE
Part A. Variation in Leaf Length
1. Examine the conifer twig. Each needlelike structure is a leaf.
2. Remove 10 leaves from the twig as shown in Figure 1.
3. Measure the length of each leaf.
4. Record your results in Table 1 by making a check below the millimeter length for each leaf.
5. Count the marks and enter the number of leaves for each length.

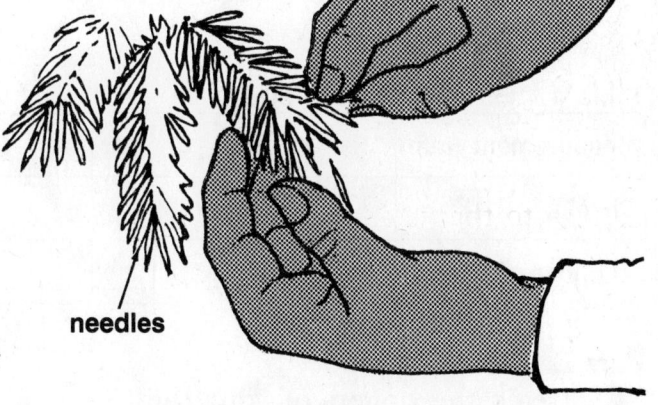

FIGURE 1. Removing conifer needles

Table 1. Variation in Leaf Lengths

Measurement (mm)	5	6	7	8	9	10	11	12	13	14	15	16	17	18	19	20	21	22	23	24
Checks																				
Number of leaves																				

Part B. Variation in the Number of Seeds in Bean Pods
1. Examine 10 opened bean pods. Each pod has seeds inside.
2. Count the number of seeds in each pod.
3. Record your results in Table 2 by making a mark below the number of seeds for each pod.
4. Count the marks and enter the number of pods for each seed number.

Table 2. Variation in Seeds in Pods

Number of seeds	5	6	7	8	9	10	11	12
Checks								
Number of pods								

Part C. Variation in Seed Coats
1. Examine 10 pinto seeds for variations in color: light, medium, and dark brown.
2. Group the seeds according to their color.
3. Count the number of seeds in each group and record this in the table.

Table 3. Variation in Seed Coat Color

Color	Light brown	Medium brown	Dark brown
Number of seeds			

Part D. Variation in Seed Size
1. Measure the length of each pinto bean seed as shown in Figure 2.
2. Record your results in Table 4 by making a mark below the length for each seed.

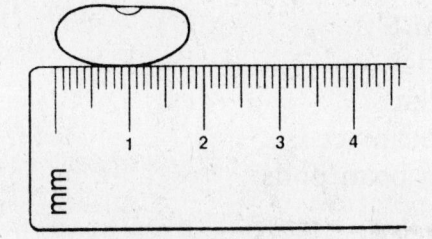

FIGURE 2. Measuring bean

Table 4. Variation in Seed Size

Measurement (mm)	5	6	7	8	9	10	11	12	13	14	15	16	17	18	19	20
Checks																
Number of beans																

Part E. Hand Spread
1. Stretch the fingers of your right hand out flat on your desk top.
2. Measure the distance from the tip of your thumb to the tip of your little finger in cm (Figure 3).
3. Write the measurement on the chalkboard.
4. Record the measurements of the class in Table 5.

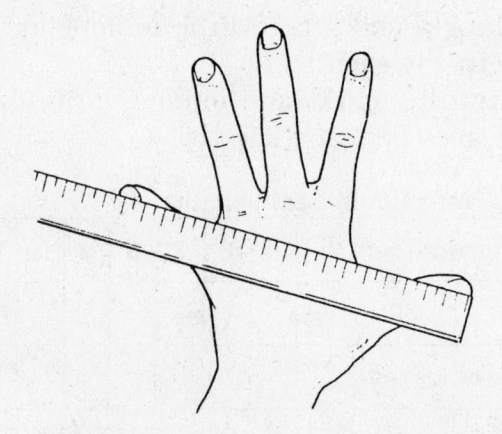

FIGURE 3. Measuring hand spread

Name _____ Class _____ Period _____

Table 5. Variation in Hand Spread

Measurement (cm)	13	14	15	16	17	18	19	20	21	22	23	24	25	26	27	28
Number of hands																

5. Plot the information from Table 5 onto the graph in Figure 4.

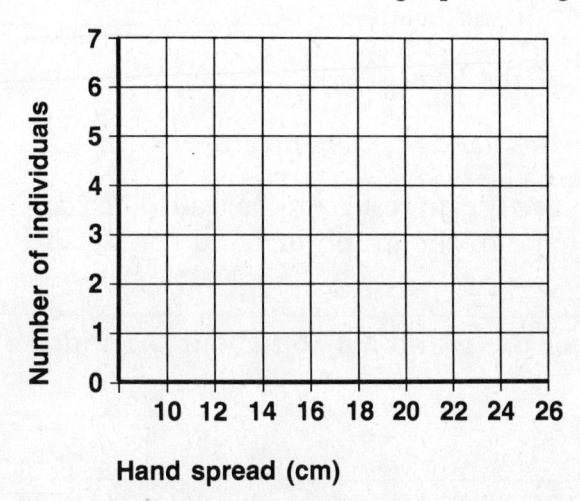

FIGURE 4. Variation in hand sizes within a class of students

QUESTIONS

1. Describe the variation that you saw in

   a. leaf length. _____

   b. seed number in each pod. _____

   c. seed coat color. _____

   d. seed size. _____

   e. hand spread. _____

2. The living things that you observed varied in ways other than those you noted in this activity. In Table 6, list another trait for which the living thing varies. You may want to look at the samples again. The table is started for you.

Table 6. Other Traits of Living Things

Sample	Trait examined	Other traits
leaves	length	
pods	seed number	
seeds	color	
hands	size	

3. In your leaf sample what is:

   a. the length of most of the coniferous needles?_____

   b. the length of the shortest needle?_____

   c. the length of the longest needle?_____

4. In your bean sample:

   a. how many seeds do most of the bean pods have?_____

   b. what color are most of the bean seeds?_____

   c. what size are most of the bean seeds?_____

5. How wide do most of the hand spreads measure?_____

6. In which width range are the fewest hand spreads?_____

7. What is the general shape of the graph of hand spread in Part E?

   _____

8. What does the shape of the graph tell you about the width of the hand spreads in your class?_____

   _____

9. What advantage could the following traits be to the organism:
   a. longer tree leaves? (HINT: What is the main job of leaves?)

   _____

   b. more seeds in a pod?_____

   c. larger bean seeds?_____

   d. larger hand spreads?_____

Name _____ Class _____ Period _____

29–2 How Do Fossils Show Change?

Most organisms live, die, and decompose. They leave no traces of having lived. Under certain conditions, an organism's remains or tracks may be preserved as a fossil. Fossils give clues about how an organism looked and where it lived. They are often used by scientists as evidence of change.

A fossil is any remains of a once-living thing. Fossils may only be the outline of some plant, animal, or other organism that is preserved in rock. Sometimes, entire skeletons of animals that lived millions of years ago are found.

INTERPRETATION

OBJECTIVES
In this activity, you will:
a. examine diagrams of fossil horses and present-day horses shown in their surroundings.
b. examine diagrams of the structure of the front foot of fossil horses and present-day horses.
c. note the changes in horses that have taken place over time.

KEYWORDS
Define the following keywords:

adaptation _____

Equus _____

fossil _____

Hyracotherium _____

natural selection _____

_____

MATERIALS
metric ruler
colored pencils: red, blue, green, and yellow

PROCEDURE
Part A. Change in Size With Time
1. Examine the diagrams in Figure 1 of *Hyracotherium, Miohippus, Merychippus,* and *Equus.*
2. Use the diagrams to fill in Table 1.

247

Table 1. Evolution in the Horse

Horse	*Hyracotherium*	*Miohippus*	*Merychippus*	*Equus*
Size				
Type of surroundings				

Hyracotherium
55 million years ago
38 cm

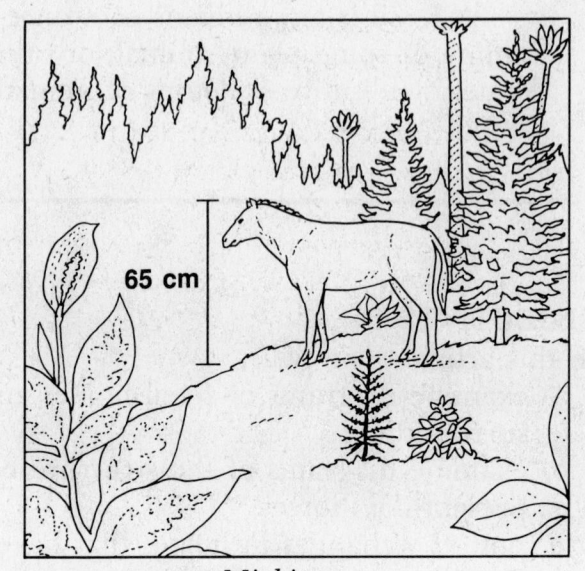

Miohippus
30 million years ago
65 cm

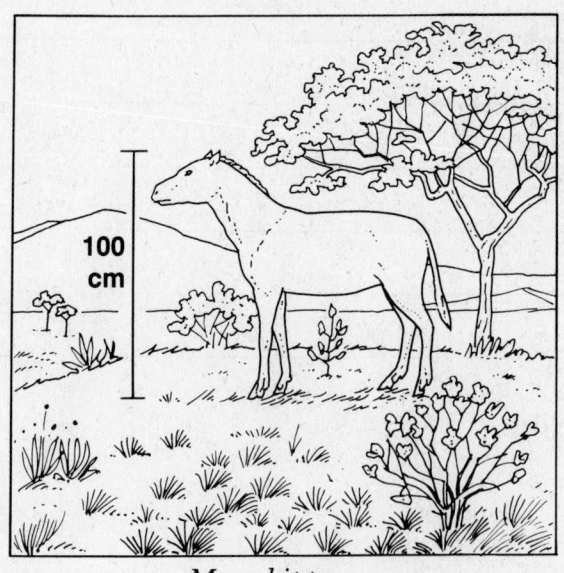

Merychippus
13 million years ago
100 cm

Equus
Today
140 cm

FIGURE 1. Evolution of the horse

Name _____ Class _____ Period _____

Part B. Changes in Bone Structures With Time

The changes in horses over the last 55 million years have been shown by studies of large numbers of fossils. The earliest kind of horse was small and had teeth that were adapted to browsing on young shoots of trees and shrubs. The present-day horse is much larger and has larger teeth that are adapted to grazing on the tough leaves of grasses. Early horses were adapted to living in wooded, swampy areas where more toes were an advantage. The single-hoofed toes of the present-day horse allow it to travel fast in the plains.

1. Examine the diagrams in Figure 2. They show fossils of the front foot bones and the teeth of horses. The foot bones at the upper right of each diagram indicate the relative bone sizes of each kind of horse.

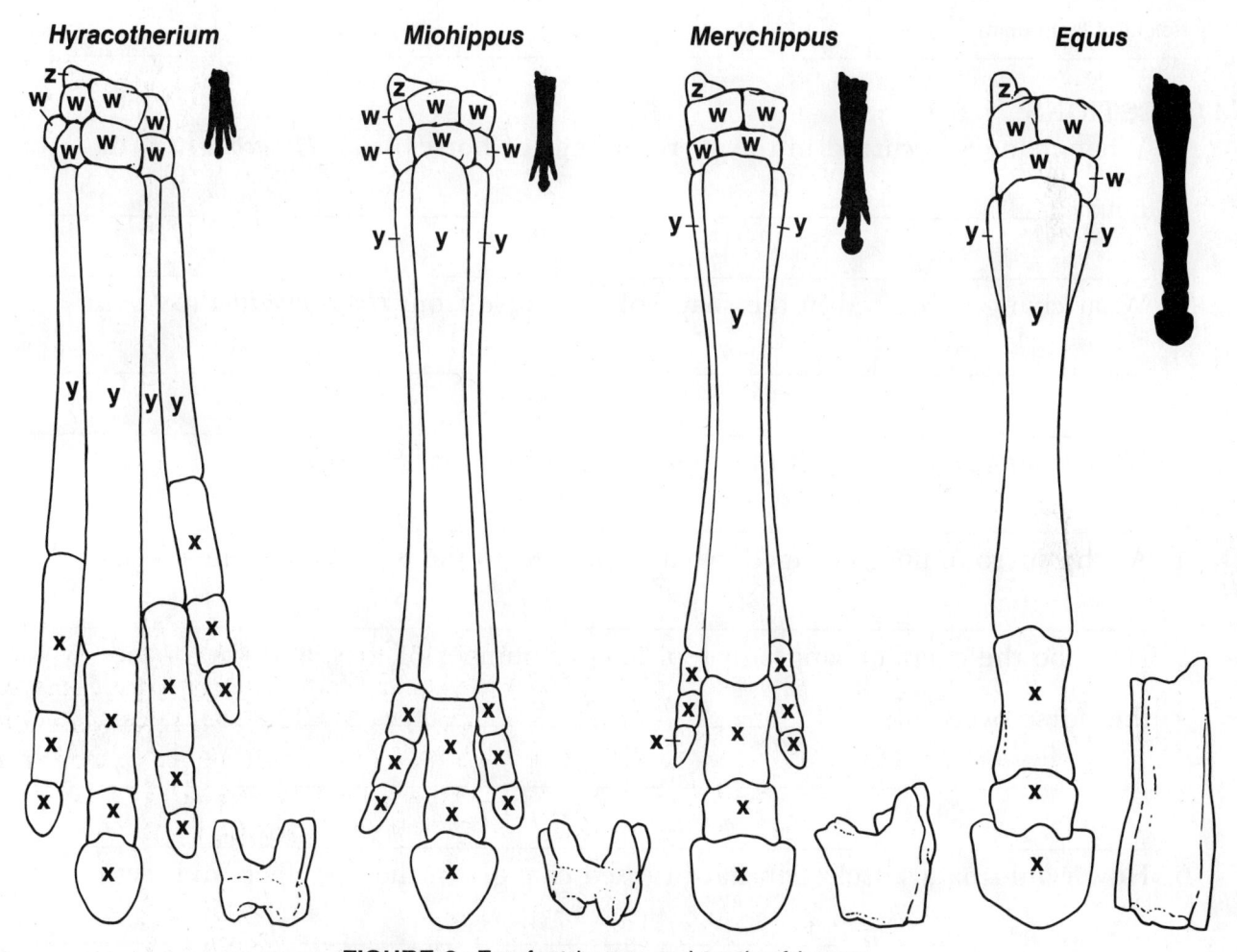

FIGURE 2. Forefoot bones and teeth of horses

2. Look for and color the following kinds of bones for each fossil horse.
    a. Color the toe bones red. These are marked for you with an *x*.
    b. Color the foot bones blue. These are marked with a *y*.
    c. Color the ankle bones green. These are marked with a *w*.
    d. Color the heel bones yellow. These are marked with a *z*.
3. Using the diagrams in Figure 2, make measurements to fill in Table 2.

249

Table 2. Evolution of the Horse

Kind of horse	Hyracotherium	Miohippus	Merychippus	Equus
Number of toes				
Number of toe bones				
Number of foot bones				
Number of ankle bones				
Number of heel bones				
Total number of foot bones				
Length of foot (measure inset diagrams) (mm)				
Height of teeth (mm)				

QUESTIONS

1. What changes occurred in the surroundings of horses from *Hyracotherium* to *Equus*? _____

2. What change occurred in the shape of the horse from *Hyracotherium* to *Equus*? _____

3. What changes occurred in the size of the horse from *Hyracotherium* to *Equus*? _____

4. As the surroundings changed, what happened to the teeth of the horse? _____

5. Describe the overall changes in foot length, number of toes, and size of toes in the horse over time. _____

6. How would natural selection have caused changes in the size, feet, and teeth of the horse? _____

Name _____ Class _____ Period _____

30-1 What Are Some Parts of a Food Chain and a Food Web?

Plants use light energy of the sun to make food. The food is stored in the cells of the plant. Plants are called producers because they make food. Some of the stored energy in the food that plants make is passed on to the animals that eat the plants. Plant-eating animals are called primary consumers. Some of the energy is passed on to the animals that eat primary consumers. Animals that eat other animals are called secondary consumers.

The pathway that food energy takes through an ecosystem is called a food chain. A food chain shows the movement of energy from plants to plant eaters and then to animal eaters. An example of a food chain can be written as follows:

(producer)	(primary consumer)	(secondary consumer)
seeds ⟶	sparrow ⟶	hawk

Some of the food energy in the seeds moves to the sparrow that eats them. Some of the food energy then moves to the hawk that eats the sparrow.

Because a hawk eats animals other than sparrows, you could make a food chain for each animal the hawk eats. If all the food chains were connected, the result is a food web. A food web is a group of connected food chains. A food web shows many energy relationships.

INTERPRETATION

OBJECTIVES
In this exercise, you will:
  a. determine what different animals eat in several food chains.
  b. build a food web that could exist in a forest ecosystem.

KEYWORDS
Define the following keywords:

consumer _____

food chain _____

food energy _____

food web _____

producer _____

MATERIALS
colored pencils   metric ruler

251

PROCEDURE
Part A. Examining Food Chains
1. Read the introduction and examine the food chains given below.

(producer)	(primary consumer)	(secondary consumers)		
plant roots	→ rabbit	→ fox		
plant seeds	→ mouse	→ fox		
plant leaves	→ earthworm	→ robin	→ snake	
plant leaves	→ rabbit	→ snake		
plant leaves	→ cricket	→ robin	→ fox	
plant stems	→ earthworm	→ snake	→ hawk	→ fox
plant stems	→ rabbit	→ hawk		
plant stems	→ small insects	→ mouse	→ owl	
plant leaves	→ rabbit	→ owl	→ fox	
plant leaves	→ cricket	→ mouse	→ hawk	
plant fruits	→ mouse	→ snake	→ owl	
plant fruits	→ small insects	→ robin	→ snake	

2. Answer the questions that follow:

   a. List the organisms that you think are producers. _____

   b. Why are they called producers? _____

   c. List the organisms that you think are primary consumers. _____

   d. Why are they called primary consumers? _____

   e. List the organisms that you think are secondary consumers. _____

   f. Why are they called secondary consumers? _____

   g. Herbivores are organisms that eat plants. List the herbivores in the food chains. _____

   h. How does your list of herbivores compare with your list in question c? _____

   i. Carnivores are organisms that eat other animals. List the carnivores in the food chains. _____

   j. How does your list of carnivores compare with your list in question e? _____

   k. Make two food chains using animals not listed in the above food chains. _____

Name _____ Class _____ Period _____

Part B. Making a Food Web
1. Use the information in Part A on the previous page to complete Figure 1.
2. Draw lines from each organism to other organisms that eat it.
3. Show which organism gets the energy by making an arrow pointing in the direction of energy flow from producers to primary consumers, to secondary consumers. One food chain has already been done for you.
4. Draw your lines with different colored pencils for different food chains. To make it easier to read when finished, do not draw through the circles.

FIGURE 1. A food web in a forest ecosystem

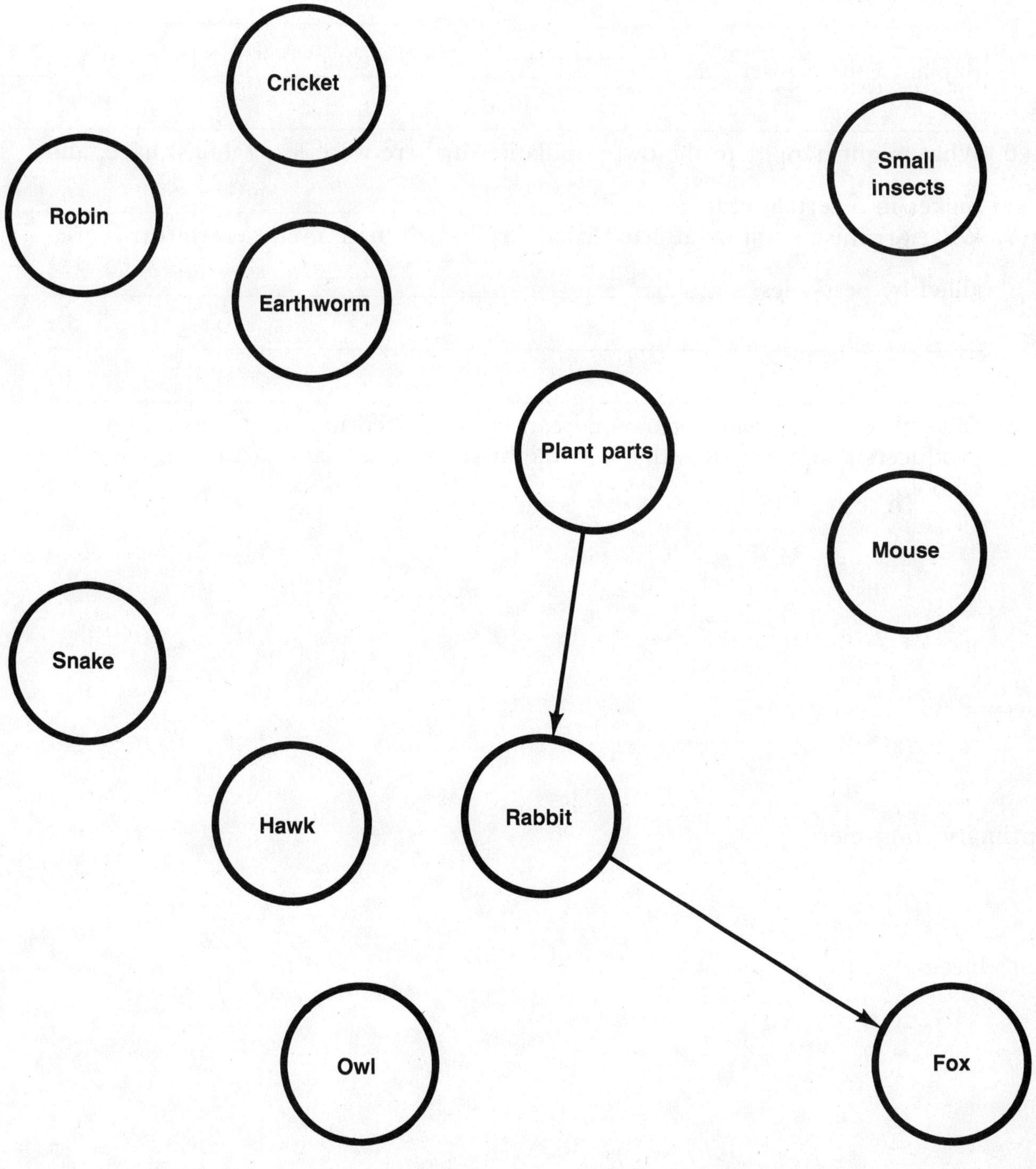

253

QUESTIONS

1. How many of the food chains you made in Figure 1 include the following animals? hawk _____ earthworm _____ fox _____ small insects _____ owl _____ snake _____

2. How many of the food chains include plant parts? _____

3. Give the names of the producers that are in the food web. _____

4. Give the names of the consumers that eat both plants and animals.

   _____

5. What would happen to the food web if all the plants were removed?

   _____

   Explain your answer. _____

   _____

6. What might happen to the owl population if there were less rabbits, mice, and snakes in a certain year? _____

7. What organisms will be affected if crickets, small insects, and earthworms are killed by pesticides? _____

   _____

   _____

8. Draw three food chains below that can be connected in a food web. Show producers and consumers that you might see in your backyard or on your way to school.

secondary consumers

primary consumers

producers

254

Name _____ Class _____ Period _____

30–2 How Do Predator and Prey Populations Change?

A predator (PRED ut ur) is an animal that kills and eats another animal. A fox is an example of a predator. The prey (PRAY) is the animal killed by a predator. A rabbit is an example of an animal that is prey for the fox.

The sizes of predator and prey populations can change with the seasons. Biologists sometimes need to know the sizes of certain predator and prey populations. They can sample the population by trapping and/or counting the animals. The results of the samplings change as the populations change.

INTERPRETATION

OBJECTIVES
In this exercise, you will:
  a. set up a model of predator and prey populations.
  b. observe changes in the results you get from sampling as the populations change.
  c. construct a graph showing your results.

KEYWORDS
Define the following keywords:

population change _____

population sampling _____

predator _____

prey _____

MATERIALS
101 brown beans   small paper bag   graph paper
13 white beans   colored pencils

PROCEDURE
Part A. Sampling a Population
1. Read this report about the animals on the abandoned James Hyde farm.

> The James Hyde farm has not had people living on it since June of 1979. An interstate highway was put through the middle of the farm. Now there are only 100 acres of land left on this farm. In April of 1986, two biologists wanted to find out how the fox and rabbit populations were changing on the farm. They counted rabbits by trapping them and releasing them. They counted foxes by looking for them with field glasses because the foxes would not go near the traps. They trapped and released 23 rabbits. They saw two foxes.

255

2. Put 92 brown beans and 8 white beans into a bag. Assume brown beans are rabbits and white beans are foxes. This number of beans is four times the sample size in the example above. This will represent the numbers in the actual populations of rabbits and foxes.
3. Shake the beans in the bag. Pick a bean without looking as shown in Figure 1. Put a strike mark in Table 1 in the correct column. If you picked a brown bean, put a mark in the rabbit column. If you picked a white bean, put a mark in the fox column.
4. Return the bean to the bag. Repeat the picking, returning the bean each time. Record the result by a mark in the table after each selection. Pick a total of 25 beans (25% of the actual numbers in the population). Total your results in the table.

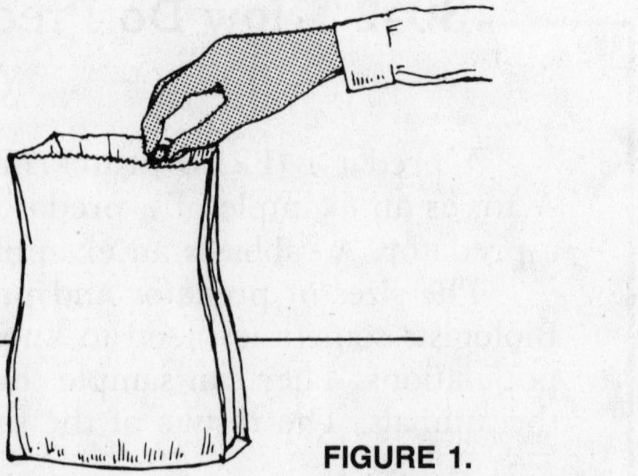

FIGURE 1.

Table 1. Recording Data in a Table

Data	Rabbits (brown beans)		Foxes (white beans)	
	Marks	Total	Marks	Total
April 1986				

Part B. Recording Changes in Populations

1. Examine Table 2 that explains how to change your numbers of beans to show how the rabbit and fox populations changed at later dates.

Table 2. Population Changes

Sampling date	Rabbit population	Fox population
October 1986	Remove 10 brown beans. (Winter was harsh and food was low. Many rabbits died.)	Add 2 white beans. (Foxes also ate pheasants. Fox numbers increased.)
October 1987	Add 15 brown beans. (Food was plentiful. More rabbits moved into the area.)	Add 2 white beans. (Foxes had larger litters than usual.)
April 1988	Remove 8 brown beans. (Many rabbits died from disease.)	Remove 3 white beans. (Food was low. Some foxes left the area.)
October 1988	Add 12 brown beans. (Spring came early. Rabbits could breed earlier.)	Remove 4 white beans. (Rabbits were fewer from disease. Foxes decreased.)
April 1989	No change.	Add 8 white beans. (Food was plentiful. Foxes moved into the area.)
October 1989	Remove 14 brown beans. (Hunters killed pheasants. Foxes ate more rabbits.)	Remove 2 white beans. (Hunters shot some foxes.)

Name _____ Class _____ Period _____

2. Using Table 2 and the sampling method in Part A, sample the populations of rabbits and foxes nine more times to fill in the data for Table 3.
   a. Compare the dates in Tables 2 and 3. For each date in Table 3, sample beans 25 times. Make marks and fill in the totals of brown and white beans.
   b. When you come to a date in Table 2 that indicates a change in population size, follow the directions as to adding or removing beans from the bag. Record this data in the same date listed in Table 3.

Table 3. Population Sampling

Date	Rabbits (brown beans)		Foxes (white beans)	
	Marks	Total	Marks	Total
October 1986				
April 1987				
October 1987				
April 1988				
October 1988				
April 1989				
October 1989				
April 1990				
October 1990				

Part C. Graphing the Data
1. Use a sheet of graph paper and make a graph like Figure 2. The number of animals are listed up the side, and the dates of sampling are along the bottom.
2. Fill in your graph using the data in Table 3 for your sampling of populations. Use different colors to color in the blocks for each animal. The first one is done for you with the biologists' data. (If you have trouble graphing, ask your teacher for help.)

Figure 2. Changes in rabbit and fox populations over 4 years

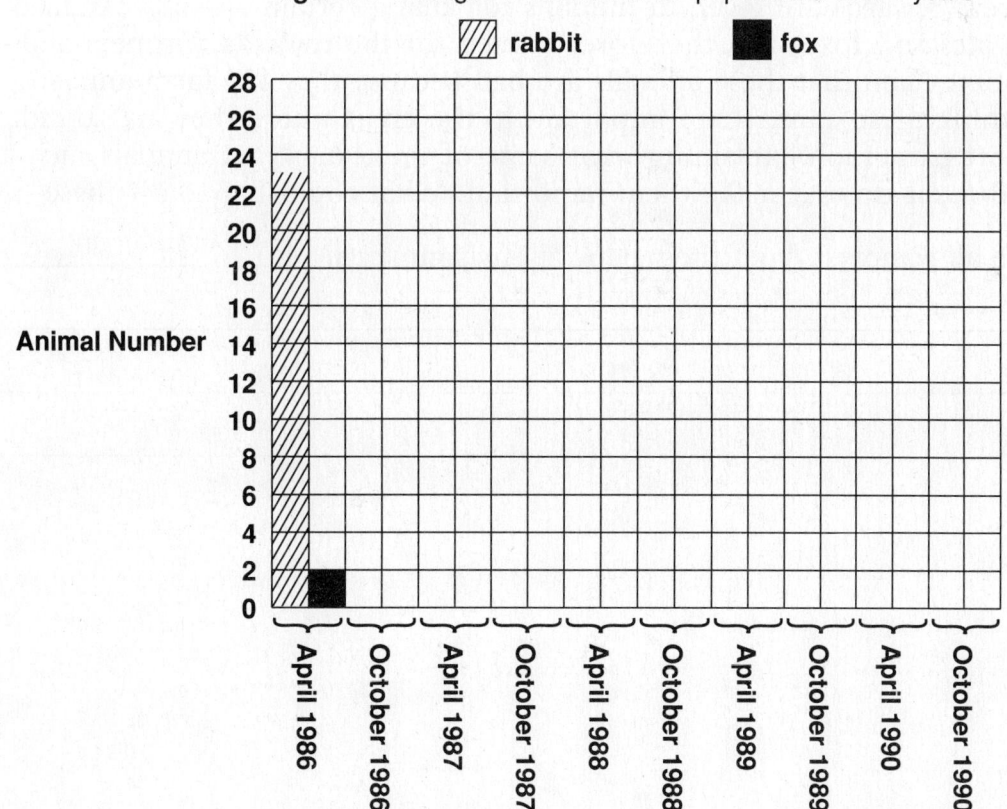

257

QUESTIONS

1. Which animal was the predator and which was the prey? _____

2. How did your sampling in Part A compare with those of the two biologists in April 1986? _____

3. Give three factors that caused a decrease in the rabbit population. _____

4. Give two factors that caused an increase in the rabbit population. _____

5. Give three factors that caused a decrease in the fox population. _____

6. Give three factors that caused an increase in the fox population. _____

7. How would the presence of pheasants affect the fox population? _____

8. What will happen to the rabbits when there is a decrease in the pheasant population? _____

9. In some areas rewards are given to humans for killing certain animals. Animals such as coyotes and foxes are, therefore, hunted for the rewards. Farmers and ranchers often claim that these animals are bad because they kill farm animals. Biologists think these animals are important to the areas where they are found. Write a short paragraph explaining what some of the things that animals such as coyotes and foxes do that make them important. What could happen if these animals are all removed from their natural environments? _____

Name _____ Class _____ Period _____

31–1 How Much Water Will Soil Hold?

Suppose you were to pick up a handful of soil and rub it between your hands. You might say that it feels like very small pieces of rock. You would be correct because soil is made of rock particles of different sizes and different shapes that are mixed together.

When you rubbed the soil between your hands, you may have noticed that it felt wet or damp. Soil particles do not fit tightly together. Rather they are loosely packed like pieces of candy in a bag. Water is in spaces between and within the soil particles. The amount of water that soil can hold is called its water-holding capacity. How can you tell what the water-holding capacity of soil is?

INVESTIGATION

OBJECTIVES
In this exercise, you will:
a. find out how much water a sample of soil will hold.
b. find out how much water a sample of sand will hold.
c. compare the water-holding capacity of soil and sand.

KEYWORDS
Define the following keywords:

balance _____

mass _____

soil _____

water-holding capacity _____

MATERIALS
2 paper cups	dry soil	pencil	large container
2 paper towels	dry sand	balance	graduated cylinder

PROCEDURE
Part A. Comparing Water-Holding Capacity of Different Soils
1. Mark the two cups with the numbers 1 and 2.
2. Using a pencil, punch several holes in the bottom of each paper cup as shown in Figure 1. **CAUTION:** *Be very careful and work slowly as you punch the holes.*
3. Line each cup with a soaked paper towel as shown in Figure 1.

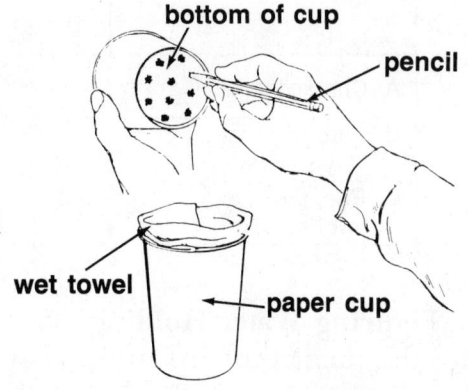

FIGURE 1. Preparing cups

4. Follow the procedure as shown in Figure 2.
   a. Find the mass of cup 1 with its towel using a balance. Record this mass in Table 1.
   b. Fill cup 1 three-fourths full with soil.
   c. Find the mass of cup 1 with the towel and soil. Record this mass in Table 1.
   d. Hold the cup over the large container. Slowly add 500 mL of water to the cup of soil. Let the water drain through the soil into the container.
   e. When no more water drips through the bottom of the cup, find the mass of the cup of wet soil. Record this mass in Table 1.
   f. Repeat steps 4a to 4e using cup 2 and sand this time in place of soil.
   g. Record your masses in Table 1.

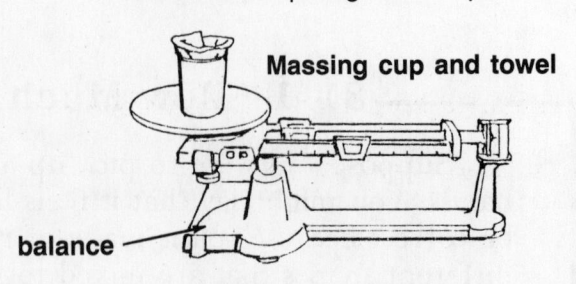

FIGURE 2. Preparing soil samples

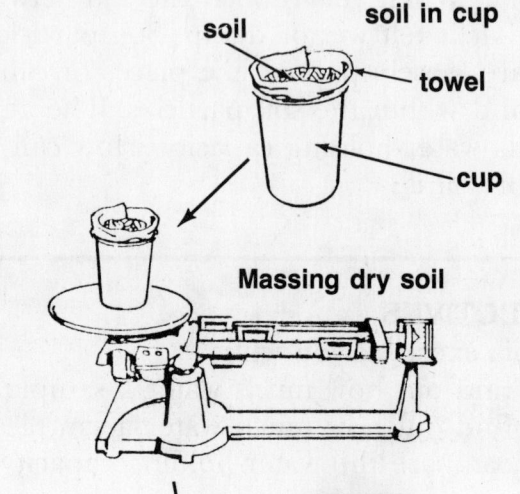

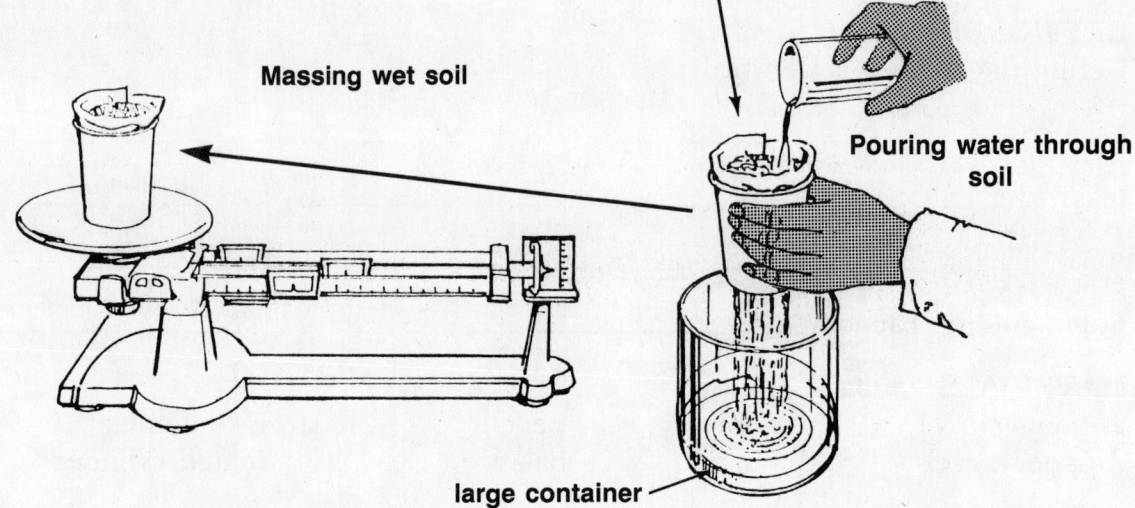

Table 1. Masses of Samples

	Mass of sample	
	Cup 1 (Soil)	Cup 2 (Sand)
A. Cup and towel		
B. Cup, towel, and dry sample		
C. Cup, towel, and wet sample		

Part B. Figuring Water-Holding Capacity of Soils
1. Using the data recorded in Table 1, complete the boxes in Table 2.
2. Follow the steps set out in Table 2 to calculate the water-holding capacity of each soil sample.

Name _____ Class _____ Period _____

Table 2. Procedure for Calculating Water-Holding Capacity

Cup 1 (soil)	Using Table 1	Cup 2 (sand)
☐	Mass *B* of cup, towel, and dry sample	☐
− ☐	Subtract mass *A* of cup and towel	− ☐
☐	Gives mass of dry sample *D*	☐
☐	Mass *C* of wet sample	☐
− ☐	Subtract mass *D* of dry sample from above	− ☐
☐	Gives gain in mass of soil *E*	☐
= ☐	Multiply *E* by 100 for percentage gain *F* in mass of soil	= ☐
☐	Enter *F* again	☐
☐	Enter here the mass *C* of wet sample	☐
= ☐	For percentage of water in the sample divide *F* by *C*	= ☐

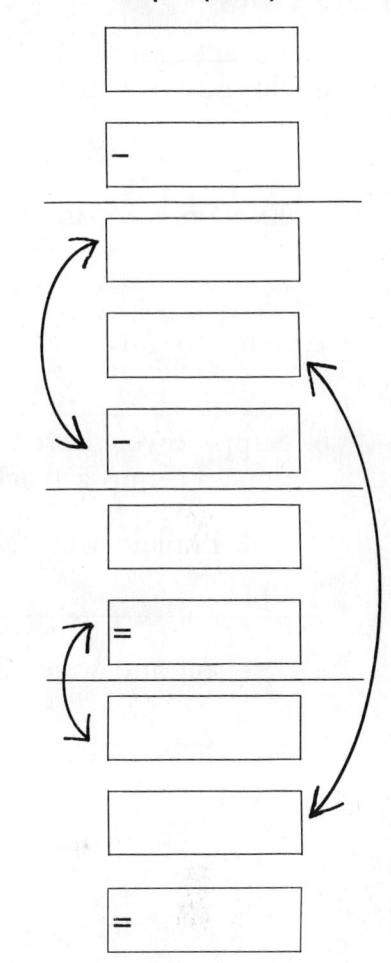

QUESTIONS

1. a. What percent of the soil sample is water? _____

   b. What percent of the sand sample is water? _____

   c. Which holds more water, soil or sand? _____

2. What determines whether soil can hold more or less water? _____
   _____

3. What does it mean when a soil sample has a mass of 40 g and a 50% water-holding capacity? _____
   _____

   Explain your answer. _____
   _____

261

5. Suppose soil particles become tightly packed together. What will happen to the water-holding capacity of the soil? _____

Explain your answer. _____

6. Suppose soil particles become more loosely packed. What will happen to the water-holding capacity of the soil? _____
Explain your answer. _____

7. Why do gardeners use a hoe to break up soil clumps in a garden? _____

8. Suppose you have a garden that stays too wet most of the time. Your friend tells you to dump a truckload of sand onto the garden and to mix it with the garden soil. Explain why this information could be good advice. _____

9. Are soil and water living or nonliving parts of an ecosystem? _____

Name _____ Class _____ Period _____

31-2 How Can a Nonliving Part of an Ecosystem Harm Living Things?

An ecosystem (EE koh sihs tum) is a place with its living and nonliving things. The living parts are the biotic (bi AHT ihk) parts. The nonliving parts are the abiotic (ay bi AHT ihk) parts.

The biotic parts of an ecosystem need the abiotic parts. Air and water are examples of abiotic parts that are needed by living things. The abiotic parts must be present in the right amounts.

One abiotic part that living things need is mineral salts. Mineral salts are chemicals that contain several different elements. Some mineral salts, for example, are made of nitrogen, calcium, sulfur, and phosphorus. Living things get their mineral salts from soil or water. Often the soil or water contains too much or too little mineral salts for living things.

INVESTIGATION

OBJECTIVES
In this exercise, you will:
a. observe what happens when different salt solutions are added to paramecia.
b. determine which solution is harmful to paramecia.

KEYWORDS
Define the following keywords:

abiotic _____

biotic _____

calcium _____

mineral salt _____

MATERIALS
Paramecium culture	cotton ball	paper towel	0.5%, 2%, and 5%
3 glass slides	3 droppers	3 coverslips	mineral salt solutions
	light microscope		Label these 1, 2, and 3.

PROCEDURE
1. Pull a small piece of cotton from the cotton ball and put it on the microscope slide. Add a drop of the *Paramecium* culture to the cotton on the slide. The strands of cotton will help keep the paramecia in the field of view.
2. Add a coverslip to the drop. Examine the slide under low power and then high power of the microscope.
3. Compare the paramecium with that shown in Figure 1.

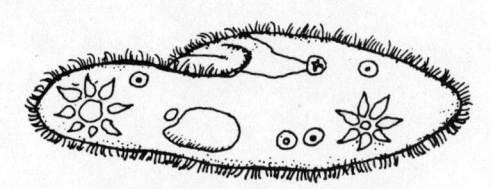

FIGURE 1. A paramecium

4. Make a drawing of your paramecium in Table 1. Draw a series of arrowed lines in the table to show how the paramecium moves.
5. Add a drop of salt solution 1 to the slide at the edge of the coverslip as shown in Figure 2. Be careful not to get the solution on top of the coverslip.
6. Place the edge of a paper towel on the other side of the coverslip so that it touches the solution. This draws the solution across the slide.
7. Use the microscope to look for the paramecia again. Record your observations of appearance and movement in Table 1.
8. Repeat steps 1 to 7 using solutions 2 and 3. Be sure to use a clean slide, coverslip, and dropper each time.

FIGURE 2. Preparing a paramecium slide

Table 1. Effects of Salt Solutions on Paramecia

Observations	No salt	Solution 1 (0.5%)	Solution 2 (2%)	Solution 3 (5%)
Appearance				
Movement				

QUESTIONS

1. Which salt solutions caused a shape change in the paramecium? _____
2. Which solutions caused a change in the movement of the paramecium?

_____

3. Which solution did not change the shape and movement of paramecium? _____
4. Which solution is the strongest salt solution? _____
5. Which solution do you think was most harmful to the paramecium? _____

Explain your answer. _____
6. Sometimes chemicals are added to streams and ponds to kill unwanted animals or plants. Not all animals die when chemicals are added to a stream. How could you explain this? _____

_____

Name _____ Class _____ Period _____

31–3 How Can a Nonliving Part of an Ecosystem Help Living Things?

In the last exercise, you found out that an abiotic part can harm living things. Abiotic parts of a ecosystem help a living thing grow and reproduce.

Duckweed is a small plant that lives in aquatic ecosystems such as ponds, lakes, or streams. The oval-shaped parts that float on the water are leaflike, but are in fact tiny flattened stems. Below the water hangs a single root. Minerals and water are taken in by the root and used for plant growth. One sign of growth is an increase in the number of leaflike parts, which will be called leaves here, and an increase in new plants. One plant can produce two or three new plants every 24 hours. The right amount of minerals can cause this good growth. Fertilizers contain large amounts of minerals that can be used by plants for growth.

INVESTIGATION

OBJECTIVES
In this exercise, you will:
  a. find out how abiotic parts help a living thing.
  b. determine which solution has the proper amount of minerals to cause growth in duckweed.

KEYWORDS
Define the following keywords:

duckweed _____

ecosystem _____

fertilizer _____

MATERIALS
9 duckweed plants
100 mL cow manure solution
100 mL 0.1% liquid plant food solution
small culture dish
100 mL distilled water
3 small jars
hand lens

PROCEDURE
1. Half fill a culture dish with water and put a duckweed plant in the dish. Look at the duckweed plant with a hand lens.
2. Look at the leaflike part and the root. Compare your duckweed plant with those shown in Figure 1.

FIGURE 1. Duckweed plants

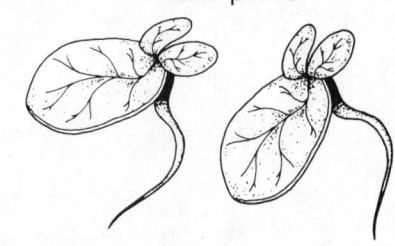

3. Label three small jars with the numbers 1, 2, and 3. Nearly fill each jar with the solutions as follows:

>jar 1—cow manure solution
>jar 2—0.1% liquid plant food solution
>jar 3—distilled water

4. Put three duckweed plants into each of the jars. Place the jars in a lighted area. Count the total number of leaves that you see in each jar. Record this number in Table 1 under Day 1.
5. Record the total number of leaves each day for the next four days in each jar.
6. At the same time look at the color of the leaves. If they are dark green, record this in Table 1 with a D. If they are pale green, record this with a P.

Table 1. Growth in Duckweed Plants

Jar number	Number and color (D or P) of leaves each day				
	Day 1	Day 2	Day 3	Day 4	Day 5
1					
2					
3					

QUESTIONS

1. Which jar had the most duckweed leaves after Day 5? _____
2. Which jar had the least duckweed leaves after Day 5? _____
3. Dark green leaves tell us that these plants are growing best. In which jar did the best growth take place? _____
4. Pale green leaves are a sign of poor growth. In which jar did the poorest growth take place? _____
5. Are your answers to questions 1 and 3 the same? _____

   Explain your answer. _____
6. Are your answers to questions 2 and 4 the same? _____

   Explain your answer. _____
7. What was the biotic part that was a limiting factor in this exercise?

   _____
8. Explain how an abiotic part of an ecosystem can help a living thing.

   _____

Name _____ Class _____ Period _____

32–1 How Do Chemical Pollutants Affect Living Things?

Chemical pollutants get into water and soil from farms and factories. Sometimes factories dump their chemical wastes into holding ponds. If it rains before these chemicals can be hauled away, the ponds may overflow. The chemicals are then carried to fields and streams by the rainwater.

Farmers spray chemicals on their crops to kill bacteria, fungi, insects, and weeds. They put fertilizer on the newly seeded fields to help plants grow and to produce a good crop. Chemical sprays and fertilizer are washed into streams during rainy weather. These chemicals are called pollutants because they can kill or harm living things. Unclean or discolored water is called polluted. Living things that live in the water are limited to where they can live because of the pH of the water. Some can live in water that is slightly acidic or slightly basic. Very few can live in water that is strongly acidic or strongly basic. Chemicals in the rain or soil enter freshwater ponds and streams. Sometimes these chemicals change the pH of the water to that of an acid. The rain falling on most of the United States has a pH in the acid range.

INVESTIGATION

OBJECTIVES
In this exercise, you will:
 a. compare the effects of two chemical pollutants on *Euglena*.
 b. determine how different amounts of chemical pollutants affect *Euglena*.
 c. determine if an acid or base added to the water can affect *Daphnia*.

KEYWORDS
Define the following keywords:

chemical wastes _____

pH _____

pollutant _____

MATERIALS
Euglena culture
10 mL fertilizer solution
10 mL chemical solution
10 test tubes
test-tube rack
hand lens

Daphnia culture
3 droppers
4 small jars
aged tap water
marking pencil
plastic spoon

acid solution
base solution
pH paper with pH color chart
applicator stick
forceps

PROCEDURE

Part A. Observing the Effect of Pollution on *Euglena*

1. Use the marking pencil to label ten test tubes 1 to 10.
2. Two-thirds fill each tube with *Euglena* culture as shown in Figure 1.
3. Notice that the color of the culture in each tube is green. This color is caused by the green color of the euglenas that are present throughout the water in each tube.

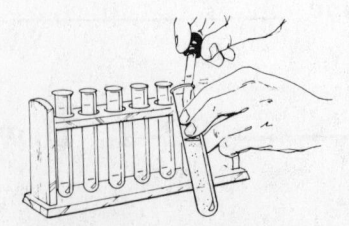

FIGURE 1. Adding *Euglena* culture to a test tube

4. Look at Table 1. Add the amounts of fertilizer and chemical solutions to the numbered test tubes as listed in Table 1.
5. Place the test tubes in the light and leave them undisturbed.
6. After 2 days observe the test tubes. Compare the colors of tubes 2 through 5 with tube 1. Describe the colors in Table 1. Note if they are dark green, pale green, or colorless. If the euglenas have reproduced, the color of the solution will appear a darker green than in the culture solution. If the euglenas have died, they will have settled on the bottom giving the solution a colorless appearance.
7. In the same way compare the colors of tubes 7 through 10 with tube 6.

Table 1. Differences in Color of *Euglena* Cultures

Test tube	Amount of fertilizer	Color at end of two days
1	0 drops	
2	5 drops	
3	10 drops	
4	15 drops	
5	20 drops	

Test tube	Amount of chemical	Color at end of two days
6	0 drops	
7	5 drops	
8	10 drops	
9	15 drops	
10	20 drops	

Part B. Studying the Effect of pH on *Daphnia*

1. Label four small jars 1 to 4. Nearly fill each jar with aged tap water.
2. Use pH paper to find the pH of the water. To do this place a drop of water on a small section of the pH paper as shown in Figure 2.
3. Look at the pH color chart and compare and match the color of your pH paper with one on the chart. This will give the pH of the water.
4. Adjust the pH of each of the four jars of water by adding drops of acid or base solutions until they match the following:
   jar 1—pH 10 (base)
   jar 2—pH 8 (base)
   jar 3—pH 6 (acid)
   jar 4—pH 4 (acid)

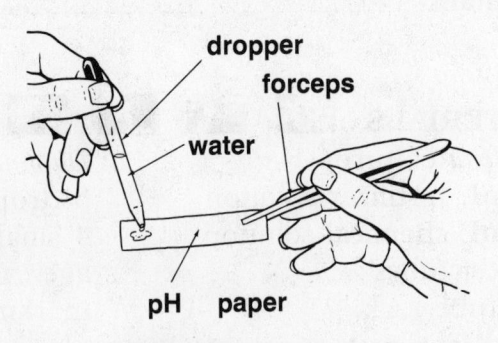

FIGURE 2. Testing pH

Name _____ Class _____ Period _____

5. To make the water in jars 1 and 2 more basic, add drops of base solution as shown in Figure 3. To make the water in jars 3 and 4 more acidic, add drops of acid solution.

6. Add a few drops of acid or base at a time and stir with a stick. Test the pH of the water in the jars after each stir as shown in Figure 2. Continue this procedure until the correct pH for each jar is reached.

7. Add 6 *Daphnia* to each jar as shown in Figure 4.
   a. First take a spoonful of culture.
   b. With an inverted dropper pick up the *Daphnia* from the spoon.
   c. Carefully drop them into the jar.
   d. Record this starting number of *Daphnia* in Table 2.

8. Add 3 droppersful of *Euglena* culture to each jar. This will provide the *Daphnia* with food.

9. Place the jars on a shelf in the lab.

10. Add 3 droppersful of *Euglena* culture to each jar for the next 2 days.

11. After three days observe each jar and record numbers of living *Daphnia* in Table 2.

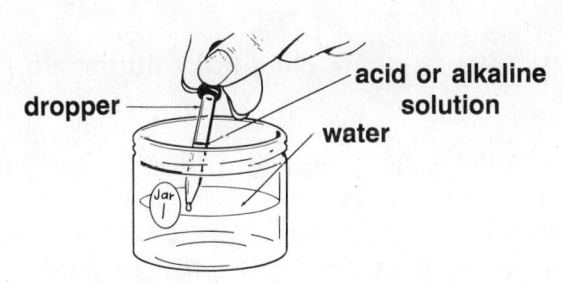

FIGURE 3. Changing pH

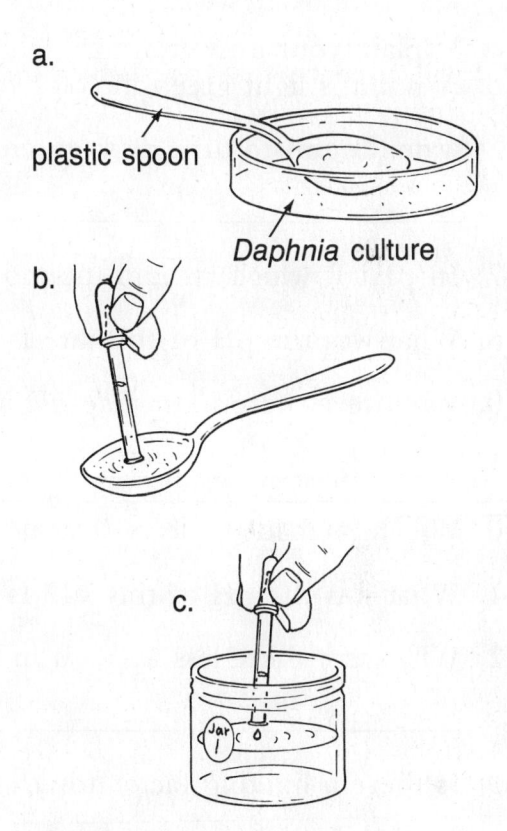

FIGURE 4. Transferring *Daphnia*

Table 2. Changes in Numbers of *Daphnia*

Jar	Numbers of *Daphnia* at start	pH of water	Numbers of living *Daphnia* after 3 days
1			
2			
3			
4			

QUESTIONS

1. In part A what did the fertilizer solution do to the euglenas? _____
_____

2. What did the chemical solution do to the euglenas? _____
_____

3. Why were fertilizer and chemical solutions not added to test tubes 1 or 6?
_____

4. Why did some of the test tubes of euglenas in part A become a darker green?
_____

5. Which tubes had pollutants in them? _____
   Explain your answer. _____

6. A pond is light green in color for several weeks. Several days after a rain the water begins to turn dark green. How can you explain this? _____
_____

7. In part B which jar had the most *Daphnia* at the end of three days? _____

8. What was the pH of this jar? Is this pH acidic or basic? _____

9. Why were there more *Daphnia* in this jar than in other jars? _____
_____

10. Which jar had the least *Daphnia* at the end of three days? _____

11. What was the pH of this jar? Is this pH acidic or basic? _____

12. Why were there less *Daphnia* in this jar than in other jars? _____
_____

13. Is there a limiting factor for *Daphnia*? _____
    Explain your answer. _____

14. Farmers put fertilizer on their soil, which can make the soil acidic. Before they plant crops, farmers usually add lime to the soil. Lime is a base. Why do farmers add lime to the soil? _____
_____

15. A chemical factory cleaned all of its large storage tanks on a weekend. Explain why many living things in a nearby stream died within a few days.
_____

Name _____ Class _____ Period _____

32–2 How Does Thermal Pollution Affect Living Things?

Sometimes the environment becomes too warm for living things. Thermal pollution is heat that is discharged into the soil, water, or air of a biological community. This heat can harm or kill living things.

Some industries heat water during the process of cooling their electric generators. While still warm, the water is sometimes dumped into small streams or ponds. Many of the organisms that make up the food chains and food webs in these water biomes may be killed.

INVESTIGATION

OBJECTIVES
In this exercise, you will:
a. find out if heated water can kill or stop the growth of living things.
b. determine if yeast cells can live in heated water for a short time.

KEYWORDS
Define the following keywords:

community _____

environment _____

food chain _____

food web _____

thermal pollution _____

MATERIALS
4 test tubes   toothpick   5 droppers   yeast suspension
test-tube rack   glass beaker   glass slide   blue stain
test-tube holder   hot plate   coverslip   microscope
clock with second hand marking pen

PROCEDURE
Part A. The Effect of Heat on Yeast
1. Label four test tubes 1 to 4. Place the test tubes in the rack.
2. Fill each test tube 1/3 full with tap water.
3. Add five drops of yeast suspension to each test tube as shown in Figure 1.
4. Shake the tubes back and forth to mix the yeast cells in the water.

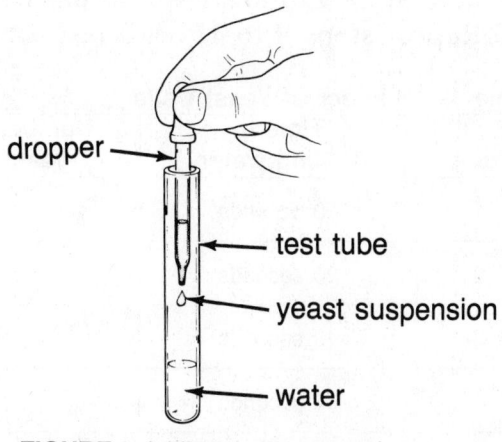

FIGURE 1. Adding yeast suspension

271

5. Heat a beaker of water to boiling over a hot plate.
6. Attach a test-tube holder to tube 2 and hold it in the boiling water for 20 seconds as shown in Figure 2. Then return it to the rack. **CAUTION:** *Always use the test-tube holder when placing the test tubes in or out of the boiling water.*
7. Repeat this process for test tube 3 but keep the test tube in the water for 40 seconds.
8. Repeat this process for test-tube 4 for 60 seconds.
9. Stir up the yeast cells in test tube 1 by filling a dropper with the yeast solution and squirting it back into the tube three times.
10. Place one drop of yeast solution from test tube 1 on a clean slide.
11. Using a clean dropper, add a drop of blue stain to the drop of yeast.
12. Use a toothpick to mix the stain with the drop of yeast solution.
13. Add a coverslip. Locate the yeast cells on low power and then turn to high power. Yeast cells will appear as small dots on low power. Look at Figure 3 to see their appearance at high power.

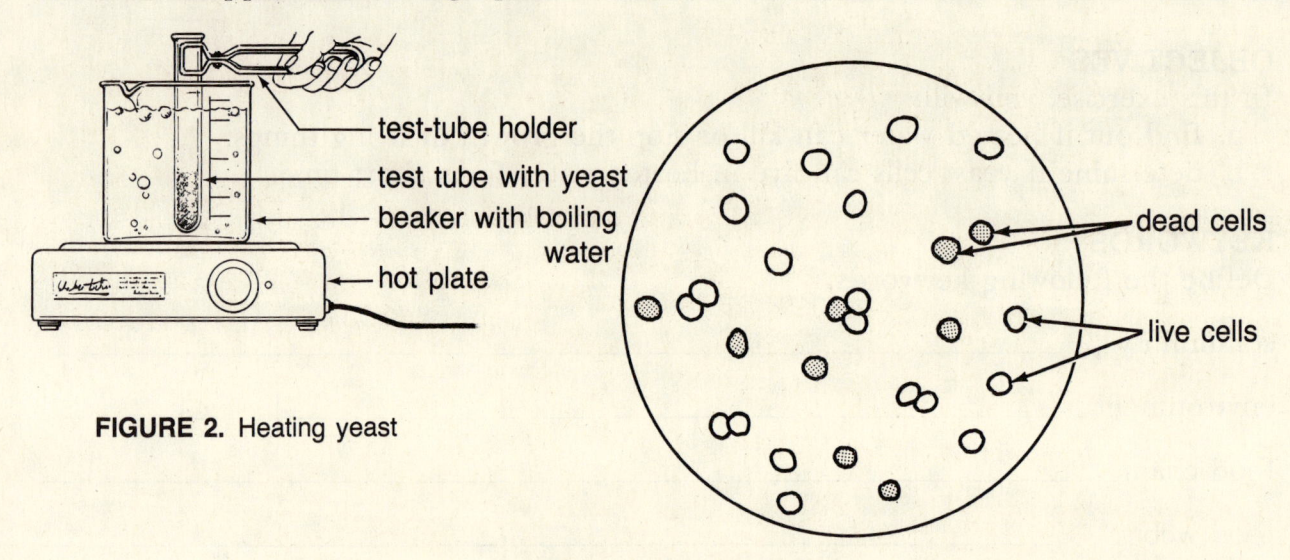

FIGURE 2. Heating yeast

FIGURE 3. Yeast cells

14. Look for live yeast cells. These will appear very light blue in color. Look for dead yeast cells. These will have the same dark blue color as the stain on the slide.
15. For each yeast cell in one field of view, make a mark in Table 1 to show if it is alive or dead. Continue counting until 50 cells have been recorded. If there are less than 50 cells in one field of view, move to another area of the slide and continue counting until 50 has been reached.
16. Repeat steps 9 to 15 with test tubes 2, 3, and 4.

Table 1. Number of Yeast Cells

Test tube	Time in boiling water	Number of live yeast cells	Number of dead yeast cells	Total number of cells counted
1	0 seconds			
2	20 seconds			
3	40 seconds			
4	60 seconds			

Name _____ Class _____ Period _____

Part B. Plotting the Data
Using the data recorded in Table 1, plot two bar graphs in Figure 4 to show the number of live and dead cells in each tube.

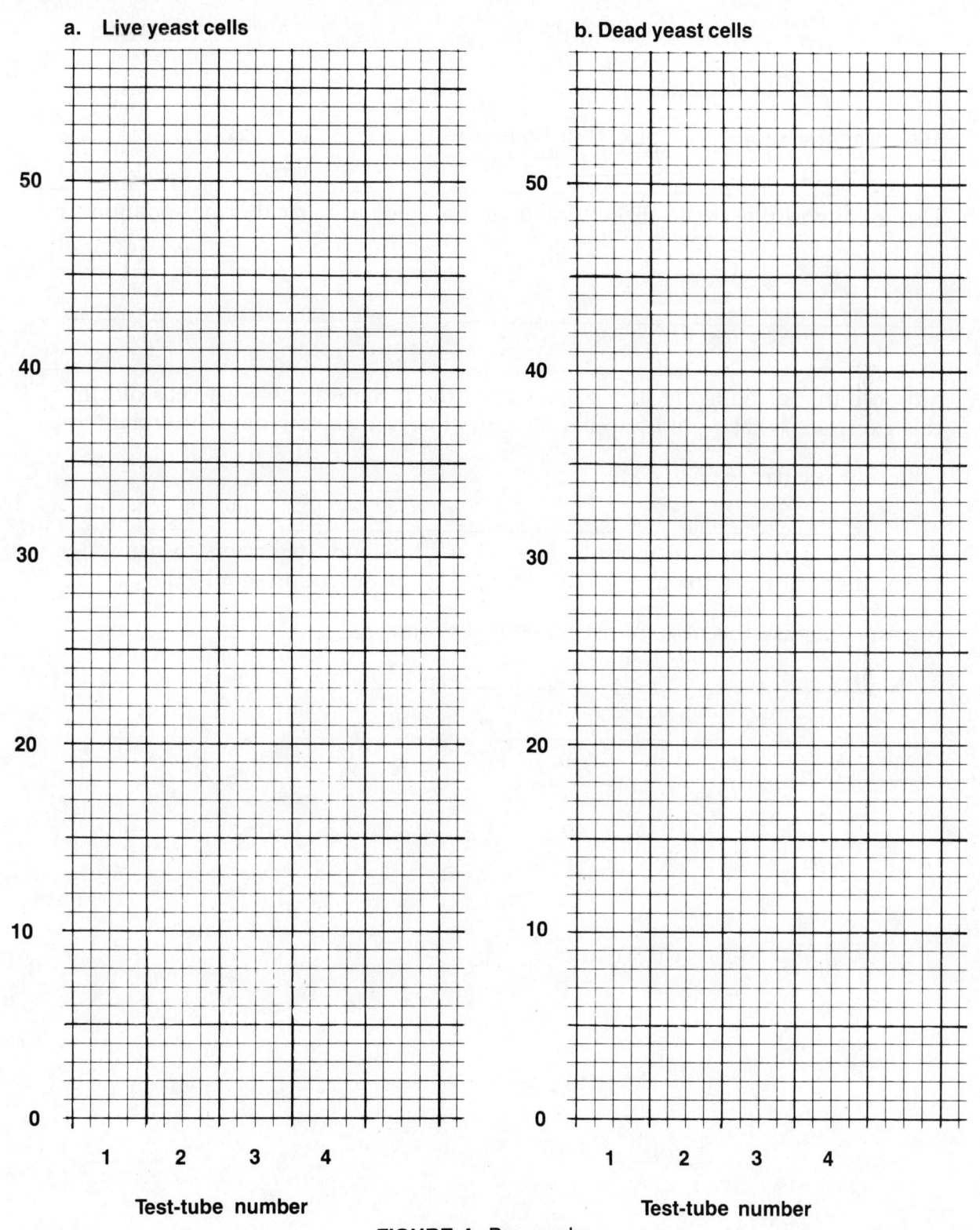

a. **Live yeast cells**

b. **Dead yeast cells**

Test-tube number

Test-tube number

FIGURE 4. Bar graphs

QUESTIONS

1. Which tube contained the most live cells? _____

2. Why were so many cells in this tube alive? _____
   _____

3. Which tube contained the most dead cells? _____

4. Why were so many dead cells in this tube? _____
   _____

5. Why was test tube 1 not placed in boiling water? _____
   _____

6. Using your results, write three sentences that explain what thermal pollution is.
   _____
   _____
   _____

7. Suppose that algae living in a stream react the same way as yeast cells did in this exercise. What would happen to food chains in the stream if thermal pollution occurred? _____
   _____

8. A new industry wants to move to your town. This industry wants to use water from the local river for its production line. What questions should the townspeople ask the new industry about the water? _____
   _____
   _____

Glossary
Pronunciation Key

You may need to review this key with students to aid them in their pronunciations.

a ... b**a**ck (bak)
ay ... d**ay** (day)
ah ... f**a**ther (fahth ur)
ow ... fl**ow**er (flow ur)
ar ... c**ar** (car)
e ... l**e**ss (les)
ee ... l**ea**f (leef)
ih ... tr**i**p (trihp)
i(i+con+e) ... **i**dea, l**i**fe (**i** dee uh, l**i**fe)

oh ... g**o** (goh)
aw ... s**o**ft (sawft)
or ... **or**bit (or but)
oy ... c**oi**n (coyn)
oo ... f**oo**t (foot)
ew ... f**oo**d (fewd)
yoo ... p**u**re (pyoor)
yew ... f**ew** (fyew)
uh ... comm**a** (cahm uh)
u(+con) flow**er** (flow ur)

sh ... **sh**elf (shelf)
ch ... na**t**ure (nay chur)
g ... **g**ift (gihft)
j ... **g**em (jem)
ing ... si**ng** (sing)
zh ... vi**s**ion (vihzh un)
k ... **c**ake (kayk)
s ... **s**eed, **c**ent (seed, sent)
z ... **z**one, rai**s**e (zohn, rayz)

A

abiotic (ay bi AHT ihk): nonliving parts of an ecosystem

acid: chemical such as vinegar that has a sour taste; it neutralizes antacids

adaptation (ad ap TAY shun): variation that helps a living thing survive

afterbirth: placenta and cord that come through the vagina after the birth of a baby

ankle: joint at bottom of leg and top of foot

antacid: drug used against acid; it neutralizes acid

antennae: sense appendages on the head

aorta (ay ORT uh): body's main artery

appendages (uh PEN dihj uz): arms or legs

artery: carries blood away from the heart

asexual (ay SEK shul) **reproduction:** form of reproduction that uses one parent, no sex cells, and cell division by mitosis

atrium (AY tree um): small, upper chamber of heart

B

bacteria: small one-celled organisms that lack a nucleus and belong to the Kingdom Monera

balance: instrument used to measure mass

bark: rough, outer covering of a woody stem

base: bottom of microscope; a chemical with a slippery feel and bitter taste, such as soap

behavior: way an animal acts

biotic (bi AHT ihk): living parts of an ecosystem

Biuret (bi yoo RET) **solution:** solution that is used to test for proteins

blood: liquid tissue in an organism that carries oxygen and carbon dioxide

blood type: trait of blood; one of four kinds, A, B, AB, or O

blue-green bacteria: microscopic organisms that can make their own food by photosynthesis, belong to the Kingdom Monera

body: main part of a sponge

breathing: moving air into and out of lungs

breeding: mating of two living things

breeding season: time of year certain animals can reproduce

brood (BREWD) **pouch:** body part of a water flea that holds eggs

budding: asexual reproduction in which a part of the parent body separates and forms a new organism

buffer (BUHF ur): solution that resists changes in pH when acid or base is added

C

caecum (SEE kum): special digestive tube that helps in the breakdown of plant foods

caesarean: birth of a baby in which the uterus must be cut open

caffeine: a drug found in foods such as coffee, cola, and chocolate

calcium: mineral salt needed by living things in an ecosystem

cambium (KAM bee um): layer of cells that forms new xylem and phloem

carbohydrates (kar boh HI drayts): nutrients that can provide energy and are contained in foods, such as bread, spaghetti, and cereal

carbon dioxide: gas given off by living things during cellular respiration

carnivore: animal that eats meat

cartilage: tough, flexible tissue that supports and shapes the body

cell division: process by which one cell becomes two; mitosis

cell membrane (MEM brayn): outer covering of a cell that controls what enters and leaves the cell

cells: tiny units of living material of which living things are made

cellular respiration: the process by which all living things release energy from food

cellulose: compound in wood

cell wall: thick outer covering of a plant cell

centimeter (SENT uh meet ur): unit of length (One hundred centimeters make one meter.)

cerebellum: brain part that makes our movements smooth and graceful

cerebrum: brain part that controls thought, reason, and senses

chamber: space within the heart that gathers and pumps blood

chemical change: change when small food pieces are turned into a form that cells can use

chemical wastes: chemicals no longer needed by humans

chlorophyll: green pigment that helps the plant make food

chloroplast (KLOR uh plast): cell part that contains chlorophyll

chordate (KOR dayt): animal that, at sometime in its life, has a stiff cord along its back

chromosome (KROH muh sohm): cell with information that determines traits of a living thing

cilia: short, hairlike parts that cover the cell membrane

citrus fruit: fruits such as oranges, grapefruit, and lemons that contain ascorbic acid

class: largest group of a phylum

classify: to put into groups

colony: group of similar cells that are attached to each other

color blindness: condition in which some colors are not seen as they should be

community: living things interacting in an area

compound light microscope: microscope that has two or more lenses

cone-bearing plant: conifer; evergreen plant

conifer (KAHN uh fur): cone-bearing plant

consumers (kun SEW murz): living things that eat other living things

contract: to shorten or squeeze together as the heart pumps

contracted: muscle that is working (shortening)

contractions: muscles of the uterus shortening during birth

control: part of an experiment in which nothing is changed; your standard for comparison

cord: stringlike part that attaches a developing baby to the placenta within the mother

cork: outer ring of cells on a woody stem

coronary (KOR uh ner ee) **artery:** blood vessel that supplies the heart muscle with blood

cortex (KOR teks): cells that store food

cross: mate; joining of the genes of two parents in offspring

coverslip: small, thin glass put over an object when mounted on a slide

cytoplasm (SITE uh plaz um): clear, jellylike substance that makes up most of the cell

D

data: observations you record

dehydrate: to remove water from a substance

development: change in form or appearance as an animal grows

diagnosis: act of distinguishing or discovering what problem a person might have

diaphragm (DI uh fram): sheetlike muscle at the bottom of the chest that helps in breathing

diffusion (dihf YEW zhun): movement of a substance from where there is a large amount to where there is a small amount

digestive system: group of organs that take in food and change it into a form the body can use

disc: raised, central portion of a starfish from which arms radiate

dissolve: to be in solution

distilled water: water that has had the impurities and bacteria removed

dividing cell: cell that is forming two cells from one

DNA: molecule that makes up genes and controls traits of organisms

dominant (DAHM uh nunt): that which controls what you do, or having more effect

dominant brain side: side which controls handedness; side having more effect on how body acts

donor: person who gives or donates their blood

dorsal: top side

dosage: how much and how often to take a drug

drug: a chemical substance that has a physical effect on living things

duckweed: small plant that lives in an aquatic ecosystem

E

ecosystem (EE koh sihs tum): community interacting with the environment

egg: female reproductive cell

embryo (EM bree oh): new undeveloped living thing; young animal that begins developing after fertilization

environment: surroundings of living things

epidermis (ep uh DUR MUS): protective layer of cells on the outside of a leaf or root

ethyl alcohol: a drug produced by plants when no oxygen is present

Equus: present-day horse

Euglena: a kind of protozoan that makes its own food

exhalation: breathing out

expected results: traits or results that can be predicted

exterior: outside

external: outside the body

external fertilization: joining of an egg and a sperm outside the body

F

false feet: parts that extend from the body of an amoeba that help the amoeba move and get food

fats: nutrients used as a source of energy by your body

femur (FEE mur): thigh bone

ferns: certain vascular plants that grow in damp areas and reproduce by spores

fertilization: joining of egg and sperm cell

fertilizer: nutrients needed by plants for growth

fetus (FEET us): embryo that has all of its body systems

field of view: lighted area you see through a microscope

filaments: long, threadlike chains of cells

flagellum: long, whiplike part

flower: reproductive part of a flowering plant

flowering plants: certain vascular plants that have flowers; reproduce from seeds

food chain: pathway of food through an ecosystem

food energy: energy an organism gets from the food it eats

food web: interconnecting food chains in an ecosystem

fossil: remains of a once-living thing

free-living: animal that does not live as a parasite

fungi (FUN ji): plantlike protists that cannot make their own food

G

gene (JEEN): chromosome part that determines the trait of an organism

genetic (juh NET ihk) **disorder:** health problem that occurs in humans and is caused by genes

genetics: study of how certain traits are passed from parents to children

gills: organs of respiration in fish and amphibians

glass slide: piece of glass on which objects are mounted for examination with the microscope

gonad: reproductive organ

H

heart value: doorlike part that opens in only one direction, keeping blood flowing in only one direction

hemoglobin: protein in red blood cells that carries oxygen around the body

hemophilia (hee muh FIHL ee uh): genetic disease in which the person's blood doesn't clot

herbivore: animal that eats plants

heterozygous (het er o ZY gus): trait controlled by a dominant and a recessive gene

hydra: freshwater coelenterate that can asexually reproduce by budding

Hyracotherium: fossil horse

I

illusion (ihl EW zhuhn): mistaken idea that you get

inhalation: breathing in

innate (ihn AYT) **behavior:** way of responding that does not require learning

interior: inside

internal: inside the body

internal fertilization: joining of an egg and a sperm inside the body

intestine: digestive organ in which much absorption of food occurs

involuntary: you have no control over the part

involuntary actions: actions you do not have to think about

isopod: animal belonging to the phylum Arthropoda, the jointed-leg phylum, class Crustacea

J

jelly layer: clear outer covering surrounding many blue-green bacteria

jointed-leg animal: invertebrate with jointed appendages

joints: place where bones come together

K

key: set of directions to identify something

kingdom: largest group of living things

L

labor: contractions of the uterus during birth

learned behaviors: behaviors that must be taught

left cerebrum side: left side of cerebrum that controls right body side's function

length: measure of how long something is

lens: curved piece of glass that makes objects look larger

leukemia (lew KEE mee uh): blood cancer

life cycle: changes that occur as an organism grows from an egg to a young animal to an adult

ligaments: tough fibers that hold bones together

light: growth requirement needed by plants

living: having the characteristics of life

M

magnifying power: how many times larger a microscope makes something look

mammal (MAM ul): any animal that has hair and can produce milk

mantle: body covering of a squid

mass: amount of matter in an object

maze (MAYZ): network of paths from start to finish; a pathway puzzle having many "blind ends"

medulla: brain part that controls heartbeat, breathing, and blood pressure

meiosis: cell division that forms egg or sperm cells; part of sexual reproduction

menstrual cycle: monthly changes that take place in female reproductive system

meter: basic unit of length

microscopic: too small to be seen with the naked eye

milliliter (MIHL ul leet ur): 0.001 of a liter; (mL)

mineral salt: chemical made of several different elements

mitosis (mi TOH sus): cell division that forms body cells; part of asexual reproduction

mold: fungus that can grow on food

mouth: in a sea anemone, the opening that leads to the pharynx; in a starfish, the opening in the center of the disc that is surrounded by spines

mRNA: molecule that carries the DNA code from the nucleus to the ribosome

muscular system: made up of all the muscles found in an animal's body

muscle contraction: shortening or working of a muscle

mutation (myew TAY shun): when the DNA code makes up a gene that has an error in it

N

natural selection: process in which something in a living thing's surroundings determines if it will or will not survive

neutral: exactly between an acid and an antacid

neutralize (NOO truh liz): to change a liquid from acid or base to neutral

nicotine: a drug found in tobacco products

nonliving: not having the characteristics of life

nonvascular (nahn VAS kyuh lur) **plants:** plants that do not have tubelike structures in roots, stems, and leaves

nucleolus: cell part that helps make ribosomes

nucleus (NEW klee us): control center of the cell

nutrient (NEW tree unt): chemical or chemicals in food

O

observed results: traits or results that are actually seen

offspring: young produced from reproduction

optical (AHP tih kul) **illusion:** mistaken idea because of what you see

organ: group of tissues that work together to do a job

organism (OR guh nihz um): living thing

osculum: large opening at the top of a sponge

osmosis (ahs MOH sus): movement of water across the cell membrane

ovary (OHV ree): female sex organ that produces eggs

oviduct: tubelike organ that connects the ovaries to uterus

ovule: tiny, round parts within ovary that contain egg cells

oxygen: (O_2); gas in the air needed by living things for cell respiration; given off in photosynthesis

P

palisade (pal uh SAYD) **layer:** layer of long, closely packed cells on inside of leaf

paramecium (par uh MEE see um): a protozoan that is slipper-shaped and has cilia

parasite (PAR uh site): animal that lives on or in another living organism

parental care: behavior in adults of giving care to eggs and offspring

pelvis (PEL vus): bones of the hip

pen: tough, transparent shell of a squid

perspiration: waste water lost through the skin

pH: scale used to tell if a substance is an acid or a base

pharynx: the part of the body lined with cilia in a sea anemone

phloem (FLOH em): cells that carry food

photosynthesis (foht oh SIHN thuh sus): process in which plants make food

phototropism: response of a plant to light

phylum (FI lum): largest group of a kingdom

physical change: breaking of large food pieces into smaller pieces

pistil: female reproductive organ of a flower

pith: cells in the center of a woody stem that store food

placenta (pluh SENT uh): organ that connects the embryo to the mother's uterus

platelets: blood cell parts that aid in forming blood clots

pollen: male reproductive part that forms sperm cells

pollutant: anything that causes pollution

pollution: anything that makes the surroundings of living things unhealthy or unclean

population change: increase or decrease in population size

population sampling: way to find the size of any population

pores: many tiny holes in the sponge

predator (PRED ut ur): animal that kills and eats another animal

premature (pree muh CHOOR): baby that is born before 38 weeks of development

prey (PRAY): animal eaten by a predator

primary consumers: animals that eat plants

producers: living things that make their own food by photosynthesis

proteins: nutrients used to build body parts; building material of living things

protist: one-celled organism that has a nucleus and other cell parts with membranes

Punnett square: way to show which genes will combine when egg and sperm join

pure dominant: combination of two dominant genes for a trait

pure recessive: trait controlled by two recesive genes

R

receptors: cells that receive information from the environment

recessive (rih SES ihv) **genes:** genes that do not show if only one is present

recipient: person who receives blood from a donor

red blood cells: cells of blood that carry oxygen to tissues

regeneration (rih jen uh RAY shun): asexual reproduction in which broken pieces of the parent body form new organisms

relaxed: muscle that is not working (returns to normal size)

repeating or repetition: doing something over and over

reproduction (ree pruh DUK shun): production of offspring

respiration: process in living organisms that produce CO_2

response: action of an animal as a result of a stimulus

resting cell: cell that is not forming new cells by cell division

reward: something given after a response to help in the learning of that response

right cerebrum side: right side of cerebrum that controls left body side's function

root hairs: long, thin cells on epidermis that absorb water

root tip: part of a root where cell division occurs

S

sacrum (SAY krum): part of pelvis made of spinal bones; base of spine

saprophyte: organism that feeds on living things

secondary consumers: animals that eat other animals

seed: part formed in the flower that contains a new plant and stored food

seed coat: outer covering of a seed

segment: body section

sense: ability to receive and react to stimuli

sense organ: cells that carry information from the receptors to the brain

sensory neurons: parts of the nervous system that tell an animal what is going on around it

sex chromosomes: chromosomes that determine sex

sexual reproduction: form of reproduction that uses two parents, egg and sperm cells, and cell division by meiosis

shade: condition in which some plants can grow best

sickle (SIHK ul) **cell anemia** (uh NEE mee uh): disorder in which the red blood cells are sickle-shaped

sickle-cell trait: trait that is carried by a person but doesn't show in that person

sickled red blood cells: red blood cells that do not have the normal shape and do not carry very much oxygen as a result; collapsed red blood cells

SI measurements: International System of Units or measurements; for example, meter

skeletal system: body framework made of bone

skeleton: framework of bones in the body which aids in support

sodium bicarbonate: baking soda; an antacid

soil: mixture of mineral particles, living matter, and dead matter

species: smallest division of living things

sperm: sex cell produced by a male

spicules (SPIHK yewlz): structures that provide support

spines: hard structures that rise through the skin from the platelike skeleton of a starfish

sponge: simple invertebrate that has pores and lives in water

spongy layer: layer of round, loosely packed cells on inside of leaf

spore: special cell that develops into a new living thing

sprain: injury that occurs to ligaments

stage: part of a microscope that holds the slide

stamen: male reproductive organ of flower

starch: complex kind of carbohydrate, such as plant starch

sterile: having all living things destroyed that may have been present

stimulus (STIHM yuh lus): something that causes a response

stoma: opening between guard cells in the epidermis

stomach: baglike organ that holds and helps digest food

style: tube-shaped part in the center of a flower which leads to the ovary

T

tendon: tough tissue fibers that connect muscle to bone

tentacle: structure with stinging cells that captures food

test cross: cross between an unknown and an organism pure recessive for the trait

thermal pollution: unneeded heat discharged into a biological community

thigmotropism: response of a plant to touch

trait: feature or characteristic of a thing

transfusion (trans FYEW zhun): blood received from another person

translucent (trans LEWS unt): spot on paper through which light can pass

trial and error: method of trying to solve something without knowing the solution in advance

tRNA: molecule that carries protein parts to the mRNA on the ribosomes

tropism: response of a plant to a stimulus

tube feet: parts that are like suction cups and help a starfish move

tubelike cells: structures that carry food and water

U

urea (yoo REE uh): waste formed by the cells of some living things

urine (YOOR un): mixture of urea, salt, and water; a waste chemical

uterus (YEWT uh rus): muscular organ in which a fertilized egg develops

V

variable: part of an experiment that is changed

variation (ver ee AY shun): difference in a trait

vascular (VAS kyuh lur) **plants:** plants that have tubelike structures in roots, stems, and leaves

vein: brings blood to the heart

ventral: bottom side

ventricle (VEN trih kul): large, lower chamber of heart

vertebrate (VURT uh brayt): an animal in which the backbone has replaced the cord (notochord)

vitamin: chemical needed in very small amounts for growth and tissue repair

vitamin C: ascorbic acid, a vitamin found in citrus fruits and many vegetables

voluntary: you have control over the part

voluntary actions: actions you can control

W

waste chemicals: chemicals not needed by and body, such as urea, water, and carbon dioxide

water flea: a small, clear arthropod that contains a brood pouch

water-holding capacity: amount of water that soil can hold

water soluble vitamin: a vitamin that will dissolve in water

wax layer: outer coating on leaves

wet mount: object placed on a slide with a drop of water and a coverslip

white blood cells: cells of blood that destroy harmful microbes and remove dead cells

X

xylem (ZI lum): cells that carry water

Before doing this lab, have students read section 1:1 in the text.
Name _____ Class _____ Period _____

Not all microscopes have magnification power stamped on the eyepiece. You may have to tell your students what it is. It is usually 10×.

1-1 How Is the Light Microscope Used?

A microscope is a tool used to look at very small things. "Micro" means small and "scope" means to look at. The microscopes that you will use in class have two or more lenses. A lens is a curved piece of glass. The lenses inside your microscope make the objects you look at appear larger. They are located in the eyepiece and in the objectives.

You may wonder how much larger your microscope can make something look. The magnifying power of a microscope is how many times larger a microscope makes something look. The eyepiece of your microscope probably makes things look ten times larger. If so, it has 10× written on it. Each objective lens also has a power written on it. To find the magnification for your microscope, multiply the eyepiece power by the power of the objective lens you are using.

OBJECTIVES   Process Skills: observing, recognizing and using spatial relationships, using a microscope, using numbers

In this exercise, you will:
a. learn the names and jobs of microscope parts.
b. learn how to use and care for the microscope.
c. determine the magnification of your microscope.

KEYWORDS*
Define the following keywords:

compound light microscope _____ a microscope that has two or more lenses

field of view _____ the lighted area you see through a microscope

lens _____ a curved piece of glass that makes objects look larger

stage _____ the part of a microscope that holds the slide

*Refer students to the glossary at the back of the lab manual.

MATERIALS
light microscope   prepared slide of
lens paper   insect leg

PROCEDURE
1. The microscope should always be handled with care. Use one hand to hold the arm. Place the other hand under the base. Move the microscope to your table and gently set it down. (The arm should be toward you.)
2. Use of the microscope is easy if you know the parts. Find the parts listed in Table 1 on page 2 on your microscope.

Remind students to carry the microscope in an upright position and not swing or tilt it while carrying it.

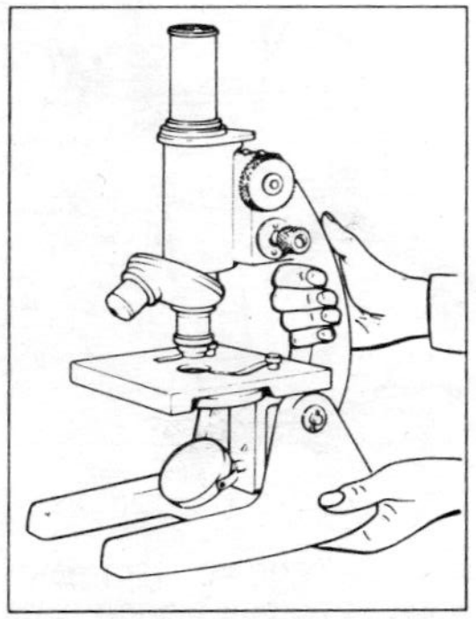

EXPLORATION

You can use prepared slides of the letter e, diatoms, pollen, or other things if insect legs are not available. See page 19T for information on making prepared slides. Legs of the fruit fly (*Drosophila*) are excellent because of small size and availability. Flies can be preserved in alcohol prior to use.

FIGURE 1. Carrying a microscope

1

Table 1. Microscope Parts and Their Jobs

Part	Name	Job
A	Eyepiece	Holds top lens, usually 10×
B	Body tube	Holds top lens certain distance from lower lenses
C	Arm	Supports body tube
D	Nosepiece	Holds lower lenses, turns to change objectives
E	High power objective	Contains 43× lens
F	Low power objective	Contains 10× lens
G	Coarse adjustment	Moves body tube up and down, brings objects into focus
H	Fine adjustment	Moves body tube up and down slightly, brings objects into focus
I	Stage	Supports slide
J	Stage clips	Holds slide in place
K	Diaphragm	Controls amount of light entering microscope
L	Light or mirror	Sends light through microscope
M	Base	Supports microscope

Magnification of these lenses may vary.

If your microscopes have mirrors, describe to the students how to use them to direct light.
CAUTION: *Do not allow students to focus on sunlight; this can cause eye damage.*
Have students direct light from the overhead lights into the microscope.

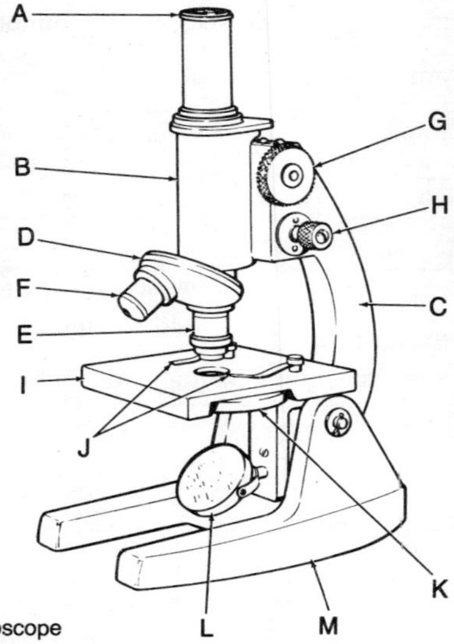

A useful technique is to put a picture of a microscope on an overhead projector. As you point out parts students find the parts on their microscopes. You can go over each part, and then ask students for part names as you point to the microscope parts on the screen.

FIGURE 2. Parts of the microscope

3. Before using the microscope, make sure the lenses are clean. Use lens paper *only*. Any other kind of paper may scratch the lenses. Wipe the eyepiece and objective lenses gently. Avoid xylene as a cleaning agent.*
4. Look through the eyepiece. Turn the diaphragm so that the most light comes through the opening in the stage. The circle of light that you see through the microscope is called the field of view.
5. Turn the nosepiece so that the low power (10×) objective is in place. Put a prepared slide of an insect leg on the stage under the clips. A prepared slide is a slide made to last a long time. Keep the slide clean by holding it by the edges.

*Supply houses have a cleaning solvent which does not harm lenses.

2

283

Name _____ Class _____ Period _____

You may want to go step by step with the entire class. Students should probably not go ahead by themselves.

6. Always find an object first on low power. Move the slide until the leg is directly over the hole in the stage. Then use the coarse adjustment knob to make what you see clear. Look to the side of your microscope when turning the coarse adjustment to keep from hitting the slide with the objective. Turn the coarse adjustment slowly. When the object is clear, we say it is in focus.

7. Move the slide to the left. Which way does the leg move as you look through the microscope? It moves to the right.

8. Move the slide away from you. Which way does the leg move as you look through the microscope? It moves toward you.

9. Draw the insect leg in the circle in Figure 4 as it appears under low power. Then turn the nosepiece carefully until the high power objective clicks into place. Observe and bring the object into focus by turning *only* the fine adjustment. It may be necessary to move the slide slightly to center the object. Draw the leg in the circle in Figure 4 as it appears under high power.

If students have trouble finding the objects, have them recenter their slides.

FIGURE 3. Using coarse adjustment

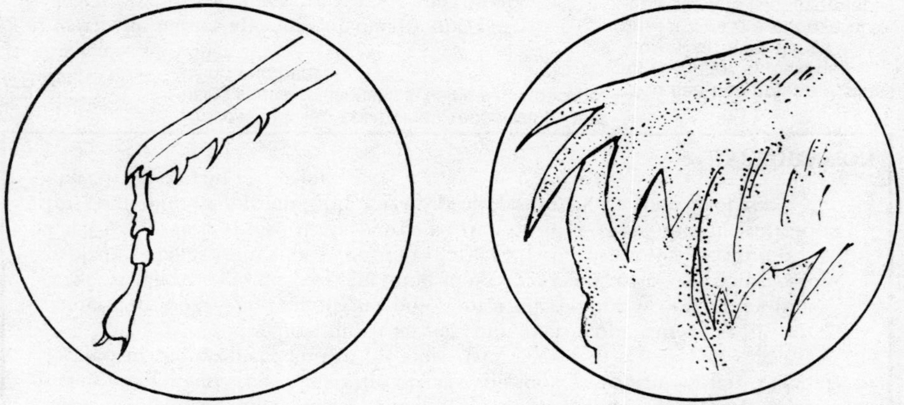

Low power   High power

FIGURE 4.

10. Switch back to low power. Remove the slide and put it away. Answer the questions on the next page. Then put your microscope away.

You may wish to have students practice proper microscope use by providing them with a second or third prepared slide.

3

QUESTIONS

1. Fill in the chart below to show the total magnification of your microscope on low and high power.

	Eyepiece magnification	×	Objective magnification	=	Total* magnification
Low power	10		10		100
High power	10		43		430

*Magnification may vary depending on magnification of the lenses. Many high power lenses magnify 45X, for example.

2. How does the leg look under high power that differs from how it looks under low power? _Under high power the leg is larger and more details show. Less of the leg can be seen._

3. When you moved the slide to the right, which way did the insect leg move? _The insect leg moved to the left._

4. Is the field of view brighter or dimmer under high power? _dimmer_

5. How should you carry a microscope? _You should carry a microscope with one hand under the base and the other on the arm._

6. Why should lenses be cleaned only with lens paper? _Lens paper will not scratch the lenses; other papers will._

7. A compound microscope has two or more lenses. Is the light microscope you used in class a compound light microscope? _yes_ Explain. _It has at least two lenses, an eyepiece and at least one objective._

8. When using any piece of laboratory equipment, what should you always do? _Always follow directions and handle the equipment with care. Learn the parts of the equipment and what they do._

9. How is the light microscope used? _It is used to magnify very small things._

4

284

Name _____ Class _____ Period _____

Before doing this lab, have students read section 1:2 in the text.

1-2 How Are SI Length Measurements Made?

Often measurements are made to learn more about biological problems. The International System of Units, or SI, is a system of measurements you will become more familiar with this year. The measurements you will make in this exercise are SI measurements.

In this exercise, you will make length measurements. The basic unit of length is the meter. The meter is divided into one hundred smaller units called centimeters. Smaller measurements are made with millimeters. Ten millimeters equal one centimeter.

When measurements are made, you should write them down. Data are observations you record—in this case, the measurements you write down. The data will be written in a table to help you keep them organized.

INTERPRETATION

OBJECTIVES
In this exercise, you will:
a. compare hand and foot bones.
b. record your data in tables and draw conclusions.

Process Skills: measuring in SI, using numbers, recognizing and using spatial relationships

KEYWORDS
Define the following keywords:

data — observations you record

length — a measure of how long something is

meter — the basic unit of length

SI measurements — International System of units measurements, for example, meter

MATERIALS
metric ruler

Review with students which markings on the metric ruler are centimeters and which are millimeters. Check by having each student show you a millimeter on his/her ruler before starting the procedure. A clear-plastic ruler can be projected on the overhead projector to show the students differences between a centimeter and a millimeter.

PROCEDURE
1. Look at the diagram of the hand in Figure 1 on page 6. Count the number of bones present in the thumb, fingers, palm, and wrist. (They are shaded in different ways in the diagram to help you.) Record your counts in Table 1.

Table 1. Bone Counts

Part	Number of bones	Part	Number of bones
Thumb	2	Big toe	2
Fingers	3 in each finger	Other toes	3 in each toe
Palm of hand	5	Center of foot	5
Wrist	8	Ankle and heel	7

Make sure students move their metric rulers on the bones so they measure the greatest length of each bone. See Bone A.

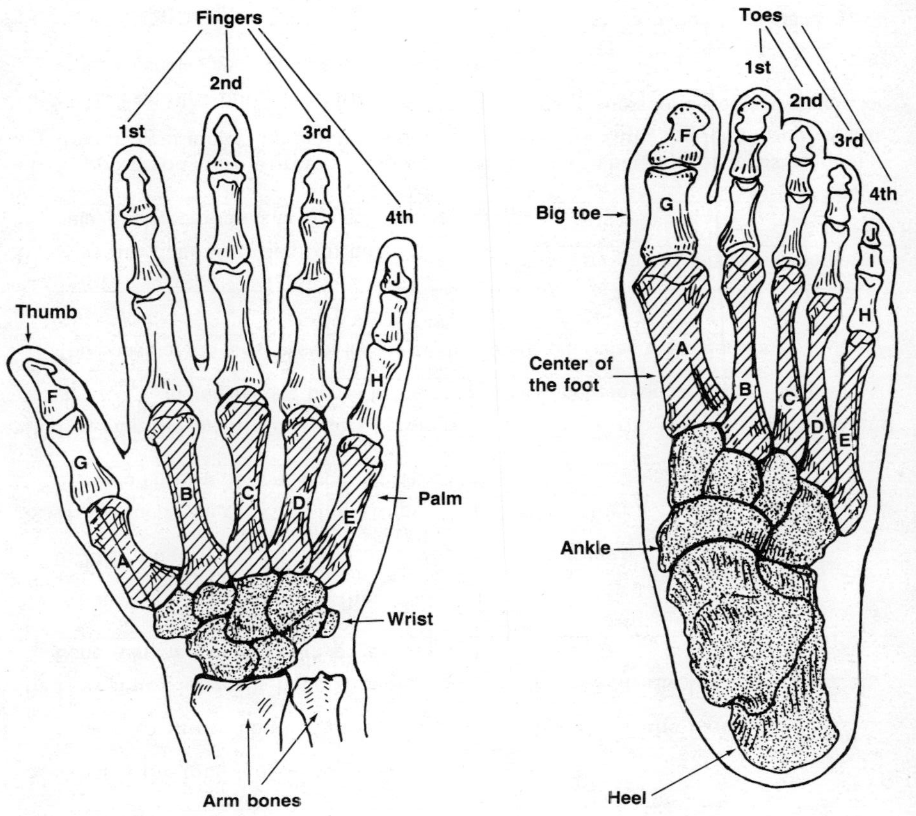

FIGURE 1.

Thumb, fingers, and toes are white in these drawings. Palm bones and bones in the center of the foot are striped. Wrist, ankle, and heel bones are spotted.
Since the fingers have three bones, students will confuse bone A as the third bone of the thumb, even though it is shaded as a palm bone.

Name _____ Class _____ Period _____

2. Measure in millimeters the lengths of the bones marked A, B, C, D, and E in the hand diagram. Record your measurements in Table 2.

3. Measure in millimeters the lengths of the bones marked A, B, C, D and E in the foot diagram. Record your measurements in Table 2.

4. Measure the length of the thumb (F+G) and record the number in the table.

5. Measure the length of the big toe (F+G) and record it in the table.

6. Measure the lengths of the smallest finger and toe (H+I+J). Record these data in the table.

7. Change all the millimeter measurements to centimeter measurements in the table. Recall that there are ten millimeters in one centimeter.

Students may need to be told that they must *divide* the millimeter number by 10 to get centimeters. Suggest that an easy way to do this division is to move the decimal point one place to the left. Give a few examples.

Measurement errors will result if the flesh outline around the bones is used in the measurements.

FIGURE 2. Measuring bone length

Table 2. Bone Lengths

Bone	Hand Millimeters	Hand Centimeters	Foot Millimeters	Foot Centimeters
Bone A	23	2.3	33	3.3
Bone B	37	3.7	40	4.0
Bone C	37	3.7	37	3.7
Bone D	33	3.3	38	3.8
Bone E	30	3.0	40	4.0
Thumb or big toe bones (F+G)	31	3.1	30	3.0
Smallest finger or toe bones (H+I+J)	36	3.6	21	2.1

QUESTIONS

1. What is the total number of bones:
   a. in the hand? __27__   b. in the foot? __26__

2. How do the total number of bones in the hand and foot compare? The hand has one more bone than the foot.

3. What is the total number of bones in the:
   a. palm? __5__   b. center of foot? __5__

4. How do the total number of bones in the palm and foot center compare? They are the same in number.

5. What is the total length of:
   a. bone A in the hand? __23 mm__   b. bone A in the foot? __33 mm__

6. How much longer is bone A in the foot than bone A in the hand? Bone A in the foot is 10 mm longer (43%).

7. What is the total length of the:
   a. little finger? __36 mm__   b. little toe? __21 mm__

8. How much longer is the little finger than the little toe? The little finger is 15 mm longer (42%).

9. Describe the main differences between the lengths of the bones in the hand and the foot. In general, the bones in the center of the foot are longer than in the palm. The bones of the toes, except for the big toe, are shorter than the finger bones.

10. Why are data often kept in tables? Answers may vary. Use of tables helps organize data. Tables provide data in a concise manner.

11. Suppose you were working in a department store. What unit of measurement (meter, centimeter, millimeter) would you use to measure the length and width of shoes and window curtains? You would use centimeters for shoes and meters for window curtains because of the objects' sizes.

12. How are SI length measurements made? SI length measurements are made in millimeters, centimeters, and meters.

Name _____ Class _____ Period _____

Before doing this lab, have students read section 2:3 in the text.

2-1 What Are Diffusion and Osmosis?

Cells have an outer covering called the cell membrane. The cell membrane controls what moves into and out of cells. Food and oxygen move into cells through the cell membrane. They move by diffusion. The movement of a substance from where there is a large amount to where there is a small amount is called diffusion.

The movement of water across the cell membrane is called osmosis. Osmosis is a special kind of diffusion.

INVESTIGATION

OBJECTIVES
In this exercise, you will:
a. observe diffusion of food coloring in water.
b. observe osmosis across the membrane of an egg.
c. measure the amount of water that moves across the egg membrane.

Process Skills: experimenting, inferring, interpreting data, predicting

This exercise could be done as a teacher demonstration if time or supplies are limited. It is, however, more beneficial to have the students do both parts working in groups.

KEYWORDS
Define the following keywords:

cell membrane — the outer covering of a cell that controls what enters and leaves the cell

diffusion — the movement of a substance from where there is a large amount to where there is a small amount

milliliter (mL) — $\frac{1}{1000}$ liter

osmosis — the movement of water across the cell membrane

MATERIALS

beaker, 400 mL
water
dropper
food coloring — grocery store items
3 glass jars with lids, 500 mL
colored pencils: blue, red, and green
wax pencil
graduated cylinder, 100 mL
200 mL white vinegar
200 mL clear syrup — Corn syrup works well.
raw egg in shell
Students can bring in clean 1 pint peanut butter or mayonnaise jars from home to use.

Students should work in groups of four or five.

PROCEDURE
Part A. Diffusion
1. Add water to the beaker until it is three-fourths full. Let the beaker stand until the water is very still.
2. Carefully drop one drop of food coloring to the surface of the water (Figure 1). Observe what happens.

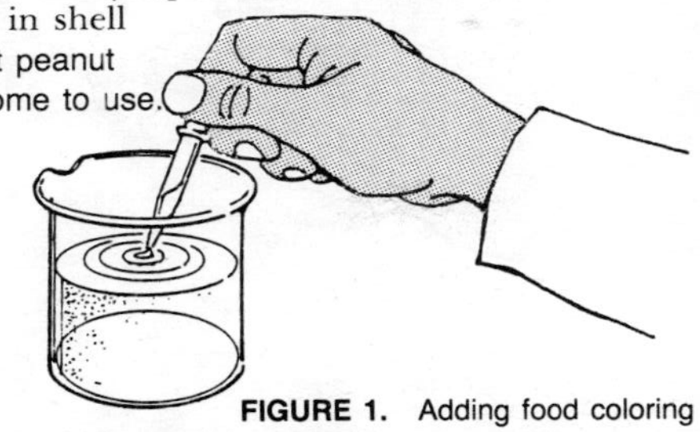

FIGURE 1. Adding food coloring

Part A could be done as a demonstration on the overhead projector. Use a large petri dish half instead of a beaker because it projects better. Also, the heat from the lamp will speed up the rate of diffusion of the food coloring.

9

3. Draw what happens in the beakers in Figure 2.

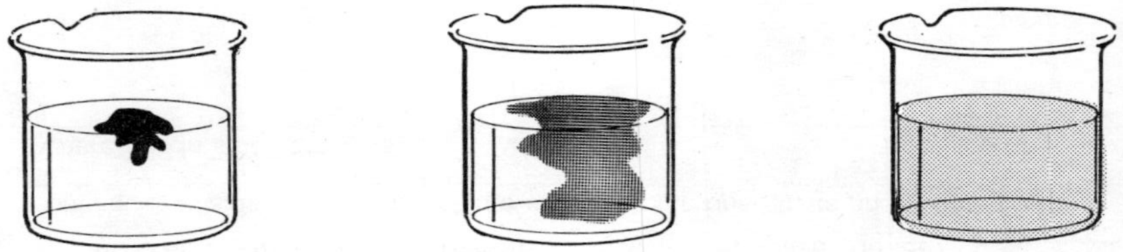

Early   Middle (after 10 minutes)   Late (after 1 hour)

FIGURE 2. Diffusion in beakers

4. Empty the water from the beakers into the sink.

Part B. Osmosis
1. Label the glass jars with the wax pencil. Write *vinegar* on one jar, *syrup* on the second jar, and *water* on the third.
2. Look at the raw egg. Note the appearance of the eggshell. Use your pencil and trace the egg in the space provided.
3. Use the graduated cylinder to measure 200 mL of vinegar. Put the vinegar in the vinegar jar. Place the raw egg into the vinegar jar. The vinegar should cover the egg. Cover the jar with the lid. Leave it undisturbed for two days (Figure 3).
4. After two days, observe what has happened. Write your results in Table 1.*

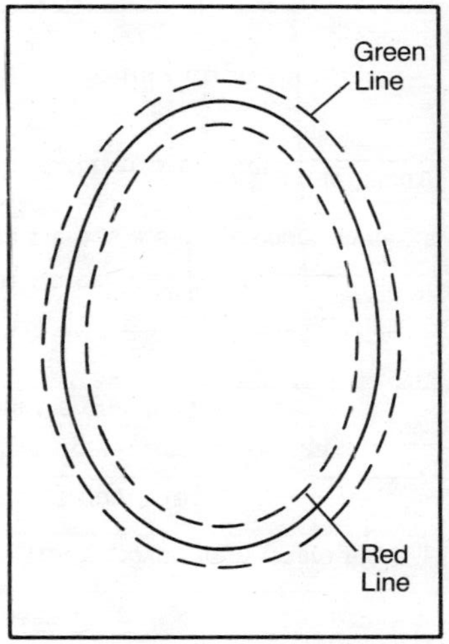

*Calcium carbonate in the shell reacts with vinegar (acetic acid). Calcium acetate is formed, which is water soluble so shell disappears but the membrane remains. Egg contents harden because acetic acid denatures egg protein.

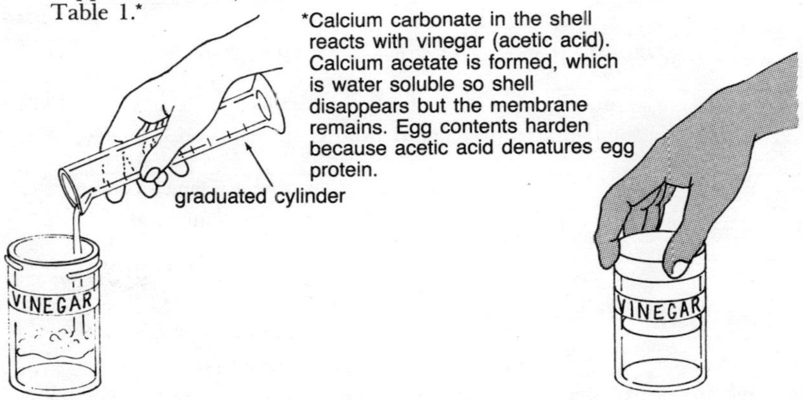

FIGURE 3. Adding vinegar and egg

10

Name _____ Class _____ Period _____

Caution students to handle the egg gently to avoid puncturing the membrane.

5. Put 200 mL syrup in the syrup jar. Place the egg in the syrup jar. Leave it for one day.
6. Measure the amount of vinegar remaining in the vinegar jar. Write the amount in the table. Then empty the vinegar into the sink.
7. The next day measure 200 mL of water and add it to the water jar.
8. Carefully remove the egg from the syrup jar. Make observations and record them. Remove the egg from the syrup and carefully rinse it with water. Use the red pencil and trace the egg over the second drawing of the egg. Place the egg in the water jar.
9. Measure the amount of syrup that remains in the syrup jar. Write the amount in the table. Empty the syrup into the sink.
10. After one or two days, remove the egg from the water. Use the green pencil and trace the egg over the third drawing of the egg. Also record your observations of the egg. Measure the amount of water that is left in the water jar. Write the amount in the table.

Students may need guidance as to what to look for—size, firmness, etc.

Table 1. Results of Osmosis

Jar	Amount present when egg was put in	removed	Observations
Vinegar	200 mL	150 mL *	Egg shell dissolved. Egg became larger.
Syrup	200 mL	250 mL *	Egg became small. Egg membrane became loose.
Water	200 mL	150 mL *	Egg became larger again.

* Student data will vary.

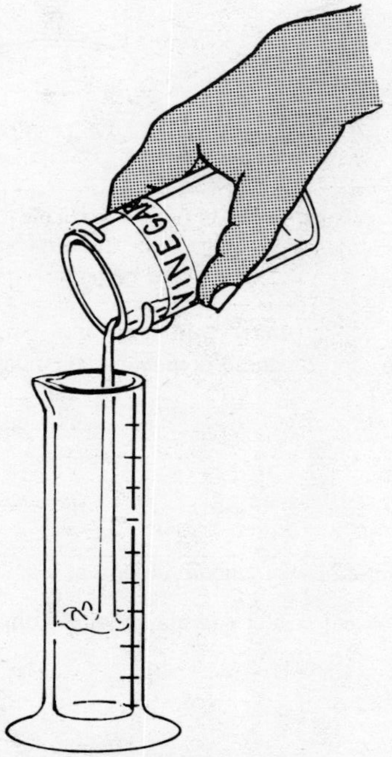

FIGURE 4. Measuring vinegar

QUESTIONS

1. What did you observe when food coloring was dropped on the water in Part A?
   Food coloring spread evenly through the water.

11

2. What is this process called? diffusion
3. In Part B, what happened to the shell of the egg? Shell dissolved.
4. Vinegar is made of acetic acid and water. Which part of the vinegar dissolved the shell? acetic acid
5. a. What happened to the size of the egg after remaining in vinegar? increased
   b. Was there more or less liquid left in the jar? less
   c. Did water move into or out of the egg? How do you know? into—egg became larger
6. a. What happened to the size of the egg after remaining in syrup? decreased
   b. Was there more or less liquid left in the jar? more
   c. Did water move into or out of the egg? How do you know? out—egg became smaller
7. a. What happened to the size of the egg after remaining in water? increased
   b. Was there more or less liquid left in the jar? less
   c. Tell why water moved into or out of the egg. Water moved from where there is a larger amount (jar) to where there is a smaller amount (egg).
8. Was the egg larger after remaining in water or after remaining in vinegar? Answers may vary. Usually they are about the same, but the one in water may be a little larger.
9. Would a cell placed in syrup probably lose or gain water? lose
   Why? Water moves from large amount (cell) to where there is a small amount (syrup).
10. Why are fresh fruits and vegetables sprinkled with water at a market? to keep them from drying out
11. Roads are sometimes salted to melt ice. What does this salting do to the plants along the roadside? It kills the plants.
    Why? Water moves out of plant cells (large amount water) by osmosis.
12. Why do dried fruits and dried beans swell when they are cooked? Water moves from liquid outside (where it is in large amounts) into dried fruits and vegetables (where it is in small amounts).

12

Name _____ Class _____ Period _____

Before doing this lab, have students read section 2:2 in the text.

2–2 What Cell Parts Can You See With the Microscope?

Living things are made of cells. All cells have parts that do certain jobs. Cells have an outer covering called the cell membrane. Cell membranes give cells their shapes and control what enters and leaves the cells. The clear, jellylike material inside the cell is the cytoplasm. The nucleus is the control center of the cell. Plant cells have a thick outer covering called the cell wall. It is on the outside of the cell membrane.

Cell parts can be studied by making wet mount slides. A wet mount slide is a temporary slide. It is not made to last a long time. You can make wet mount slides of living and once-living materials to study cell parts.

Process Skills: observing, using a microscope, recognizing and using spatial relationships, classifying

EXPLORATION

OBJECTIVES

In this exercise, you will:
a. make wet mount slides for examination under the microscope.
b. study four cell parts—the cell wall, cytoplasm, nucleus, and cell membrane.

KEYWORDS

Define the following keywords:

cell wall ___ the thick outer covering of a plant cell

cytoplasm ___ jellylike substance inside the cell

nucleus ___ the control center of the cell

wet mount ___ an object placed on a slide with a drop of water and a coverslip

MATERIALS

glass slide
coverslip
light microscope
water
dropper
forceps
iodine stain
elodea leaf
*cork shaving
*bamboo shaving
onion skin
prepared slide of frog blood

See annotation on page 14.
Use frog skin slide if frog blood is unavailable.

PROCEDURE

Part A. Making a Wet Mount

1. Get a clean glass slide and coverslip. Handle the slide and coverslip by the edges to keep them clean (Figure 1).

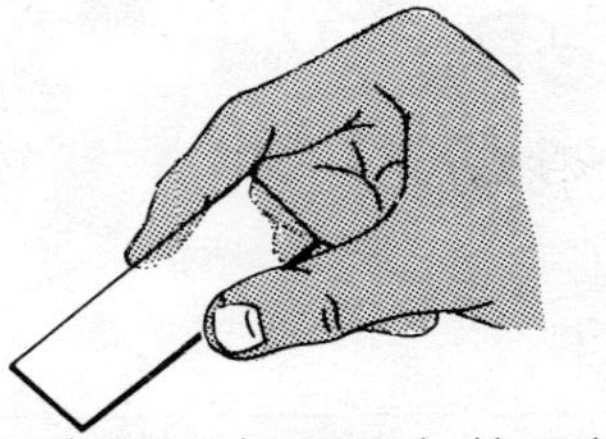

FIGURE 1. Holding slide by edges

2. Use a dropper to put a drop of water in the center of the slide.
3. With forceps, place the object to be examined in the drop of water.
4. Hold the coverslip at an angle. Gently lower it onto the drop of water (Figure 2). Forceps can be used to lower the coverslip.

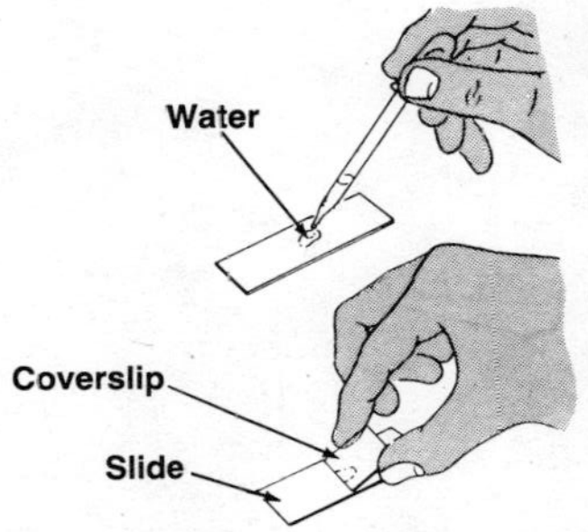

FIGURE 2. Making a wet mount

Part B. Looking at Cells

1. Prepare a wet mount of the cork shaving. Follow the steps given in Part A. Only cork cell walls can be seen since cells are dead.
2. Examine the slide of cork under low power of the microscope. Switch to high power. Draw the cells you see in the circle in Figure 3. Label the cell wall.
3. Prepare a wet mount of a bamboo shaving. Examine the bamboo under low and then high power of your microscope. Draw several dead bamboo cells in the circle in Figure 3. Label the cell wall.

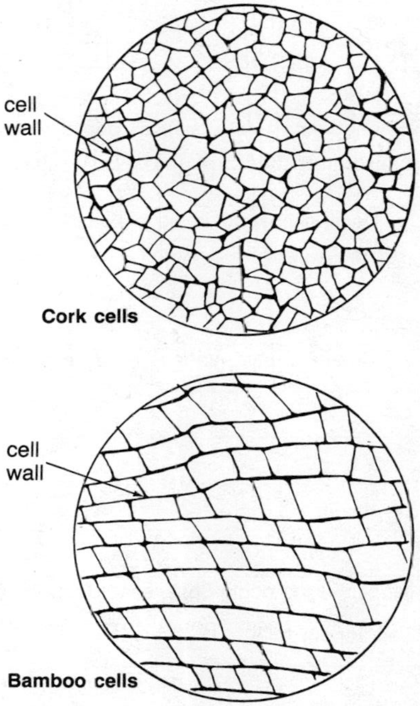

FIGURE 3. Looking at cells

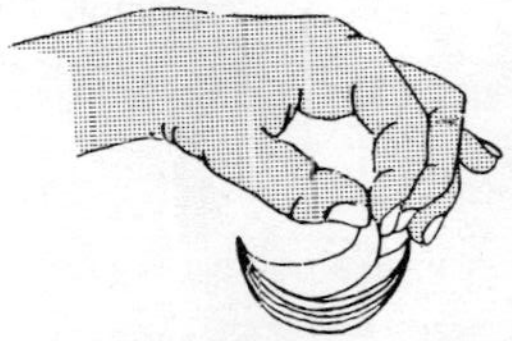

4. Peel the thin layer of cells from the inside of an onion as shown in Figure 4. Make a wet mount of the onion skin cells.* Add one drop of stain in place of water.
5. Examine the onion slide under low and high power of your microscope.
6. Find the cell wall, nucleus, and cytoplasm. Draw living onion cells that you see in the circle in Figure 5. Label the parts.
7. Prepare a wet mount of an elodea leaf.
8. Examine the elodea leaf under low and then high power of your microscope. Find the cell wall. Try to find the nucleus. Also find the parts that make the leaf green. Draw some elodea cells in the circle in Figure 5. Label the parts you see.

FIGURE 4. Peeling layer of cells

* Iodine stain can be used to make cells more visible. Prepare by dissolving 1.5 g potassium iodide and 0.3 g iodine in 1 L water. Store in dark room or brown bottle. Students can add a drop to the onion tissue before adding coverslip. CAUTION: *Iodine is poisonous and can burn the skin.*

* Prepare shavings in advance by scraping a cork and a saxophone reed with a sharp razor blade. Only very thin sections are desirable. Discard thicker sections. Wood shavings from the school woodshop are excellent sources of plant cells that are similar to bamboo if reeds are not available.

Name _____ Class _____ Period _____

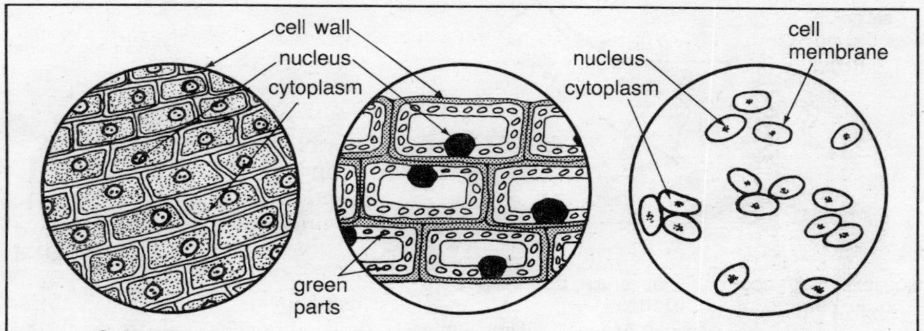

FIGURE 5. Examining cells under the microscope

Onion skin cells — cell wall, nucleus, cytoplasm
Elodea cells — green parts
Frog blood cells — cell membrane, nucleus, cytoplasm

10. Examine a prepared slide of frog blood with low and then high power. In the circle in Figure 5, draw some frog blood cells. Label the nucleus, cytoplasm, and cell membrane.

11. Complete Table 1.

Table 1. Parts of Cells

Cell type	Cell wall present? (yes or no)	Nucleus present? (yes or no)	Cytoplasm present? (yes or no)	Shape of cell?	Cell living or dead?
Cork	yes	no	no	rectangular	dead
Bamboo	yes	yes	no	rectangular	dead
Onion	yes	yes	yes	rectangular	living
Frog blood	no	yes	yes	oval	dead*
Elodea	yes	yes	yes	rectangular	living

Shapes may be drawn in table

*because of preserved slide

QUESTIONS

1. What is the name of the small units that make up cork? __cells__

2. Are the cork cells filled with living material or are they empty? __empty__

15

3. Describe how the small units of cork look. __heavy, outer cell wall present; insides are empty; no nuclei or cytoplasm present__

4. Are bamboo cells living or dead? __dead__

5. How are cork cells and bamboo cells alike? __Both have cell walls and similar shape; both are empty and are plant cells.__

6. How are onion cells different from the cork cells? __Onion cells are living; cork is dead. Onion cells have nuclei and cytoplasm; cork cells are empty.__

7. Compare the onion skin cells and the frog blood cells. __Onion cells are rectangular and have cell walls. Frog blood cells are oval and lack cell walls.__

8. Compare the cell parts seen in the elodea and the frog blood cells. __Elodea cells have cell walls, green color, nuclei, and cytoplasm that moves. Frog blood cells lack cell walls, and cytoplasm does not move.__

9. a. Are onion cells plant or animal? __plant__
   b. Are elodea cells plant or animal? __plant__
   c. Are frog blood cells plant or animal? __animal__

10. Why do cells have different shapes? __Cells with different shapes have different jobs; cell shape is related to cell function.__

11. Skin cells seem to fit together or overlap. How is this cell arrangement helpful to the organism? __The cells protect; overlapping helps make a better protective layer.__

12. If blood cells were box-shaped, like onion cells, why would they be unable to do their job as well? __Since blood cells move through blood vessels, they would get stuck along the sides of the vessels.__

16

290

Name _____ Class _____ Period _____

Before doing this lab, have students read section 3:1 in the text.

3-1 How Can Paper Objects Be Grouped?

You group, or classify, many things every day. The phone book, the library, and the grocery store are all classified to make finding things easier. In any classification, things with similar traits are grouped together. Scientists divide living things into groups. The main groups are called kingdoms. Kingdoms are divided into phyla. Phyla are divided into classes. The other groups are order, family, genus, and species. The more similar two living things are, the more groups they share.

INTERPRETATION

Process Skills: classifying, recognizing and using spatial relationships, inferring

OBJECTIVES
In this exercise, you will:
a. group paper objects.
b. use the words *kingdom*, *phylum*, and *class* in your classification.
c. decide what traits were used in the classification.

KEYWORDS
Define the following keywords: Refer students to glossary for help.

class ____ the largest group of a phylum
classify ____ to group
kingdom ____ the largest group of living things
phylum ____ the largest group of a kingdom
trait ____ a feature a thing has

MATERIALS
scissors
Provide copies of Figure 2 for each student.

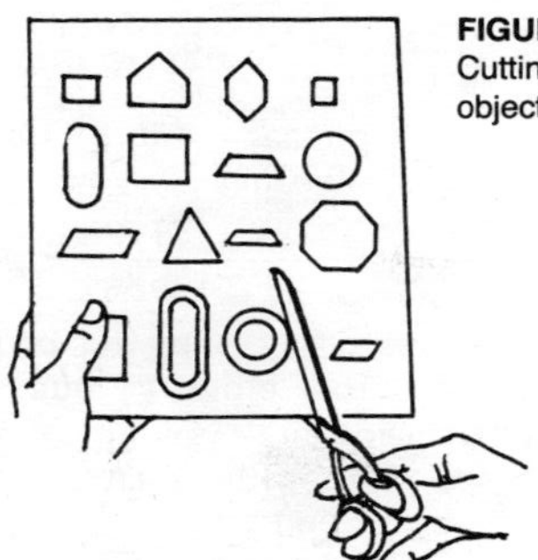

PROCEDURE
1. Get a copy of the paper objects in Figure 2 on the next page from your teacher.
2. Cut out the objects as shown in Figure 1. **CAUTION:** *Use extreme care with the scissors.*
3. Place the objects on your desk. Divide them into two groups as follows:
   a. Put objects 1, 4, 6, 7, 9, 11, 13 and 16 into one group. These will represent the classification level of one kingdom. What trait do all of these objects have in common? ____ four sides

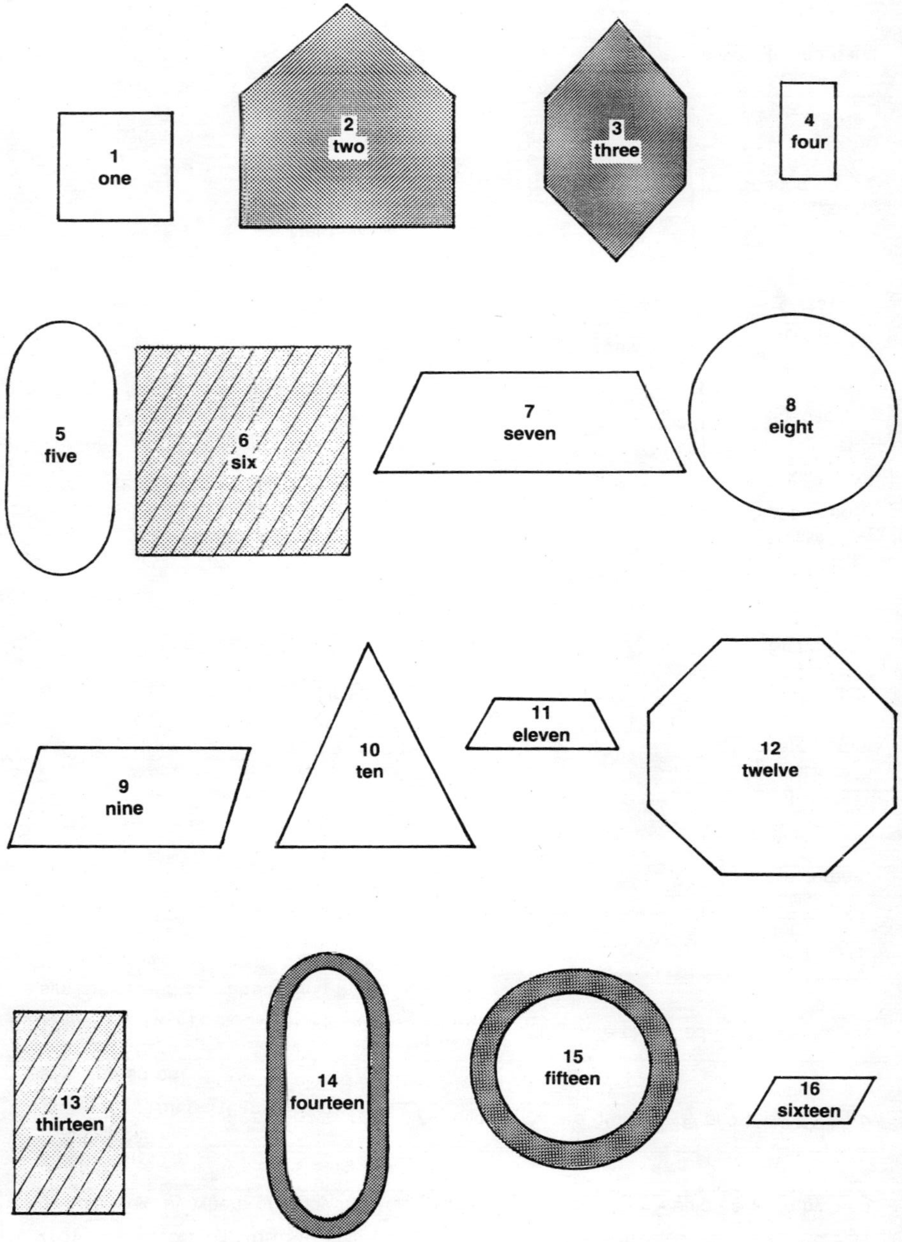

FIGURE 1. Cutting out objects

FIGURE 2. Paper objects

Name _____ Class _____ Period _____

b. Put objects 2, 3, 5, 8, 10, 12, 14, and 15 into a second group. These will represent a second kingdom. What trait do all these objects have in common? **They do not have four sides.**

c. Write a good kingdom name for each group in Chart 1 on the next page.

4. a. Divide the objects from the first kingdom only in this way. Put objects 1, 4, 11, and 16 into one group. Put objects 6, 7, 9, and 13 into a second group. This represents the next classification level called Phylum.

   What trait was used to separate the two groups? **size**

   b. Write a good phylum name for each group in Chart 1.

5. a. Divide the objects from the second kingdom in this way. Put objects 2, 3, 10, and 12 into one group. Put objects 5, 8, 14, and 15 into a second group.

   What trait did you use to separate the objects into two groups? **shape**

   b. Write a good phylum name for each group in Chart 1.

6. a. Use objects 1, 4, 11, and 16. Separate them into two groups as follows. Put objects 1 and 4 into one group and objects 11 and 16 into a second group. This represents the next classification level called Class.

   What trait did you use to separate the objects into two groups? **square corners**

   b. Write the class level name for these groups in Chart 1.

7. a. Separate out objects 2, 3, 10, and 12. Then group them into two groups.

   What traits did you use to group them as you did? **shaded**

   b. Write the objects' classification level and group names in Chart 1.

8. a. Repeat Step 7 for objects 6, 7, 9, and 13. Complete Chart 1.
   b. Repeat Step 7 for objects 5, 8, 14, and 15. Complete Chart 1.

QUESTIONS

1. Underline the correct word choice in each of these sentences.

   a. Objects in the same class also belong to the same (phylum, genus, family).

   b. Objects in the same phylum also belong to the same (family, order, <u>kingdom</u>).

2. List the classification levels in order from largest to smallest. **kingdom, phylum, class, order, family, genus, species**

3. Classifications are not all alike. Suppose objects 14 and 15 were put in one group. What trait do they have in common? **They each have an outer band.**

4. Write a name below for each of these groups.*

   Objects 1 and 4 **small, four sided, square-cornered object group**

   Objects 2 and 3 **large, shaded, 5- or 6-sided object group**

   Objects 10 and 12 **3- or 8-sided object group**

   Objects 8 and 15 **round object group**

   * Answers will vary.

292

---

5. What are used to determine if living things belong to a particular group? **traits**

6. How are pieces of clothing classified in a department store? *Clothes are grouped by size, sex, by where on body they are worn, by age group they would be worn by, and so on.

7. How are athletic teams grouped? * Athletic teams are grouped by sport, by ability, by sex, and so on.

8. What are two reasons for classifying things? **Things are classified to show similarities and to make finding things easier.**

* Answers will vary.

```
                                                    ┌─ 1,4 ─ Class: Square corners
                                     ┌─ 1,4,11,16 ──┤
                                     │  Phylum      └─ 11,16 ─ Class: Not square corners
                                     │  Small size
                  ┌─ 1,4,6,7,9,11,13,16 ─┤
                  │  Kingdom              │                   ┌─ 6,13 ─ Class: Striped
                  │  Four sides           └─ 6,7,9,13 ────────┤
                  │                          Phylum           └─ 7,9 ─ Class: Not striped
1,2,3,4,5,6,7,8,  │                          Large size
9,10,11,12,13,14, │
15,16             │                                           ┌─ 2,3 ─ Class: Shaded
All paper objects │                       ┌─ 2,3,10,12 ───────┤
                  │                       │  Phylum           └─ 10,12 ─ Class: Not shaded
                  │                       │  Straight sides
                  └─ 2,3,5,8,10,12,14,15 ─┤
                     Kingdom               │                  ┌─ 14,15 ─ Class: Shaded band
                     Not four sides        └─ 5,8,14,15 ──────┤
                                              Phylum          └─ 5,8 ─ Class: Not shaded band
                                              Curved sides
```

Name _____ Class _____ Period _____

Before doing this lab, have students read section 3:3 in the text.

3–2 How Can Living Things Be Grouped?

Living things can be grouped or classified just like nonliving things. Scientists classify living things to put them in order and to show how they are alike. This makes the study of the many different living things easier. One way of classifying is by placing all living things into five main groups. Each group is called a kingdom. The five kingdoms are the Moneran kingdom, the Protist kingdom, the Fungus kingdom, the Plant kingdom, and the Animal kingdom.

Process Skills: classifying, inferring, interpreting data, scientific terminology

INTERPRETATION

OBJECTIVES
In this activity, you will:
a. observe and record the traits of living things.
b. use the observed traits to see similarities and differences in living things.
c. group living things into kingdoms.

KEYWORDS
Define the following keywords:

classify ____ to put into groups

kingdom ____ the largest group of living things

trait ____ a feature or characteristic of a living thing

MATERIALS
specimens of living things
microscope
hand lens

Organisms such as *Nostoc*, paramecium, mushroom, fish, geranium, earthworm, or others may be used. CAUTION: *If you use wild mushrooms instruct students not to eat them.*

PROCEDURE
Part A. Describing Organisms
1. Look at Figures 1 and 2 of living things on the next page. Use Figure 3-10 on page 63 of your text.
2. Fill in the blanks for each organism in the space provided as follows:
   a. For 'number of cells' write *one* or *many*.
   b. For 'cell nucleus' write *yes* or *no*.
   c. Describe the color of the organism.
   d. For 'makes food' write *yes* or *no*.
   e. For 'can move' write *yes* or *no*.
   f. Determine and write the kingdom to which it belongs.

21

FIGURE 1

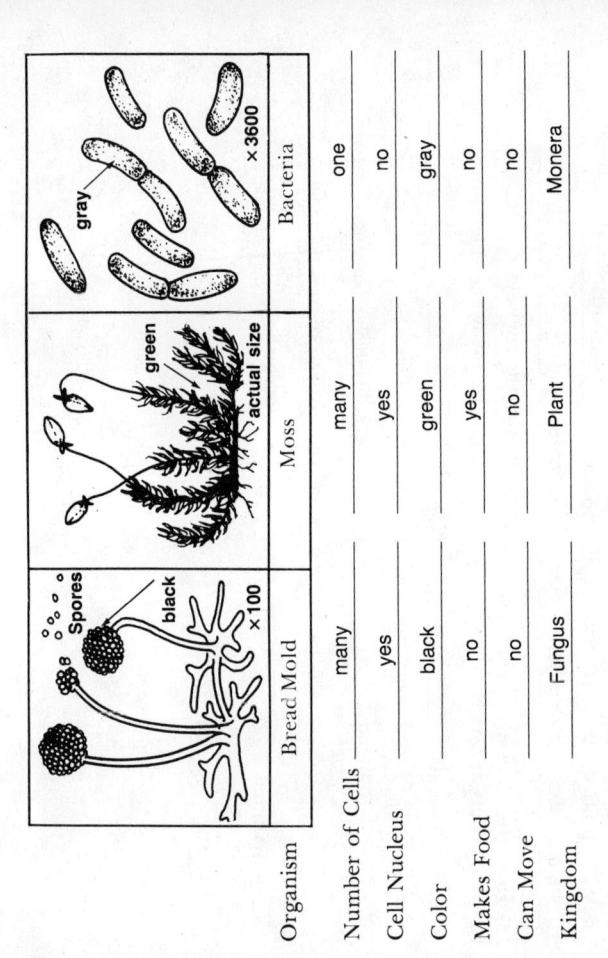

Organism	Bread Mold	Moss	Bacteria
Number of Cells	many	many	one
Cell Nucleus	yes	yes	no
Color	black	green	gray
Makes Food	no	yes	no
Can Move	no	no	no
Kingdom	Fungus	Plant	Monera

FIGURE 2

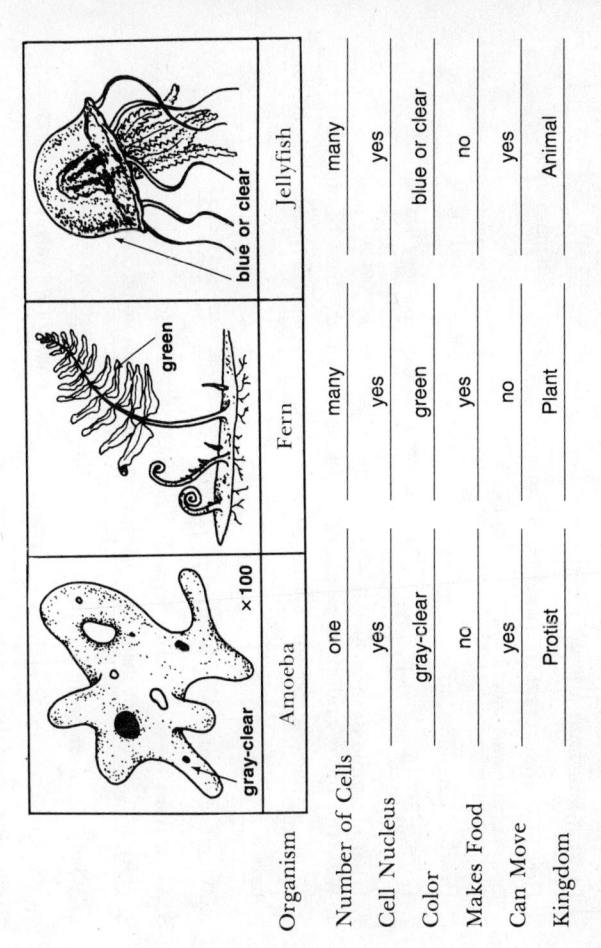

Organism	Amoeba	Fern	Jellyfish
Number of Cells	one	many	many
Cell Nucleus	yes	yes	yes
Color	gray-clear	green	blue or clear
Makes Food	no	yes	no
Can Move	yes	no	yes
Kingdom	Protist	Plant	Animal

22

293

Name _____ Class _____ Period _____

Part B. Classifying Organisms

1. Look at the six living specimens provided for you to study.
2. Examine each one carefully and make drawings of the organisms in the spaces provided.
3. Fill in the blanks as for Part A.
4. Determine the kingdom to which each organism belongs.

Organism	Paramecium	Earthworm	Mushroom
Number of Cells	one	many	many
Cell Nucleus	yes	yes	yes
Color	gray	tan	white
Makes Food	no	no	no
Can Move	yes	yes	no
Kingdom	Protist	Animal	Fungus

Organism	Geranium	Fish	Nostoc
Number of Cells	many	many	one
Cell Nucleus	yes	yes	no
Color	green	brown	blue-green
Makes Food	yes	no	yes
Can Move	no	yes	no
Kingdom	Plant	Animal	Monera

23

QUESTIONS

1. In what kingdoms are one-celled organisms placed? **Monera and Protist**
2. How are producers different from consumers? **Producers can make their own food. Consumers cannot make their own food.**
3. In what kingdoms do you find producers? **Monera, Protist, and Plant**
4. How do fungi get their food? **Fungi absorb food from what they live on.**
5. What are some examples of fungi? **bread mold, and mushrooms**
6. Which of the organisms were placed in a kingdom on the basis of one trait? Explain. **Bacteria and Nostoc were placed in the kingdom Monera because they do not have a nucleus.**
7. What organisms were classified as plants? **geranium, fern, and moss** Answers will vary, but will include green with chlorophyll, many cells, cannot move about, and are producers.
8. What traits do these plants have in common?
9. What organisms were classified as animals? **jellyfish, earthworm, and fish**
10. What traits do these animals have in common? Answers will vary, but will include can move about, have many cells, and are consumers.
11. What organisms are classified as protists? **paramecium and amoeba**
12. What traits do these protists have in common? Answers will vary, but will include are one-celled, the cells have a nucleus, and move about.
13. How are these protists different? **They move in different ways.**
14. What traits are used to place living things into the following kingdoms?
    a. Monera **one-celled, does not have a nucleus**
    b. Protist **one-celled, has a nucleus and other cell parts**
    c. Fungus **many cells, absorbs food from its surroundings, does not move**
    d. Plant **many cells, has chlorophyll and can make food, does not move**
    e. Animal **many cells, cannot make food, can move about**

Name _____ Class _____ Period _____
Before doing this lab, have students read section 4:2 in the text.

4-1 How Can You Test for Bacteria in Foods?

Bacteria are living things too small to be seen without a microscope. Bacteria are said to be microscopic. Often a test can be done to show the presence of microscopic living things. The test can be done with a chemical indicator.

As its name says, a chemical indicator is a chemical that "indicates" something. One type of chemical indicator will tell you if carbon dioxide is present. The chemical turns from blue to yellow or green if carbon dioxide is present.

Respiration in bacteria and other living things produces carbon dioxide. Thus, this chemical can tell if bacteria or other living things are present in a sample being treated.

Bacteria are used to make many dairy products. The samples you will be testing in this exercise are different dairy products.

INVESTIGATION

Process Skills: predicting, experimenting, separating and controlling variables, interpreting data

OBJECTIVES
In this exercise, you will:
a. learn how to do a chemical test for the presence of carbon dioxide.
b. test dairy products to determine if they contain bacteria.

KEYWORDS
Define the following keywords:

control   _part of an experiment that is used to compare changes_

microscopic   _too small to be seen with the naked eye_

respiration   _process in living organisms that produces CO_2_

variable   _part of an experiment that is changed_

MATERIALS
6 test tubes   wax pencil
6 cork stoppers 2 droppers
test tube rack   3 wooden splints
150 mL bromthymol blue solution
dairy products—milk, buttermilk, yogurt, cottage cheese, sour cream
See pages 17T–19T for preparation instructions.

PROCEDURE
1. Label 6 test tubes 1 through 6.
2. Fill each test tube with the blue chemical solution.

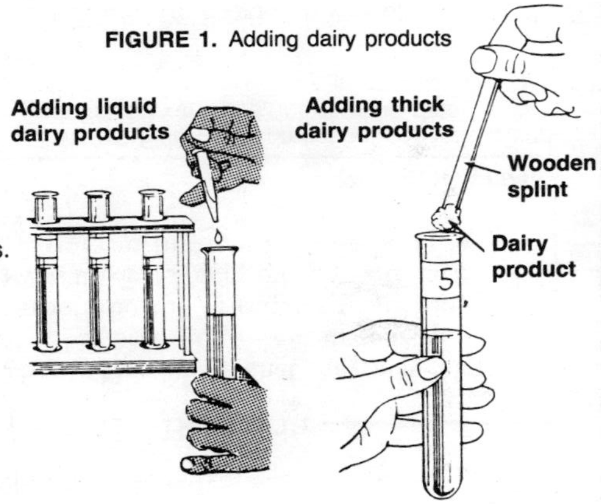

FIGURE 1. Adding dairy products

Adding liquid dairy products

Adding thick dairy products

Wooden splint
Dairy product

3. Use a clean dropper to add one drop of the following to each tube. (Do not let the dairy products go down the sides of the test tubes.) Figure 1 shows you how.
   Tube 1–nothing   Tube 2–milk   Tube 3–buttermilk
   Tube 4–plain yogurt   Tube 5–cottage cheese   Tube 6–sour cream
4. Using a wooden splint, add an amount about the size of a green pea of the following to each tube.
5. Stopper each tube with a cork. Record in Table 1 the color of the tube contents. Leave the tubes undisturbed. Do not shake.
6. Observe and record the color of tube contents at the bottom of each tube, both at the end of the class period and the next day.

Table 1. Test for Bacteria in Food

Tube	Contents	Color at start	Color at class end	Color 1 day later	Carbon dioxide gas present?	Bacteria present?
1	Nothing	blue	blue	blue	no	no
2	Milk	blue	blue	blue or* blue-green	no or* yes	very little, if any
3	Buttermilk	blue	blue-green*	yellow	yes	yes
4	Plain yogurt	blue	blue-green*	yellow	yes	yes
5	Cottage cheese	blue	blue-green*	yellow	yes	yes
6	Sour cream	blue	blue-green*	yellow	yes	yes

*Answers will vary.

QUESTIONS

1. Was there a color change in the contents of Tube 1 one day later? _no_
   What was the purpose of Tube 1? _to compare changes in other test tubes_
   What name is given to this part of the experiment? _control_
2. In which tubes was a color change detected one day later? _3, 4, 5, 6_
   What foods were added to these tubes? _dairy products_
   Do these foods contain bacteria? _yes_
   What is your proof? _Bacteria changed the color of bromthymol blue._
3. What is the variable in this experiment? _Tubes with dairy products._
4. Why are bacteria important to the dairy industry? _Bacteria are used to add flavor and texture to dairy foods._
5. Are bacteria in dairy products helpful or harmful? _They are both helpful and harmful. Certain bacteria give products flavor and texture, while others cause food to spoil._

Name _____ Class _____ Period _____

Before doing this lab, have students read sections 4:2 in the text.

4-2 Does Soil Contain Bacteria

Bacteria are present almost everywhere. This means that they may be found in soil. How can you find out if bacteria are present in soil? Most living things use oxygen and give off carbon dioxide in cellular respiration. If living bacteria are present in soil, they will use oxygen and give off carbon dioxide. By carrying out an experiment, you can find out if living bacteria are present in soil.

INVESTIGATION

Process Skills: experimenting, separating and controlling variables, interpreting data

OBJECTIVES
In this exercise, you will:
a. test soil samples and measure the amount of carbon dioxide given off during 24 hours by the different soil samples.
b. compare the amount of carbon dioxide given off by different soil samples.
c. explain why the amount of carbon dioxide may differ in each sample tested.

KEYWORDS
Define the following keywords:

bacteria _small, one-celled organisms that lack a nucleus and belong to Kingdom Monera_

carbon dioxide _a gas given off by living things during cellular respiration_

cellular respiration _the process by which all living things release energy from food_

distilled water _water that has had all the impurities and bacteria removed_

sterile _having all living things destroyed that may have been present_

MATERIALS
4 large beakers, 500 mL[1]
4 small beakers, 250 mL[1]
4 plastic stirring sticks
sterile soil[2]
Chemical A[3]
8 labels
4 circles of filter paper

distilled water
teaspoon
100 mL graduated cylinder
4 funnels
aluminum foil
glucose or dextrose
2 droppers
Chemical B[3]

Have students work in groups of 2 or 3.

[1] Use jars, milk cartons, or plastic cups.
[2] Do not use clay or playground soil. For sterile soil, purchase from a garden supply store or prepare your own. Heat soil for 1 hour at 300°F in kitchen oven. CAUTION: gives off odor
[3] See pp. 17T-18T for preparation of solutions.

PROCEDURE
Part A. Preparing Soil Samples
1. Label 4 large beakers 1 through 4. Add the following as shown in Figure 1 and use the teaspoon to mix the substances in each beaker:

FIGURE 1. Preparation of beakers 1 through 4

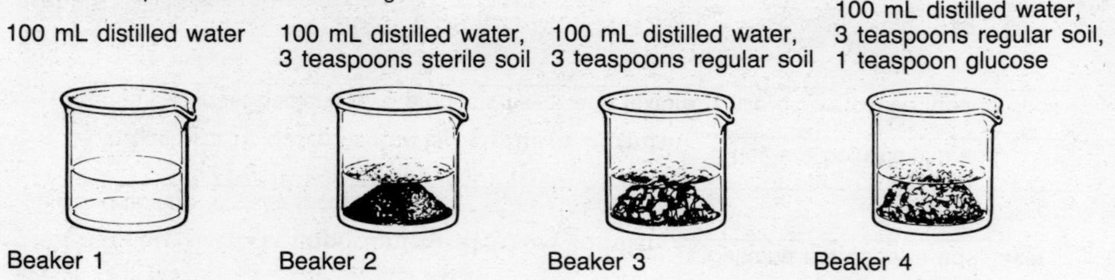

Beaker 1: 100 mL distilled water
Beaker 2: 100 mL distilled water, 3 teaspoons sterile soil
Beaker 3: 100 mL distilled water, 3 teaspoons regular soil
Beaker 4: 100 mL distilled water, 3 teaspoons regular soil, 1 teaspoon glucose

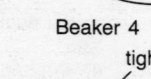

a. Measure 100 mL of distilled water with the graduated cylinder. Add it to Beaker 1.
b. Add 100 mL of distilled water and 3 teaspoons of sterile soil to Beaker 2.
c. Add 100 mL of distilled water and 3 teaspoons of regular soil to Beaker 3.
d. Add 100 mL of distilled water, 3 teaspoons of regular soil, and 1 teaspoon of glucose to Beaker 4.

2. Cover each beaker tightly with aluminum foil as shown in Figure 2. Allow the beakers to remain overnight.

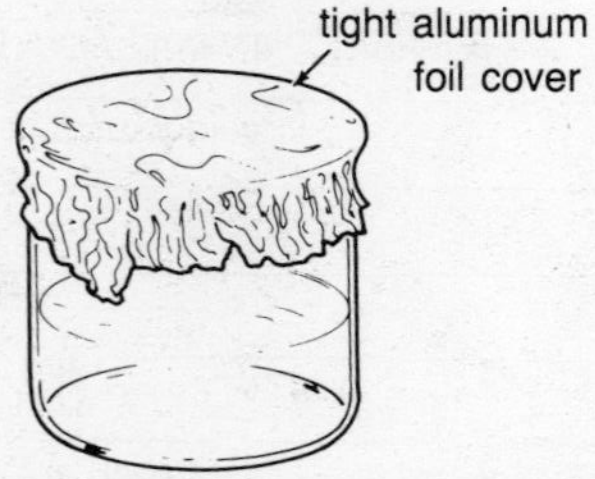

FIGURE 2. Beaker covers

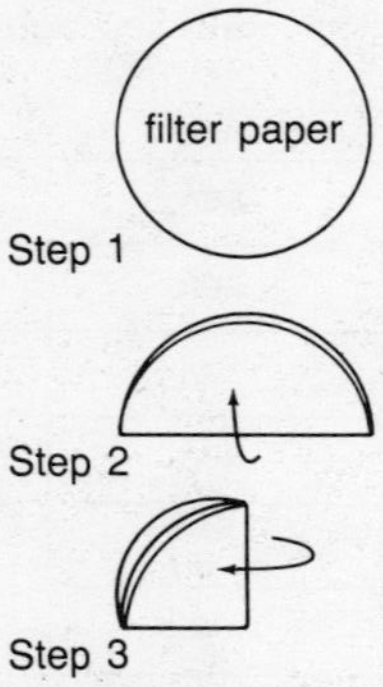

FIGURE 3. Making a funnel
Step 1, Step 2, Step 3

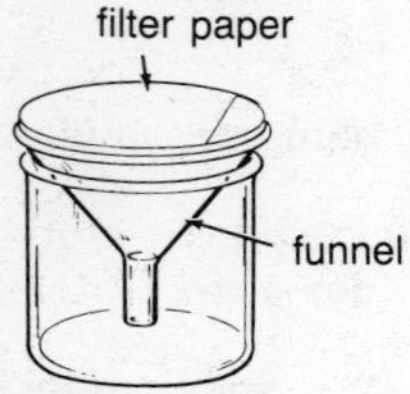

FIGURE 4. Using a funnel

Part B. Filtering the Soil Samples
1. Label four small beakers 1 through 4.
2. Prepare four circles of filter paper as shown in Figure 3. Fold each one in half. Fold in half again. Place each folded paper into the funnel.
3. Stand each funnel with filter paper into the beakers as shown in Figure 4.
4. Pour the contents of the large beakers into the small beakers.
   a. Pour the liquid from large Beaker 1 into the funnel resting in small Beaker 1. Do not let the liquid overflow onto the sides of the filter paper.
   b. Repeat step a for Beakers 2 through 4.

Name _____ Class _____ Period _____

5. Remove the filter paper from each small beaker when filtering is completed.

Part C. Measuring the Amount of Carbon Dioxide Formed by Soil Samples

1. Use a dropper and add three drops of Chemical A to each small beaker.
2. Read all of step 2 before actually doing it.
   a. Use a different dropper for Chemical B. Add Chemical B one drop at a time to the water in beaker 1. **CAUTION:** *Rinse Chemical B with water if it spills on skin or clothes.*
   b. Use a stirring stick to swirl the water in the beaker.
   c. Wait 15 seconds.
   d. Continue to add Chemical B one drop at a time until a light pink color *remains after swirling and waiting 15 seconds.*
   e. Count the number of drops of Chemical B needed to turn the water a *lasting light pink color.*
   f. Record the number of drops needed to turn the water in Beaker 1 a lasting light pink color in Table 1. **NOTE:** The more drops of Chemical B needed to make a lasting light pink color, the more carbon dioxide is present in the water. The fewer drops of Chemical B needed to make a lasting light pink color, the less carbon dioxide is present in the water.
3. Repeat step 2 for Beakers 2 through 4.

*Call attention to this paragraph during any prelab discussion.

Table 1. Drops of Chemical B used in each beaker

Beaker	Contents	Number of drops of Chemical B used
1	water	1–3**
2	water and sterile soil	1–3**
3	water and regular soil	15
4	water, soil, and glucose	30

**Student data will vary. However, Beakers 1 and 2 should contain very little carbon dioxide. Beaker 3 should have a high amount of Carbon dioxide and Beaker 4 should have the highest amount of carbon dioxide.

The procedure used here is a titration technique. Carbon dioxide combines with water and forms a weak acid. Adding sodium hydroxide neutralizes the acid using phenophthalein as a pH indicator.

QUESTIONS

1. a. What was the source of bacteria in the experiment? _soil_
   b. What was used to trap or catch the carbon dioxide given off? _water_
   c. What was the purpose of Beaker 1? _to show that water does not give off carbon dioxide_

2. a. What was being measured by the number of drops of Chemical B added to the water? _the amount of carbon dioxide given off during the 24 hours of waiting_
   b. What do a few drops of Chemical B show about carbon dioxide amount present in the water? _little carbon dioxide present_
   c. What do many drops of Chemical B show about carbon dioxide amount present in the water? _much carbon dioxide present_

3. a. How many drops of Chemical B were used to give a pink color to Beaker 1? _2–3 drops (will vary)_
   b. Beaker 3? _14–15 drops (will vary)_

4. a. Which beaker, 1 or 3, had more carbon dioxide present? _3_
   b. On the first day, what was put into Beaker 1? _distilled water only_
   c. On the first day, what was put into Beaker 3? _distilled water and regular soil_
   d. Where did most of the carbon dioxide come from, the bacteria in the soil or the bacteria in the water? _bacteria in the soil_

5. a. Which beaker, 2 or 3, had more carbon dioxide present? _3_
   b. On the first day, what was put into Beaker 2? _sterile soil and distilled water_
   c. Does sterile soil contain any bacteria? _no_
   d. Where did most of the carbon dioxide come from, the bacteria in the soil, or the soil itself? _bacteria in the soil_

6. a. Which beaker, 3 or 4, had more carbon dioxide present? _4_
   b. What do bacteria do with glucose? _use it for food_
   c. On the first day, what was put into Beaker 4? _distilled water, regular soil and glucose_
   d. Which beaker, 3 or 4, carried on more cellular respiration? _4_

7. Why was a beaker with only water used in this experiment? _to show that it was not the water that produced the carbon dioxide; water was the control_

8. Why was sterile soil used in this experiment? _to show that it was not the soil that produced the carbon dioxide, but the bacteria in the soil_

9. What are your conclusions from this experiment? _The regular soil sample and the soil sample with glucose contained bacteria that carried on cellular respiration. Sterile soil and water only contained no bacteria and were therefore unable to carry on cellular respiration._

Name _____ Class _____ Period _____

Before doing this lab, have students read section 4:2 in text.

4-3 What Are the Traits of Blue-green Bacteria?

Living things can be classified into five kingdoms. One kingdom, Monera, includes bacteria and blue-green bacteria. Organisms in this kingdom do not have a nucleus or other cell parts, such as chloroplasts. Blue-green bacteria are different from other bacteria. They have pigments inside their cells and can make their own food. Most are blue-green in color and have the pigment chlorophyll. Blue-green bacteria live as single cells, in colonies or in filaments. The cells are often surrounded by a jelly layer.

EXPLORATION

OBJECTIVES
In this exercise, you will:
a. observe the traits of two blue-green bacteria.
b. draw and label the parts observed in each blue-green bacterium.
c. record the traits of blue-green bacteria.

Process Skills: observing, using a microscope, making and using tables, classifying

KEYWORDS
Define the following keywords:

blue-green bacteria microscopic organisms that can make their own food by photosynthesis, belong to the Kingdom Monera

colony a group of similar cells that are attached to each other

filament long, threadlike chain of cells

jelly layer clear outer covering surrounding many blue-green bacteria

*Dissolve 1.5 g methylene blue in 100 mL ethyl alcohol. Dilute by adding 10 mL solution to 90 mL of water.

MATERIALS
microscope
glass slide
coverslip
pencil, green
* methylene blue stain
  Gloeocapsa living culture
  Lyngbya living culture
  dropper

CAUTION: *Methylene blue will stain clothing.*

PROCEDURE
1. Place a drop of Gloeocapsa on a microscope slide. Add one drop of stain to the slide. Add a coverslip.
2. Observe the Gloeocapsa under low and high power. The stain will form a dark background. This will allow you to determine if the cells are surrounded by a clear jelly layer.

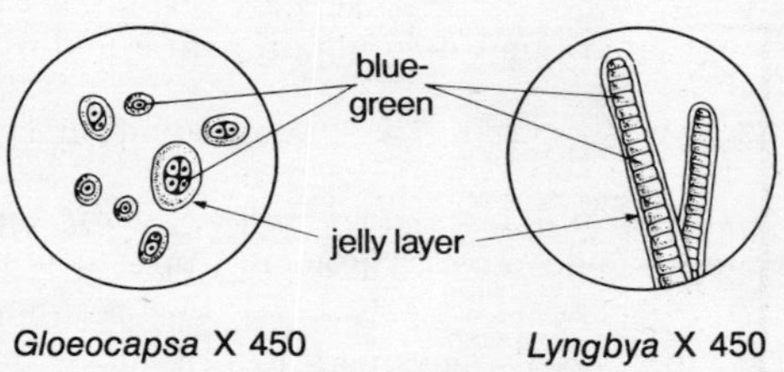

FIGURE 1. Circles for drawing blue-green bacteria

Gloeocapsa X 450   Lyngbya X 450

(labels: blue-green, jelly layer)

3. Make a drawing of Gloeocapsa in the circle provided in Figure 1. Note the color and location of the chlorophyll. Color those parts that are blue-green.
4. Complete the first column in Table 1 for Gloeocapsa.
5. Repeat steps 1 through 4 for Lyngbya.

Table 1. Comparison of blue-green bacteria

	Gloeocapsa	Lyngbya
Kingdom	Monera	Monera
Filament or colony	colony	filament
Nucleus present?	*no	no
Chlorophyll present?	*yes	yes
Chloroplasts present?	*no	no
Makes own food?	*yes	yes
Jelly layer present?	*yes	yes

*Refer students to introduction and keywords sections for help with table.

QUESTIONS

1. What is the largest number of cells that you can count within one jelly layer of Gloeocapsa? Usually 4, students' answers will vary

2. a. Is the color of Gloeocapsa the same or different from that of Lyngbya? the same
   b. What chemical provides the color? chlorophyll
   c. What does this chemical allow these organisms to do? make their own food through photosynthesis

3. Compare the jelly layer of Gloeocapsa and Lyngbya. Which is thicker? Gloeocapsa

4. What are the traits of blue-green bacteria? They contain chlorophyll and other pigments, they do not have a nucleus or cell parts, they live as single cells in filaments and colonies, they have special cells for fixing nitrogen, and special cells for reproduction.

You may expand this lab through the use of other samples of blue-green bacteria. Many are easily collected locally or can even be found on the inside glass surface of classroom aquaria.

Name _____ Class _____ Period _____

Before doing this lab, have students read section 5:1 in the text.

5-1 Where Are Protists Found?

Almost all protists are single-celled organisms that have a nucleus and other cell parts. Protists must have water in which to live. Most move around in order to search for food, darkness or other, suitable living conditions. There are three kinds of protists: animal-like protists, plantlike protists and fungilike protists. Protists can be found wherever water is found and many protists actually live inside other animals.

EXPLORATION

OBJECTIVES

In this activity, you will:
a. observe and identify protists that can be found in water samples, a termite, diatomaceous earth and scouring powder.
b. determine where animal-like and plant-like protists are likely to live.

Process Skills: using a microscope, observing, classifying, inferring

KEYWORDS

Define the following keywords:

cellulose   a compound in wood

cilia   short, hairlike parts that cover the cell membrane

false feet   parts that extend from the body of an amoeba that help the amoeba move and get food

flagellum   long, whiplike part

protist   a one-celled organism that has a nucleus and other cell parts with membranes

MATERIALS

microscope
5 glass slides
5 coverslips
forceps
prepared slide of diatomaceous earth (or materials to make a wet mount of diatomaceous earth)
marking pen
5 droppers
water samples
probe
household cleaning powder
*buffer solution
1 to 3 living termites
Zootermopsis angusticallis and *Reticulitermes flavipes* can be purchased from a biological supply house. It is less costly to collect *R. flavipes*, found in many areas in the U. S.

PROCEDURE

Part A. Examination of Water Samples

1. Obtain 3, labeled water samples from your teacher. Label a slide with the letter A as shown in Figure 1.

2. Prepare a wet mount of sample A on the slide you marked with an 'A'.

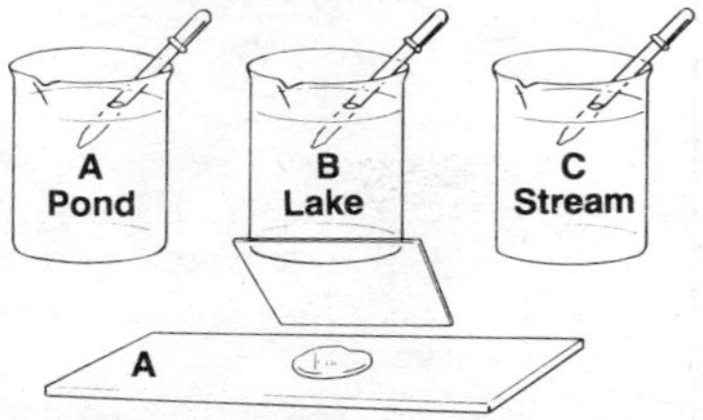

FIGURE 1. Examining water samples

* The best buffer to use for termite flagellates is Trager's solution A or U (see page 19T). If you don't have the chemicals for preparing Trager's solution, a 0.60–0.65 percent saline solution can be used. Plain water can also be used. The protists will not live as long.

3. Observe the slide under low power and search for protists. Watch for movement. Look around pieces of debris. You may need to reduce the amount of light passing through the slide in order to locate some protists that look like amoebas*.

4. When you find some protists that are not moving or are moving only very slowly, switch over to high power.

5. Draw, in the proper column in Figure 3, the protists you observe under high power.

6. Use your text and other sources to help you to identify the protists. Look for flagella, cilia, false feet, pigments and other noticeable features.

7. Repeat steps 1 through 6 for water samples B and C.

* You may wish to add a drop of methyl cellulose solution to the water or slide. See page 19T for preparation.

Part B. Examination of the Inside of a Termite

Not all protists living inside other animals are parasites. Flagellates live in the intestines of termites. The termite cannot digest the wood that it eats. The flagellates make an enzyme that digests the cellulose in the wood particles.

1. Place a drop of buffer solution on a clean slide.

CAUTION: *Handle buffer solution with care.*

2. Use forceps to place a termite on the slide near the drop of buffer.

3. Cover the termite and the buffer with a coverslip and a second clean slide.

4. With the end of a probe, gently press on the second slide in order to squash the termite. *The second slide prevents breaking coverslips.*

5. Remove the top slide. Observe the slide under both low and high power.

6. Make a drawing (or drawings) in Figure 4 of the protists that you observe in the termite. The largest protist you will see is called *Trichonympha*.

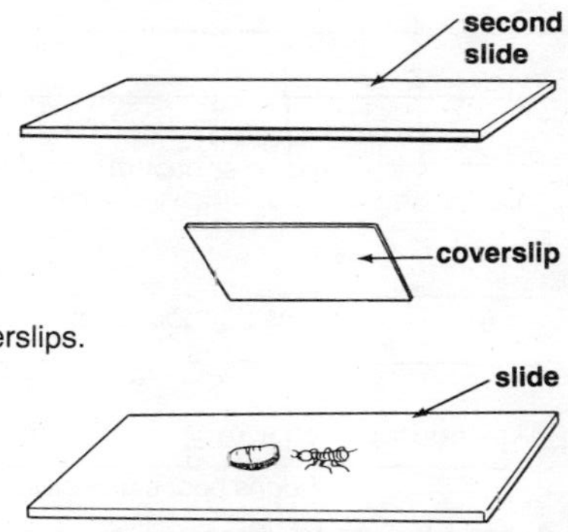

FIGURE 2. Preparing termite slide

Part C. Examination of Diatomaceous Earth and Scouring Powder.

1. Obtain a prepared slide of diatomaceous earth or prepare a wet mount of the powdered diatomaceous earth given to you.

2. Observe the slide under both low and high power.

3. Make drawings of what you observe under high power in Figure 5.

4. Use forceps to place a small amount of scouring powder on a clean microscope slide. Add a drop of water and a coverslip.

5. Observe the slide under both low and high power.

6. Make drawings of what you observe under high power in Figure 5.

Name _____ Class _____ Period _____

FIGURE 3. Protists observed in water samples

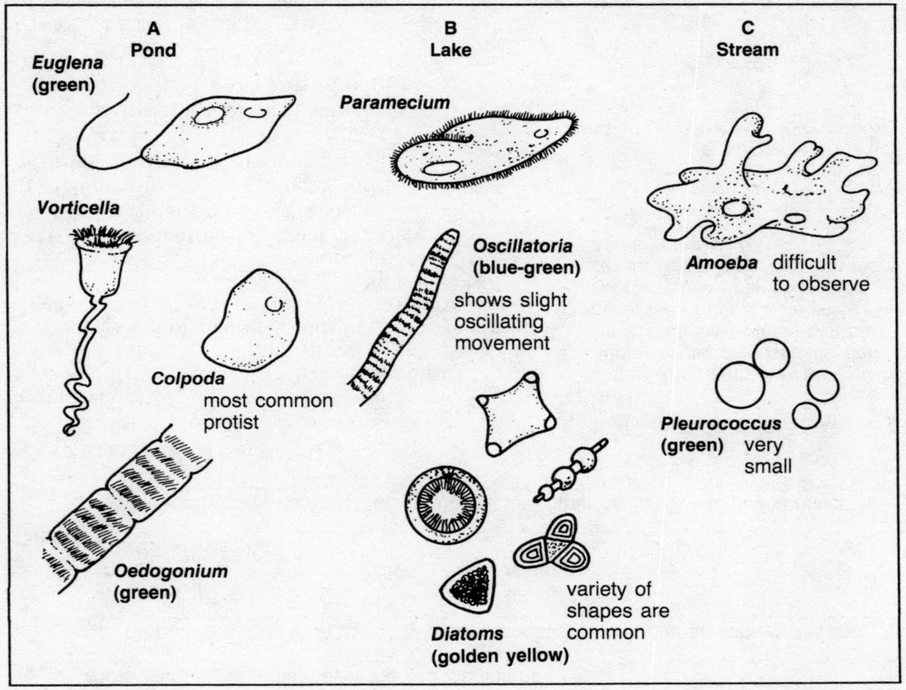

A Pond — *Euglena* (green), *Paramecium*, *Vorticella*, *Colpoda* most common protist, *Oedogonium* (green)

B Lake — *Oscillatoria* (blue-green) shows slight oscillating movement, *Diatoms* (golden yellow) variety of shapes are common

C Stream — *Amoeba* difficult to observe, *Pleurococcus* (green) very small

FIGURE 4. Termite protists

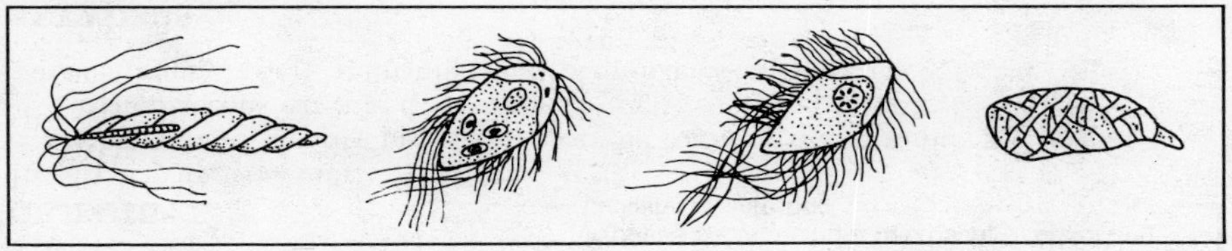

FIGURE 5. Diatomaceous earth and scouring powder

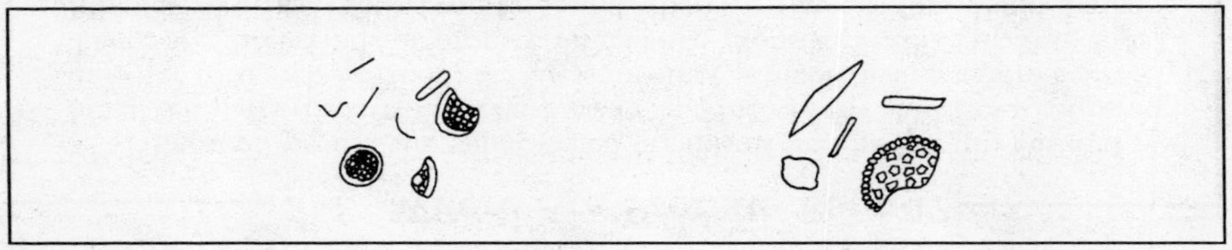

QUESTIONS

1. Which of the types of protist (animal-like or plant-like) did you find in the greatest numbers in:
   a. pond water __Answers will vary. Animal-like, ciliates.__
   b. lake water __Answers will vary.__
   c. stream water __Answers will vary.__
2. Which water sample had the least number of protists? __pond or lake__
3. Which water sample had the greatest number of protists? __stream__
4. What is the most common type of protist in your water samples? __probably ciliates__
5. How is *Trichonympha* different from the other protists that you have observed? __It has many flagellae and lives inside another organism.__
6. What does the termite use for food? __wood__
7. In what ways do the termite protists depend upon the termites? __Termite protists depend upon the termite for a place to live and a food supply.__
8. Wood is cellulose. Cellulose is made up of simple sugars. Into what compounds do the protists probably digest the wood? __simple sugars__
9. Why is the termite dependent upon the protists? __It cannot digest the wood it eats.__
10. What are the glasslike parts that you see in the diatomaceous earth. __The glasslike parts are diatoms or parts of diatoms.__
11. Can you see similar parts in the scouring powder? __yes__
12. What might be the reasons for adding diatomaceous earth to scouring powder? __Diatomaceous earth may be added to make it more abrasive and a better cleaner.__

Name _____ Class _____ Period _____

Before doing this lab, have students read section 5.2 in the text. You will need an incubator and a refrigerator. Substitute a heat vent for the incubator.

5-2 What Do Molds Need in Order to Grow?

Molds are a group of living things in the Kingdom Fungi. Many molds are called saprophytes because they feed on things that were once living. They also feed on and destroy many foods that humans eat.

When a mold spore that is carried in the air lands on food, it sprouts and sends out rootlike parts into the food. These parts break down the food on which they are growing. The food becomes a source of energy for the molds. In a short time, the mold* can completely cover the food that the spore landed upon.

INVESTIGATION

Process Skills: predicting, inferring, experimenting, separating and controlling variables

OBJECTIVES

In this exercise, you will:
a. observe the growth of molds on different kinds of food.
b. determine what conditions promote the growth of molds.
c. determine how mold growth can be slowed or stopped.

*To prepare mold, place bread in a small container. Use a dropper to moisten the bread and leave it uncovered for about an hour. Cover the container and place in a dark room. Set this up a week in advance of the lab; observe after three days.

KEYWORDS

Define the following keywords:

dehydrate _____ removing water from a substance

mold _____ a fungus that can grow on food

saprophyte _____ an organism that feeds on once-living things

spore _____ a special cell that develops into a new living thing

Have students work in groups of four or five.

MATERIALS

bleach
alcohol
instant potatoes
bread, 3 days old — cut into 3 cm squares prior to class
raisins
5 droppers
spoon
marking pen
household cleaner
household ammonia
3 brown paper bags
**16 small jars with covers
6 cotton swabs

**Small paper, plastic, or foam cups may be used in place of glass jars. Adding aluminum foil or plastic wrap covers to each cup will replace caps. This will also facilitate clean up. Keep one or two jars of fungal growth for use next year. This will provide a handy spore source. Spores will remain viable for years, even when dried.

PROCEDURE

Part A. Growth of Mold on Foods

1. Label 6 small jars 1 through 6. Place your name on each label.
2. Add the following materials to the jars.
   Jars 1 and 2–1 piece of bread, cut to fit into the jar.
   Jars 3 and 4–1 spoonful of instant potatoes.
   Jars 5 and 6–1 spoonful of raisins.
3. Add 20 drops of water to jars 2, 4 and 6.

37

4. Complete the first three columns in Table 1.
5. Get a cotton swab with mold spores on it from your teacher. Rub the swab across the surface of the food in all 6 jars. Figure 1 shows you how.
6. Place the covers on the jars and store them on a shelf for 3 days.
7. After 3 days, observe the jars for any growth of mold.
8. Complete the last column in Table 1.

2–3 bread slices with mold growing on them will be needed to inoculate the swabs. Rub swabs on spores prior to class. Place swabs on aluminum foil.

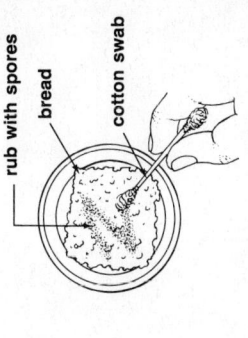

FIGURE 1. Top view of jar

Table 1. Growth of Molds

Jar number	Type of food	Water added?	Is mold growing on food?
1	bread	no	no
2	bread	yes	yes
3	potatoes	no	no
4	potatoes	yes	yes
5	raisins	no	no
6	raisins	yes	yes

Part B. Examination of Growing Conditions for Molds

1. Label 6 jars with numbers from 7 through 12.
2. Cut 6 pieces of bread and place one into each jar.
3. Add 20 drops of water to jars 7, 8 and 9.
4. Get a cotton swab with mold spores on it from your teacher. Rub the swab across the surface of the bread in each of the 6 jars.
5. Write your name on 3 brown paper bags. Put jars 7 and 10 into one paper bag. Place it in a refrigerator.
6. Put jars 8 and 11 into a second paper bag. Place this bag in an incubator at 30°C or leave it in a warm place, perhaps near a radiator.
7. Put jars 9 and 12 into the third paper bag. Place this bag on a shelf at room temperature.
8. Record the conditions for growth (use the words moist or dry, cold, warm or room temperature and light or dark) in the proper column in Table 2.
9. Begin checking for mold growth on the third day. Continue to check the jars each day for 6 more days.
10. Record in Table 2 the amount of mold you find each day in each jar. Use the following terms: none, little, some and much.

38

Name _____ Class _____ Period _____

Table 2. Growing Conditions for Molds

Jar number	Conditions	Day 3	Day 4	Day 5	Day 6	Day 7	Day 8	Day 9
7	moist, cold, dark	none	none	none	none	none	little	little
8	moist, warm, dark	none	none	little	some	some	much	much
9	moist, room temp, dark	none	none	none	some	little	some	some
10	dry, cold, dark	none	none	none	none	none	none	none
11	dry, warm, dark	none	none	none	none	none	may have a little	may have a little
12	dry, room temp, dark	none	none	none	none	none	none	none

Part C. Slowing or Stopping Mold Growth

1. Label 4 jars with numbers 13 through 16.
2. Cut 4 pieces of bread and place one into each jar.
3. Add 20 drops of water to each jar.
4. Get a cotton swab with mold spores on it from your teacher. Rub the swab across the surface of the bread in all 4 jars.
5. Add nothing to jar 13.
6. Place 10 drops of bleach across the surface of the bread in jar 14.
   CAUTION: *Avoid spilling or inhaling bleach.*
7. Place 10 drops of household cleaner across the surface of the bread in jar 15.
8. Place 10 drops of household ammonia across the surface of the bread in jar 16.
   CAUTION: *Avoid spilling or inhaling ammonia.*
9. Record in the table below the substances you have added to each jar.
10. Place the jars in a warm, dark area.
11. Observe the jars for mold growth after 3 days. Record your observations in the proper column in Table 3.
12. Continue to check on the jars each day for 6 more days.
13. Record in Table 3 your observations of mold growth for each of these days.

Table 3. Slowing or Stopping Mold Growth

Jar number	Substance added	Day 3	Day 4	Day 5	Day 6	Day 7	Day 8	Day 9
13	nothing	none	none	none	none	little	some	some
14	bleach	none	none	none	none	none	none	none
15	cleaner	none	none	none	none	little	little	little
16	ammonia	none	none	none	none	none	little	little

Caution students not to mix the bleach with the ammonia. A dangerous chemical reaction results, releasing poisonous fumes.

QUESTIONS

1. Consider the observations you made in Part A of this exercise.
   a. Does mold grow on moist food? __yes__ dry food? __no__
   b. Which jars show mold growth under these conditions?
      dry __none__ moist __2, 4, 6__
   c. What types of food does mold grow on? __all types__
2. Consider the observations you made in Part B of this exercise.
   a. What conditions caused the most growth of mold? __moist, warm__
   b. In order to prevent mold growth, how should food be stored? __cool and dry__
   c. Why does bread become moldy faster than crackers? __Bread has more moisture than crackers.__
3. Consider the observations you made in Part C of this exercise.
   a. What was the purpose of jar 13? __control__
   b. What substances seem to slow mold growth? __best; cleaner and ammonia some.__ __Bleach slowed mold growth the best; cleaner and ammonia some.__
4. What three things are needed for mold growth? __Warmth, moisture, food, and some may say darkness.__
5. Foods are often packaged several days or weeks before they are sold. Food companies add preservatives to keep molds from growing on many of these foods. How could you tell if the preservatives that are added are able to keep the molds from growing? __Put food in jars similar to 2, 4, and 6 in Part A; observe to see if mold grows.__
6. Many dehydrated foods do not contain chemical preservatives. What prevents molds from growing on these foods? __The water has been removed, and mold needs moisture.__

Name _____ Class _____ Period _____

Before doing this lab, have students read section 6:1 in the text.

6-1 What Are the Traits of Vascular and Nonvascular Plants?

EXPLORATION

Roots, stems and leaves are special plant structures. Roots anchor plants in the soil and take in water and minerals. Stems carry water to all parts of the plant, support the plant and hold the leaves up to the sunlight. Leaves make food. These structures are traits of plants such as ferns, conifers and flowering plants. These are vascular plants. Many simple plants do not have these parts. Vascular plants have special tubelike structures in the roots, stems and leaves to carry food and water. Nonvascular plants do not have the tubelike structures.

Process Skills: using the microscope, observing, classifying, scientific terminology

OBJECTIVES
In this exercise you will:
a. use models to determine the differences in vascular and nonvascular plants.
b. examine and compare vascular and nonvascular plants under the microscope.

KEYWORDS
Define the following keywords:

nonvascular plant ___ plants that do not have tubelike structures in roots, stems, and leaves ___

tubelike cells ___ structures that carry food and water ___

vascular plant ___ plants that have tubelike structures in roots, stems, and leaves ___

MATERIALS
2 drinking straws   liverwort
green pencil
scissors   lettuce   microscope
razor blade single edge celery   2 glass slides
CAUTION: *Students should be extremely careful with razor blades.*

PROCEDURE
Part A. Constructing Plant Models.
1. The diagrams in Figure 2 on the last page of this lab are outline models of two plants. One is marked vascular plant and the other is marked nonvascular plant. Color parts A and B green. These are the parts that contain chlorophyll.
2. Tape properly cut straws along the dashed lines *on only one model.* The straws represent tubelike cells. Review the definitions of vascular and nonvascular plants in order to determine on which model to place the straws. CAUTION: *Use care with scissors.*

3. Plant parts that appear to be leaves but lack the tubelike cells are said to be leaflike. Label parts A and B as either leaf or leaflike.
4. Plant parts that appear to be stems but lack the tubelike cells are said to be stemlike. Label parts C and D as either stem or stemlike.
5. Plant parts that appear to be roots but lack the tubelike cells are said to be rootlike. Label parts E and F as either a root or rootlike. *These leaflike, stemlike and rootlike parts are features of nonvascular plants.*

Part B. Examination of vascular and nonvascular plants.
1. Obtain a piece of lettuce. Use a razor blade to cut out a small section (Figure 1). CAUTION: *Use extreme care with razor blades.*
2. Place the section on a glass slide. Add two or three drops of water.
3. Place a second slide over the first slide. Squash the slides together (Figure 1).
4. Examine the lettuce under low power magnification only.
5. Look for long, tubelike cells. They may appear as spirals or train tracks. Draw a diagram of what you see in the appropriate place in Table 1.
6. Obtain a piece of liverwort from your teacher. Repeat steps 1 through 5. Draw a diagram in Table 1 of what you see.
7. Obtain a piece of celery from your teacher. Prepare the celery as shown in Figure 1. Cut off a 1 cm piece of the string. Repeat steps 2 through 5. Draw a diagram in Table 1 of what you see.
8. Complete the last two columns in Table 1.

Table 1. Comparison of Vascular and Nonvascular plants

Plant	Diagram	Tubelike cells present or absent	Vascular or nonvascular
Lettuce		present	vascular
Liverwort		absent	nonvascular
Celery		present	vascular

CAUTION: *Students should not use high power because of the double slides.*

Name _____ Class _____ Period _____

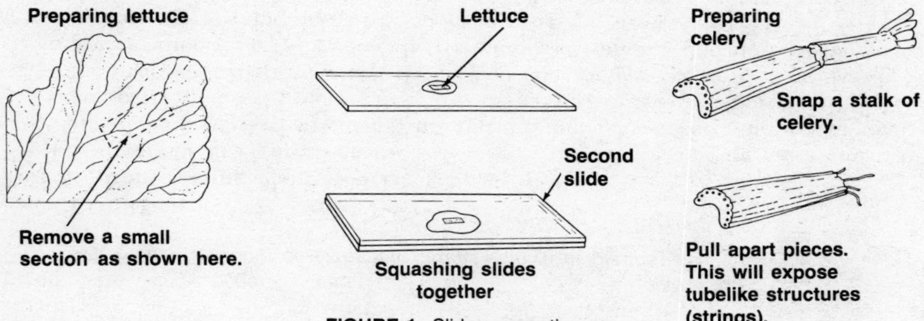

Preparing lettuce

Preparing celery
Snap a stalk of celery.

Pull apart pieces. This will expose tubelike structures (strings).

Remove a small section as shown here.

Lettuce
Second slide
Squashing slides together

FIGURE 1. Slide preparation

QUESTIONS

1. What does the green color used in coloring the plant models represent? chlorophyll

2. What do the straws used in the model represent? tubelike structures

3. List one way in which vascular and nonvascular plants are alike. Both have chlorophyll.

4. List three ways in which vascular and nonvascular plants are different. Use your model labels for help. Vascular plants have roots, stems, and leaves. Nonvascular plants have rootlike, stemlike, and leaflike parts.

5. Explain how tubelike cells in vascular plants are used. These parts carry food and water.

6. Explain the difference between stems and stemlike parts. Stemlike parts look like stems but do not have tubelike parts inside.

7. Explain the difference between roots and rootlike parts. Rootlike parts look like roots but do not have tubelike parts inside.

8. Consult the models. Are tubelike cells continuous throughout the plant if the plant is vascular? yes How is this helpful to the plant? Food and water are carried to all parts.

9. Would you expect to find roots, stems, and leaves in lettuce, liverwort, and celery? lettuce and celery only Why? You can see the tubelike parts in the lettuce leaves and in the celery. You cannot see tubelike parts in liverworts.

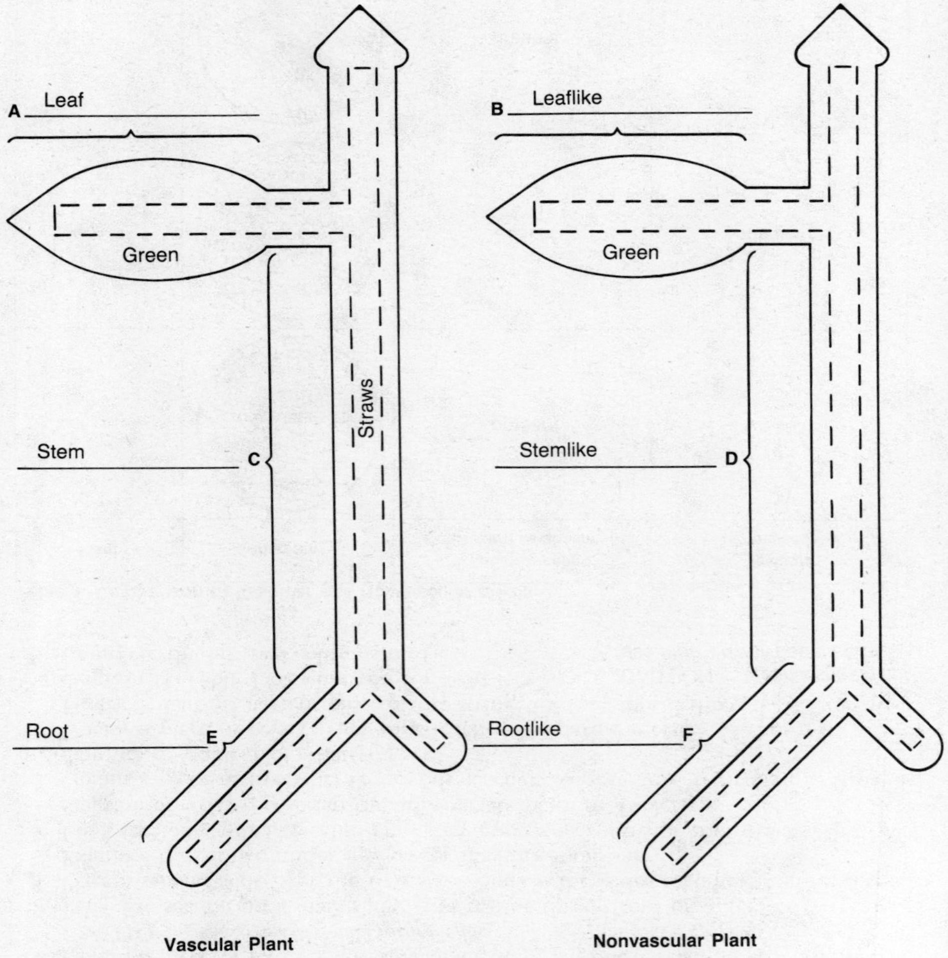

43

A — Leaf
Green
B — Leaflike
Green
C — Stem
Straws
D — Stemlike
E — Root
F — Rootlike

Vascular Plant   Nonvascular Plant

FIGURE 2.

44

304

Name _____ Class _____ Period _____

Before doing this lab, have students read section 6:3 in the text.

6-2 How Can Conifers Be Identified?

Perhaps you have followed a set of directions to get to some place that you wanted to go. Biologists use a set of directions to identify the many kinds of living things. The set of directions used to identify something is called a key. If you follow directions and make correct choices with a key, you can identify living things.

A group of plants called conifers can be identified by their leaves. Conifers are cone-bearing plants. Leaf shape and how the leaves are arranged on stems are features used to identify the different conifers.

EXPLORATION

Process Skills: observing, classifying, making and using keys, recognizing and using spatial relationships

OBJECTIVES
In this exercise, you will:
a. examine the leaves of some conifers.
b. use a key to identify the conifers from which the leaves came.

KEYWORDS
Define the following keywords:

conifer _a cone-bearing plant_

key _a set of directions to identify something_

MATERIALS
twigs of various conifers
hand lens   metric ruler

PROCEDURE
1. Examine the numbered twigs of the conifers carefully with the hand lens.
2. Record the features of the conifers in Table 1. Use the diagrams shown in Figure 1 to help you.
3. Pick one twig to key. Read question 1 of the key in Table 2. Answer the question, yes or no, using the twig you selected.

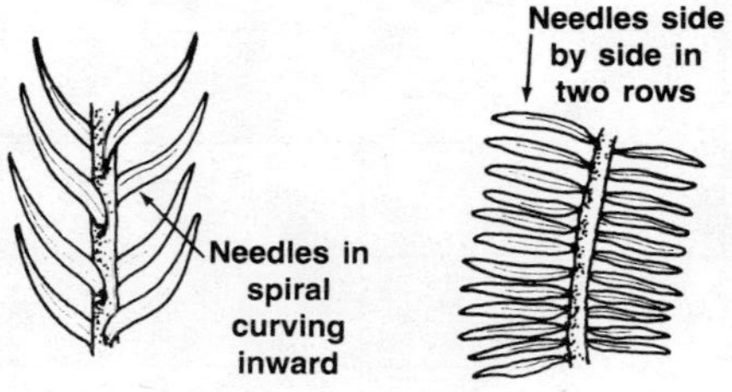

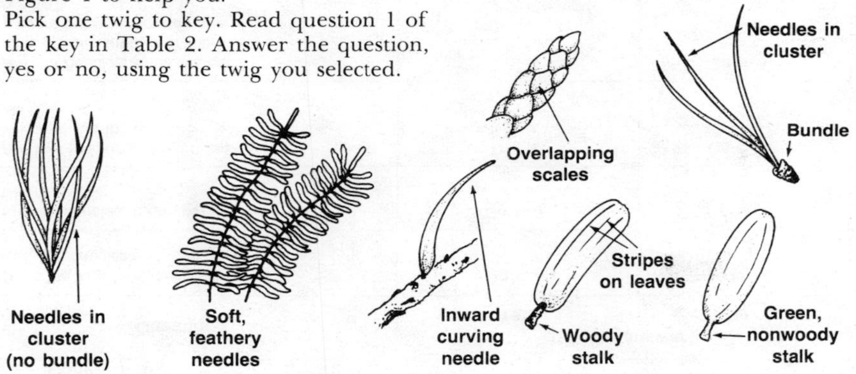

FIGURE 1. Characteristics of conifers

Answers will vary depending on the conifers examined. Sample data are given.

Table 1. Identifying Conifers

Plant	Observations	Kind of Conifer
1	leaves like needles in clusters of three	pine
2	needles soft and flat / needles growing in spiral pattern	Douglas fir
3	two kinds of leaves (scales and needles)	juniper
4	needles that are flat and rounded on end; gray stripes on bottom of leaf	hemlock
5	leaves not needlelike but like overlapping scales	cedar
6	needles with four sides, sharp pointed needles	spruce
7	light green leaves, soft and featherlike	bald cypress
8	needles over 5 cm long, 6 or more leaves in a bundle	larch

305

Name _____ Class _____ Period _____

You may need to go step-by-step with students so they will remember how to use a key.

Table 2. Conifer Key

Questions	If you answer Yes	No
1. Does the twig have leaves that look like needles?	Go to 2.	Go to 12.
2. Are the needles in groups of two or more?	Go to 3.	Go to 5.
3. Are the needles at least 5 cm long and are there 2 to 5 enclosed in a bundle?	This is a pine.	Go to 4.
4. Are the needles at least 5 cm long and are there 6 or more present but not enclosed in a bundle?	This is a larch.	Go to 5.
5. Do the needles curve inward and grow in a spiral pattern?	Go to 6.	Go to 8.
6. Are the needles soft and flat?	This is a Douglas fir.	Go to 7.
7. Do the needles have four sides? Are the needles sharp?	This is a spruce.	Go to 8.
8. Do the needles grow straight and side by side in two rows on the twig?	Go to 9.	Go to 10.
9. Are the leaves light colored, soft, and featherlike?	This is a bald cypress.	Go to 10.
10. Do the needles have gray stripes on the bottom and are they attached to the stem by a short woody stalk?	This is a hemlock.	Go to 11.
11. Do the needles have two silver stripes on the bottom and are they attached to the stem by a short green stalk?	This is a fir.	Go to 12.
12. Do the twigs have only one kind of leaf that looks like overlapping fish scales?	This is a cedar.	Go to 13.
13. Do the twigs have two kinds of leaves—one that looks like overlapping scales and the other that looks needlelike?	This is a juniper.	This is the end. If you haven't named the plant, go to 1 again.

4. Based on your answer, follow what the key says to do. Move to the next question where the key says to go. Answer this question, yes or no.
5. Based on your answer, do what the key says to do. Follow the key until it tells you the identity of your conifer twig. Record your answer in the last column of Table 1.
6. Repeat steps 3 through 5 with the other twigs. Record your results.

QUESTIONS

1. List two shapes of conifer leaves. __needles and scalelike__
2. Which conifers have leaves that are flat? __Douglas fir, bald cypress, hemlock, fir__
3. Using the traits in the key, list two traits of the bald cypress. __light green leaves that are soft and featherlike__
4. List the kinds of conifers that have leaves that look like overlapping scales. __cedars__
5. Which conifers have leaves in clusters? __pine and larch__
6. List two ways in which hemlock and fir leaves are alike? __both are striped and have stalks__
7. Describe a spruce leaf. __needlelike with four sides__
8. A tree has leaves that have two stripes. What else do you need to know before you can identify the tree? __if the stalk is woody or green__
9. Below are some descriptions of different conifer leaves. If you can identify the tree from the description, write the name on the blank. If you cannot identify the conifer, write "cannot identify" on the blank.
   a. needles present __cannot identify__
   b. leaves like overlapping fish scales __cedar__
   c. needles flat __cannot identify__
   d. needles in bundles of 2 to 5 __pine__
   e. leaves light colored and featherlike __bald cypress__
   f. needles with four sides __spruce__
   g. leaves grow in spiral on twig __Douglas fir__

Name _____ Class _____ Period _____

Before doing this lab, have students read section 7:2 in the text.

7-1 What Traits Does a Sponge Have for Living in Water?

Sponges are simple animals that live in water. They do not move about. Sponges do not have the same traits that other animals in the animal kingdom have. They are divided into classes based upon structures that support the sponge body.

Sponges do not have real skeletons. Many have tiny glasslike parts inside their bodies that give support. These glasslike parts are called spicules. Sometimes people refer to the spicules of a sponge as a skeleton.

EXPLORATION

Process Skills: observing, recognizing and using spatial relationships, inferring, scientific terminology

OBJECTIVES
In this exercise you will:
a. observe and compare two different sponge types, *Grantia* and *Spongia*.
b. observe the spicules that support each sponge.

KEYWORDS
Define the following keywords:

body _____ main part of a sponge _____

osculum _____ large opening at top of the sponge _____

pores _____ many tiny holes in the sponge _____

spicules _____ structures that provide support _____

sponge _____ a simple invertebrate that has pores and lives in water _____

MATERIALS
probe
scissors
10 mL bleach
  in bottle with dropper
hand lens
microscope
water
  in bottle with dropper
2 glass slides
2 coverslips
petri dish
Grantia sponge
Spongia sponge

Grantia is available from biological supply companies.
Spongia is available in paint stores as commercial sponge.

PROCEDURE
Part A. *Grantia* Sponge

1. Place the sponge in a small glass dish. Examine the sponge with a hand lens. Observe the many tiny holes. These bring water into the sponge. Food is present in the water that enters the sponge. The food is trapped by cells which line the hollow inside of the sponge. Observe the larger hole. The larger hole allows the water to leave the sponge. It is called the osculum.
2. Label the body, pores and osculum of the sponge in Figure 1.

49

3. Carefully place the probe into one of the small holes. Determine if the probe enters the center of the animal.
4. Place the probe in the larger hole called the osculum. Determine if the probe enters the center of the animal.
5. Insert the point of the scissors into the osculum of the sponge. Carefully cut the sponge lengthwise. Use the hand lens to examine the inside of the sponge. Sponges have only two cell layers. **CAUTION:** *Use care with scissors.*
6. Locate several canals passing through the body wall. The canals are lined with flagella. The flagella set up currents of water that cause the water to move through the sponge.
7. In Figure 2 draw arrows to show how water enters and leaves the sponge. Refer to steps 1 and 2 if necessary.

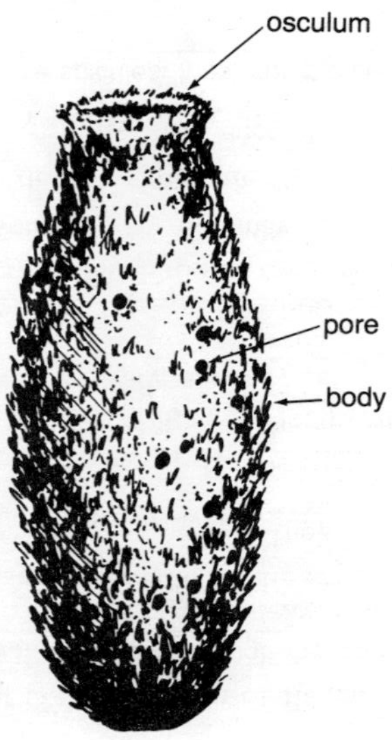

FIGURE 1. Sponge body parts (osculum, pore, body labeled)

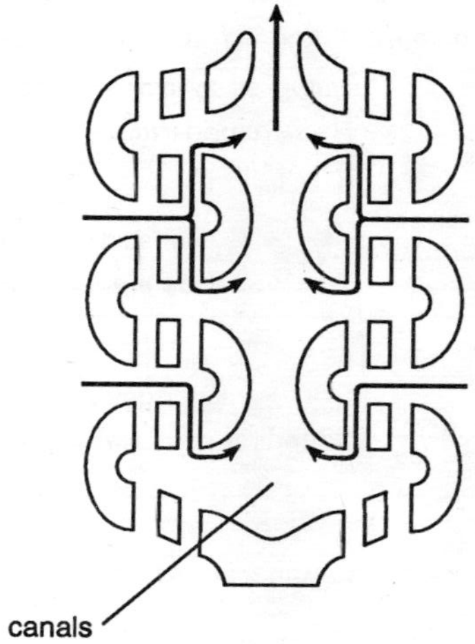

FIGURE 2. Water flow in a sponge (canals labeled)

8. Add two drops of bleach to a microscope slide. **CAUTION:** *Be careful not to spill the bleach on your hands or clothing. Avoid inhaling.*
9. Put a small piece of *Grantia* in the bleach. When a sponge is placed in bleach the animal's cells begin to break apart while the spicules remain unchanged.
10. Add a coverslip. Observe *Grantia* using low power. The structures that you see are called spicules. They should look similar to the ones shown in Figure 3. *Grantia* spicules are made of calcium carbonate.

50

307

Name _____ Class _____ Period _____

11. Draw some spicules in the space provided.

FIGURE 3. *Grantia* spicules

Part B. *Spongia* Sponge

1. Examine the *Spongia*. Note the many pores present that allow food and water to enter the animal.
2. The *Spongia* has a hard support system that can be observed with the microscope. Add two drops of water to a microscope slide.
3. Pull off a small piece of *Spongia* and place it in the water.
4. Add a coverslip. Using low power, observe the *Spongia*.
5. Draw what you see in the space provided. The spicules that you see are composed of spongin. Spongin is made of protein fibers.

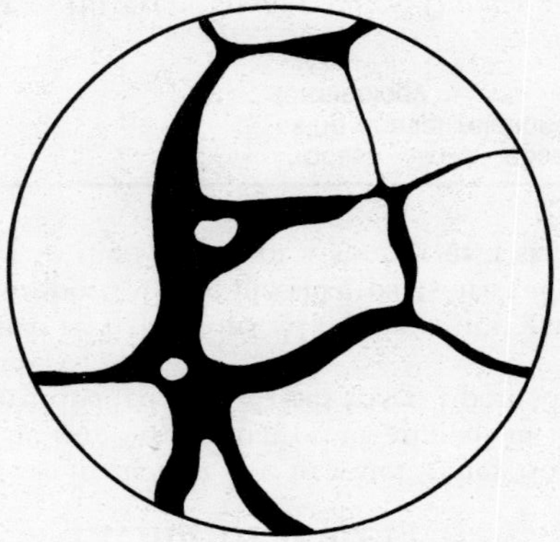

QUESTIONS

1. Where do sponges live? __in water__
2. From what you have observed do sponges have any parts that might help them move about? __no__
3. What are three traits of sponges? __Sponges have pores, spicules, two cell layers, and live in water.__
4. How does water enter and leave the sponge? __Water enters through the pores and leaves through the opening in the top center of the sponge.__
5. How do sponges get food? __Food is in the water that enters the pores. The food is trapped by cells that line the hollow inside of the animal.__
6. What is the function of flagella? __The flagella set up currents of water that cause the water to move through the sponge.__
7. Are spicules made of the same material as the rest of the sponge? __no__ Explain your answer. __They are made of a harder substance.__
8. What do spicules do? __provide support__
9. Describe the shape of the spicules from the *Grantia* sponge. __Answers will vary, and may be: like spikes, needles, pins, etc.__
10. Describe the shape of the spicules from the *Spongin* sponge. __Answers will vary, and may be: tough fibers, not as needlelike as Grantia.__
11. Why do you think the *Grantia* sponge is different from the kinds used to wash cars or walls? __The kinds of sponges used to wash cars or walls do not have glasslike spicules. Most commercial sponges contain spongin or are synthetic.__
12. What traits does a sponge have for living in water? __The sponge has pores, an opening in the top center, spicules for support, and flagella.__

Name _____ Class _____ Period _____

Before doing this lab, have students read section 7:4 in the text.

7-2 What Are the Parts of the Squid?

Squid are soft-bodied animals that live in the ocean. They are among the quickest members of the invertebrates. Some can swim as fast as 20 km per hour and leap as much as 3 m from the surface of the water. They vary in size from 5 cm to 6 m in length. The giant squid is the largest of all invertebrates. It has tentacles that range from 9 to 12 m in length.

Squid feed on smaller invertebrates. They use two methods to avoid being eaten, themselves. They release an inky fluid that hides them and allows them to swim away unseen, and they can disappear into their surroundings by changing color to match the background.

EXPLORATION

Process Skills: observing, recognizing and using spatial relationships, inferring, interpreting scientific illustrations

OBJECTIVES
In this exercise, you will:
a. observe the external parts of the squid.
b. dissect and locate the internal parts of the squid.

KEYWORDS
Define the following keywords:

external _____ outside the body _____

gonad _____ reproductive organ _____

internal _____ inside the body _____

mantle _____ the body covering of the squid _____

pen _____ tough, transparent shell of the squid _____

MATERIALS
squid
scissors
dissecting pan
forceps
hand lens
colored pencils: red, green, gray, blue, yellow, and purple

Frozen squid can be purchased at large food chain stores (in large quantities if sufficient notice is given). This avoids handling specimens preserved in formaldehyde.

PROCEDURE
Part A. External Parts of Squid

1. Place the squid right side up in the dissecting pan. Stretch out the two tentacles and eight arms.
2. Examine one tentacle with the hand lens. Note the small suction cup parts on the tentacle.
3. The body is the largest part of the animal. Find the head. It appears loosely attached to the body. The head has eyes on each side.

53

4. Behind the head the body is covered by the mantle. The two small flaps at the tail end of the body are fins.
5. The opening to the body just behind the head is the collar. Locate the water jet on the underside where the collar opening and head meet. Squid swim by squirting water away from their water jet. The direction the water jet faces allows the animal to swim backward or forward.
6. Separate the arms and find the mouth. Notice the dark spot in the center of the mouth. These are the jaws. With the forceps, carefully pull on the upper jaw and remove it. Use the same procedure to remove the lower jaw. Place it with the upper jaw and observe how they work together.
7. Label the tentacle, suction-cup part of the tentacle, arm, body, mantle, head, eye, fin, collar, water jet, and mouth on Figure 1.

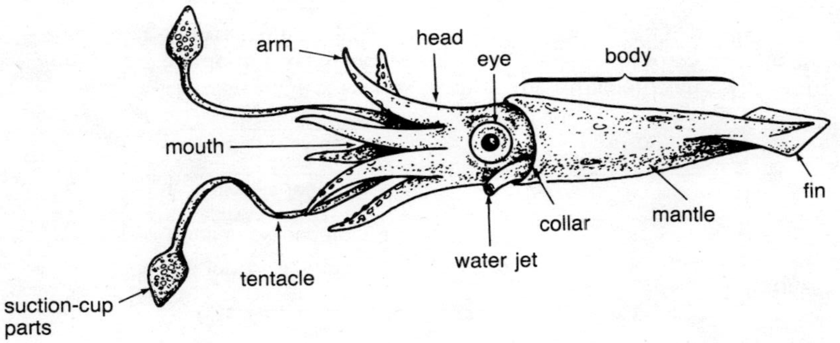

FIGURE 1. Squid external anatomy

Part B. Internal Parts of the squid

1. Turn the squid over so that it is on its back. Use forceps to hold the mantle up away from the internal organs. With scissors, carefully cut through the mantle. CAUTION: *Use care with scissors.* Place the blade of the scissors under the collar and cut toward the fins all the way to the end.
2. Lay the mantle open. Use care in finding the internal organs. Find the silvery black ink sac.
3. Locate the esophagous, the tube that leads from the mouth. Food passes through the esophagous.
*4. Locate the stomach, the white organ attached to the esophagous at the end opposite the mouth.
5. Locate the liver. It is located beneath the place where the ink sac was found.
6. The gonad is the reproductive organ. It is a mass near the far end of the mantle. The gonad is white in the male and orange in the female.
7. Locate the gills, two long organs that look like feathers. The gill hearts are located at the base of the gills.
*It is difficult to distinguish the stomach and liver but students can find the general area.

54

Name _____ Class _____ Period _____

8. Probe the mantle near the fins with the scissors. Find a hard object. Grasp this hard object with the forceps and pull straight out from the body. This tough, transparent part is called the pen. It is the shell of the squid.
9. Remove an eye. Make a small cut in the center and find a round object. This is the lens.
10. In Figure 2 label the water jet, ink sac, stomach, liver, gonad, gill and gill heart.

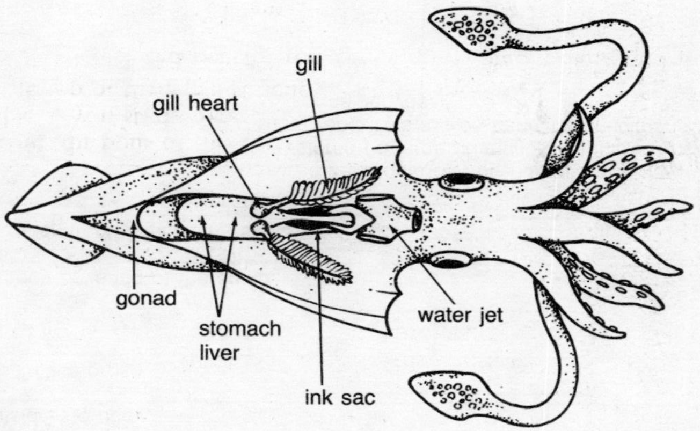

FIGURE 2. Internal parts of the squid

QUESTIONS

1. Describe the shape of the squid's body. cylindrical, ends with two fins

2. How might the body shape help it as it swims? It helps it to move through the water faster.

3. What do squid feed upon? smaller invertebrates

4. How might the eyes, tentacles and arms help the squid to capture food? Eyes help find food, tentacles and arms help grasp food and take it into the body.

5. How does the squid use its ink sac? It releases an inky fluid that hides the squid until it can swim away.

55

6. What is the function of the squid shell? to provide support
7. Figure 3 shows several squid parts.
   Color red the part used for finding food.
   Color green the part used for holding food.
   Color gray the part used for tearing food.
   Color blue the part used to support the body.
   Color yellow the part used to help the squid swim forward or backward.
   Color purple the part that forms the body covering.

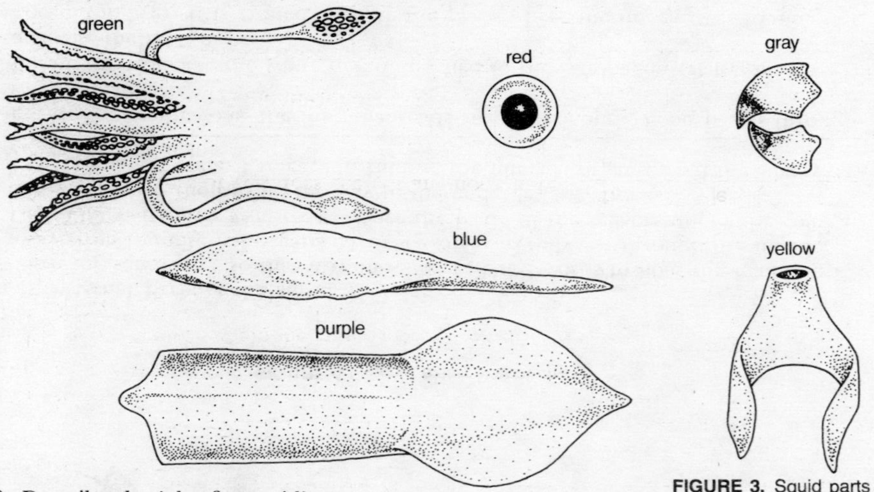

FIGURE 3. Squid parts

8. Describe the job of a squid's
   a. gills. takes in oxygen from the water
   b. stomach. digests food
   c. gonad. used in reproduction

9. Shade in the correct arrow ends to show the direction water will squirt out from each jet. Shade in the correct arrow ends to show the direction the animal moves as water squirts from its jet.

FIGURE 4. Squid and water directions

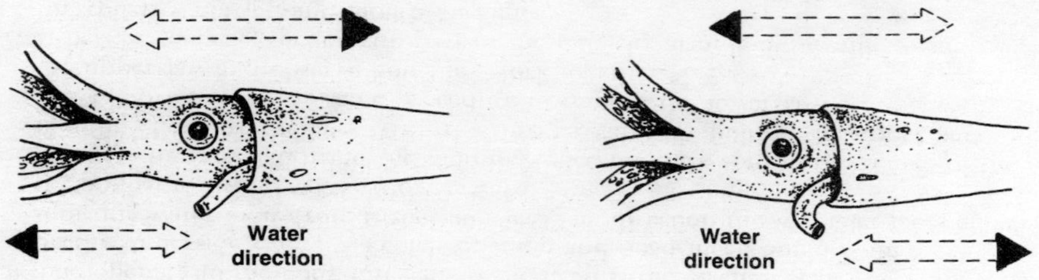

56

310

Name _____ Class _____ Period _____

Before doing this lab, have students read sections 7:2 in the text.

7-3 What Are the Parts of a Stinging-cell Animal?

Stinging-cell animals have two basic body forms. One is free-swimming and the other remains attached to one place. The basic difference between the two is that the free-swimming form has a mouth and tentacles facing downward. The form that remains attached has a mouth and tentacles that face upward. The sea anemone is an example of the attached type. The sea anemone is so named because it resembles a flower called the anemone.

Process Skills: observing, inferring, making and using tables, interpreting scientific illustrations

EXPLORATION

OBJECTIVES
In this exercise, you will:
a. observe the external parts of a sea anemone.
b. dissect and locate the internal parts of a anemone.

KEYWORDS
Define the following keywords:

mouth _____ opening that leads to the pharynx

pharynx _____ the part of the body lined with cilia

tentacle _____ structure with stinging cells that captures food

MATERIALS
forceps
scalpel
scissors
hand lens
dissecting pan
preserved sea anemone

PROCEDURE
1. Place the sea anemone in the dissecting pan.
2. Examine the sea anemone and find each of the parts labeled in Figure 1. Write the functions of the parts in the table as you study them.
3. Look at the top of the sea anemone and find the mouth and tentacles. The bottom of the sea anemone is the disc by which the animal can attach itself to a rock or other surface.
4. Use the scalpel to cut through the center of the sea anemone from mouth to disc, to make two halves. CAUTION: *Always be careful when using a scalpel.*
5. Use the hand lens to examine the outer and inner layers. Find the jellylike layer between them.
6. Find the mouth with its ringlike canal. The mouth leads into the pharynx. The pharynx extends from on-half to two-thirds the way down into the body cavity. Use the hand lens to find the grooves lined with cilia that run lengthwise in the pharynx. These cilia beat downward to create an incoming current of water that brings in oxygen and small particles of food. Other cilia beat upward to remove water from the anemone. This water takes with it carbon dioxide and other wastes.
7. Find the reproductive structures located on the sides of the body cavity below the pharynx.
8. Use the scalpel to open a tentacle and look inside it.
9. Dispose of the dissected sea anemone as your teacher directs. Wash your equipment and your hands.

Table 1. Stinging-cell Animal Parts

Mouth	food enters and wastes leave
Tentacles	capture food
Body	support
Disc	attachment
Pharynx	lined with cilia that bring in oxygen and food and remove wastes
Body cavity	removes wastes; has digestive and reproductive tissues
Reproductive structures	produces eggs and sperm

QUESTIONS
1. What does the sea anemone remove from the water it takes in? _____ oxygen _____ and small particles of food
2. What is removed with the water that leaves a sea anemone's body? _____ carbon dioxide and other wastes
3. How many layers of cell does the sea anemone have? _____ two, with a jellylike layer between them
4. How does the sea anemone capture food? _____ with its tentacles
5. How does the pharynx create an incoming current of water? _____ the cilia beat downward
6. How does the sea anemone attach to a surface? _____ with its disc
7. Describe the inside of a tentacle. _____ The center is hollow with two layers of cells surrounding it.

FIGURE 1. Sea Anemone

(labels: tentacles, mouth, body cavity, disc, pharynx, body, reproductive structures)

57

58

311

Name _____ Class _____ Period _____

Before doing this lab, have students read section 8:1 in the text.

8-1 What Are The Traits Of Certain Jointed-leg Animals?

Ticks, fleas, bedbugs, and lice belong to the phylum of jointed-leg animals. Jointed-leg animals share certain traits that separate them from other animal groups. Which traits do they share?

Each jointed-leg animal has a life cycle in which it changes from an egg to a young animal to an adult. The life cycle of a louse is typical of other jointed-leg animals.

EXPLORATION

OBJECTIVES
In this exercise, you will:
a. observe four jointed-leg animals and determine the traits of jointed-leg animals.
b. complete the life cycle of a louse.
c. determine what diseases are carried by certain jointed-leg animals.

Process Skills: classifying, sequencing, making and using keys

KEYWORDS
Define the following keywords:

antennae _____ sense appendages on the head

jointed-leg animal _____ an invertebrate with jointed appendages

life cycle _____ the changes that occur as an organism grows from an egg to a young animal to an adult

segment _____ a body section

MATERIALS
scissors   ruler
tape   preserved tick, flea, bedbug and louse
petri dish   stereoscopic microscope   substitute hand lens
forceps

Use fowl tick *Argas* or *Dermacentor andersoni* (*D. andersoni*), dog or cat flea *Ctenocephalides*, human head louse, and *Cimex* bedbug. Prepared slides may be used for the flea and the louse.

PROCEDURE
Part A. Traits of Four Jointed-leg Animals
1. Obtain a glass dish that contains a tick.
2. Examine the tick using a stereoscopic microscope.
3. Use forceps to turn the animal over so that you can observe both sides.
4. Measure the length of the body of the tick by placing a metric ruler under the glass dish alongside the animal. You can read the millimeters on the ruler through the stereoscopic microscope.
5. Complete the column in Table 1 for the tick. You will notice that one feature is already filled in.
6. Repeat steps 1 through 5 using a flea, a bedbug, and a louse.

Table 1. Traits of Four Jointed-leg Animals

Trait	Tick	Flea	Bedbug	Louse
Wings present?	No	No	No	No
Antennae present?	No	No	Yes	Yes
Antennae in segments?	—	Yes	Yes	Yes
Three body regions easily seen?	No	Yes	Yes	Yes
Abdomen in segments?	No	Yes	Yes	Yes
Legs in segments?	Yes	Yes	Yes	Yes
Number of legs?	8	6	6	6
Claw-like ends on legs?	No	Yes	No	Yes
Feed on human	Blood	Blood	Blood	Blood
Body length in mm	4 mm*	2 mm	6 mm	2 mm

*may vary

Part B. Life Cycle of a Louse
Locate Figure 1 of the life cycle of a louse. Complete the life cycle by following the steps below.
1. Note the drawings numbered 1 through 4 below the life cycle (Figure 1). Cut each one out along the solid lines that enclose the drawing. **CAUTION:** *Use care with scissors.*
2. Arrange them on the life cycle figure according to the position of the numbers.
3. Tape them in place.
4. Also beneath the life cycle are boxes containing a series of facts. Cut these boxes of facts out along the solid lines.
5. Place the facts onto the figure so that they best describe the correct diagram or event. The facts should be placed in the spaces outlined by the dashes.
6. Tape the facts in place.

Part C. Diseases Carried by Jointed-leg Animals
Each of the four animals studied in Part A can carry diseases to humans. The animal itself is not the cause of the disease. The animals carry certain disease-causing organisms in or on their bodies. When they bite a person, they may pass the disease-causing organism on to that person.
1. Table 2 lists five diseases. Complete the table by writing in the name of the jointed-leg animal that can carry that disease to a human.
2. The disease-carrying animal is pictured within Table 2.

Name _____ Class _____ Period _____

FIGURE 1. Life cycle of head louse

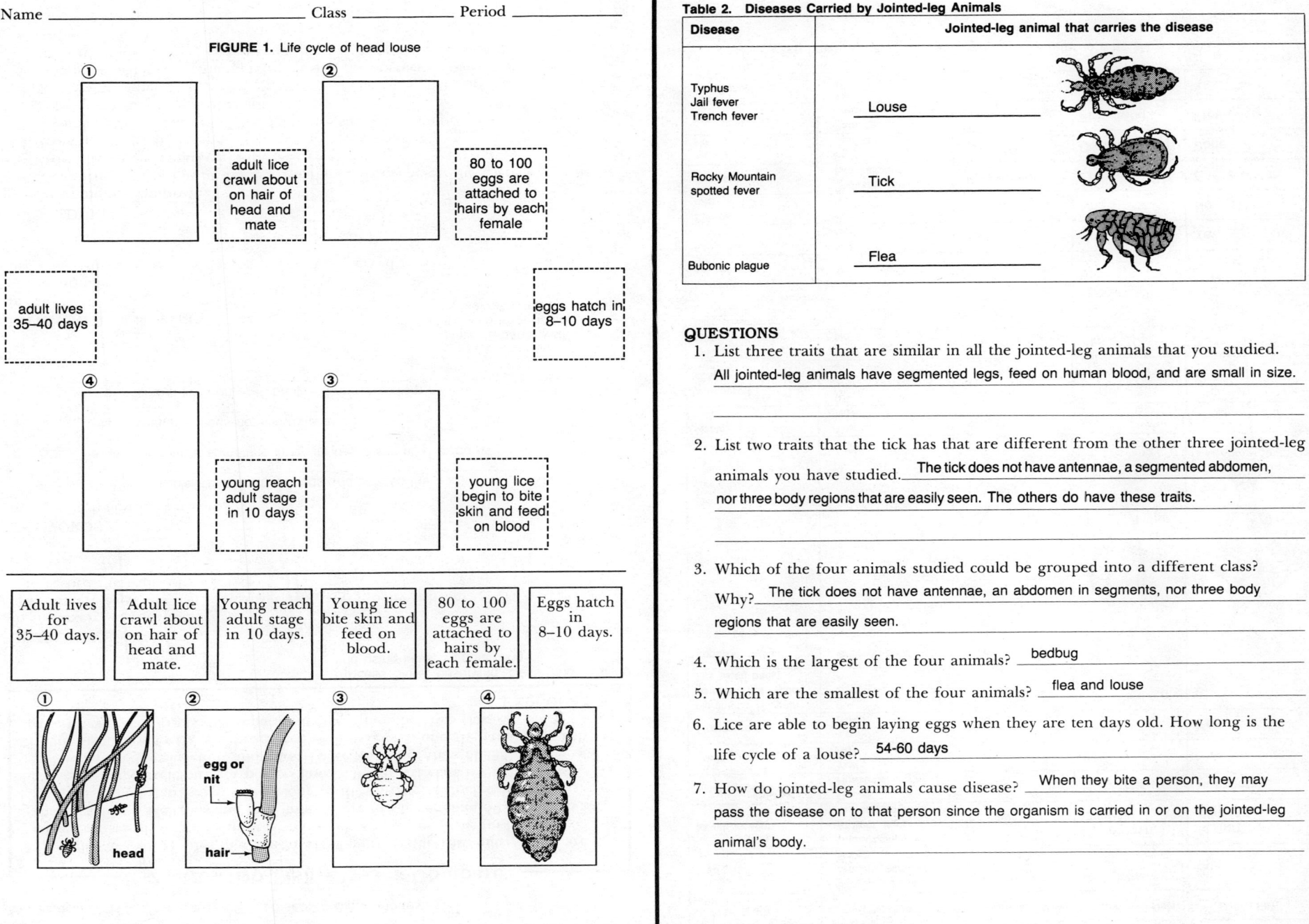

① [ ]   adult lice crawl about on hair of head and mate
② [ ]   80 to 100 eggs are attached to hairs by each female
         eggs hatch in 8-10 days
③ [ ]   young lice begin to bite skin and feed on blood
④ [ ]   young reach adult stage in 10 days
         adult lives 35-40 days

① Adult lives for 35-40 days. | Adult lice crawl about on hair of head and mate. | Young reach adult stage in 10 days. | Young lice bite skin and feed on blood. | 80 to 100 eggs are attached to hairs by each female. | Eggs hatch in 8-10 days.

① head
② egg or nit / hair
③
④

Table 2. Diseases Carried by Jointed-leg Animals

Disease	Jointed-leg animal that carries the disease
Typhus Jail fever Trench fever	Louse
Rocky Mountain spotted fever	Tick
Bubonic plague	Flea

QUESTIONS

1. List three traits that are similar in all the jointed-leg animals that you studied. All jointed-leg animals have segmented legs, feed on human blood, and are small in size.

2. List two traits that the tick has that are different from the other three jointed-leg animals you have studied. The tick does not have antennae, a segmented abdomen, nor three body regions that are easily seen. The others do have these traits.

3. Which of the four animals studied could be grouped into a different class? Why? The tick does not have antennae, an abdomen in segments, nor three body regions that are easily seen.

4. Which is the largest of the four animals? bedbug

5. Which are the smallest of the four animals? flea and louse

6. Lice are able to begin laying eggs when they are ten days old. How long is the life cycle of a louse? 54-60 days

7. How do jointed-leg animals cause disease? When they bite a person, they may pass the disease on to that person since the organism is carried in or on the jointed-leg animal's body.

Name _____ Class _____ Period _____

Before doing this lab, have students read section 8:2 in the text.

8-2 How Do Fish Classes Compare?

Fish are vertebrates that live in water and breathe with gills. There are three classes of fish. The most primitive class is jawless fish. Jawless fish have cartilage skeletons and lack jaws. There are fewer than 50 species in this class. The sharks and rays belong to the cartilage-fish class. They also have cartilage skeletons but possess jaws. The largest class is the bony fish. They have bony skeletons and jaws. There are more than 30,000 species of bony fish. Besides the characteristics of the skeleton and the jaws, each fish class has certain traits that separate it from the other two classes.

EXPLORATION

Process Skills: observing, classifying, making and using keys, summarizing

OBJECTIVES
In this exercise, you will:
a. observe a jawless fish, a cartilage fish, and a bony fish.
b. record the traits of each fish.

KEYWORDS
Define the following keywords:

cartilage _____ a tough, flexible tissue that supports and shapes the body

chordate _____ an animal that at sometime in its life has a stiff cord along its back

free-living _____ an animal that does not live as a parasite

parasite _____ an animal that lives on or in another living organism

MATERIALS
probe
dissecting pan
preserved shark, preserved lamprey, and a preserved perch

Have students wear gloves to keep from getting preservative on their hands.

PROCEDURE
1. Examine the lamprey.
2. Complete the column in Table 1 for the lamprey. Select the correct answers from the column marked "choices."
3. Examine the shark.
4. Complete the column in Table 1 for the shark by using the correct answers from the column marked "choices."
5. Examine the perch.
6. Complete Table 1 for the perch by using the correct answers from the column marked "choices."

314

Table 1. Comparison of Jawless, Cartilage, and Bony Fish

Trait	Choices	Jawless fish	Cartilage fish	Bony fish
Skin	Scales / Smooth, no scales / Rough, no scales	Smooth, no scales	Rough, no scales	Scales
Mouth location	Under chin, jaws / At front, jaws / At front, no jaws	At front, no jaws	Under chin, jaws	At front, jaws
Mouth shape	Round / Horizontal	Round	Horizontal	Horizontal
Teeth	Yes / No	Yes	Yes	Yes
Fins present	Yes / No	Yes	Yes	Yes
Number of body fin pairs (do not include fins along back)	0 / 2 / 3	0	2	2
Location of fins	Draw on body			
Nostrils present	Yes / No	Yes	Yes	Yes
Number of nostrils	0 / 1 / 2	1	2	2
Gills present	Yes / No	Yes	Yes	Yes
Gills covered by a flap of skin	Yes / No	No	No	Yes
Number of gill openings	1 / 5 / 7	7	5	1
Eyes present	Yes / No	Yes	Yes	Yes
Eyelids present	Yes / No	No	No	No
Skeleton	Endoskeleton / Exoskeleton	Endoskeleton	Endoskeleton	Endoskeleton
Kind of skeleton	Bone / Cartilage	Cartilage	Cartilage	Bone
Way of life	Parasite / Free-living	Parasite	Free-living	Free-living
Backbone	Yes / No	Yes*	Yes	Yes

*Jawless fish have a notochord. For simplicity, we refer to it as a backbone.

64

63

Name _____ Class _____ Period _____

QUESTIONS

1. To what phylum do jawless fish, cartilage fish and bony fish belong? They belong to the chordate phylum (Chordata).

2. List two traits of all chordates. Chordates have an endoskeleton and a backbone.

3. Are jawless fish, cartilage fish, and bony fish invertebrates or vertebrates? vertebrates
   Explain your answer. The jawless fish, cartilage fish, and bony fish all have a backbone.

4. List three ways in which all fish are alike. All fish have fins, gills, gill openings, an endoskeleton, and a backbone.

5. List three ways that jawless fish and cartilage fish are alike. Jawless fish and cartilage fish do not have scales nor gills covered by a flap of skin. They have a skeleton of cartilage.

6. List three ways in which jawless fish and cartilage fish are different. Jawless fish have smooth skin, a round mouth at the front, no jaws, no body fins, 7 gill openings, and live as parasites. Cartilage fish have rough skin, a horizontal mouth under the chin, jaws, 2 pairs of fins, 5 gill openings, and are free-living.

8. List three ways that cartilage fish and bony fish are different. Cartilage fish do not have scales, have a mouth under the chin, do not have gills covered with skin, have 5 gill openings, and a skeleton of cartilage. Bony fish have scales, a mouth at the front, gill covers, 1 gill opening, and a skeleton of bone.

65

9. Explain the meaning of the following terms:
   a. jawless fish Jawless fish do not have jaws and have a skeleton of cartilage.
   b. cartilage fish Cartilage fish have a skeleton of cartilage but have jaws.
   c. bony fish Bony fish have a skeleton of bone and have jaws.

10. Complete the classification for the following three animals. Write the correct name on the lines provided below each fish.

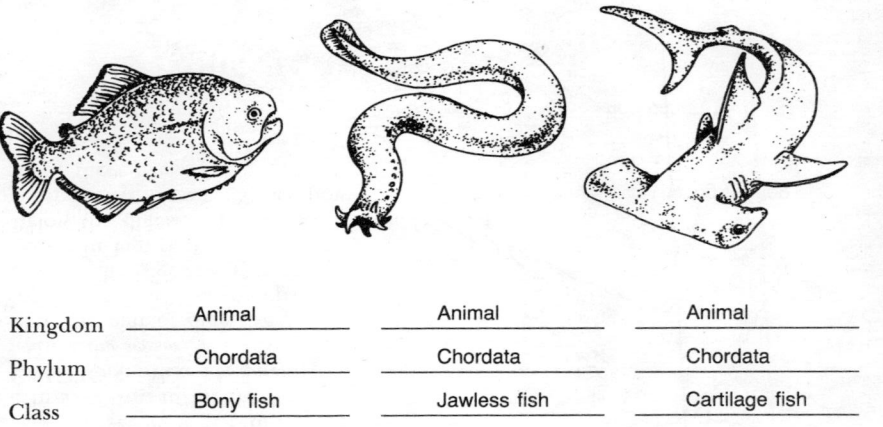

Kingdom __Animal__
Phylum __Chordata__
Class __Bony fish__

Kingdom __Animal__
Phylum __Chordata__
Class __Jawless fish__

Kingdom __Animal__
Phylum __Chordata__
Class __Cartilage fish__

Kingdom __Animal__
Phylum __Chordata__
Class __Bony fish__

Kingdom __Animal__
Phylum __Chordata__
Class __Cartilage fish__

Kingdom __Animal__
Phylum __Chordata__
Class __Bony fish__

66

Name _____ Class _____ Period _____

Before doing this lab, have students read sections 8:1 in the text.

8-3 What Are the Parts of a Starfish?

EXPLORATION

Starfish are perhaps the best known spiny-skin animals. They live all over the world in coastal waters and along rocky shores. They may be seen in various colors such as red, purple, green, blue, and yellow. They range in diameter from about 2 centimeters to almost a meter. These slow moving animals are economically important because they prey on oysters, clams, and other organisms that are used by people for food.

Process Skills: observing, inferring, interpreting scientific illustrations, using a microscope

OBJECTIVES
In this exercise, you will:
a. study the internal and external structures of a starfish.
b. observe how a starfish is adapted for living in water.

KEYWORDS
Define the following keywords:

disc _____ raised central portion from which the arms radiate

dorsal _____ top side

mouth _____ opening in the center of the disc surrounded by spines

spines _____ hard structures that rise through the skin from the platelike skeleton

tube feet _____ parts that are like suction cups and help a starfish move

ventral _____ bottom side

MATERIALS
preserved starfish
dissecting pan
scissors
forceps
stereomicroscope or hand lens

PROCEDURE

Part A. External Structure of the Starfish

1. Place the starfish in the dissecting pan with the dorsal side up.
2. Use the information in Figure 1 to find the external parts of the starfish.
3. Locate the disc, the center raised portion from which the arms radiate.

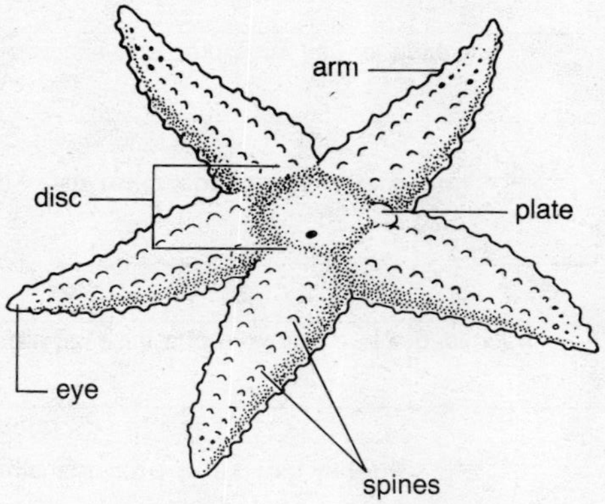

FIGURE 1. Dorsal

4. Find the small, round plate on one side of the starfish near its center. The plate takes in water.
5. Find the red spot at the tip of each arm. These are the eyes.
6. Examine the skin of the starfish. Look at the spines under the stereomicroscope or with the hand lens.
7. Turn the starfish so that the ventral side is up. Use Figure 2 to find the mouth in the middle of the starfish. Examine the ring of small spines around the mouth.
8. Find the groove that extends from the mouth to the tip of each arm.
9. Locate the tube feet that line the groove. Observe the tube feet under the stereomicroscope or with the hand lens.

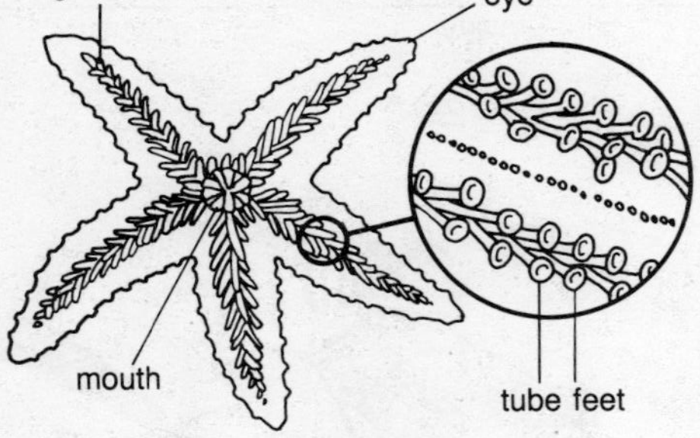

FIGURE 2. Ventral

Part B. Internal Structure of the Starfish

1. Place the starfish dorsal side up in the dissecting pan.
2. Use the scissors to remove the top skin and skeleton carefully from one arm as shown in Figure 3. CAUTION: *Always be careful when using scissors.*
3. Find the pair of digestive glands inside the arm. Use forceps to remove the digestive glands as shown in Figure 4.
4. Below the digestive glands are the reproductive organs. Notice how many there are.

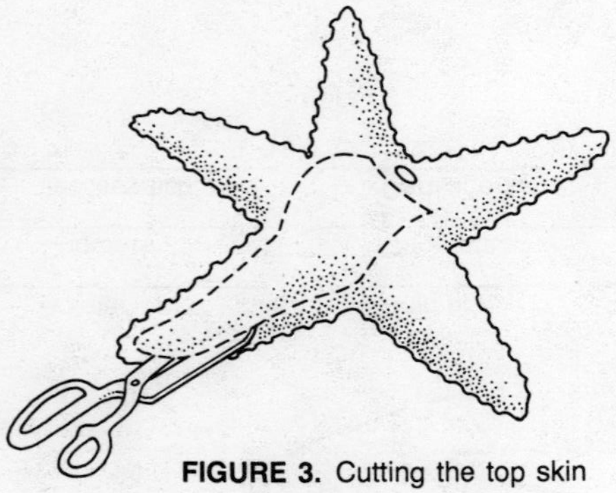

FIGURE 3. Cutting the top skin

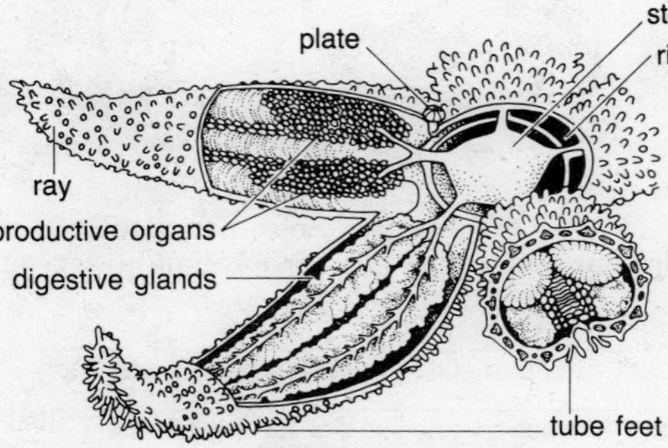

FIGURE 4. Internal structure

Name _____ Class _____ Period _____

5. Remove the skin and skeleton from a second arm.
6. Carefully remove the remaining skin and skeleton from the central disc. The part directly below is the stomach.
7. Remove all organs of the digestive system and reproductive system so that you can see the water vascular system as shown in Figure 5.
8. Dispose of the dissected starfish as your teacher directs. Wash your equipment and your hands.

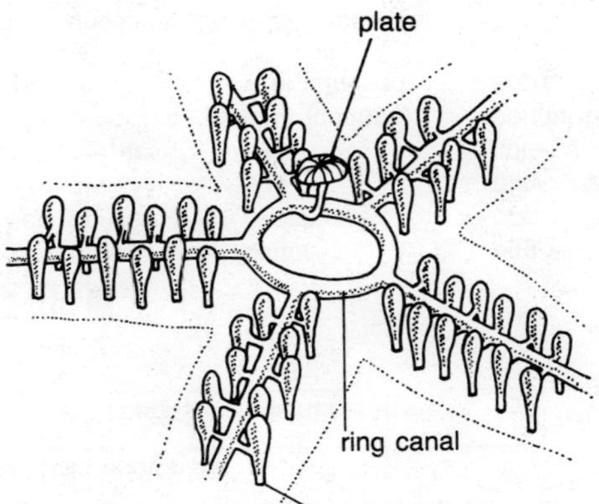

FIGURE 5. Water vascular system

QUESTIONS

1. How many arms does the starfish have? five
2. What kind of symmetry does the starfish have? radial symmetry
3. Why can the starfish move equally well in any direction? It has radial symmetry, so there is no distinct anterior or posterior.
4. Describe the surface of the starfish? It has spines that come through the skin. It is hard and rough.
5. How many rows of tube feet does the starfish have? generally 2 or 4 rows per arm
6. What are the functions of the tube feet? to grip objects and to move the animal
7. Where is the mouth on the starfish? on the ventral side
8. Where is the mouth located in relation to the stomach? directly below the stomach

9. How many digestive glands does the starfish have? If it has 5 arms, it has 5 pairs of digestive glands (10 glands).
10. How many reproductive organs does the starfish have? one pair per arm; if it has 5 arms, it has 10 reproductive organs
11. How is the starfish adapted for living in shallow marine waters? It has a water vascular system with tube feet for movement.

Name _____ Class _____ Period _____

Before doing this lab, have students read section 9:1 in the text.

9-1 What Are the Tests for Fats and Proteins?

All foods contain some of the six nutrients. However, you cannot tell what nutrients are in food by looking at the food. You can tell if certain nutrients are present by doing certain tests. This activity will let you test a number of different foods to find out if fats and proteins are present.

INVESTIGATION

OBJECTIVES

Process Skills: experimenting, making and using tables, separating and controlling variables

In this exercise, you will:
a. test a food containing fat to observe the test results when fat is present.
b. test a food containing protein to observe the test results when protein is present.
c. test a variety of foods to see if fat and protein are present.

*Caution students regarding spillage of Biuret solution—it is caustic. For preparation, see p. 18T.

KEYWORDS

Define the following keywords:

control ___ that part of an experiment in which nothing is changed—your standard for comparison

fat ___ nutrient used as a source of energy by your body

protein ___ nutrient used to build body parts

translucent ___ a spot on paper through which light can pass

MATERIALS

brown paper***	8 test tubes	**butter
cooking oil	labels or wax pencils	bacon
8 droppers		lard
test tube rack	food samples**	water
8 stoppers	sticks, wood or plastic	tuna (canned)
Biuret solution*		veal (strained)
		egg white (solid) hard-boiled
		egg white (liquid)
		mayonnaise
		cottage cheese
		peanut butter
		apple

***grocery bags cut up in advance, into 20 X 20 cm sections

PROCEDURE

Part A. Testing a Fat and a Not-fat

1. Use a dropper to place one small drop of cooking oil (fat) onto a piece of brown paper. Label the drop *fat*.

FIGURE 1: Rubbing foods on brown paper

2. Rub the drop around on the paper, making a circular area (Figure 1).
3. Use a different dropper to place one small drop of water onto the piece of brown paper away from the oil spot. Label the spot *Non-fat*.
4. Rub the drop around on the paper, making a circular area. Don't use the same finger that you used for the oil.
5. Wait 5 to 10 minutes for the liquids to dry. You can speed up the drying time by waving the paper in the air.
6. Hold the paper up toward window light.
7. Light passes through the oil spot. This spot is said to be translucent. *Fat forms a translucent spot on brown paper.*
8. Record your results in Table 1.

Part B. Testing for Fat

1. Test the following foods for fat: butter, egg white (solid), peanut butter, bacon, mayonnaise, cottage cheese, lard, and apple. Apply a small amount of each food to the large piece of brown paper. Make sure you label the food used at each circular area on the paper. **CAUTION:** *Do not taste food samples. Chemical spills may have made them unsafe.*
2. Wait 5 to 10 minutes and then check to see if a translucent spot appears.
3. Record the results of your tests in Table 1.

Emphasize to students that it is a good practice never to taste anything in the lab.

Table 1. Testing for Fat

Food	Translucent spot present?	Fat present?
Cooking oil (fat)	yes	yes
Water (non-fat)	no	no
Butter	yes	yes
Egg white (solid)	no	no
Peanut butter	yes	yes
Bacon	yes	yes
Mayonnaise	yes	yes
Cottage cheese	no (may get a slight translucent spot)	no
Lard	yes	yes
Apple	no	no

Name _____ Class _____ Period _____

Part C. Testing a Protein and a Non-protein

1. Add 10 drops of egg white liquid (a protein) to a test tube. Label this tube *protein*.
2. Add 10 drops of water to a second test tube. Label this tube *water*.
3. Label a third test tube *Biuret*.
4. Add 10 drops of Biuret solution to each of the three test tubes. **CAUTION:** *If the Biuret solution is spilled, wash it off your hands or clothing immediately.*
5. Place a stopper into the first two tubes. Shake both tubes gently for one minute.
6. Compare the color of all three tubes. Biuret solution forms a purple color when mixed with protein. Introduce in prelab as the tube to compare against. Color difference is obvious when two tubes are held next to one another. Biuret is blue; Biuret and protein is violet, lavender, or purple
7. Record your results in Table 2.

Table 2. Testing for Protein

Tube number and/or contents	Color with Biuret solution	Protein present?
Egg white liquid (protein) + Biuret	violet or purple	yes
Water (non-protein) + Biuret	blue	no
Biuret solution	blue	no
1. Biuret	blue	no
2. Biuret + cottage cheese	violet	yes
3. Biuret + lard	blue	no
4. Biuret + veal (or turkey)	violet	yes
5. Biuret + tuna (or salmon)	violet	no

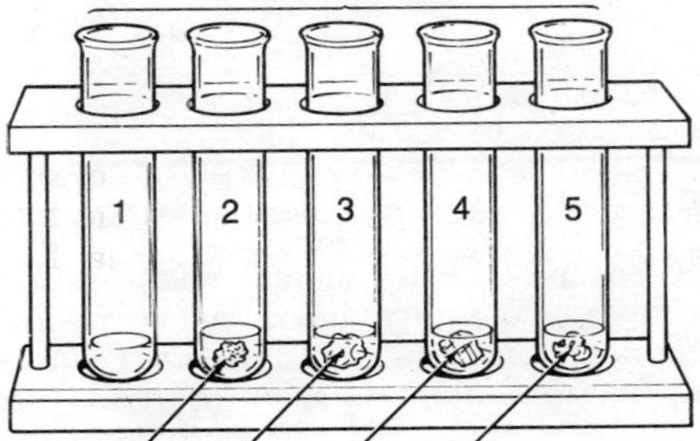

Part D. Testing Foods for Protein

1. Number five test tubes 1 through 5 and place the tubes in a test tube rack.
2. Add the following to each tube using Figure 2 as a guide. Sticks may be used to add very small amounts of each food (about the size of 1/2 a pea) to each tube.

cottage cheese — lard — veal — tuna
10 drops of Biuret

FIGURE 2. Testing for protein

73

Tube 1: 10 drops of Biuret solution.
Tube 2: 10 drops of Biuret solution and cottage cheese.
Tube 3: 10 drops of Biuret solution and lard.
Tube 4: 10 drops of Biuret solution and veal.
Tube 5: 10 drops of Biuret solution and tuna.

3. Place a stopper in tubes 2 through 5. Shake tubes 2 for 1 minute. Critical step!
4. Compare the color of the solution in tubes 2 through 5 with tube 1.
5. Record your results in Table 2.

QUESTIONS

1. Describe how you can test for the nutrient fat in foods. Rub a sample onto a piece of brown paper. Wait 10 minutes. If a translucent spot forms, the food contains fat.
2. Why was it helpful to first test a food that you were told already had fat in it? to see what a translucent spot actually looks like; a basis for comparison
3. Why was it helpful to compare the spot left by fat to the spot left by water? See your definition of *control* in the keywords section. to see how a translucent spot differs from one which is not
4. How is the nutrient fat used in the body? It is used as an energy source.
5. List those foods tested in Part B that contained fat butter, peanut butter, bacon, mayonnaise, lard
6. Your friend's brown lunch bag has a large translucent spot on the bottom. How would you explain why the spot is there? the lunch contains some food that contains fat
7. Describe how you can test for the nutrient protein in foods. Add 10 drops of Biuret solution to the food, shake, and note the color. A purple color means protein is present.
8. Why was it helpful to first test a food that you were already told had protein in it? to see what color appears in the Biuret solution if protein is present
9. Why was it helpful to compare the color in a tube that contained protein to the color in a tube that did not contain protein? so that you had a basis for color comparison between protein and no protein
10. How is the nutrient protein used in the body? it is used to help build and repair body parts (cells and tissues)
11. List those foods tested in Part B that contain protein. cottage cheese, veal (or turkey), tuna (or other fish)

74

Name _____ Class _____ Period _____

Before doing this lab, have students read sections 9:1 in the text.

9-2 How Do You Test for Vitamin C in Foods?

Vitamin C is also known as ascorbic acid. It is a water soluble vitamin. This means that it will dissolve in water. As a vitamin, it is present in many different foods. Tomatoes and salad greens are examples of foods that are good sources of vitamin C. Citrus fruits, such as lemons, limes, and oranges are excellent sources of vitamin C.

Vitamin C can be detected in foods by using a special blue testing chemical. You will use this chemical to test for the vitamin in a variety of different foods. If a food contains vitamin C, the blue color of the testing chemical will turn colorless. If no vitamin C is present, the color will remain blue.

Process Skills: experimenting, interpreting data, inferring, making and using tables

OBJECTIVES
In this exercise, you will:
a. learn how to test for vitamin C.
b. test citrus fruits to find out which ones do or do not have vitamin C.
c. find out if cooking removes vitamin C from food.
d. find out if canning food changes the amount of vitamin C present in the food.

INTERPRETATION

KEYWORDS
Define the following keywords:

citrus fruit — fruits such as oranges, grapefruit, and lemon that contain ascorbic acid

vitamin — chemical needed in very small amounts for growth and tissue repair

vitamin C — ascorbic acid, a vitamin found in citrus fruits and many vegetables

water soluble vitamin — a vitamin that will dissolve in water

MATERIALS
- 9 test tubes
- blue testing chemical*
- 10 droppers
- water
- vitamin C*
- raw cabbage juice**
- cooked cabbage juice***
- fresh lemon juice
- fresh orange juice
- fresh grapefruit juice
- fresh tomato juice
- canned tomato juice

NOTE: all liquids to be tested may be placed in small, labeled plastic dropping bottles for student use.

NOTE: pale, red color of tomato juice may cause a slight problem for detecting when color change is complete. As a comparison, prepare a tube with 10 drops of water and 4 drops of tomato juice. Have students use this tube as a comparison to the tube with indophenol.

*See page 18T for solution preparation.
**Raw cabbage juice—use 200 mL chopped cabbage soaked in 100 mL water for 30 minutes.
***Cooked cabbage juice—cook 200 mL chopped cabbage in 100 mL water for 30 minutes on low heat. Replace water periodically. Return final volume to 100 mL. Pour liquid off both mixtures into small beakers for student use.

75

PROCEDURE

Part A. Testing for Vitamin C

1. Label one test tube *Vitamin C* and a second test tube *Water*.
2. Use a dropper to add 10 drops of blue testing chemical to each test tube. NOTE: add 1 drop at a time of vitamin C
3. Using a different dropper, add one drop at a time of vitamin C to the test tube labeled *Vitamin C* until the blue testing chemical turns from blue to clear (see Figure 1a). Shake the tube after adding each drop.
4. Count the number of drops of vitamin C needed to turn the blue testing chemical from blue to clear.
5. Record this number in Table 1.
6. Repeat steps 3 through 5 for the test tube labeled *water*, only this time, use water instead of vitamin C in step 3 (see Figure 1b). add 1 drop at a time of water
7. If the blue testing chemical has not turned clear after 50 drops of water, stop and record the number *50* in Table 1. NOTE: *If a liquid has much vitamin C in it, it will take only a few drops to turn the blue testing chemical clear. If a liquid has very little vitamin C in it, it will take many drops to turn the blue testing chemical clear.*

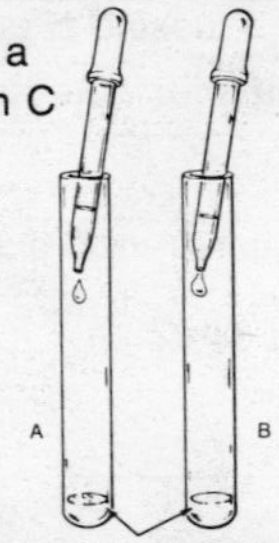

If beakers are used with individual droppers, caution students about mixing up droppers.

10 drops of blue testing chemical
FIGURE 1. Testing for vitamin C

Table 1. Testing for Vitamin C

Liquid being tested	Number of drops of liquid used
Vitamin C	1
Water	50, stopped at this point, no change

Certain foods will cause a colorless to colorless appearance; Students are looking for a colorless appearance; continue to add food until tube is indeed colorless.

Table 2. Testing Food for Vitamin C

Food Being Tested	Number of Drops of Food Used	Vitamin C Present?
1. Lemon juice	1–2	yes
2. Orange juice	1–2	yes
3. Grapefruit juice	1–2	yes
4. Raw cabbage juice	40	very little
5. Cooked cabbage juice	12–18	yes
6. Fresh tomato juice	3	yes
7. Canned tomato juice	5	yes

Student data may vary.

76

Name _____ Class _____ Period _____

Part B. Testing Citrus Fruits for Vitamin C
1. Label three test tubes from 1 through 3.
2. Use a dropper to add 10 drops of blue testing chemical to each test tube.
3. Test for vitamin C as follows:
   a. Using a different dropper, add to Tube 1 one drop of lemon juice at a time until the blue testing chemical turns from blue to clear.
   b. Shake the tube after adding each drop of lemon juice.
   c. Count the number of drops of lemon juice needed to turn the blue testing chemical from blue to clear.
4. Record this number in Table 2.
5. Repeat steps 3 through 4 for Tube 2 using orange juice instead of lemon juice.
6. Repeat steps 3 through 4 for Tube 3 using grapefruit juice instead of lemon juice.

Part C. Is Vitamin C Lost During Cooking?
1. Label two test tubes from 4 through 5.
2. Use a dropper to add 10 drops of blue testing chemical to each test tube.
3. Test for vitamin C as follows:
   a. Using a different dropper, add to Tube 4 one drop of raw cabbage juice at a time until the blue testing chemical turns from blue to clear.
   b. Shake the tube after adding each drop of cabbage juice.
   c. Count the number of drops of raw cabbage juice needed to turn the blue testing chemical from blue to clear.
4. Record this number in Table 2.
5. Repeat steps 3 through 4 for Tube 5 using cooked cabbage juice instead of raw cabbage juice.

Part D. Is Vitamin C Lost When Food is Canned?
1. Label two test tubes from 6 through 7.
2. Use a dropper to add 10 drops of blue testing chemical to each test tube.
3. Test for vitamin C as follows:
   a. Using a different dropper, add to Tube 6 one drop of fresh tomato juice at a time until the blue testing chemical turns from blue to clear.
   b. Shake the tube after adding each drop of tomato juice.
   c. Count the number of drops of fresh tomato juice needed to turn the blue testing chemical from blue to clear.
4. Record this number in Table 2.
5. Repeat steps 3 through 4 for Tube 7 using canned tomato juice instead of fresh tomato juice.

QUESTIONS
1. What was the color of the testing chemical
   a. before adding anything to it? _blue_
   b. after adding vitamin C? _clear_
   c. after adding water? _blue_
2. In Part A, which liquid
   a. took the fewest drops to turn the testing chemical from blue to clear? _vitamin C_
   b. took the most drops to turn the testing chemical from blue to clear? _water_
   c. contains little or no vitamin C? _water_
   d. was your control? _water_
3. a. In Part B, which food contained vitamin C? _lemon, orange, grapefruit_
   b. Recheck the meaning of the word _citrus fruit_ in the Keywords section.
      Were all the foods tested in Part C citrus fruits? _yes_
   c. If you wanted to increase the amount of vitamin C in your diet, what food type would be helpful to you? _citrus fruits_
4. In Part C, which liquid
   a. took the fewest drops to turn the testing chemical from blue to clear? _cooked cabbage juice_
   b. took the most drops to turn the testing chemical from blue to clear? _raw cabbage juice_
   c. contains the most vitamin C? _cooked cabbage juice_
   d. contains the least vitamin C? _raw cabbage juice_
5. a. Based on your results from Part C, does cooking raise or lower the amount of vitamin C in food? _lowers_
   b. Does cooking raise or lower the amount of vitamin C in the cooking liquid? _raises_
   c. If you wanted to increase the amount of vitamin C in your diet, should you throw away the liquid in which foods like cabbage have been cooked? _no_
   d. Recheck the meaning of _water soluble vitamin_ listed in the Keywords section. Explain why cooking lowers the amount of vitamin C in foods. _Vitamin C dissolves in water. When cooked, the vitamin leaves the cabbage and enters the water._
6. In part D, which food
   a. took the fewest drops to turn the testing chemical from blue to clear? _fresh tomato juice_
   b. took the most drops to turn the testing chemical from blue to clear? _canned tomato juice_
   c. contains the most vitamin C? _fresh tomato juice_
   d. contains the least vitamin C? _canned tomato juice_
7. a. Based on your results from part D, tell what happens to the amount of vitamin C in foods when they are canned. _It decreases._
   b. If you wanted to increase the amount of vitamin C in your diet, should you eat more canned or raw foods? _raw foods_

Name _____ Class _____ Period _____

Before doing this lab, have students read section 10:2 in the text.

10-1 How Do Digestive System Lengths Compare?

EXPLORATION

You know that the diet of different animals may vary. You can buy cat food, dog food, and bird food in most supermarkets.

The length of the digestive system may also vary. Animals that eat plants usually have longer digestive systems than animals that eat meat.*
*This is the key phrase to the activity.

OBJECTIVES

In this exercise, you will:
a. measure the length of the digestive system in three animals.
b. compare these lengths with the type of food eaten.

Process Skills: recognizing and using spatial relationships, measuring in SI, inferring, scientific terminology

KEYWORDS
Direct students to the text or the lab glossary for help.
Define the following keywords:

caecum ___ special digestive tube that helps in the breakdown of plant foods _____

carnivore ___ animal that eats meat _____

digestive system ___ a group of organs that take in food and change it into a form the body

_____ can use _____

herbivore ___ animal that eats plants _____

MATERIALS
string   metric ruler   scissors   tape

A transparency of page 74 placed on an overhead and a demonstration with string may help get students started.

PROCEDURE

1. Place a piece of string down on the outline drawing of the rabbit digestive system in Figure 2 on the next page. Figure 1 shows you how.
2. Tape the end of the string in place at the label marked "start" on the stomach of the rabbit.
3. Position the string only over the entire length of the *unshaded* organs. It must match, exactly, the many twists and turns of the stomach, the small intestine and the large intestine (the unshaded organs).
4. When you reach the anus, cut the string, remove it from the drawing, and stretch it out its full length.
   CAUTION: *Use care with scissors.*
5. Measure the length of the string in centimeters and record this number in Table 1.

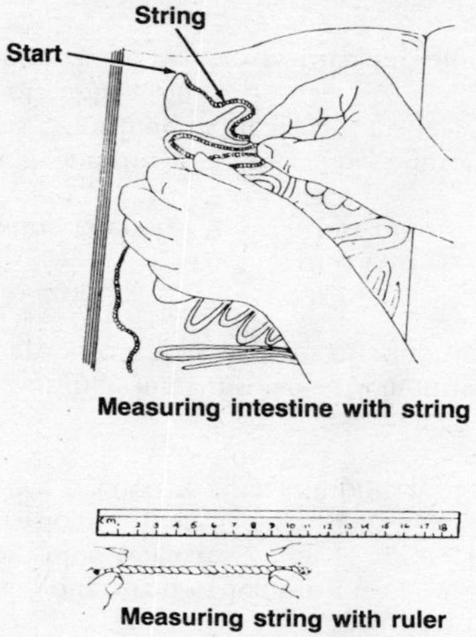

FIGURE 1. Measuring the digestive system on the string. This will enable you to use the same piece of string for all 3 animals.

*You may have students simply mark the end of the digestive system on the string. This will enable you to use the same piece of string for all 3 animals.

79

Rabbit — Start, Stomach, Small intestine, Caecum, Large intestine, Anus

Dog — Start, Stomach, Small intestine, Caecum, Large intestine, Anus

Koala — Start, Stomach, Small intestine, Caecum, Large intestine, Anus

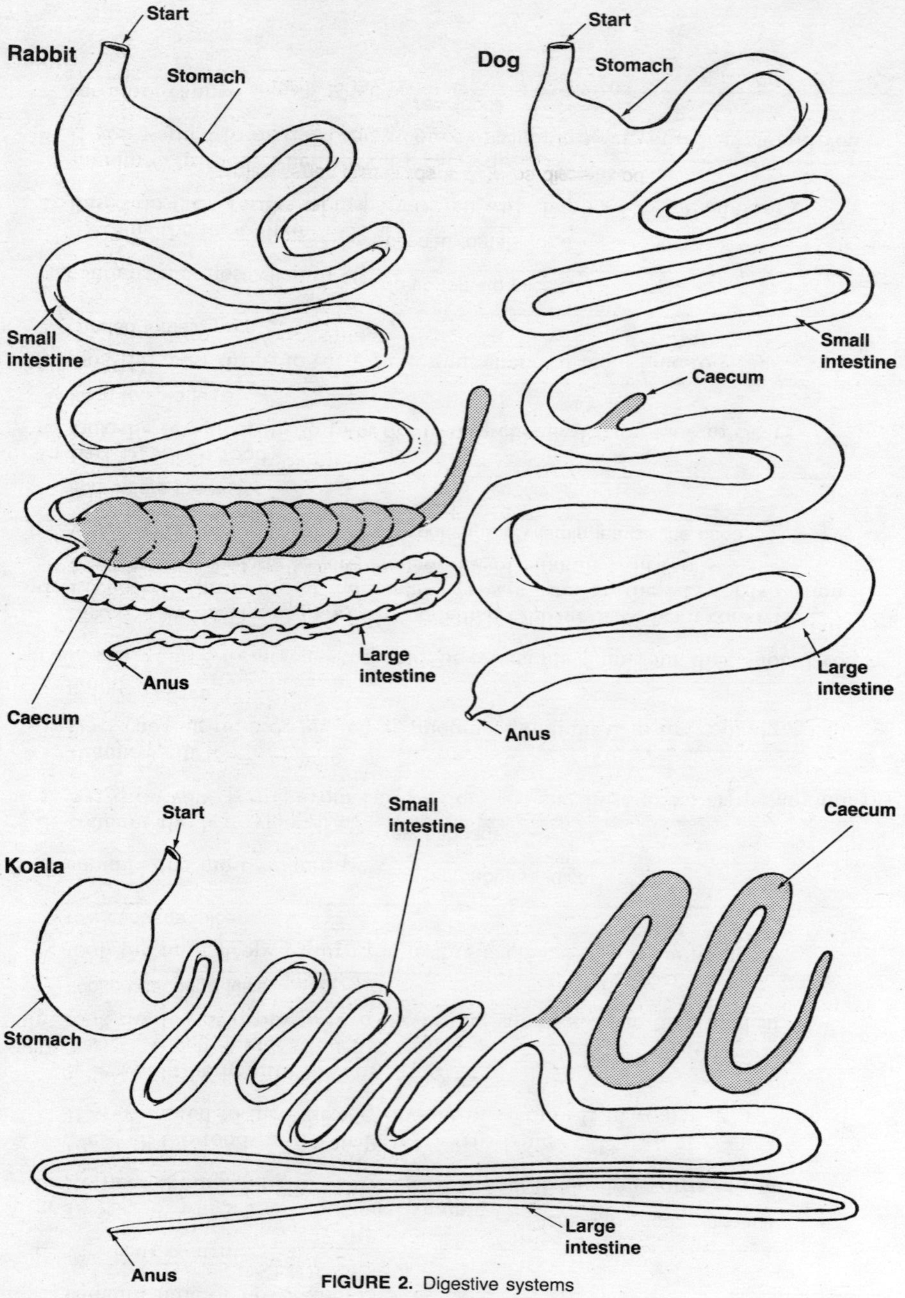

FIGURE 2. Digestive systems

80

322

Name _____ Class _____ Period _____

6. Position the string over the shaded portion of the rabbit digestive system and measure the length of the caecum. Record this measurement in centimeters in Table 1.
7. Add together the two numbers that you have now recorded in the table in order to get the total length of the digestive system. Record this number in Table 1.
8. The diagram of the rabbit digestive system is drawn ⅓ smaller than actual size. Multiply the total digestive system length by 3* to complete the first row of Table 1. This number is the actual length of the rabbit digestive system.
9. Repeat steps 1 through 8 for the digestive system of the koala and the dog.

*Applicable only in this exercise.

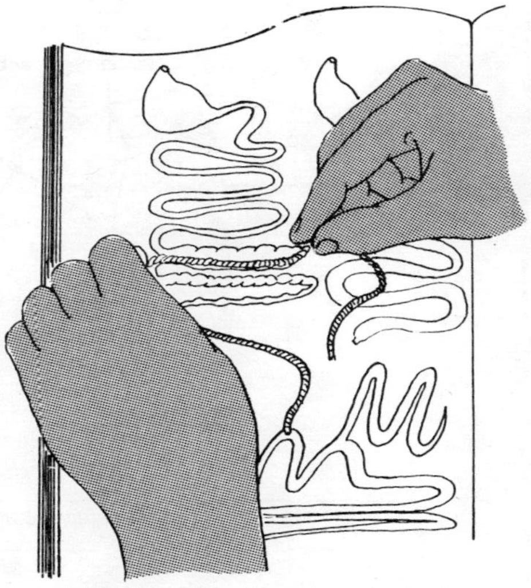

Table 1. Digestive System Measurements

Animal	Length of stomach, small intestine, large intestine		Caecum length		Total digestive system length		Multiply by 3		Actual length of digestive system
Rabbit	68 cm	+	9	=	77	×	3	=	231 cm
Koala	84 cm	+	20	=	104	×	3	=	312 cm
Dog	47 cm	+	1	=	48	×	3	=	144 cm

FIGURE 3. Measuring caeca of animals

QUESTIONS

1. Which animal has the longest actual digestive system? __koala__
2. Which animal has the shortest actual digestive system? __dog__
3. Based on what you already know, tell whether the animals used in this experiment are carnivores or herbivores. NOTE: The koala is an Australian animal that feeds only on the leaves and buds of the eucalyptus tree.

   rabbit __herbivore__   koala __herbivore__   dog __carnivore__

81

4. Circle the correct answer to the following questions:
   a. The animal that has the longest actual digestive system is a (carnivore, **herbivore**).
   b. The animal that has the shortest actual digestive system is a (**carnivore**, herbivore).
   c. The animal that has the longest caecum is a (carnivore, **herbivore**).
   d. The animal that has the shortest caecum is a (**carnivore**, herbivore).
5. By using your answers to question 4, describe how the length of the digestive system in animals seems to be related to the type of food the animals eat. __A long digestive system is found in herbivores. A short digestive system is found in carnivores.__
6. Are plants or meat more difficult to digest? __plants__
   Explain. __Plant-eating animals have a longer system so more time can be spent during digestion.__
7. Use the word *long* or *short* to describe what you think the length of the digestive system might be in the

   lion __short__   cat __short__   horse __long__   deer __long__
   panther __short__   donkey __long__   cow __long__   wolf __short__

8. Two different animals of almost the same size have digestive systems that are of the following lengths:
   Animal A—410 cm   Animal B—145 cm
   a. Which one of these animals is most likely to be a carnivore? __B__
   b. Explain your answer. __The digestive system is short.__
   c. Which one of these animals is most likely to be a herbivore? __A__
   d. Explain your answer. __The digestive system is long.__
9. Describe the function of the:
   a. stomach __baglike organ that holds and helps digest food__
   b. small intestine __helps in the digestion and absorption of food__
   c. large intestine __helps in the absorption of food and water__

82

323

Name _____ Class _____ Period _____

Before doing this lab, have students read section 10:2 in the text.

10-2 How Does the Fish Digestive System Work?

Biologists study many different animals in order to learn more about animals. These studies often help us to understand more about ourselves.

By studying the organ systems of many animals, we have learned that the organ systems of different animals are often very similar. For example, the digestive system of a frog or a fish is similar to the digestive system of a human.

By studying the digestive system of a fish, you will learn how the different parts work together to break down food. You will also see how the fish digestive system can be compared to the human digestive system.

EXPLORATION

Have students work in groups of 2 or 3 to conserve on cost of specimens.

OBJECTIVES
In this exercise, you will:
a. observe the outside and inside of a fish.
b. study the fish digestive system.
c. compare the fish digestive system to that of a human.

Process Skills: observing, recognizing and using spatial relationships, scientific terminology, measuring in SI

KEYWORDS
Direct students to the text or the lab glossary if help is needed.
Define the following keywords:

chemical change __ change when small food pieces are turned into a form that cells can use

digestive system __ a group of organs that take in food and change it into a form the body can use.

physical change __ breaking of large food pieces into smaller pieces

MATERIALS
scissors metric ruler preserved perch*
scalpel or single-edge dissecting pan colored pencils:
razor blade or paper towels red and blue
forceps
*Purchase from a biological supply house or from a supermarket.

PROCEDURE
1. Place the fish in the dissecting pan.
2. Locate the parts on the outside of the fish as shown in Figure 1.
3. Use the forceps to lift the gill cover and to open the mouth.

Caution students to use extreme care with the razor blade. Have students wear gloves to keep from getting preservative on their hands.

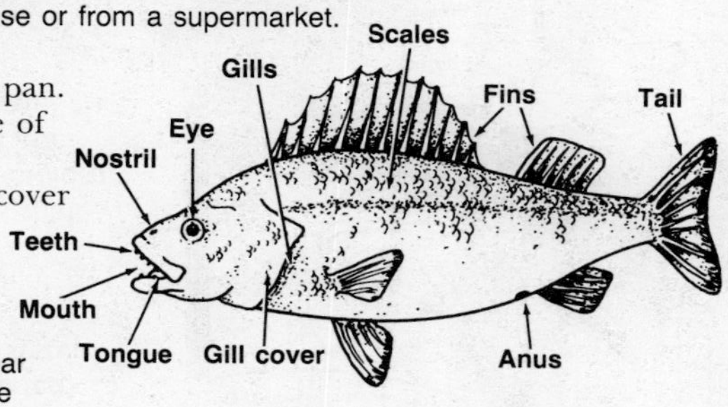

Advise students as to proper disposal of fish at the conclusion of this experiment.

FIGURE 1. Outside of a fish.

(Labels: Tail, Fins, Anus, Gill cover, Tongue, Mouth, Teeth, Nostril, Eye, Gills, Scales)

324

---

4. Complete Table 1 at this time. It may be necessary to use your textbook and the glossary at the back of this guide as reference.

Table 1. Outside Fish Parts and Their Functions

Fish part	Function	Part of what body system?*
Gills	remove oxygen from water; release carbon dioxide into water	respiratory
Eyes	see or locate food; predators.	nervous
Scales	protective cover	integumentary skeletal
Fins	aid in swimming	muscular
Nostrils	detecting odors	nervous
Mouth	taking in and chewing food	digestive
Tail	aid in swimming	muscular
Anus	allows undigested waste to leave	excretory
Teeth	help chew or grasp food	digestive
Tongue	helps to move food into esophagus	digestive

*You may have to help with this column.

5. By using a scissors, remove a section of skin and muscle from the side of your fish. CAUTION: *Use care with scissors.* Use the directions in Figure 2 as a guide to where to make the first, second, third, and fourth cuts. They are marked cuts 1, 2, 3, and 4. *The body parts shown in Figure 2 should be seen easily now.*
6. Locate and identify each of the body parts shown in Figure 2.
7. Complete Table 2 at this time. It may be necessary to refer to your textbook and the glossary in this guide.

Table 2. Parts of the Fish Digestive System and Their Functions

Students will locate organ functions for the human.
Advise them that fish functions are similar.

Digestive system organ	Function
Esophagus	short tube leading from the mouth to the stomach
Stomach	saclike organ; digests foods, mainly proteins
Liver	produces bile which aids in fat breakdown
Gallbladder	stores bile which aids in physical change of fat
Caecum	helps to digest plant material used as food
Small intestine	digests food and absorbs it into the blood

83

84

Name _____ Class _____ Period _____

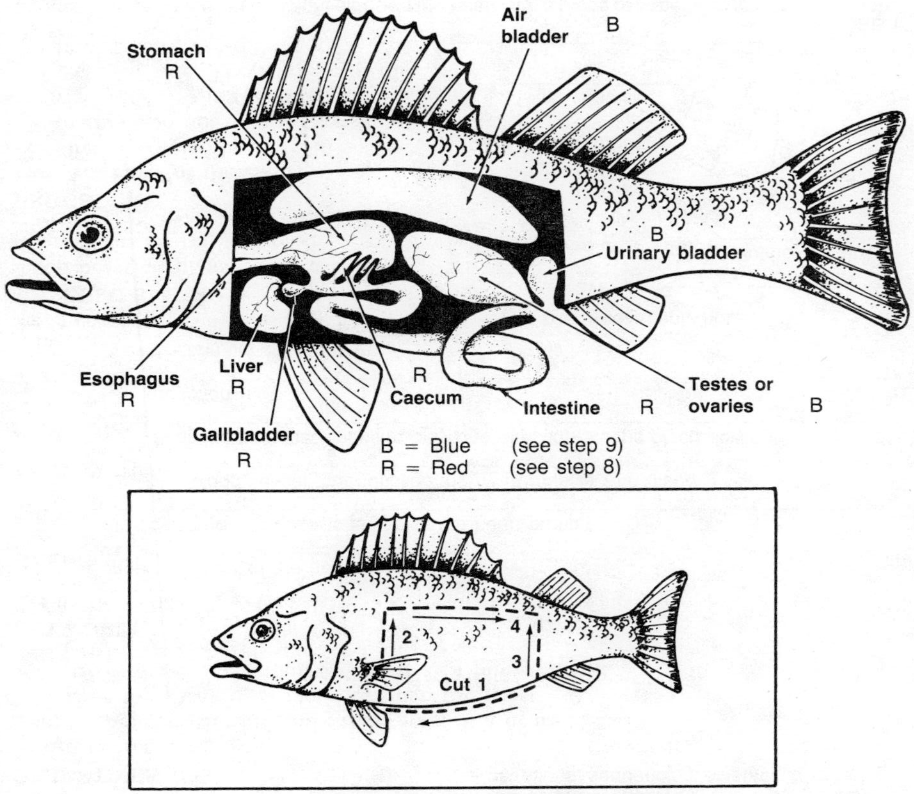

Stomach R
Esophagus R
Gallbladder R
Liver R
Caecum R
Air bladder B
Urinary bladder B
Intestine R (see step 9)
Testes or ovaries B (see step 8)

B = Blue
R = Red

FIGURE 2. Fish digestive system

8. Color red those organs in Figure 2 that are part of the digestive system.
9. Color blue all organs that are not part of the digestive system.
10. Use a scissors to cut loose and remove the entire digestive system.
11. Straighten out the digestive system as best you can and measure its length in centimeters. Record this number after step 12.
12. Measure the entire length of the fish from mouth to tail in centimeters. Record this number in the space provided here.

    Length of digestive system = __45 cm__   Answers will vary.

    Length of fish = __18 cm__

13. Use a scalpel to cut through the organs listed in Table 3 to determine if they are hollow or solid. If they are hollow, food will pass through them. If solid, food will not pass through them. **CAUTION:** *Use extreme care with a scalpel.*
14. Complete Table 3. The gallbladder has already been done for you.

85

Table 3.

Organ	Hollow or solid?	Food passes through?
Stomach	Hollow	Yes
Liver	Solid	No
Gallbladder	hollow	no
Caecum	Hollow	Yes
Small intestine	Hollow	Yes

QUESTIONS Answers may vary.

1. How do the teeth and tongue help with the digestive system? They aid in physical changes—grind food and push it into esophagus.
2. Do the teeth and tongue help with chemical or physical changes? __physical__
3. Describe the shape of the stomach. __saclike, hollow organ__
4. Is the stomach hollow or solid? __hollow__
5. How does the shape of the stomach help it to do its job? __By being saclike and hollow, the stomach can hold food while it is being digested.__
6. Describe the shape of the small intestine. __long tube__
7. Is the intestine hollow or solid? __hollow__
8. How does the shape of the intestine help it to do its job? __By being long and hollow, food can pass through and spend time as it is digested.__
9. The liver is not hollow. How can it still do its job? __It produces bile which is carried to and stored in the gallbladder.__
10. How does the length of the digestive system compare to the length of the fish? __It is longer.__
11. How does the great length of the digestive system help it to do its job? __This length allows food to remain long enough to be digested and absorbed.__
12. By using the picture of the human digestive system appearing on page 209 in your text, explain how the fish and human digestive systems
    a. are alike. __They contain many of the same organs.__
    b. are different __Humans have no caecum; humans have a large intestine.__

86

Name _____ Class _____ Period _____

Before doing this lab, have students read section 11:2 in the text.
Note: This lab can be done without the preserved animal heart being available for students.

11-1 How Does the Heart Work?

The heart is a muscular organ which pumps blood. It is divided into four chambers. The two upper chambers take in blood. The two lower chambers pump blood out of the heart. An upper chamber is called an atrium. A lower chamber is called a ventricle. Between each atrium and each ventricle there is a door that opens in only one direction. The valve acts like a door that opens in only one direction.

Blood first moves into the two upper chambers. The top chambers then pump blood through the valves into the lower chambers. As the lower chambers fill with blood, the valves close. When the lower chambers squeeze together, the blood is forced out of the heart. Blood does not move back into the top chambers.

Process Skills: observing, interpreting scientific illustrations, sequencing, scientific terminology

OBJECTIVES
In this exercise, you will:
a. examine the outside and inside parts of a heart.
b. trace the pathway of blood through the heart.
c. follow the events within the heart as it pumps blood.

KEYWORDS
Define the following keywords:

atrium _____ upper chamber of the heart

contract _____ to shorten or squeeze together as the heart pumps

coronary artery _____ blood vessel that supplies the heart muscle with blood

heart valves _____ door-like part that opens in only one direction; keeps blood flowing in only one direction

ventricle _____ lower chamber of the heart

MATERIALS
Pig or cow hearts can be used and are available from a biological supply house.
sheep heart on paper towel or dissecting tray
colored pencils: red and blue

PROCEDURE
Part A. Parts of the Heart

1. Obtain a sheep heart from your teacher. Do not turn it over. The right side of the sheep heart is on your left side. The left side of the heart is on your right side.*

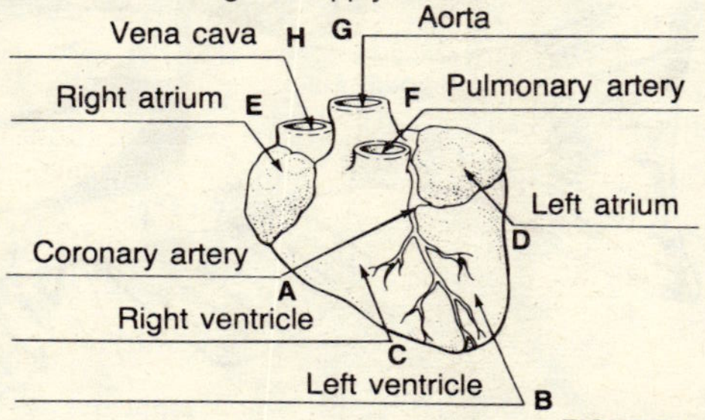

*Suggestion: Label the left and right sides on the heart with a piece of paper and pin stuck through paper into the muscle. You as the teacher must position the heart correctly on the paper towel. The easiest way to tell the left from the right is that the coronary artery divides the heart in half. The left half will be wider.

FIGURE 1.

(Labels on figure: Aorta, Pulmonary artery, Left atrium, Vena cava, Coronary artery, Right atrium, Right ventricle, Left ventricle; letters A–H)

2. On your sheep heart, find the parts listed in Table 1. Use the information in the table to help you.
3. Label the eight parts of the heart correctly on Figure 1. To help with the labels use the letters provided in the table and on the figure.

Table 1. Front Parts of the Heart

Part	Location	Traits	Name
A	across front of heart center	small blood vessel	coronary artery
B	bottom right chamber	large muscle section or chamber	left ventricle
C	bottom left chamber	large muscle section or chamber	right ventricle
D	top right chamber	small muscle section or chamber	left atrium
E	top left chamber	small muscle section or chamber	right atrium
F	top center	large blood vessel* from right ventricle	pulmonary artery
G	top center behind F	large blood vessel* from left ventricle; largest artery in body	aorta
H	top left	large blood vessel* from right atrium	vena cava

*All you will see is a hole where the blood vessel was attached to the heart.

Part B. Direction of Blood Flow Through the Heart

1. Examine Figure 2. It is a diagram of the inside of a heart. Arrows show the direction of blood flow.
2. Examine Figure 3 on the next page, which shows the inside of a sheep's heart. The arrows outlined in dashes indicate *possible* directions of blood flow. Using Figure 2 as a guide, fill in with a pencil the arrowheads that show the correct direction of blood flow.
3. Label the inside parts of this figure using Figure 2 as a guide.

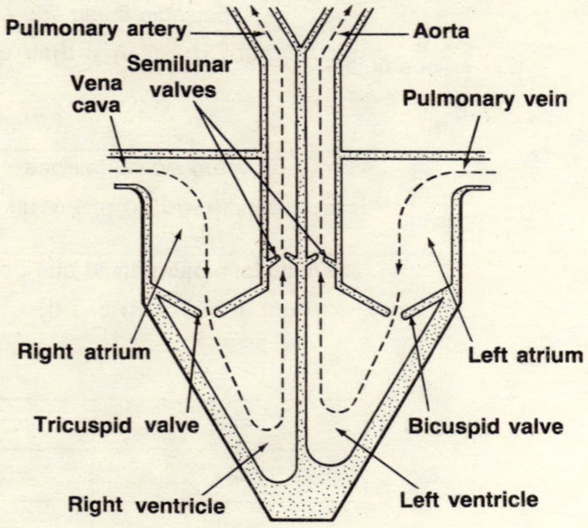

FIGURE 2. Blood flow through the heart

(Labels: Aorta, Pulmonary vein, Left atrium, Bicuspid valve, Left ventricle, Pulmonary artery, Vena cava, Semilunar valves, Right atrium, Tricuspid valve, Right ventricle)

Name _____ Class _____ Period _____

Part C. Condition of Blood in the Heart

All blood on the heart's right side has little oxygen and much carbon dioxide. Blood on the left side has much oxygen and little carbon dioxide.

1. Using colored pencils, fill in the arrows in Figure 3 to show these differences in gas content:
   a. all arrows that indicate blood with much oxygen should be colored red.
   b. all arrows that indicate blood with much carbon dioxide should be colored blue.

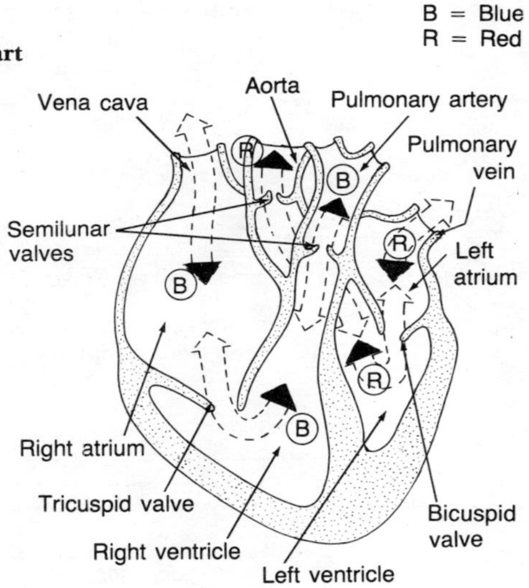

B = Blue
R = Red

FIGURE 3. Inside of a sheep's heart

Part D. Pumping Action of the Heart

Blood enters the two top chambers of the heart. Because they are made of muscle, they are able to squeeze together or contract. When this happens, blood is pumped to the two bottom chambers which are relaxed. These events are shown in Figure 4.

1. Note that certain valves in Figure 4 are open while other valves are closed. Complete the first column of Table 2. *Once blood fills the two bottom chambers they contract. Blood is then pumped out of the heart into the rest of the body.* These events are shown in Figure 5.

2. Note which valves are open or closed in Figure 5. Complete the second column of Table 2.

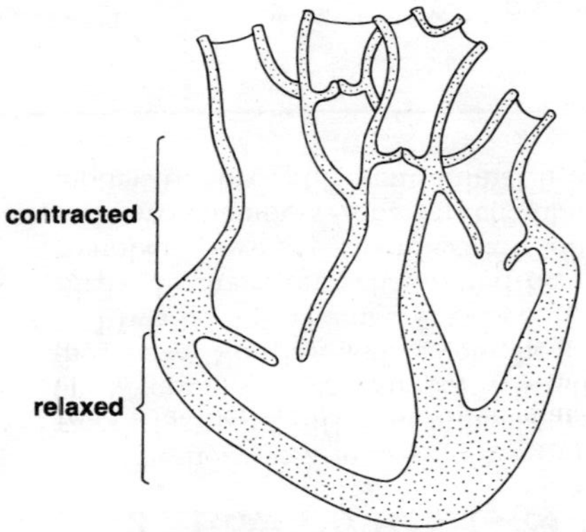

FIGURE 4. Blood entering ventricles

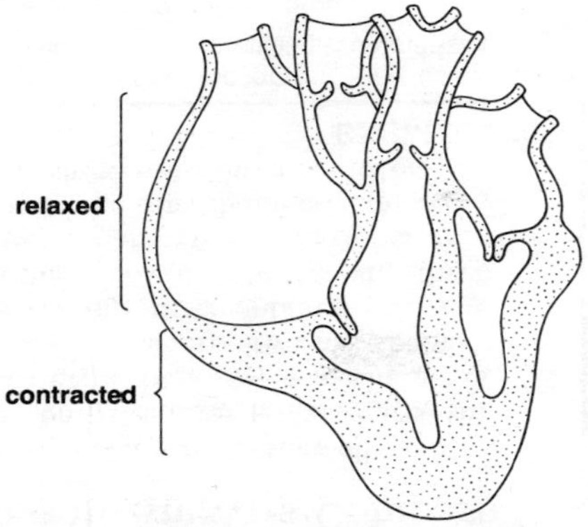

FIGURE 5. Blood leaving ventricles

Table 2. The Opening and Closing of Parts of the Heart

	Blood entering ventricles	Blood leaving ventricles
Top chambers (atria) relaxed or contracted?	contracted	relaxed
Bottom chambers (ventricles) relaxed or contracted?	relaxed	contracted
Semilunar valves open or closed?	closed	open
Bicuspid valve open or closed?	open	closed
Tricuspid valve open or closed?	open	closed

QUESTIONS

1. What is the job of the coronary artery? __to supply blood to the heart muscle__

2. Blood is pumped from the heart to the body through the aorta.
   a. Which chamber does this job? __left ventricle__
   b. Does this blood have more oxygen or more carbon dioxide? __more oxygen__
   c. Which valves are open during this process? __semilunars (Figure 5)__

3. Blood is pumped from the heart to the lungs through the pulmonary artery.
   a. Which heart chamber does this job? __right ventricles__
   b. Does this blood have more oxygen or more carbon dioxide? __more carbon dioxide__
   c. Which valves are open during this process? __semilunars (Figure 5)__

4. Trace a drop of blood through the heart by putting these heart chambers and valves in proper order: left atrium, semilunar valve, right atrium, right ventricle, bicuspid valve, tricuspid valve, left ventricle, semilunar valve.

   Begin with the right atrium. __1__ __5__ __4 or 8__ __3__
   __2__ __7__ __4 or 8__
   __6__

5. Using colored pencils, indicate if each heart chamber listed in question 4 contains blood with more oxygen (red pencil) or more carbon dioxide (blue pencil). Underline each part in your answer to question 4 with the proper color.

   Right atrium   Left atrium
   Right ventricle   Left ventricle
   _____   _____
      blue   red

Name _____ Class _____ Period _____

Before doing this lab, have students read section 11:1 in the text.

11-2 How Do Hearts of Different Animals Compare?

The heart is a muscle that pumps blood. Even though they all pump blood, hearts of different animals are different in several ways. Hearts can be different sizes and shapes. Also, they differ in the number of chambers they have. You know that your heart has four chambers—two top chambers and two bottom chambers. Other mammals have four-chambered hearts. Birds also have four-chambered hearts. Amphibians have hearts with three chambers—two top chambers and one bottom chamber. Fish have hearts with two chambers—one top chamber and one bottom chamber. Making models of three different animal hearts will show you how they differ.

EXPLORATION

OBJECTIVES Have students work in groups; assign one model per group.

Process Skills: recognizing and using spatial relationships, sequencing, summarizing

In this exercise, you will:
a. build a two-chambered heart model such as the type found in fish.
b. build a three-chambered heart model such as the type found in amphibians.
c. build a four-chambered heart model such as the type found in mammals.

KEYWORDS Direct students to the text or the lab glossary for help.
Define the following keywords:

artery _____ carries blood away from the heart
atrium _____ small chamber within the heart
chamber _____ space within the heart that gathers and pumps blood
vein _____ brings blood to the heart
ventricle _____ large chamber within the heart

MATERIALS
9 straws
labels or masking tape
heavy paper manilla folder
scissors
page of model parts on page 92
colored pencils: red, blue and purple or crayons
9 paper cups not foam cups, 7 oz.
glue inexpensive type works fine
tape
ruler

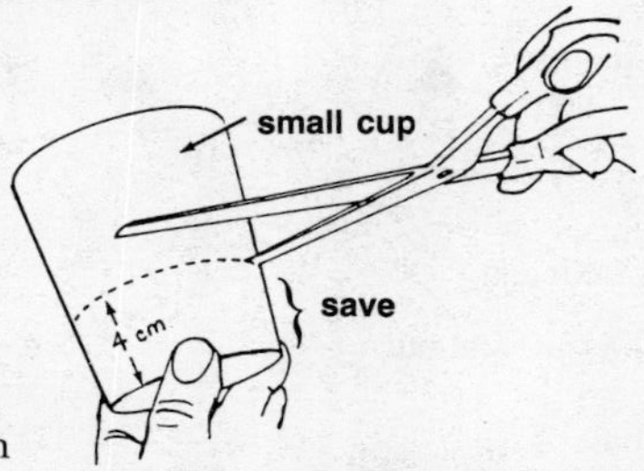

FIGURE 1. Cutting off top of cup

PROCEDURE
Part A. Building a Two-Chambered Heart Model
1. Remove the bottom of a paper cup as shown in Figure 1. **CAUTION:** *Use care with scissors.*

91

Trace three small circles and one large circle.

These circles will fit 7-oz cups. If you purchase larger cups, these circles will not fit. You will have to adjust accordingly.

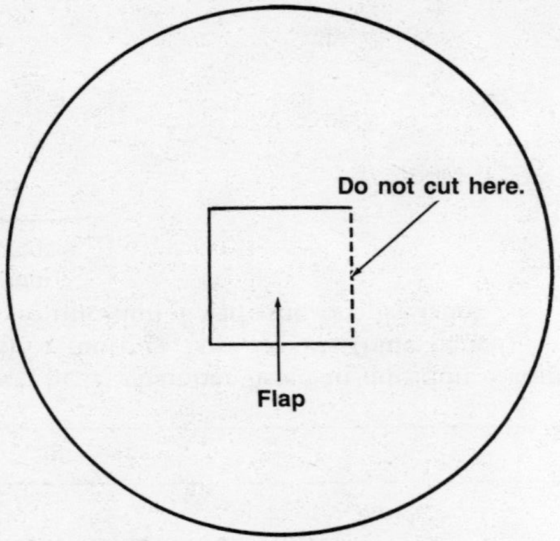

Students can use a compass to make the circles needed, or you can make a spirit master of these circles on scrap paper prior to class. Students need 3 small circles and 1 large circle.

Circles to trace

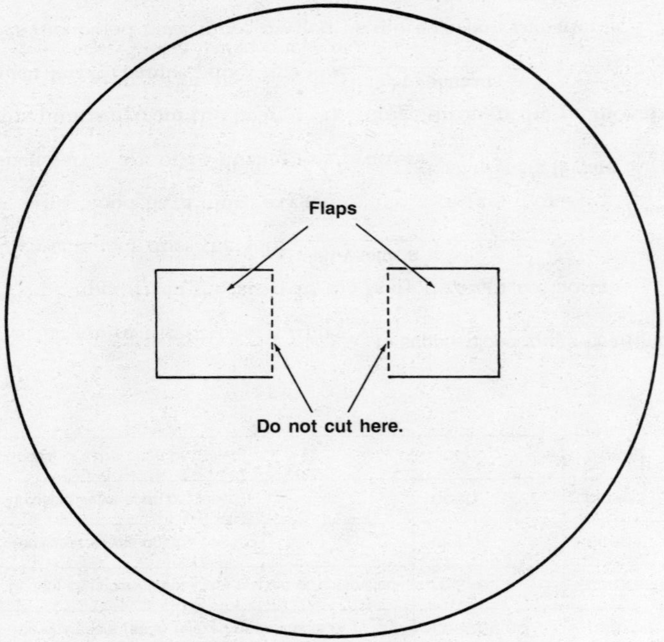

92

328

Name _____ Class _____ Period _____

Caution students on the amount of pressure exerted on the cups. Too much pressure will buckle them.

2. Cut out one small circle from the page of circles given to you.
3. Trace the circle onto heavy paper. Then, cut the circle out from the heavy paper and cut through the flap in the center.
4. Poke holes in the cups as shown in Figure 2. Use the tip end of the scissors. **CAUTION:** *Always be careful with scissors.* Push straws into these holes.
5. Glue the cup pieces and paper circle together.
*6. Color and print onto four labels the information shown in Figure 2.
7. Position the labels using Figure 2 as a guide.

Arrows on the labels indicate the direction of blood flow.

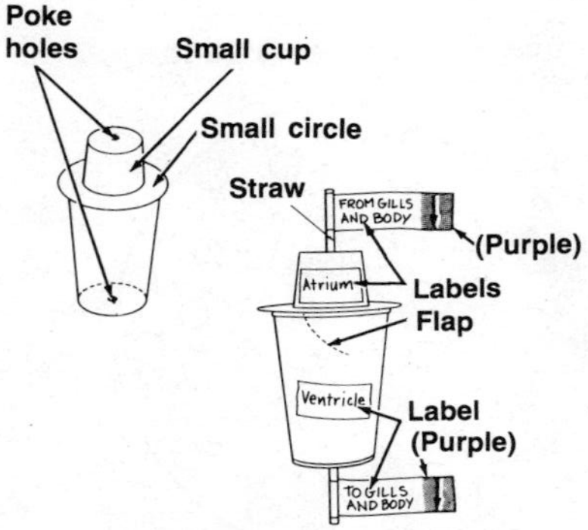

Part B. Building a Three-Chambered Heart Model

1. Prepare this model in the same way as in Part A. Make the following changes: cut off the bottom of *two* paper cups, use a large circle on *heavy* paper rather than a small one, and use *three* straws.
2. Glue the parts together and position the straws as shown in Figure 3.
3. Color and print the information onto six labels using Figure 3 as a guide.
4. Position the labels as shown.

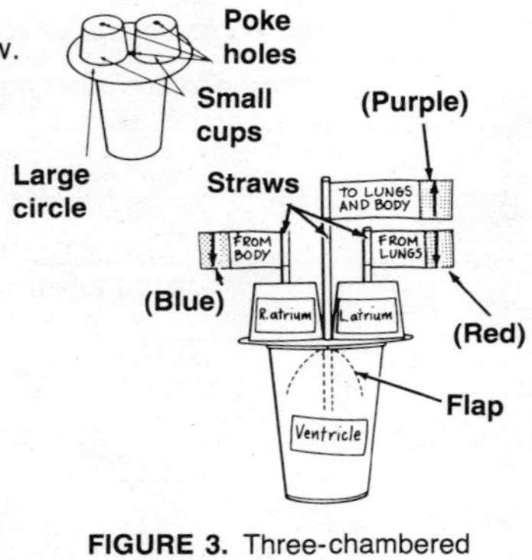

Part C. Building a Four-Chambered Heart Model

1. Prepare this model in the same way as in Part A. Make the following changes: cut off the bottom of *two* paper cups, use *two small circles* on heavy paper, and use *four* straws.
2. Glue the parts together and position the straws as shown in Figure 4.
3. Color and print the information shown in Figure 4.
4. Position the labels using Figure 4 as a guide.
5. You will have to tape the two cups together.

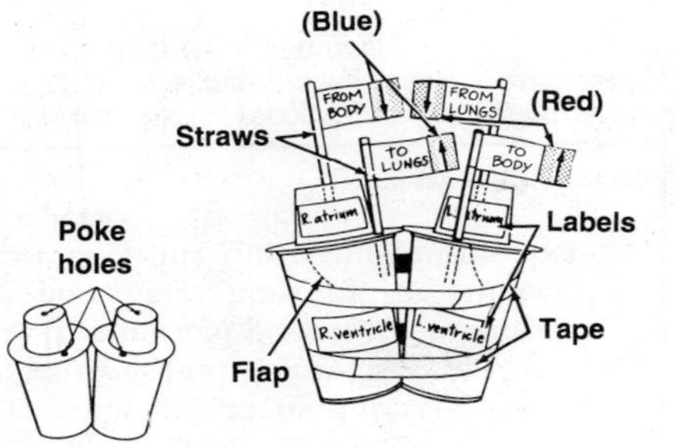

* The colors reference blood gas composition. Direct attention to statements just above question section on page 94.

FIGURE 2. Two-chambered heart model

FIGURE 3. Three-chambered heart model

FIGURE 4. Four-chambered heart model

QUESTIONS

Note: The colors used on the labels indicate the condition of the blood present. Red = blood with much oxygen, blue = blood with much carbon dioxide, purple = blood with equal amounts of oxygen and carbon dioxide.

1. What kinds of animals have two-chambered hearts? __fish__ from introduction
   a. Name the two chambers. __atrium and ventricle__ from model labels
   b. Is the straw in the small cup an artery or a vein? __a vein__
   c. Is the straw in the large cup an artery or a vein? __an artery__ Based on model label arrows and definition of artery and vein
   d. Describe the direction in which blood flows through a two-chambered heart. __enters atrium, pumped to ventricle, leaves heart__
   e. Describe the condition of the blood in each chamber. __equal amounts of oxygen and carbon dioxide__

2. What kinds of animals have a three-chambered heart? __amphibians__
   a. Name the three chambers. __right and left atria, ventricle__ Based on model label and statement above.
   b. Are the straws in the small cups arteries or veins? __veins__
   c. Is the straw in the large cup an artery or a vein? __an artery__
   d. Describe the direction in which blood flows through a three-chambered heart. __blood enters both atria, pumped to ventricle, leaves ventricle__
   e. Describe the condition of the blood in each chamber. __Right atrium contains blood with much carbon dioxide; left atrium contains blood with much oxygen. Ventricle contains equal amounts of oxygen and carbon dioxide.__

3. What kinds of animals have a four-chambered heart? __mammals__
   a. Name the four chambers. __right and left atria, right and left ventricles__
   b. Are the straws in the small cups arteries or veins? __veins__
   c. Are the straws in the large cups arteries or veins? __arteries__
   d. Describe the direction in which blood flows through a four-chambered heart. Start with the right atrium. __Right atrium to right ventricle and then out of the heart. Left atrium to left ventricle and then out of the heart.__
   e. Describe the condition of blood in each chamber. __Right atrium and right ventricle has much carbon dioxide. Left atrium and left ventricle has much oxygen.__

Name _____ Class _____ Period _____

Before doing this lab, have students read section 12:2 in the text.

12-1 How Can Blood Diseases Be Identified?

Blood is a tissue. It has many different cells with many different jobs. If you look at blood under the microscope, you will find three different cell types—red cells, white cells, and platelets. In a normal person the numbers of types of blood cells are fairly constant. Sometimes, however, the number of cells will change due to a certain disease. Noticing this change in number can help a physician in the diagnosis of a person's disease.

Process Skills: recognizing and using spatial relationships, interpreting scientific illustrations, interpreting data, inferring

INTERPRETATION

OBJECTIVES
In this exercise, you will:
a. learn how to recognize three blood cell types.
b. examine diagrams of blood samples from six hospital patients.
c. match the blood samples with certain diseases.

KEYWORDS
Define the following keywords:

diagnosis __to distinguish or discover what problem a person might have__

platelet __blood cell part that aids in forming blood clots__

red blood cell __cell of blood that carries oxygen to tissues__

white blood cell __cell of blood that destroys harmful microbes and remove dead cells__

As a prelab, you may wish to project a 35-mm color film slide of human blood on the screen and identify the three cell types present. You may wish to have students examine prepared slides of blood under the microscope as a prelab.

PROCEDURE

Part A. Normal Blood Cells

1. Examine Figure 1, which shows human blood cells magnified 1000 times.
2. Count each cell type present.
   HINT: To help avoid counting cells twice place a checkmark on each cell as you count.
 * a. red blood cells—round, very numerous, no nucleus.
 * b. white blood cells—round, larger than red blood cells, nucleus present.
 * c. platelets—dotlike, many but less than red cells, very small.
 * Students must use these traits to determine which cell is which type.

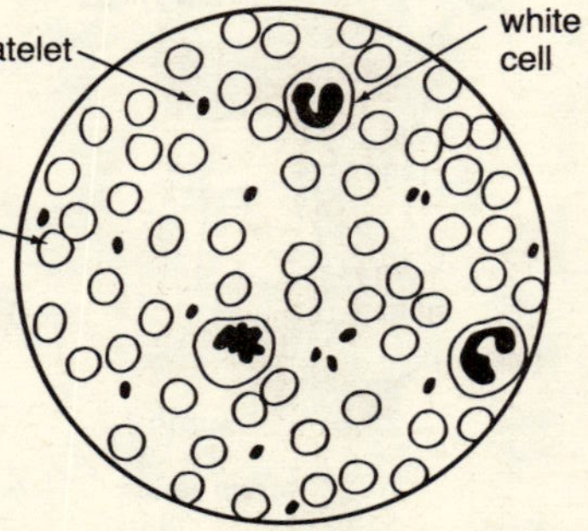

FIGURE 1. Normal blood sample (labels: white cell, platelet, red cell)

95

3. Record the number of each cell type for Figure 1 in Table 1. These numbers are for normal blood.
4. Using the numbers 1, 2, or 3, rank the cells in order from the most common (1) to the least common (3). Enter these rankings in the next column in Table 1 marked *Rank*.
Provide an example of ranking data. Use numbers and examples other than blood cells. Check student data for correct ranking. You may want to have all students check their data in Part A before continuing on with Parts B and C.

Part B. Examining Abnormal Blood Cells

1. Examine Figures 2 to 6. These represent human blood samples from people with certain diseases.
2. Count each cell type and record the number for each sample in Table 1 under the right column.
3. Complete the rank columns using the numbers 1 to 3 as with the normal blood sample.

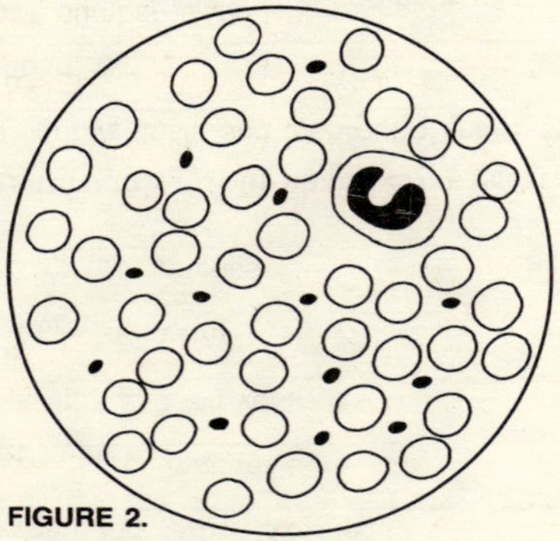

FIGURE 2.

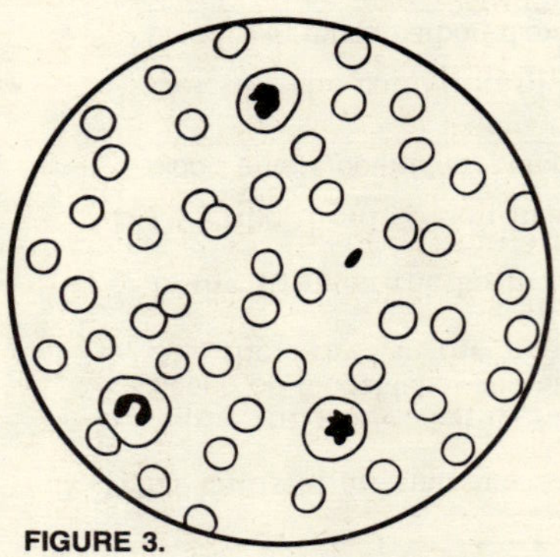

FIGURE 3.

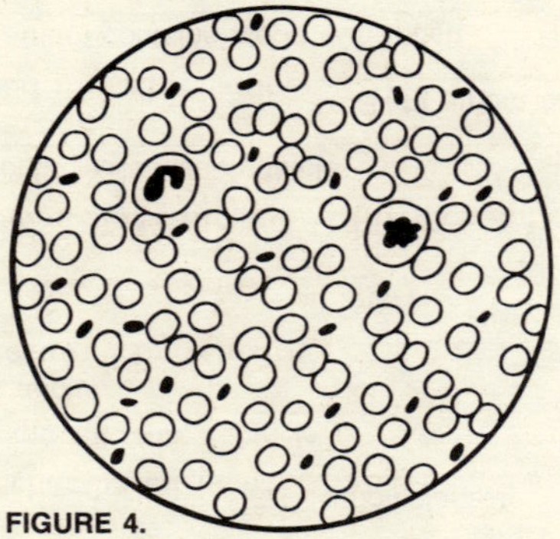

FIGURE 4.

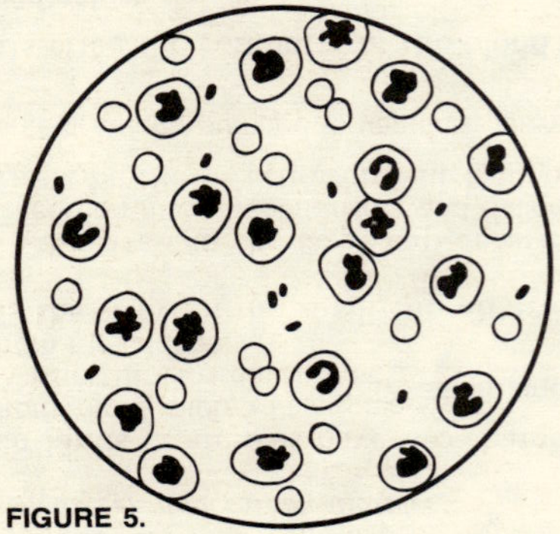

FIGURE 5.

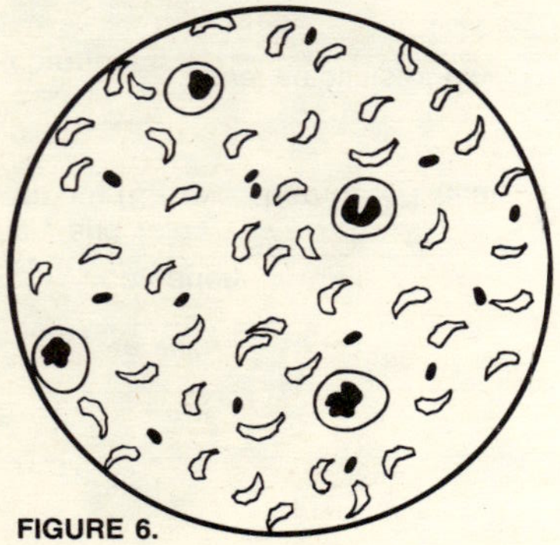

FIGURE 6.

96

330

Name _____ Class _____ Period _____
Student data may vary but ranking should be the same as the sample data.

Table 1. Blood Cell Counts

Cell type	Fig. 1 No.	Fig. 1 Rank	Fig. 2 No.	Fig. 2 Rank	Fig. 3 No.	Fig. 3 Rank	Fig. 4 No.	Fig. 4 Rank	Fig. 5 No.	Fig. 5 Rank	Fig. 6 No.	Fig. 6 Rank
Red	57	1	66	1	47	1	109	1	17	2	47	1
White	3	3	3	3	3	2	2	3	20	1	4	3
Platelet	15	2	13	2	1	3	26	2	11	3	14	2
Disease diagnosis	Normal blood		AIDS		Thrombo-cytopenia purpurea		Poly-cythemia		Leukemia		Sickle-cell anemia	

Part C. Diagnosing Blood Diseases

1. Read over the following case histories for five hospital patients.
2. Match each case history with the appropriate blood sample.
3. Record the name of the disease below each sample in Table 1 in the space provided for disease diagnosis.

Make sure that the disease diagnosis is included in the Table.

Case History: Male, white, age 28; has admitted to injecting drugs for the past 6 years, has pneumonia and skin cancer
Blood analysis: Few white cells present
Disease Diagnosis: AIDS (acquired immunodeficiency syndrome)

Case History: Male, black, age 15; is always tired and short of breath
Blood Analysis: Red cells—shaped like crescent moons
Disease Diagnosis: Sickle-cell anemia

Case History: Female, oriental, age 14; has a fever, sore throat, and frequent nosebleeds
Blood Analysis: Red cells—low in number; White cells—high in number
Blood cell rank—white = 1, red = 2, platelets = 3
Disease Diagnosis: Leukemia (leuk = white, emia = blood)

Case History: Male, white, age 68; has frequent headaches, nosebleeds, shows high blood pressure, a very red complexion
Blood Analysis: Red cells—a very high number
Disease Diagnosis: Polycythemia (poly = many, cyth = cell, emia = blood)

Case History: Female, white, age 22; has sudden appearances of purple marks under the skin, bruises easily, blood does not clot easily after a cut
Blood Analysis: Platelets—very few in number
Blood cell rank—red = 1, white = 2, platelets = 3
Disease Diagnosis: Thrombocytopenia purpurea (thrombo = platelet, cyto = cell, penia = shortage, purpurea = purple)

Other test data is used by physicians before a final diagnosis is normally made.

97

QUESTIONS

1. What is the function of
   a. red blood cells? carry oxygen to body cells
   b. white blood cells? destroy bacteria
   c. platelets? aid in forming blood clots
2. How many
   a. red blood cells are in a drop of normal blood? 5,000,000
   b. white blood cells are in a drop of normal blood? 8,000
   c. platelets are in a drop of normal blood? 250,000   Information is in student text.
3. Rank your answers given to question 2 as to the most common (1) to the least common (3). red (1), platelets (2), white (3)
4. Do your rankings for normal blood in Table 1 agree with your answer to question 3? Yes
5. Explain why a person with AIDS may also have pneumonia. (Keep in mind the main job of white blood cells). White blood cells are destroyed with AIDS. This means that the body can no longer fight off other diseases such as pneumonia as efficiently as before.
6. The rank of blood cells in a normal person and one with polycythemia is the same. How can you conclude that the person has polycythemia? The number of red cells is much higher than normal.
7. The rank of blood cells in a normal person and one with sickle-cell anemia is the same. How can you conclude that the person has sickle-cell anemia? The shape of red cells is sickle-shaped.
8. Name a blood disease that shows
   a. too many white blood cells leukemia
   b. too few platelets thrombocytopenia purpurea
   c. too few red blood cells anemia (or leukemia)
   d. too many red blood cells polycythemia
   e. two few white blood cells. AIDS
9. Explain why a person with thrombocytopenia purpurea shows many bruises or purple marks. A bruise is the result of bleeding under the skin. A person with too few platelets will have much bleeding because the platelet number is too low.
10. Explain how the counting and appearance of blood cells can help in the diagnosis of blood diseases. A physician can determine by rank, number of cells of each type, or shape of cells what disease a patient may have.

98

331

Name _____ Class _____ Period _____

Before doing this lab, have students read section 12:3 in the text.

12–2 What Blood Types Can Be Mixed?

INVESTIGATION

Sometimes patients may lose a lot of blood. In these cases blood from another person can be given to the patient. This giving of someone else's blood to a person is called a transfusion.

There are four main blood types: A, B, AB, and O. Only certain blood types can be mixed when a transfusion is made. Mixing blood types incorrectly during a transfusion can lead to serious illness or the death of a patient.

OBJECTIVES Have students work in teams of 2 or 3.
In this exercise, you will:
a. set up plastic cups filled with water and food coloring to represent the four blood types.
b. mix "blood" to see if color changes take place.
c. judge which blood types can be mixed safely.

Process Skills: observing, predicting, experimenting, interpreting data

KEYWORDS
Define the following keywords:

blood type _a trait of blood—one of four kinds, A, B, AB, or O_

donor _person who gives or donates blood_

recipient _person who receives blood from a donor_

MATERIALS
colored pencils: red, green, and black
food coloring: red and green
graduated cylinder

20 small clear plastic cups
6 droppers   1 oz dose cup ideal, available from drug store, or order from Jet Plastica, Hatfield, PA 19440

PROCEDURE
Part A. Set Up

1. Turn over the page and examine the grid in Figure 2. Note the columns marked *Recipient* and the rows marked *Donor*.
2. Place one of the small plastic cups onto each of the 20 squares as shown here in Figure 1.
*3. Fill each cup with 10 mL of water.
4. Using a dropper, add 4 drops of red food coloring to each of the four cups in the column marked *Recipient A* (red), and to the cup marked *Donor A*.

	Recipient			
Donor	A	B	AB	O
	⌴	⌴	⌴	⌴
	⌴	⌴	⌴	⌴
	⌴	⌴	⌴	⌴
	⌴	⌴	⌴	⌴

FIGURE 1. Placing cups on grid

*Students can save time by measuring only the first cup and using this volume as a standard. The setup can be done on the overhead projector as a demonstration. Because of the bright light, the amounts of food color will have to be adjusted.

99

---

5. Using a different dropper, add 2 drops of green food coloring to the four cups in the column marked *Recipient B* (green), and to the cup marked *Donor B*.
6. Add 3 drops of red food coloring and 3 drops of green food coloring to each of the four cups in the column marked *Recipient AB* (red and green), and to the cup marked *Donor AB*.
7. Note that the four cups in the column marked *Donor O* have no food coloring added to them.
8. Using colored pencils, color in Table 1 to show the colors of all 16 cups marked *Recipient*.

Make sure that students follow directions for number of drops exactly: 4 red for A, 2 green for B and 3 red plus 3 green for AB.

Table 1. Before Blood Is Mixed

Donor	Recipient			
	A	B	AB	O
A	red	green	black	clear
B	red	green	black	clear
AB	red	green	black	clear
O	red	green	black	clear

Part B. Mixing Blood Types Make sure that droppers are clean for each donor.

1. Using a clean dropper, remove "blood" from the cup marked *Donor A*. Moving across the grid, add 2 droppers full of Type A "blood" to each of the four cups in the same row. This step shows what happens when a donor gives his or her blood to a recipient.
2. Repeat step 1 for the next row, but this time use "blood" from the cup marked *Donor B*.
3. Repeat step 1 for the next row, but this time use "blood" from the cup marked *Donor AB*.
4. Repeat step 1 for the final row, but this time use "blood" from the cup marked *Donor O*.
5. Color in Table 2 to show the colors of all 16 recipient cups.

NOTE: No color change has taken place along bottom row, column marked *AB*, or reading from top left to bottom right.

Table 2. After Blood Is Mixed

Donor	Recipient			
	A	B	AB	O
A	red	black	black	red
B	black	green	black	green
AB	black	black	black	black
O	red	green	black	clear

100

Name ______________ Class ______ Period ______

To make it easier for students to use this grid and to use their lab books at the same time, you may wish to make copies of this page for them to use.

FIGURE 2. Grid for mixing food colors

Donor	Recipient			
	A	**B**	**AB**	**O**
A (red)	(red)	(green)	(red + green)	(clear)
B (green)	(red)	(green)	(red + green)	(clear)
AB (red + green)	(red)	(green)	(red + green)	(clear)
O (clear)	(red)	(green)	(red + green)	(clear)

Part C. Judging If Blood Is Safe to Mix

1. Compare Tables 1 and 2. Blood is *safe* to mix between donor and recipient if there is *no change in color* in the same cup from Table 1 to Table 2. Blood is *not safe* to mix between donor and recipient if there is *a change in color* in the same cup from Table 1 to Table 2.
2. Complete Table 3. Write the word *safe* or *unsafe* in each of the 16 squares.
   Your textbook implies that giving Type O to a Type A or Type B person will cause problems of agglutination. In theory, this is true. However, in practice, when 1 pint of Type O blood is added to a large volume of Type A or Type B blood, no agglutination occurs. Thus, Type O is called the universal donor.

Table 3. Is Blood Safe To Mix?

Donor	Recipient			
	A	**B**	**AB**	**O**
A	Safe	Unsafe	Safe	Unsafe
B	Unsafe	Safe	Safe	Unsafe
AB	Unsafe	Unsafe	Safe	Unsafe
O	Safe	Safe	Safe	Safe

QUESTIONS

1. List the blood types of people to which a Type A donor can safely donate blood. A and AB
2. List the blood types of people to which a Type B donor can safely donate blood. B and AB
3. List the blood types of people to which a Type AB donor can safely donate blood. AB
4. List the blood types of people to which a Type O donor can safely donate blood. A, B, AB, and O
5. List the blood types of people from which a Type AB recipient can receive blood. A, B, AB, and O
6. A person with Type O blood is often called a "universal donor." Why might this be a good term to use to describe such a person? They can donate blood to someone with any blood type.
   Have students look at row marked Donor O in Table 3.
7. A person with Type AB blood is often called a "universal recipient." Why might this be a good term to use to describe such a person? They can receive all blood types.
   Have students look at column marked Recipient AB in Table 3.

Name _____ Class _____ Period _____

Before doing this lab, have students read section 13:2 in the text.

13-1 How Does Breathing Occur?

If you have ever tried to hold your breath, you know that breathing is automatic. Breathing is moving air into and out of the lungs. Taking in air is called inhalation. Letting out air is called exhalation.

Your ribs and chest help with breathing. A muscle called the diaphragm also helps by contracting as you inhale and relaxing as you exhale. Let's see how these parts work together during the process of breathing.

Process Skills: observing, making and using tables, experimenting, scientific terminology

EXPLORATION

OBJECTIVES

In this exercise, you will:
a. compare a model to the human chest.
b. use the model to show how the diaphragm helps inhalation and exhalation.
c. use the model to show how the chest wall helps inhalation and exhalation.

KEYWORDS

Define the following keywords:

contracted _____ muscle that is working (shortened)

diaphragm _____ sheetlike muscle at bottom of chest that helps in breathing

exhalation _____ breathing out

inhalation _____ breathing in

relaxed _____ muscle that is not working (returned to normal size)

Students may want to do this lab as a demonstration. Students can build the model on day 1 and run the experiment on day 2. Bend glass tubing for students. Caution them on inserting glass tubing into the stopper.

This lab correlates pressure changes within the chest cavity to events occurring with lungs, diaphragm, and rib cage.

MATERIALS

model of the human chest

PROCEDURE

Part A. Model Parts and How They Work

See Figure 1 for details on building the model.

1. Obtain a model of the human chest from your teacher.
2. Note the following parts on your model and in Figure 1:
   a. rubber sheet along the bottom
   b. glass tube that has a short and a long side; tube bottom contains colored water
   c. outer plastic dome
   d. two balloons attached to an upside-down Y-shaped tube
   e. air trapped inside the plastic dome (air cannot escape from inside because of the balloons and the water inside the tube)
3. Push up gently on the rubber sheet and note the water level change in the tube.
4. Record the water level changes for both sides of the tube in Table 1.

FIGURE 1. Model of the human chest

(Y-shaped tube, 2-hole stopper, plastic dome, A #6 or #7 will fit a 3-liter bottle, air inside, colored water, rubber band, twist tie balloons, glass tube, rubber sheet)

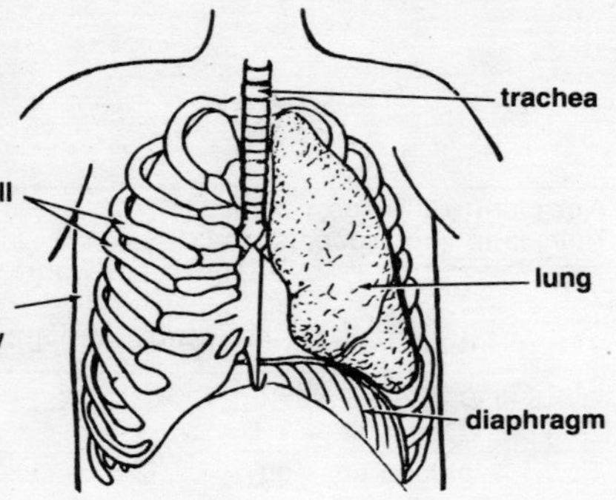

5. Pull down gently on the rubber sheet and note the water level change in the tube.

Air trapped inside the model will cause changes in air pressure when the rubber sheet is pushed up or pulled down. When the rubber sheet is pulled down, air pressure inside the chamber is decreased or low, and water rises on the short side of the tube. When the rubber sheet is pushed up, air pressure inside the chamber is increased or high, and water rises on the long side of the tube.

For plastic dome, use a three-liter plastic soft drink bottle. Rubber sheets are available from local dentist or supply house (called rubber dam).

6. Complete Table 1.

Table 1. Water Levels in the Chest Model

Rubber sheet	Water level on long side	Water level on short side	Change in inside air pressure	Air pressure in model
Pushed up	rises	falls	rises (increases)	rises
Pulled down	falls	rises	falls (decreases)	falls

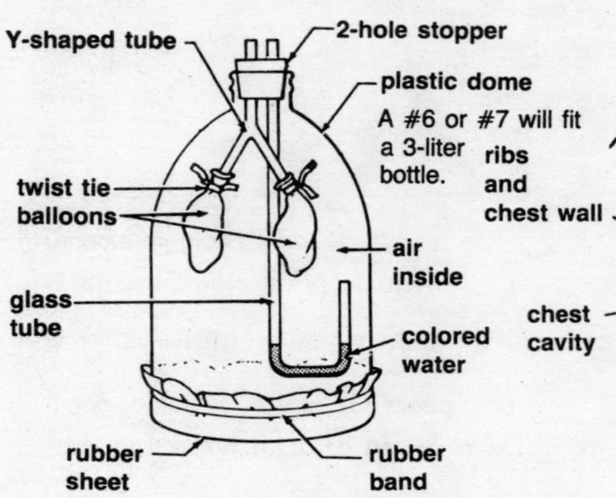

FIGURE 2. The human chest

(trachea, lung, diaphragm, chest wall, chest cavity)

Part B. Comparing Your Model With the Human Chest

1. Compare Figures 1 and 2. The model you have been working with represents the human chest.
2. Match the parts of the model (Figure 1) with the parts of the human chest (Figure 2) that are listed below.

Model Parts   Parts of the Human Chest

balloons _____ E   A. trachea and bronchi

rubber sheet _____ D   B. ribs

Y-shaped tube _____ A   C. chest cavity

air inside dome _____ C   D. diaphragm

plastic sides of dome _____ B   E. lungs (air sacs)

Name _____ Class _____ Period _____

Part C. The Role of the Diaphragm in Breathing

1. Gently push up on the rubber sheet (diaphragm) of the model.
2. Note the following:
   a. what happens to the balloons.
   b. in which side of the tube the water rises.
3. As pressure inside the chest increases, it squeezes the air sacs. This pushes the air within them out.
4. Complete the top row of Table 2. Note that the diaphragm is in a relaxed condition when it pushes up in your body.
5. Gently pull down on the rubber sheet (diaphragm) of the model.
6. Note the following:
   a. what happens to the balloons.
   b. in which side of the tube the water rises.
7. As pressure inside the chest decreases, the air sacs return to their original shape. They fill with air as you breathe in.
8. Complete the bottom row of Table 2. Note that the diaphragm is in a contracted condition when it pulls down in your body.

Table 2. Using the Chest Model

Rubber sheet	Diaphragm condition (relaxed/ contracted)	Diaphragm position (up/down)	Tube side in which water rises (long/short)	Inside air pressure (high/low)	Balloons (air sacs) (empty/fill)	Person breathing (exhale/ inhale)
Pushed up	relaxed	up	long	high	empty	exhale
Pulled down	contracted	down	short	low	fill	inhale

Part D. The Role of the Ribs and the Chest Wall in Breathing

1. Gently squeeze in the sides at the bottom of the plastic dome (chest wall).
2. Note the following:
   a. what happens to the balloons.
   b. in which side of the tube the water rises.
3. Complete the top row of Table 3. Note that the chest wall and the ribs in Figure 3A move down slightly when the human chest wall moves in.
4. Gently squeeze in the sides at the bottom of the plastic dome, then let go and note:
   a. what happens to the balloons.
   b. in which side of the tube the water rises.
5. Complete the bottom row of Table 3. Note that the chest wall and the ribs in Figure 3B move slightly up when the human chest wall moves out and that the size of the chest cavity gets larger.

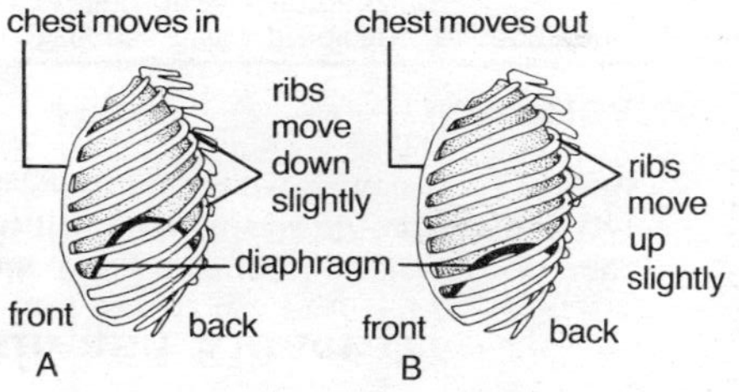

FIGURE 3. Side views of the rib cage

Table 3. The Movement of the Chest During Breathing

	Chest wall pushed in	Chest wall back to original shape
Tube side in which water rises (long/short)	long	short
Inside air pressure (falls/rises)	rises	falls
Air pressure (high/low)	high	low
Rib cage movement (up/down)	down	up
Chest cavity size (large/small)	small	large
Balloons or air sacs (empty/fill)	empty	fill
Person breathing (exhale/inhale)	exhale	inhale

QUESTIONS

1. Complete this summary table by writing in the correct word or words.

	Inhalation	Exhalation
Diaphragm pulled up or down?	down	up
Diaphragm relaxed or contracted?	contracted	relaxed
Chest wall pushed in or out?	out	in
Ribs pulled up or down?	up	down
Air pressure in chest high or low?	low	high
Pressure does or does not squeeze air sacs?	does not	does
Chest cavity size increases or decreases?	increases	decreases
Lungs filling or emptying?	filling	emptying
Breathing in or out?	in	out

105

106

335

Name _____ Class _____ Period _____

Before doing this lab, have students read section 13:4 in the text.

13-2 What Chemical Wastes Are Removed by the Lungs, Skin and Kidneys?

When foods are broken down in the body, chemical wastes are formed. Carbon dioxide and water are two such waste chemicals. Other waste chemicals given off by the body include urea and salts. Which body organs help to remove certain waste chemicals?

INVESTIGATION

Process Skills: predicting, experimenting, interpreting data, summarizing

OBJECTIVES

In this exercise, you will:
a. learn how to test for water using cobalt chloride paper.
b. learn how to test for salt using silver nitrate.
c. test to see if the lungs, skin and/or kidneys give off water.
d. test to see if the lungs, skin and/or kidneys give off salt.

KEYWORDS

Define the following keywords:

distilled water ___ water that has had all the impurities and bacteria removed

perspiration ___ waste water lost through the skin

urine ___ mixture of urea, salt, and water, a waste chemical

waste chemical ___ chemical not needed by the body, such as urea, water, and carbon dioxide

MATERIALS

* silver nitrate
plastic bag grocery store item
** cobalt chloride paper available from a biological supply house
5 test tubes
distilled water available in a grocery store
test-tube rack

salt water
urine
water dropper
cotton

18 g sodium chloride to 200 mL water (enough for 1 class)

3 drops yellow food coloring and 9 g sodium chloride to 1 L water (enough for 1 class)

PROCEDURE

Part A. Testing for Water

1. Examine a dry piece of cobalt chloride paper and record the color in Table 1.
2. Place a drop of water on the paper (by using a dropper). Record the color of the cobalt chloride paper in Table 2.

FIGURE 1.

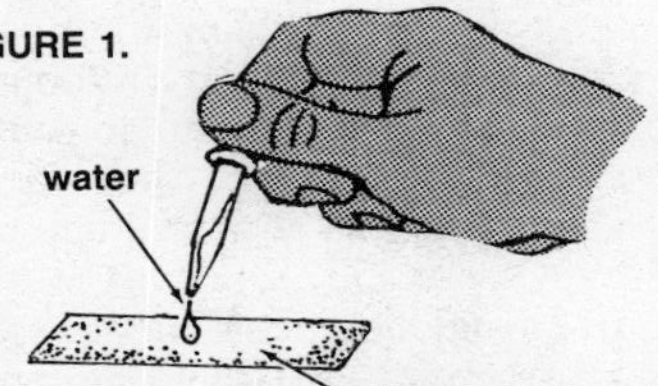

* 4 g silver nitrate to 250 mL **distilled** water. Place into Barnes dropping bottles or plastic dropping bottles for ease in dispensing. This quantity is enough for 5 classes. A chemistry teacher may be able to provide this reagent.
** You can reuse cobalt chloride paper. When allowed to dry (heat in an incubator or an oven), it returns to its blue color.

How could you test for the presence of water by using dry cobalt chloride paper? ___ Dry cobalt chloride paper turns from its blue color to a pink color if water is added.

Part B. Testing for Salt

1. Fill a clean test tube ¼ full of distilled water. Amounts are not critical.
2. Fill a second test tube ¼ full of salt water.
3. Add 1 drop of silver nitrate to each tube. CAUTION: *Wash with water immediately if silver nitrate is spilled on the skin.*
4. Record the color of both tubes in Table 1. The white cloud that forms in the tube with salt indicates that salt is present. How can you test for the presence of salt using silver nitrate? ___ Silver nitrate forms a white cloud if salt is present. (AgNO₃ + NaCl → AgCl [a white precipitate]

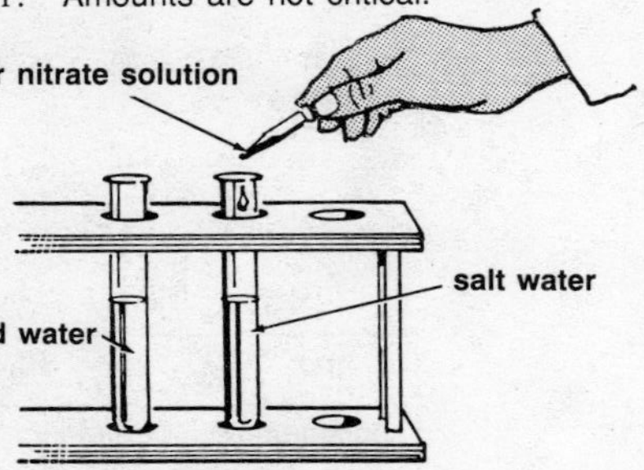

FIGURE 2.

Table 1.

Substance Tested	Color
Dry cobalt chloride	blue
Wet cobalt chloride	pink
Distilled water (no salt)	clear
Salt water	white cloud

Part C. Testing Breath from the Lungs for Water and Salt

1. Place a dry piece of cobalt chloride paper in a plastic bag.
2. Hold the plastic bag to your mouth and breathe into the bag five times.
3. Hold the bag closed for about two minutes.
4. Note and record in Table 2 the color of the cobalt chloride paper. Also record whether water is given off by the lungs in your breath in the same table.
5. Tear open the plastic bag and soak up any liquid inside the bag with a small piece of cotton. Too large a piece of cotton will stick in the test tube.
6. Roll up the cotton so that it will fit into a test tube.
7. Fill the tube ¼ full of distilled water. Do *not* use tap water. Amount is not critical.
8. Add 1 drop of silver nitrate to the tube.
9. Note and record in Table 2 if a white cloud appears in the tube. Also record whether salt is given off by the lungs.

Name _____ Class _____ Period _____

Part D. Testing Skin Perspiration for Water and Salt

1. Hold a piece of dry cobalt chloride paper in your hand with your fist closed for about 5 minutes.
2. Record the color of the cobalt chloride paper in Table 2.
3. Rub the palm of your hand at least 20 times with a clean piece of cotton, as if you were trying to dry your palm.*
4. Roll up the cotton and place in the bottom of a test tube.
5. Fill the tube ¼ full with distilled water. Do not use tap water.
6. Add 1 drop of silver nitrate to this tube.
7. Note and record in Table 2 if a white cloud appears in the tube. Also record whether salt is given off by perspiration. Cloud may be very faint.

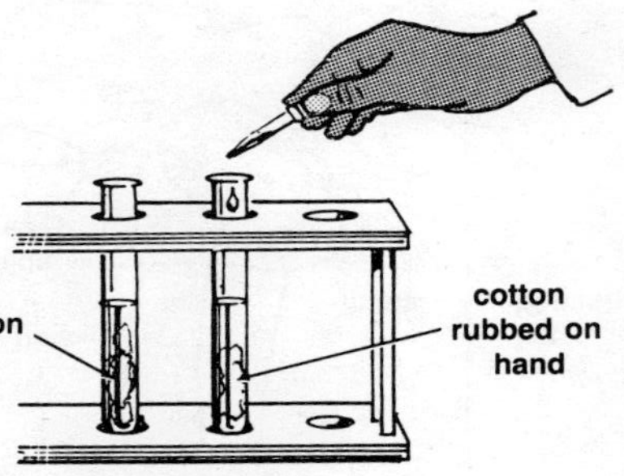

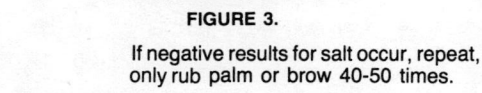

*You may want to suggest the brow instead of the palm or use both.

FIGURE 3.

If negative results for salt occur, repeat, only rub palm or brow 40-50 times.

Table 2.

	Cobalt chloride color	Water given off?	Silver nitrate color	Salt given off?
Breath from lungs	pink	yes	clear	no
Skin perspiration	pink	yes	white cloud	yes
Urine from kidneys	pink	yes	white cloud	yes

Advise students that this is artificial urine.

Part E. Testing Urine for Water and Salt

1. Fill a test tube ¼ full of "urine." Amount is not critical.
2. Tip the tube and insert a small piece of dry cobalt chloride paper so that it just touches the urine.
3. Record any color change in Table 2. Also record whether water is given off by the kidneys in urine.
4. Add one drop of silver nitrate to the contents of the test tube.
5. Note and record in Table 2 if a white cloud appears in the tube. Also record whether salt is given off by the kidneys in urine.

109

QUESTIONS

1. Name three organs that give off waste chemicals. __skin, lungs, and kidneys__
2. Do the lungs help to get rid of wastes? __yes__
   a. What waste was shown to be given off by the lungs in this experiment? __water__
   b. What waste was shown not to be given off by the lungs in this experiment? __salt__
3. Does the skin help to get rid of wastes? __yes__
   What wastes were shown to be given off by the skin in this experiment? __water and salt__
4. Do the kidneys help get rid of wastes? __yes__
   What wastes were shown to be given off by the kidneys in this experiment? __water and salt__
5. A scientist drew up the following chart based on her results from a similar experiment. The chart appears as follows:

   (a + indicates "present", a − indicates "absent")

Organ	Water	Salt	Carbon dioxide	Urea
Lungs	+	−	+	−
Skin	+	+	−	−
Kidneys	+	+	−	+

   a. Do your results agree with the scientist's for tests on water? __yes__ Student answers may vary.
   b. Do your results agree with the scientist's for tests on salt? __yes__ Student answers may vary.
6. Examine the chart in question 5.
   a. List the organ that gives off more *different* wastes. __kidney__
   b. List the only organ that gives off urea. __kidney__
   c. List the only organ that gives off carbon dioxide. __lung__
   d. List those organs that can give off water. __lungs, skin, and kidneys__
   e. List those organs that can give off salt. __skin and kidneys__

110

Name _____ Class _____ Period _____

Before doing this lab, have students read section 14:3 in the text.

14-1 What Causes Sports Injuries?

A number of different kinds of injuries can take place that involve the skeletal system or the muscular system. Many of these injuries result from everyday accidents while others may occur while participating in certain sports.

INTERPRETATION

Process Skills: scientific terminology, interpreting scientific illustrations, recognizing and using spatial relationships, inferring

OBJECTIVES

In this exercise, you will:
a. learn what the difference is between ligaments and tendons.
b. relate sprains, torn tendons, and tendonitis to certain injuries.
c. learn the names of certain body muscles, bones, and tendons.

KEYWORDS

Define the following keywords:

ankle   joint at bottom of leg and top of foot

ligament   tough fibers that hold bones together

muscular system   made up of all the muscles found in an animal's body

skeletal system   body framework made of bone

sprain   injury that occurs to ligaments

MATERIALS

#2 pencil or any numbered pencil
colored pencils: blue and red may substitute other colors but correct student instructions

PROCEDURE

Part A. Sprains

1. Examine Figure 1. This is a drawing of the bones that are a part of the human ankle.
2. Examine Figure 2. This is a similar drawing of the ankle except that three ligaments have been added. They are marked 1, 2, and 3.

Students will not have to learn bone names but will use them.

FIGURE 1. Bones of the ankle
(labels: lower leg bones, fibula, navicular, tibia, ankle area, talus, heel bone, foot)

111

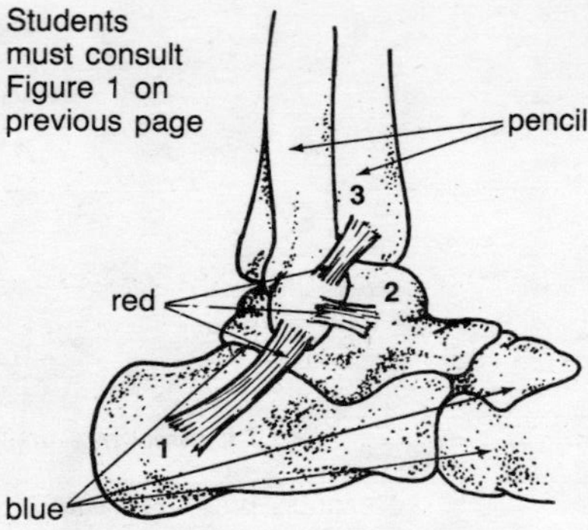

FIGURE 2. Ligaments of the ankle

Students must consult Figure 1 on previous page

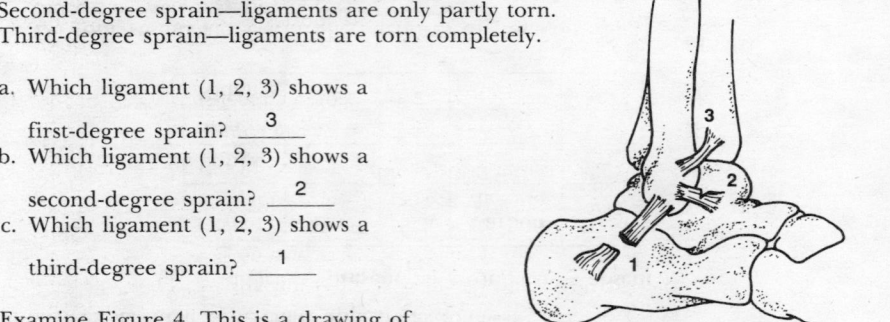

FIGURE 3. Sprained ligaments

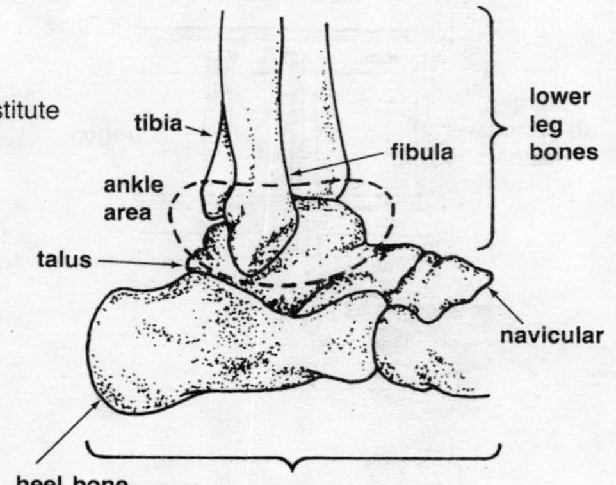

3. a. Color all leg bones in Figure 2 grey (use #2 pencil).
   b. Color all foot bones in Figure 2 blue.
   c. Color all ligaments in Figure 2 red.

4. Answer the following questions:
   a. Name the two bones held together by ligament 1.
      fibula and heel (or tibia and heel)
   b. Name the two bones held together by ligament 2.
      tibia and talus
   c. Name the two bones held together by ligament 3.
      fibula and tibia

5. Examine Figure 3 showing the three types of sprains. They are:
   First-degree sprain—ligaments are only stretched.
   Second-degree sprain—ligaments are only partly torn.
   Third-degree sprain—ligaments are torn completely.

6. a. Which ligament (1, 2, 3) shows a first-degree sprain? 3
   b. Which ligament (1, 2, 3) shows a second-degree sprain? 2
   c. Which ligament (1, 2, 3) shows a third-degree sprain? 1

7. Examine Figure 4. This is a drawing of the bones and ligaments of the shoulder. Color all shoulder bones grey. Color all upper arm bones blue. Color all ligaments red.

   Students must consult the labels.

8. a. Name the two bones held together by ligament 1.
      clavicle and scapula
   b. Name the two bones held together by ligament 2.
      clavicle and scapula

9. Examine the incomplete drawing of the shoulder in Figure 5. Finish the drawing by:
   a. drawing in a second-degree sprain of ligament 1.
   b. drawing in a third-degree sprain of ligament 2.
   c. drawing in a normal ligament holding the humerus to the scapula.

112

Name _____ Class _____ Period _____

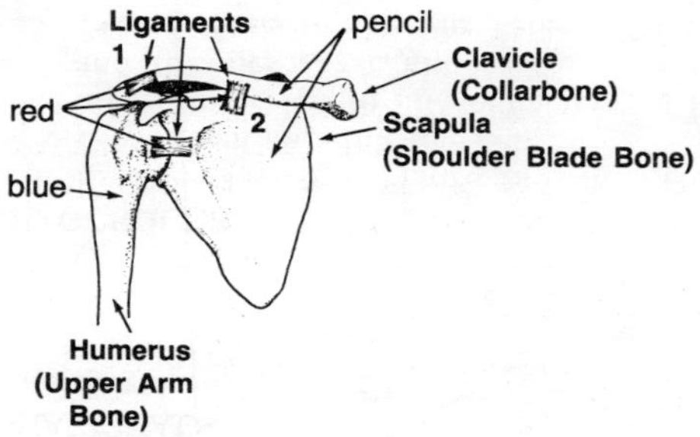

FIGURE 4. Ligaments of the shoulder

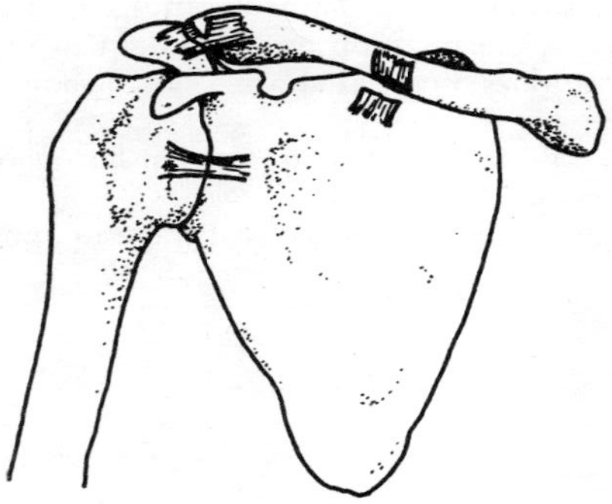

FIGURE 5. Sprains of the shoulder

Part B. Totally Torn Tendons—Tendonitis

1. Locate your calf muscle (your Gastrocnemius muscle). Run your hand down your calf until you nearly reach the back of your heel. You should now be able to feel a thick cord at the back of your heel. This cord is a tendon (your Achilles tendon.)
2. Examine Figure 6a. This drawing shows an actual view of the back of a person's leg. The skin has been removed.
3. Finish Figure 6b by showing what a totally torn Achilles tendon would look like. Draw an arrow pointing to the torn area and label it.
4. Finish Figure 6c by showing what tendonitis of the Achilles tendon would look like. Tendonitis is a soreness of the tendon. It is caused by small tears which occur along the tendon. Draw an arrow pointing to the tears and label them.

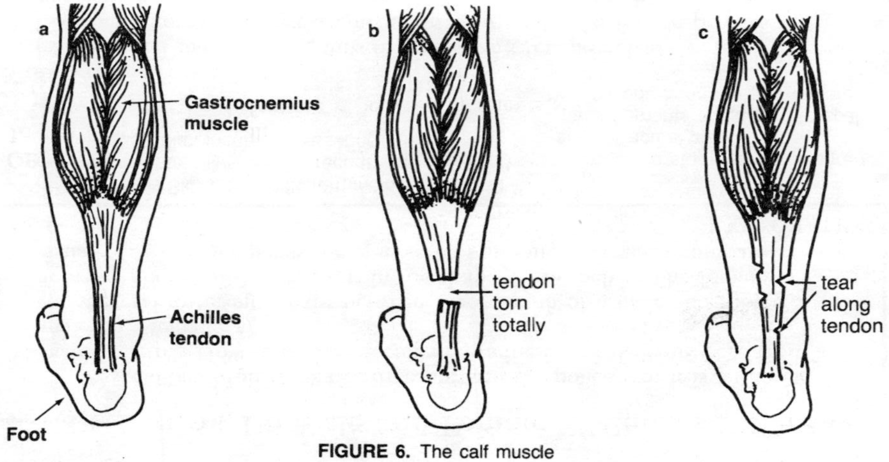

FIGURE 6. The calf muscle

113

QUESTIONS

1. What body parts are held together by ligaments? __bones__
2. Are ligaments a part of the muscular system or the skeletal system? __skeletal__ Why? __They hold bones together.__
3. Explain how a first, second, and third degree sprain differ? __first-degree sprain, ligaments are only stretched; second-degree sprain, ligaments are only partly torn; third-degree sprain, ligaments are torn completely__
4. What type of sprain probably takes the least time to heal? __1st degree__
5. What type of sprain takes the most time to heal? __3rd degree__
6. Describe what one might have to do to cause a sprain. __stretch bones too far out of normal position__
7. What body parts are connected by tendons? __muscle and bone__
8. Are tendons a part of the muscular system or the skeletal system? __muscular__ Why? __They are actually an extension of the muscle.__
9. Explain how tendons differ from ligaments. __Tendons connect muscle to bone and help move bones. Ligaments connect bone to bone and help hold them in place.__
10. Describe what one might have to do to cause a tendon to totally tear or develop tendonitis. __Too much tension or too sudden a pull on a body part such as a leg, a foot, or an arm.__
11. A totally torn tendon is a serious problem for an athlete or anyone else. A person will lose the use of the body part to which the tendon attaches. For example, a totally torn Achilles tendon will prevent a person from lowering his foot. Muscles shorten (contract) when they work. The Gastrocnemius shortens and pulls the foot down.
    a. Explain why the foot cannot be pulled down if the Achilles tendon is totally torn. __The muscle which pulls the foot down is no longer connected.__
    b. Might the foot be raised if the Achilles tendon is totally torn? __Yes__ Why? __A different muscle has this job.__
    c. Might a person with a totally torn Achilles tendon still be able to move his leg? __Yes__ Why? __Different muscles move the leg. The Achilles tendon only moves the foot.__

114

Name _____ Class _____ Period _____

Before doing this lab, have students read section 14:1 in the text.

14-2 How Do Male and Female Skeletons Differ?

A skeleton is found. A doctor reports to the police that it is an adult male skeleton. How could the doctor determine if the skeleton were from a male or a female?

Several differences exist between the skeleton of a male and that of a female. The main difference is in the shape of the pelvis. The female usually has a wider pelvis. Let's see how some measurements compare.

Process Skills: scientific terminology, interpreting scientific illustrations, recognizing and using spatial relationships, measuring in SI

Point out to students that diagrams are *not* actual size. The measurements are drawn to scale when comparing male to female.

EXPLORATION

OBJECTIVES

In this exercise, you will:
1. examine and measure diagrams of a male and female pelvis.
2. determine how these measurements differ in male and female pelvises.
3. use your data to determine if a third pelvis is male or female.

An excellent follow-up to this lab would be to bring a human skeleton into class if available. Ask students to determine if it is male or female, and explain why.

KEYWORDS

Define the following keywords:

femur _____ the thigh bone

pelvis _____ bones of the hip

sacrum _____ part of pelvis made of spinal bones (base of spine)

skeleton _____ framework of bones in the body which aids in support

MATERIALS
metric ruler

PROCEDURE

1. Examine Figure 1. Figure 1a is the pelvis from an adult male. Figure 1b is the pelvis from an adult female.
2. Measure the length (in millimeters) of the following dashed lines on Figures 1a and 1b: lines *a*, *b*, *c*, and *d*.
3. Record these numbers in Table 1. Note that lines b and c are part of the sacrum bone (shaded). This bone is found at the back of the pelvis and does not block the pelvis opening. It does, however, appear to block the opening in the figures.

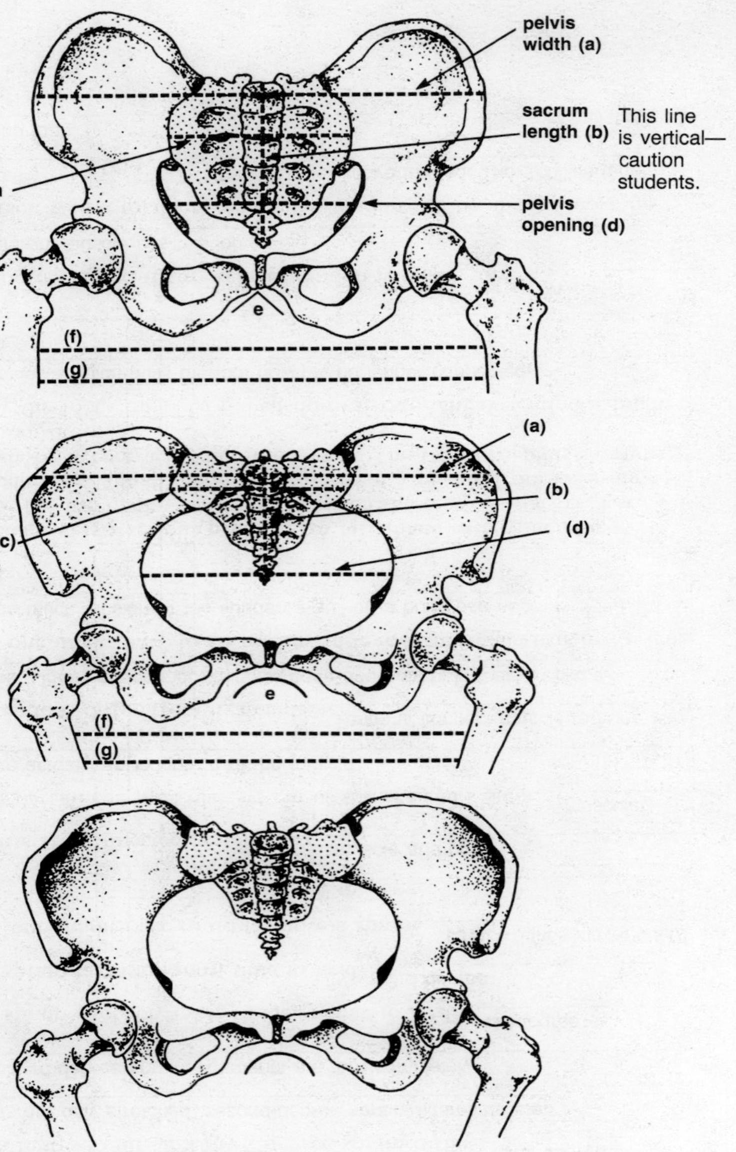

---

a

pelvis width (a)
sacrum length (b) — This line is vertical—caution students.
pelvis opening (d)
sacrum width (c)

b

(a)
(b)
(d)
(c)
(f)
(g)

The slant inward of femur bones in the female is also obvious in a person. Have a female and a male with jeans on stand in front of the class. The difference is most obvious.

c

Students should record actual data rather than merely say "female" because of its shape.

FIGURE 1. Human pelvises

Name _____ Class _____ Period _____

4. Locate letter *e* on Figures 1a and 1b. Note that the bottom of each pelvis is either round or pointed at this location. Record in Table 1 if the bottom is pointed or round.

5. Measure and record in Table 1 the lengths of dotted lines *f* and *g* on Figures 1a and 1b. If the femur bones hang *straight down*, the lengths of lines *f* and *g* will be the same. If the femur bones *slant inward*, the lengths of lines *f* and *g* will differ. This position of femur bones (thigh bones) provide a clue as to whether the skeleton is male or female.

6. Record your measurements and position of the femurs in Table 1.
Note that now you have a way of telling male from female skeletons by using all the data in Table 1.

7. Measure and record all the lengths of the pelvis and femur parts in Figure 1c just as you did for Figures 1a and 1b. The dashed lines are not included in the figure. Record your data in Table 1.

8. Indicate in Table 1 if Figure 1c represents a male or female skeleton.

Table 1. Pelvic Bone Measurements

Figure	Sex	Pelvis width line a	Sacrum length line b	Sacrum width line c	Pelvis opening line d	Bottom shape	Line f	Line g	Position of femurs
1a	Male	85	32	35	36	Pointed	84	84	Straight
1b	Female	88	26	32	47	Round	74	70	Slant in
1c	Female	89	27	33	48	Round	75	70	Slant in

Student data for Figure 1c may vary.

QUESTIONS

1. Explain how each of the following differs in adult male and adult female skeletons:

   a. pelvis width   pelvis is wider in female than in male _____

   b. sacrum   length and width are less in female than in male _____

   c. pelvis opening width   wider in female than male _____

   d. bottom shape of pelvis   round in female, pointed in male _____

   e. position of femur bones   slant inward in female, hang straight in male _____

2. Figure 1c is from a   female   (male or female.) List three things that helped you with your answer.   bottom of pelvis is round, pelvis opening is wide, legs slant in
Student answers may vary.

117

---

3. The approximate age of a skeleton can be told by measuring the length of the femur bone. The graph shown in Figure 4 gives you these measurements. By using this graph, determine the approximate age of a skeleton whose femur measures

   a. 200 millimeters.   5 years old

   b. 300 millimeters.   10 years old

   c. 350 millimeters.   13 years old

4. Explain why the graph in Figure 2 cannot be used to determine the age beyond 18 years.   growth of the femur stops after this age

FIGURE 2. Age of human skeletons

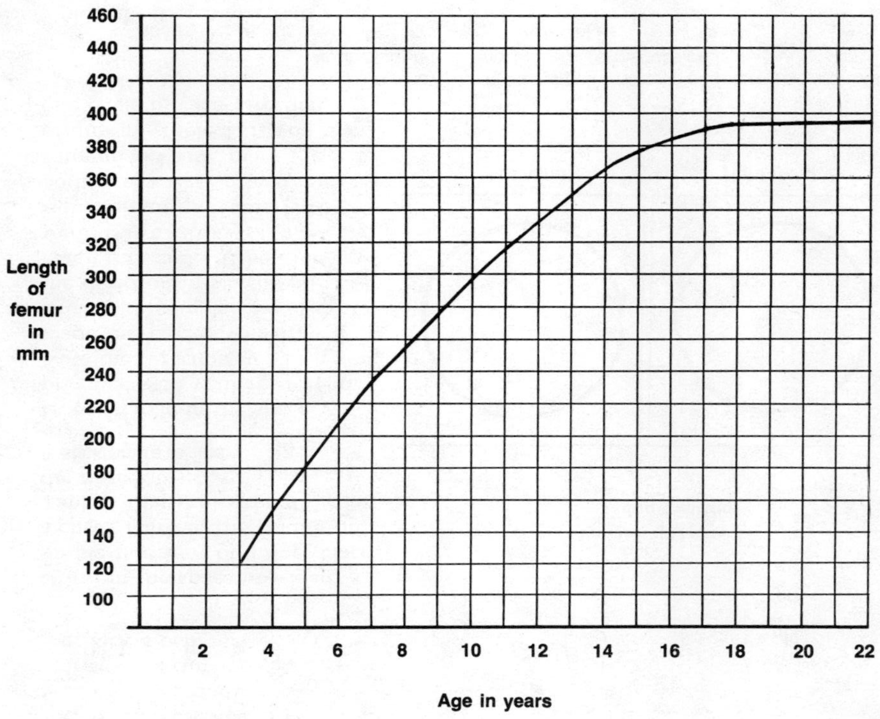

118

Name _____ Class _____ Period _____

Before doing this lab, have students read section 15-2 in the text.

15-1 Which Brain Side Is Dominant?

The human brain is divided into a left and a right side. Many things that you do with the right side of your body are controlled by your brain's left side. Many things that you do with the left side of your body are controlled by your brain's right side. If much of what you do is done by your body's right side, your dominant brain side is the left side. If much of what you do is done by your body's left side, your dominant brain side is the right side.

INVESTIGATION

Process Skills: experimenting, inferring, observing, interpreting scientific illustrations

OBJECTIVES
In this exercise, you will:
a. check to see how many activities you do using your left hand or your right hand.
b. check how many activities you do using your left foot or your right foot.
c. find out if you draw or see objects more to the right side or the left side.
d. find out if the left side or the right side of your brain is dominant.

KEYWORDS
Define the following keywords:

cerebrum ____ brain part that controls thought, reason, and senses

dominant ____ that which controls what you do; having more effect

left cerebrum side ____ side of cerebrum that controls right body side's function

right cerebrum side ____ side of cerebrum that controls left body side's function

MATERIALS
paper   red pencil

PROCEDURE
1. Place a check mark in the proper column in Table 1 to show which hand you usually use to do the following tasks. Note: If you use either hand just as often, then check both columns.
Tell which hand you use to
   a. write your name.
   b. wave "hello."
   c. bat while playing baseball.
   d. which thumb is on top when folding your hands.
   e. hold your spoon or fork while eating.

119

2. Place a check mark in the proper column in Table 1 to show which foot you usually use to do the following tasks. Note: If you use either foot just as often, check both columns. Tell which foot you use to
   a. start down a flight of stairs.
   b. start up a flight of stairs.
   c. catch yourself from falling as you lean forward.
   d. start skipping.
   e. place most weight on when you are standing.
   f. start to run.
   g. kick a ball.

Students may have to actually try these activities in order to mark the table correctly. One is usually not aware of these actions.

Example

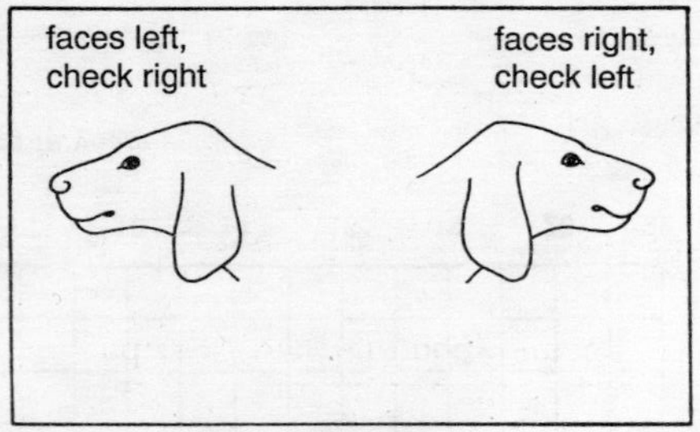

faces left,   faces right,
check right   check left

3. Draw, in the space provided, a simple side view of a dog. Place a check mark in the column in Table 1 that shows the direction the nose faces.
A stick figure is fine.

dog drawing

4. Draw a circle in the space provided with your *right* hand. Note the direction in which you made this circle. Now draw a circle with your left hand. Note the direction in which you made this circle. If both circles were drawn clockwise, mark the right column in Table 1. If both circles were drawn counterclockwise, mark the left column in Table 1. If you drew one circle in each direction, check both columns.

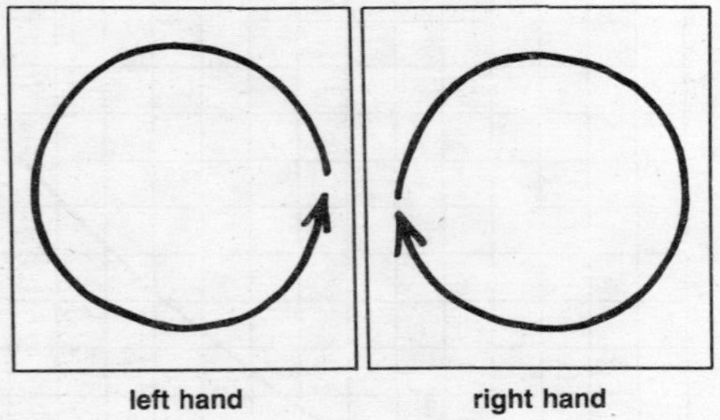

left hand   right hand

5. Roll a sheet of paper into a tube. Look through the tube at some distant object with both eyes open as shown in Figure 1. Then while looking through the tube at that distant object, close one eye and then the other. The eye that sees the object through the tube is your dominant eye. Place a check mark in the proper column in Table 1.

You may have to review clockwise and counterclockwise with your students during the prelab discussion.

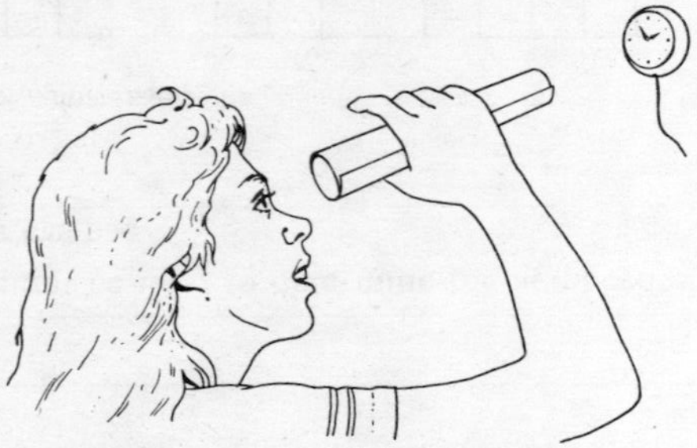

FIGURE 1. Finding your dominant eye

120

342

Name _____ Class _____ Period _____

6. Total up the check marks for each column of Table 1 and place the total at the bottom of the columns.

Table 1. Finding Your Dominant Side

Task	Left side	Right side
Write name		✓
Wave "hello"		✓
Bat		✓
Thumb position		✓
Hold spoon		✓
Walk down stairs		✓
Walk up stairs		✓
Catch from falling		✓
Skipping		✓
Standing	✓	
Start to run		✓
Take off shoe		✓
Leg on top		✓
Kick		✓
Dog drawing		✓
Circle drawing	✓	
Dominant eye	✓	
Totals =	3	13

Answers will vary. Sample data given.

QUESTIONS

1. Which column in Table 1 has the most check marks? __Student answers will vary.__
2. Which column in Table 1 has the fewest check marks? __Student answers will vary.__
3. Which body side seems to be your dominant side? __Student answers will vary.__
4. The human cerebrum is divided into left and right sides.
   a. Which brain side controls the left side of your body? __right__
   b. Which brain side controls the right side of your body? __left__
5. The brain side that you use the most is said to be your dominant brain side. Which is your dominant brain side? __Answers will vary.__ (HINT: The answer will be the opposite side from your answer to question 3.)
6. Look at Figure 2. It shows a top view of the brain. Label the following parts: left cerebrum side, right cerebrum side. Use a red pencil to shade in your dominant brain side.

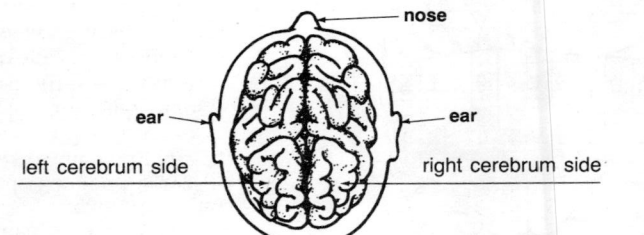

FIGURE 2. Top view of brain
(labels: nose, ear, ear, left cerebrum side, right cerebrum side)

7. Your teacher will ask for a class survey of certain results. Complete the following data for your class:
   a. number of students who are right-handed and show the right body side as dominant. __Student answers will vary.__
   b. number of students who are right-handed and show the left body side as dominant. __Student answers will vary.__
   c. number of students who are left-handed and show the right body side as dominant. __Student answers will vary.__
   d. number of students who are left-handed and show the left body side as dominant. __Student answers will vary.__
*8. Using your results from question number 7,
   a. does a person who uses his or her right hand for writing always show a dominant right body side? __Most students will say no.__
   b. does a person who uses his or her left hand for writing always show a dominant left body side? __Most students will say no.__

*Dominant body side will usually correlate with handedness but may not. Students may have been *taught* to use their right hand when actually they are left-handed.

Name _____ Class _____ Period _____

Before doing this lab, have students read section 15:2 in the text.

15-2 The Brain and Its Functions

The human brain is divided into three different parts. Each part is specialized. Each part has a job to perform that is different from the other parts. The brain is even more specialized in that specific brain sides control only specific body sides.

INTERPRETATION

Process Skills: interpreting scientific illustrations, scientific terminology, recognizing and using spatial relationships

OBJECTIVES

In this exercise, you will:
a. identify and label the three brain areas.
b. determine the jobs of certain brain areas.
c. match brain areas with their corresponding areas of body control.

KEYWORDS

Define the following keywords:

cerebellum brain part that makes our movements smooth and graceful

cerebrum brain part that controls thought, reason, and senses

involuntary you have no control over the part

medulla brain part that controls heartbeat, breathing, and blood pressure

voluntary you have control over the part

MATERIALS

#2 pencil   colored pencils: red, green, blue, gray, and yellow

PROCEDURE

Part A. Control Areas of the Brain

1. Examine Figure 1. This shows a side view of the human brain. Label the brackets correctly to show the brain's three parts or areas. Use the following labels: medulla, cerebrum, and cerebellum. Students will need to consult their text.

2. Label the functions of certain brain parts by using the following labels:
   A. vision center   F. heartbeat center
   B. speech center   G. coordination center for body muscle
   C. sensation of body pain H. smell center
   D. muscle control of body I. personality center
   E. hearing center

3. Still using Figure 1, color in the voluntary parts of the brain with a red pencil. Color the involuntary parts blue. Students will need to consult their text for help with this.

123

FIGURE 1. Side view of the human brain

cerebrum (red) {
 — sensation of body pain
 — muscle control of body
 — hearing center
 — personality center
 — speech center
 — smell center

cerebellum (blue) {
 — coordination center for body muscle

medulla (blue) {
 — heartbeat center

vision center

FIGURE 2. The cerebrum

left side of cerebrum   right side of cerebrum

A—solid red   A'—stripe red   a—crisscross red   a'—dot red
B—solid blue   B'—stripe blue   b—crisscross blue   b'—dot blue
C—solid green   C'—stripe green   c—crisscross green c'—dot green
D—solid yellow   D'—stripe yellow   d—crisscross yellow d'—dot yellow
E—solid gray   E'—stripe gray   e—crisscross gray   e'—dot gray

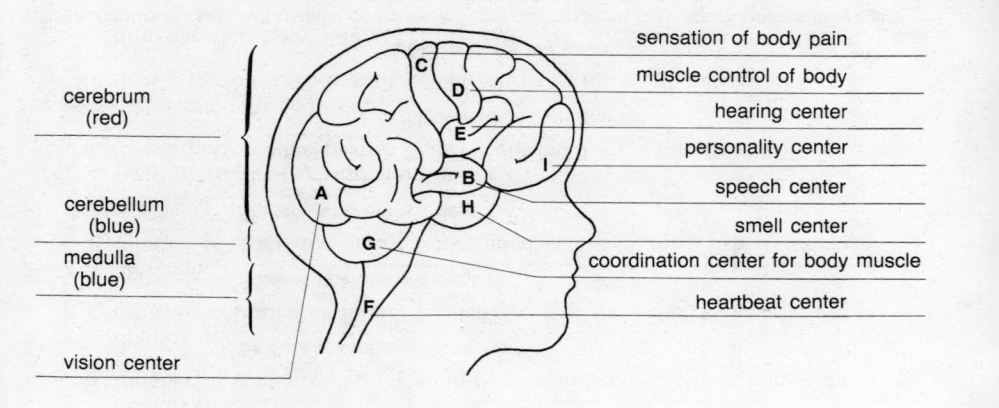

Part B. How the Brain Controls the Body

1. Examine the top view of the cerebrum in Figure 2. Note that the cerebrum is divided into left and right sides. Each side has been marked for you.

2. Locate and examine the two front views of the body in Figure 3. Note that in these views the left and right sides are reversed (this is because they are front views). The body views are marked either "sensation of body pain" or "muscle control of body." Muscle control and body pain are controlled by certain brain areas. The brain area controlling this and the corresponding areas on the body are marked with similar letters.

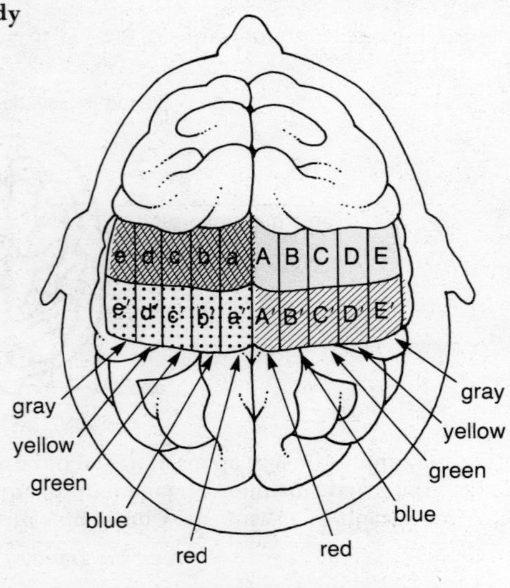

3. Match the brain areas of Figure 2 with their corresponding body areas of Figure 3 by coloring in parts of the figures as follows:

124

Name _____ Class _____ Period _____

FIGURE 3. Front views of the human body

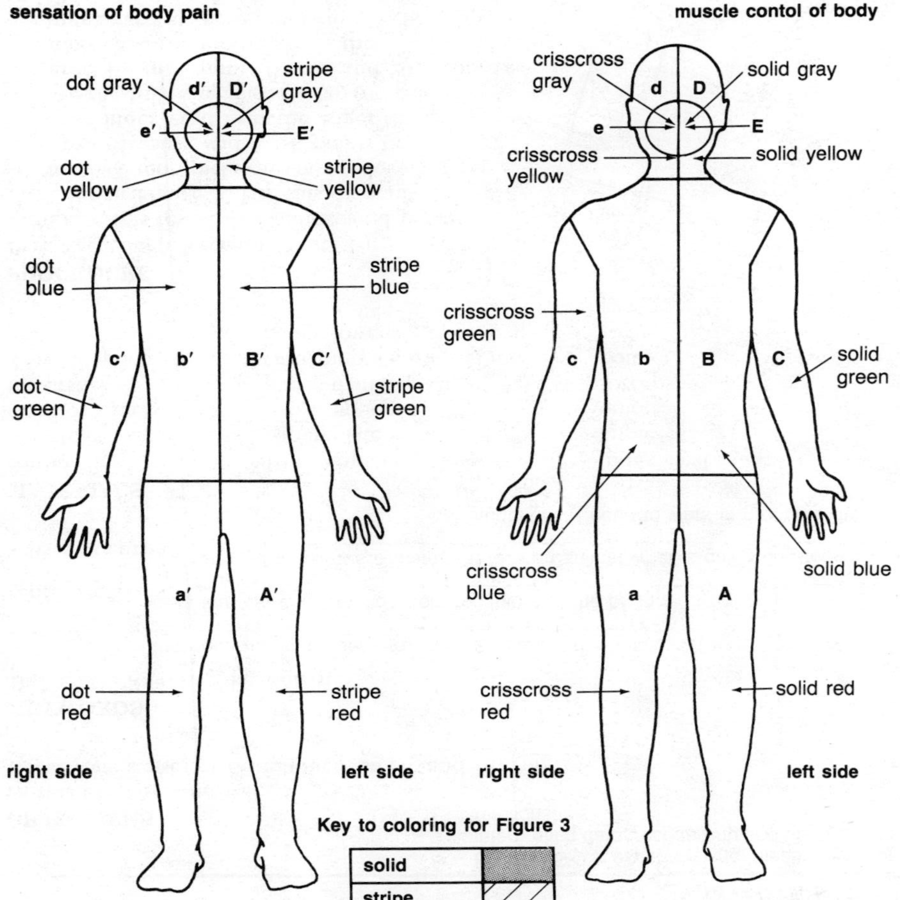

sensation of body pain | muscle control of body

right side | left side | left side | right side

Key to coloring for Figure 3:
| solid |
| stripe |
| crisscross |
| dots |

QUESTIONS

1. Which brain area is the largest? **cerebrum**
2. What side and functions are part of body areas
   a. A–E? **muscle control of left body side**
   b. a–e? **muscle control of right body side**

3. On what brain side of the cerebrum do the following appear:
   a. A–E? **right side**
   b. a–e? **left side**

4. By using your answers to questions 2 and 3, explain how brain side and body muscle control are related. **The brain's right side controls muscles on the left side of the body. The brain's left side controls muscles on the right side.**

5. What side and function are part of body areas
   a. A'–E'? **sensation of body pain on left side**
   b. a'–e'? **sensation of body pain on right side**

6. On what brain side of the cerebrum do the following appear:
   a. A'–E'? **right side**
   b. a'–e'? **left side**

7. By using your answers to questions 5 and 6, explain how brain side and body sensation of pain are related. **The brain's right side controls sensation on the left side of the body. The brain's left side controls sensation on the body's right side.**

8. Circle the answer that correctly completes the following statements:
   a. Body areas A–E for muscle movement go from (top to bottom, **bottom to top**) of the body.
   b. Brain areas A–E for control of muscle movement go from (**center to right side**, right side to center) of the brain.
   c. Body areas A'–E' for sensation of pain go from (top to bottom, **bottom to top**) of the body.
   d. Brain areas a'–e' for sensation of pain go from (**center to left side**, left side to center) of the brain.

9. A stroke or cardiovascular accident results when blood vessels in the brain burst. This results in damage to the brain area near the broken blood vessel. Using Figure 2 as a guide, predict how a person would be affected if they suffer a stroke in
   a. area A. **unable to move left leg muscles**
   b. area C. **unable to move left arm muscles**
   c. area E. **unable to move left side of mouth**
   d. area D'. **unable to feel pain on left side of head and face**
   e. area b. **unable to move right chest muscles**
   f. area e'. **unable to feel pain on right side of mouth**

Name _____ Class _____ Period _____

Before doing this lab, have students read section 16:2 in the text.

16–1 How Can You Test Your Senses?

INVESTIGATION

Information about your surroundings reaches the brain through sensory neurons. The sensory neurons get information from the receptors. Some of these receptors for pain, pressure, and touch are scattered throughout the skin and other parts of the body. Other receptors for taste, smell, sight, and hearing are contained in special organs. We are dependent, in part, on our sense receptors for our understanding of the world.

OBJECTIVES

In this exercise, you will:
test your senses of taste, touch, smell, and hearing.

Process Skills: experimenting, measuring in SI, interpreting data, recognizing and using spatial relationships

KEYWORDS

Define the following keywords:

receptor ___ cell that receives information from the environment

sense organ ___

sensory neuron ___ part of the nervous system that tells an animal what is going on around it

MATERIALS

Students will need to work in groups of two.

cotton
blindfold
clock or timer
meterstick
2 metal spoons
glue

10 toothpicks
paper plate
plastic spoon
paper drinking cup
cinnamon chewing gum
paper towel

index card
apple
onion
forceps
potato

PROCEDURE

Part A. Touch Receptor Test

Complete this test with a blindfolded partner.

1. Look at Figure 1. Use this diagram to place toothpicks on your index card. Measure and mark points on your index card near the edges that are the following distance apart: 2 mm, 4 mm, 8 mm, 10 mm, and 20 mm. Glue the toothpicks to the index card at these points to match those shown in Figure 1.

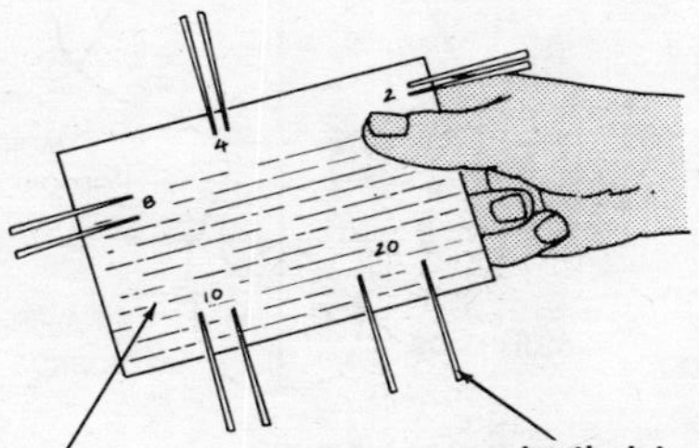

FIGURE 1. Touch test card

2. Gently touch your partner's fingertips (palm side) with the points of the toothpicks that are 2 mm apart. Ask your partner to report how many points are touching the skin. If your partner can feel two points, put a check mark in Table 1 under the proper column. If your partner cannot feel two points, put a minus mark in the table. **CAUTION:** *Do not apply heavy pressure when touching your partner's skin.*

3. Repeat step 2 touching your partner's palm, back of the hand, inside of the forearm, back of the neck, and lips. Record your results in Table 1.

4. Repeat steps 2 and 3 using each pair of toothpicks.

Table 1. Testing the Sense of Touch   Answers will vary. Sample data given.

Body part	Distance between two points of touch				
	2 mm	4 mm	8 mm	10 mm	20 mm
Fingertip (palm side)	✓	✓	✓	✓	✓
Palm of hand	–	–	–	✓	✓
Back of hand	–	–	✓	✓	✓
Forearm (inside)	–	✓	–	✓	✓
Back of neck	–	–	✓	✓	✓
Lips	–	✓	✓	✓	✓

Part B. Food Test

Complete this test with a blindfolded partner.

1. Using forceps, place a small amount of chopped potato, onion, and apple on a paper plate.

2. Instruct your partner to hold his or her nose with thumb and forefinger so that he or she cannot smell.

3. Using a plastic spoon, give your partner one of the foods to taste. (Do not let your partner swallow the food.) Be sure the nostrils are tightly closed while tasting foods. Ask your partner to tell you what food is being tasted. Record in Table 2 what food your partner thought he or she tasted. Dispose of the food in a paper towel. Have your partner rinse his or her mouth with water.

4. Repeat steps 2 and 3 with the remaining food samples. Be sure to record the information in Table 2. Also be sure to rinse the mouth after each food test.

5. Have your partner rinse his or her mouth after the foods have been tasted. Have your partner hold his or her nose closed again. Give your partner a piece of cinnamon chewing gum. Ask your partner to describe the taste. Record the answer in Table 2. Then have your partner release the nostrils and describe the taste.

Table 2. The Sense of Taste

Food sampled	Tasted like
Potato	Answers may vary. bland
Onion	Not a strong onion flavor
Apple	sweet
Cinnamon gum	sweet

They may not be able to distinguish the onion from the potato.

Name _____ Class _____ Period _____

Part C. Sound Test

1. Hold a ticking clock or timer at the opening of your partner's left ear. Slowly back away until your partner can no longer hear the ticking. Measure and record in Table 3 the distance at which the sound can no longer be heard.
2. Repeat step 1 with your partner's right ear.
3. Have your partner block his or her right ear with cotton (or your partner could use a finger.) Repeat step 1. Record the results in Table 3.
4. Have your partner block his or her left ear and repeat. Record the results in Table 3.
5. Blindfold your partner. Strike two metal spoons together at a distance of 1 meter from the front of the person's head. Have your partner indicate the direction of the sound. Place a check mark in the proper place in Table 4 if your partner is correct and a minus sign if your partner is not correct. Then strike the spoons together on the right side, the left side, in back of, and above the head at a distance of 1 meter. Record the results of each of these tests in Table 4.

Table 3. The Sense of Hearing

Sound source	Distance from	
	Left ear	Right ear
Ticking clock	3 m	3 m
Ticking clock with ear blocked	2 m	2 m

Answers will vary depending on an individual's quality of hearing.

Table 4. How Distance Affects the Sense of Hearing

Location	Ears open		Left ear blocked		Right ear blocked	
	1 m	3 m	1 m	3 m	1 m	3 m
Front of head	✓	✓	✓	✓	✓	✓
Left side of head	✓	✓	✓	–	✓	–
Right side of head	✓	✓	✓	–	✓	–
Back of head	✓	✓	✓	–	✓	–
Above the head	✓	✓	✓	–	✓	–

Most students will accurately locate sound in front of the head.

6. Repeat step 5 at a distance of 3 meters. Record the results in Table 4.
7. Block only your partner's right ear (as before.) Repeat steps 5 and 6. Record the results in Table 4.
8. Block only your partner's left ear (as before.) Repeat steps 5 and 6. Record your results in Table 4.

129

QUESTIONS

1. Which area of the body as tested in Part A seems to have the greatest number of touch receptors? _The fingertips seem to have the greatest number of receptors._
2. How does this help you? _You feel things with your fingertips._ Most students will say the back of the hand has the least number of receptors.
3. Which area seems to have the least number of touch receptors?
4. What do the answers to questions 1 and 3 tell you about the distribution of touch receptors in the skin? _they are not evenly distributed_
5. How many basic tastes can your tongue sense? _four_
   What are they? _bitter, sour, salty, and sweet_
6. a. Is it possible to detect correctly the three foods tested in part B with the nose closed? _usually no_
   b. Would it be possible to detect the three foods correctly if the nose were not closed? _yes_
7. What does the gum taste like with the nostrils closed? _sweet_
   With the nostrils open? _cinnamon_ Why? _smell is needed to distinguish flavor_
8. How can you tell the difference between two things that taste sweet? _the sense of smell helps you_
9. When you eat vanilla ice cream, what senses are you using? _taste and smell_
10. Why do you lose the ability to taste foods when you have a cold? _when you have a cold, mucus interferes with the smell receptors_
11. In Part C, do you hear sounds that originate at greater distances better with one ear or two ears? _you hear better with two ears_
12. Which sense organ was being used in Part A? _the skin_
13. Which sense organ was being used in Part B? _the tongue_
14. Which sense organ was being used in Part C? _the ears_

130

Name _____ Class _____ Period _____
Before doing this lab, have students read section 16:2 in the text.

16-2 How Are Your Senses Sometimes Fooled?

What you see and what you think you see are not always the same. Seeing is done with your eyes. The message your eyes pick up is sent to the brain where it is interpreted. The brain may then "tell" you that you see something that is not present. The mistaken idea that you get is an illusion. If the mistaken idea is because of what you see, or your eyes, it is called an optical illusion.

INVESTIGATION

Process Skills: observing, experimenting, recognizing and using spatial relationships, interpreting scientific illustrations

OBJECTIVES
In this exercise, you will:
a. look at several diagrams and record what you see.
b. compare your results with what you know is present.

KEYWORDS
Define the following keywords:

illusion   a mistaken idea that you get

optical illusion   a mistaken idea because of what you see

sense   the ability to receive and react to stimuli

MATERIALS
ruler   red marker
file card   yellow marker
blank paper

PROCEDURE

Part A. Triangle Illusion
1. Examine Figure 1 for a white triangle. Record in Table 1 on page 127 if you see it.
2. Now record in the table if the white triangle is really present. Is there a white triangle drawn on the diagram? __no__

Part B. Bent Line Illusion
1. Examine Figure 2. Record in Table 1 if you think the lines across the diagram could meet without the resulting line being bent.

FIGURE 1.

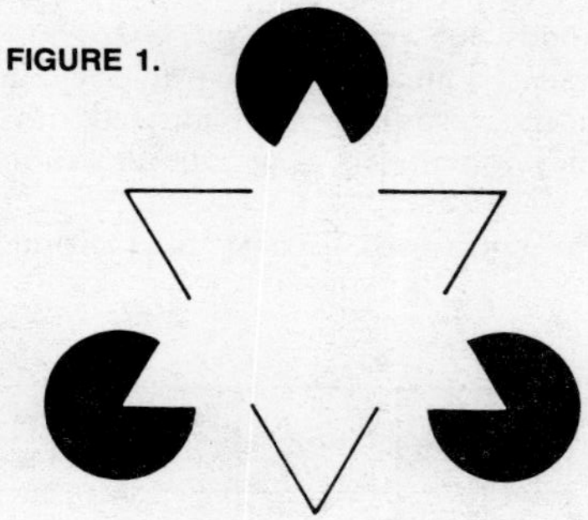

FIGURE 2.

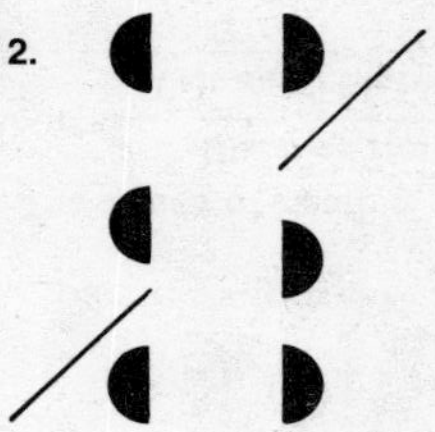

131

FIGURE 3.

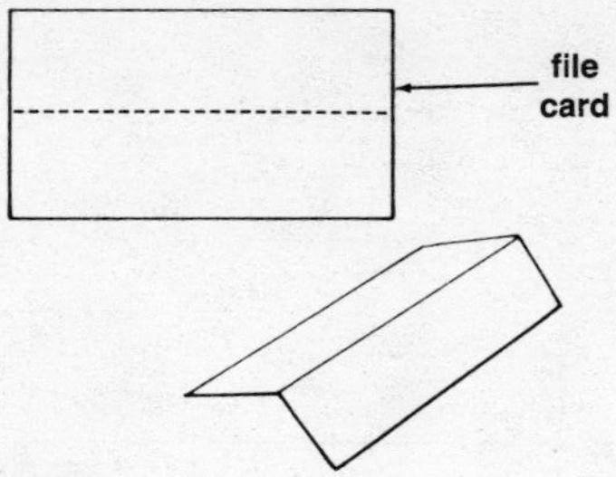

FIGURE 4.

FIGURE 5.

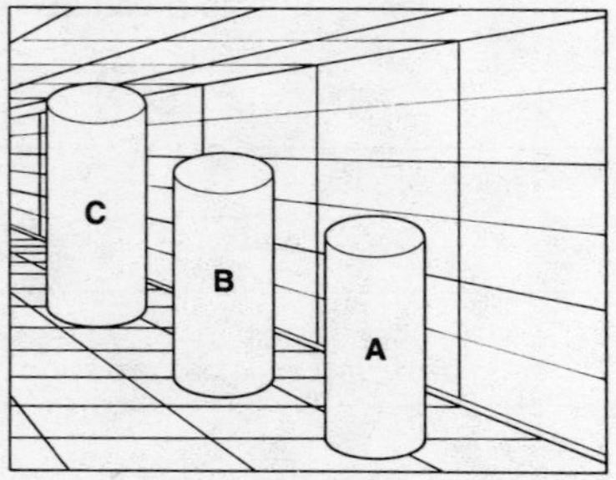

FIGURE 6.

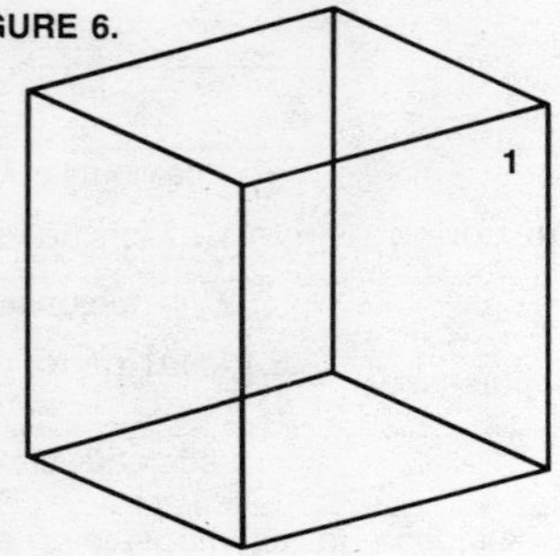

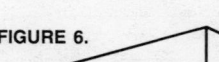

2. Lay a ruler on Figure 2 next to both lines. Does it now appear that the lines would meet in the center without the resulting line being bent? __yes__
3. Record your results in Table 1.

Part C. Paper Fold Illusion
1. Fold a file card as shown here in Figure 3. Lay it on your desk about 20 cm from your eyes.
2. Stare at the figure for about 20 seconds with one eye closed. Does the fold appear to always be in the top of the figure? __no__
3. Record your answer in Table 1. Also record where you know the fold really is.

Part D. Flower Illusion
1. Color Figure 4 so that the center is red and the petals are yellow.
2. Stare at the diagram for 30 seconds. Then stare at a blank piece of paper.
3. Record in Table 1 what you see. Also record what you know is on the blank paper.

Part E. Cylinder Illusion
1. Look at Figure 5.
2. Which cylinder appears largest? __C__
3. Record your answer in Table 1.
4. Measure the cylinders with a ruler. Was your answer correct? __no__
5. Record your correct answer in Table 1.

Part F. Cube Illusion
1. Stare at the number 1 in Figure 6 for at least one minute.
2. What appears to happen to corner 1 when you gaze at it steadily? __it alternates from front to back.__
3. Record this in Table 1.

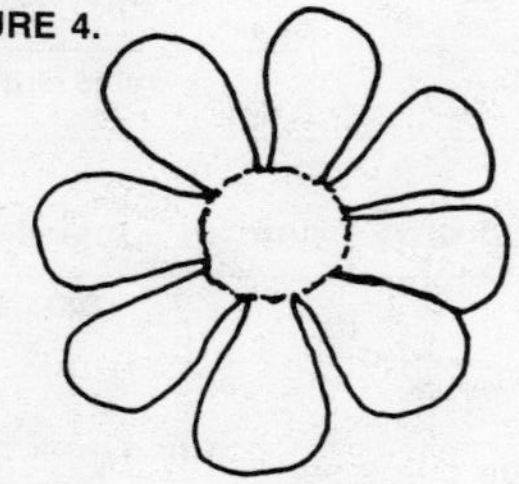

132

348

Name _____ Class _____ Period _____

4. Also record in Table 1 what Figure 6 really shows.

Table 1. Observing Illusions

Illusion	Appears	Really Is
Triangle	White triangle appears to be present.	No white triangle is present.
Bent line	The line appears bent.	The line is straight.
Paper fold	The paper appears folded two different ways.	The paper is folded only one way.
Flower	Color appears on the blank paper.	The paper is blank.
Cylinders	C appears larger.	It is the same size.
Cube	Corner 1 alternates from front to back.	It remains in the same place.

FIGURE 7.

Part G. Figure Reversals

1. Examine Figure 7. What do you see? a seated woman

2. When you look a second time, do you see anything different? a face

3. Do both figures ever appear at the same time? no

4. Examine Figure 8. This figure was used in 1900 by psychologist Joseph Jastrow. What do you see when the face looks to the left? a duck

5. What do you see when the face looks to the right? a rabbit
Can you see both at the same time? no

FIGURE 8.

133

FIGURE 9.

6. Examine Figure 9. What do you see? a goblet

7. Look at the figure again. Do you see anything different? What do you see? a pair of faces
Can you see both things at the same time? no

Part H. Creating Illusions

1. Examine the figure of the two girls below. They are both exactly the same size. Without changing their size, create the illusion that one girl is taller than the other.

A hat on one girl may create the illusion of tallness. Vertical striped clothing or background may also give the illusion of tallness.

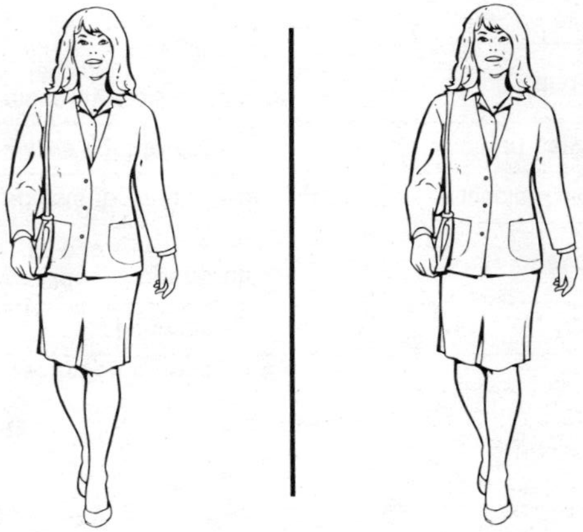

A horizontal background or horizontal striped clothing may make one girl appear shorter.

FIGURE 10. Make your own illusion

QUESTIONS

1. Which of the illusions in this activity were optical illusions? All of the illusions in this activity were optical illusions.

2. Sometimes a driver sees water on a dry highway. Why is this water an optical illusion? There is no water present on a dry highway.

3. Name some jobs people have in which they must be aware of optical illusions. Fabric designer, wallpaper designer, airplane pilot, and truck driver. Students may give other answers. Have them explain why each must be aware of optical illusions.

134

Name _____ Class _____ Period _____

Before doing this lab, have students read sections 16:2 in the text.

16-3 What Are the Parts of an Eye?

EXPLORATION

The eye is a most complex sense organ. It can detect differences in light intensity, color, and motion. It's a perfect example of a body organ where form follows function. This means that each eye part (form) is perfectly designed to carry out its specific job (function).

Process Skills: observing, communicating, classifying, modeling

OBJECTIVES
In this exercise, you will:
a. observe and identify the parts on the exterior of a pig eye.
b. build an eye model to study the action of four eye muscles.
c. observe and identify the parts on the interior of a pig eye.

KEYWORDS
Define the following keywords:

exterior ____ outside

interior ____ inside

muscle contraction ____ shortening or working of a muscle

MATERIALS
Have students work in groups of 2 or 3.

preserved pig eye
colored pencils: blue, black
polystyrene ball (7.5 cm diameter)
dissecting pan or paper towel
large paper clip
4 cloth strips

scissors
ruler
paper
pen
tape

Sheep or cow eyes may be substituted. Eyes are available from biological supply houses. Cow eyes have fat attached to the back of the eye making location of muscles and optic nerve difficult.

PROCEDURE
Part A. The Exterior Parts of an Eye

1. Examine a preserved pig eye that has been placed in a dissecting pan.
2. Locate the following parts using Figure 1 as a guide: sclera, iris, pupil, eye muscles, optic nerve, cornea.
3. Describe the job or function of these eye parts. Use your text or other references for help in completing the column marked "Function" in Table 1.
4. Complete the column marked "Description." As an example, the iris can be described as "part of the eye that gives it color."

The text describes functions of all structures except eye muscles and ciliary muscles.

350

NOTE: This diagram is of a human eye. Students may have difficulty locating these same parts on the pig eye. Point out that: muscles are pink while fat is yellow; cornea and lens may be opaque due to preservative; pupil is more egg shaped than round.

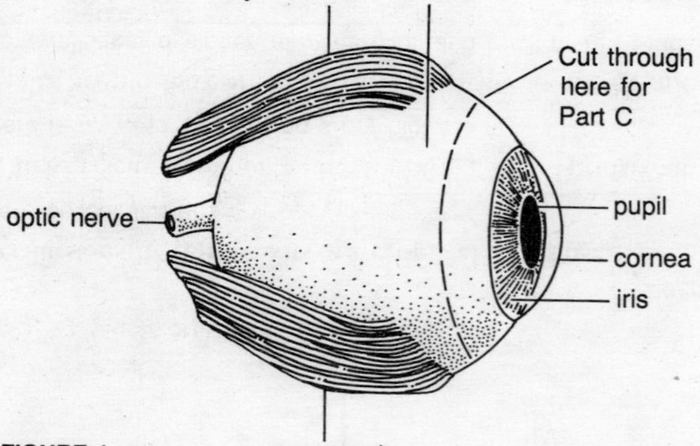

FIGURE 1.

Table 1. Function and Description of Exterior Eye Parts

Part	Function	Description
sclera	provides shape, protection to eye	white of eye, outer covering
iris	adjusts for bright, dim light	muscle, colored part of eye
pupil	allows light to enter eye	hole in center of iris
eye muscles	move eyeball up, down, left, right	tough band-like tissue
optic nerve	carries messages to brain	tube-like part found at back of eye
cornea	bends light, protects eye	clear front part of eye

Parts A and C can be completed without doing Part B.

Part B. Eye Muscles and Movement

1. Trace Figure 2 onto a sheet of paper. Color the pupil black and the iris blue. Cut the traced drawing out and tape it onto a polystyrene ball (eye model) as shown in Figure 3.
2. Use scissors to cut four cloth strips, each measuring 2 x 8 cm.
3. Use a pen to number the strips 1–4.

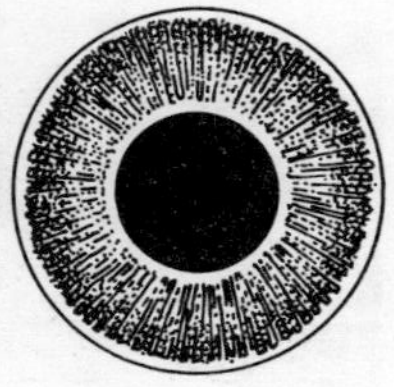

FIGURE 2.

136

Name _____ Class _____ Period _____

4. Tape the cloth strips onto the eye model as shown in Figure 3. These strips will represent eye muscles.
5. Determine the action of muscles 1 and 2 as follows:
   a. Stick the eye model onto a large paper clip as shown in Figure 4.
   b. Push the paper clip about 1 cm into the ball.
   c. Hold the paper clip between the fingers of one hand and gently pull cloth strip 1 toward the back of the eye with your other hand. The pulling of the cloth shows what happens in the eye as this muscle contracts or shortens.
   d. Note the direction that the eye model turns and record it in Table 2.
   e. Repeat steps c and d, only this time pull on cloth strip 2.
6. Determine the action of muscles 3 and 4 as follows:
   a. Stick the eye model onto the open end of a large paper clip as shown in Figure 5.
   b. Push the paper clip about 2 cm into the ball.
   c. Hold the paper clip between the fingers of one hand and gently pull cloth strip 3 toward the back of the eye with your other hand.
   d. Note the direction that the eye model turns and record it in Table 2.
   e. Repeat steps c and d, only this time pull on cloth strip 4.

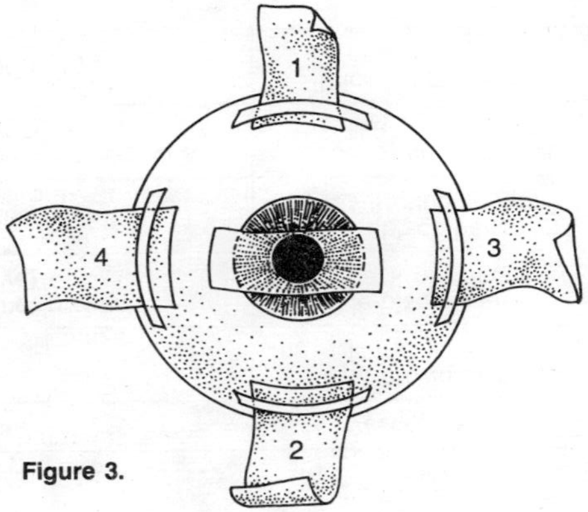

NOTE: Muscle 3 is on the eye side next to the ear while muscle 4 is on the eye side next to the nose.

Students may have to rock the round end of the paper clip slightly within the ball in order to show muscle action properly in the model.

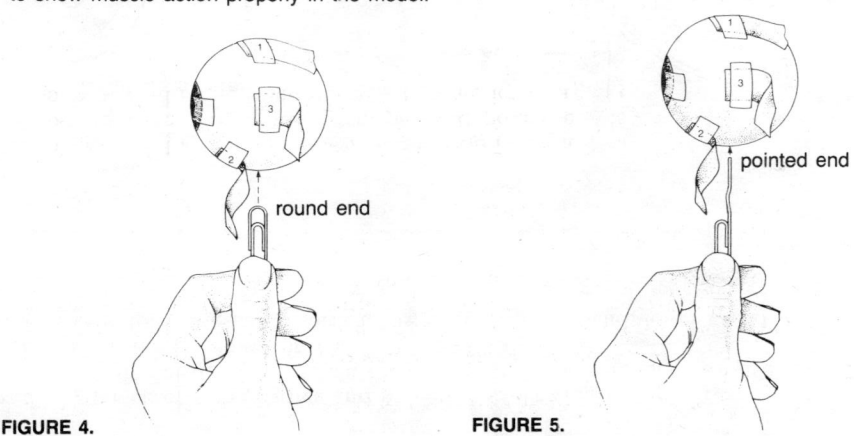

Figure 3.

FIGURE 4. round end

FIGURE 5. pointed end

137

7. Complete Table 2.

Table 2. Eye Muscle Action

Muscle	Muscle Name	Direction Eye Moves as Muscle Contracts	Location on Eye
1	Superior rectus	up	across top of eyeball
2	Inferior rectus	down	along bottom of eyeball
3	Lateral rectus	toward right	along side next to ears outer side
4	Medial rectus	toward left	along side next to nose inner side

Advise students to cut through the sclera, but not through the cornea area.

Part C. The Interior of an Eye

1. Using the preserved eye, make a circular cut with your scissors through the sclera. Use Figure 1 as a guide.
2. Separate the eye into two halves. A jellylike material and marble-shaped part may fall out of the cut eye.
3. Locate the following eye parts using Figure 6 as a guide: retina, tapetum, ciliary muscles, lens, vitreous humor.
4. Describe the job or function of these eye parts. Use your text or other references for help in completing the first column of Table 3.
5. Complete the column marked "Description." As an example, the lens can be described as "clear, marble-shaped."

NOTE: It may be difficult for students to differentiate between iris and ciliary muscles. Point out that the lens is normally clear.

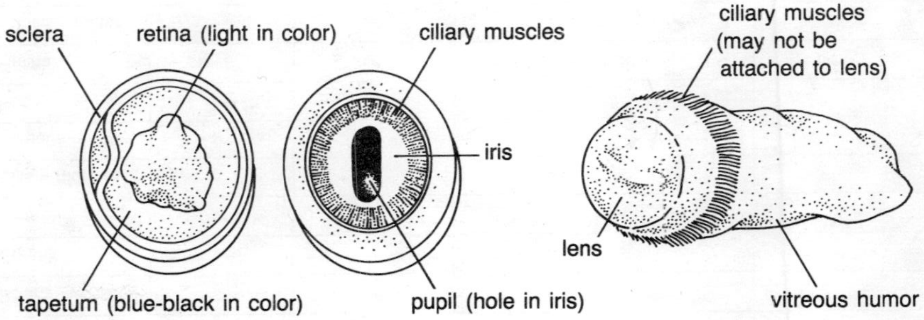

sclera — retina (light in color) — tapetum (blue-black in color) — ciliary muscles — iris — pupil (hole in iris) — ciliary muscles (may not be attached to lens) — lens — vitreous humor

FIGURE 6.

138

351

Name _____ Class _____ Period _____

6. To better observe the cornea and iris, do the following:
   a. use a single-edge razor to cut the front half of the eye, as shown in Figure 7.
   b. look at the cut edge and compare it to Figure 8.
7. Label the following parts: cornea, iris, sclera, upper eyelid, lower eyelid, pupil.

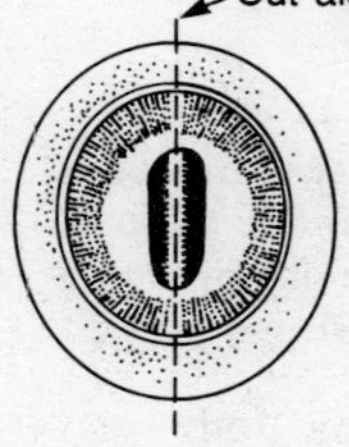

FIGURE 7.

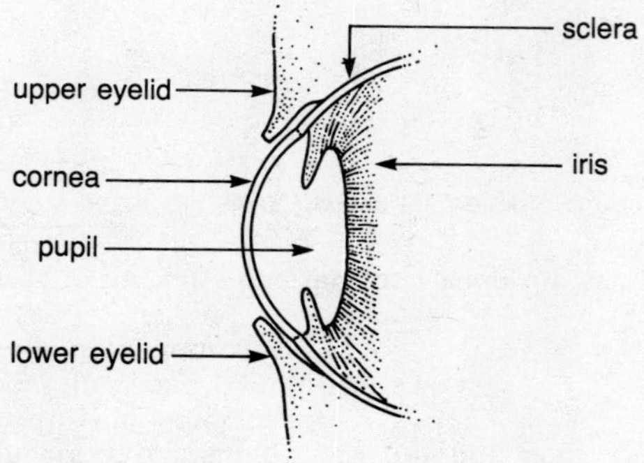

FIGURE 8.

Table 3. Function and Description of Interior Eye Parts

Part	Job or Function	Description
retina	receives light	light brown, thin film
tapetum	reflects light (not present in human eye)	blue-black layer in back of retina
ciliary muscles	adjusts lens shape	thin, black bands surrounding lens
lens	changes shape for close up and distant viewing	round, clear, marble-like
vitreous humor	provides shape to eye	thick liquid, almost gelatin-like

139

QUESTIONS

1. Complete the following chart. Use check marks. Some eye parts may be marked more than once in a row.

Eye part	Light does pass through	Light does not pass through	Muscle tissue	Can change shape	Made of nerve cells
Sclera		✓			
Cornea	✓				
Lens	✓			✓	
Ciliary muscles		✓	✓	✓	
Vitreous humor	✓				
Pupil	✓			✓	
Retina		✓			✓
Superior rectus		✓	✓	✓	
Lateral rectus		✓	✓	✓	
Iris		✓	✓		
Optic nerve		✓			✓

2. Using your results with the eye model and muscle action, predict the meaning of the following terms:
   a. superior _____ top _____
   b. inferior _____ bottom _____
   c. lateral _____ to side, toward ear _____
   d. medial _____ toward center, toward nose _____

You may wish to have students cut the rear half of the eyeball in half, as was done in Part C with the front half. This will allow students to see the continuation of the retina and its blind spot with the optic nerve.

140

Name _____ Class _____ Period _____

Before doing this lab, have students read sections 17:2 in the text.

17–1 What Self-protecting Behaviors Do Isopods Show?

INVESTIGATION

How would an animal behave if given a choice of spending time in a light or a dark place? The answer to this question will be different depending upon the kind of animal. The way that each animal responds to its surroundings is part of its behavior. Usually, most behavior in animals will help rather than harm them.

Process Skills: observing, experimenting, prediction, interpreting data

OBJECTIVES
In this exercise, you will:
a. experiment with an animal called an isopod.
b. test the animal's behavior in response to four different conditions.
c. determine how the isopod behaves under these four conditions.

KEYWORDS
Define the following keywords:

behavior   the way an animal acts

innate behavior   a way of responding that does not require learning

isopod   an animal belonging to the phylum Arthropoda, the joined leg phylum, class Crustacea

learned behavior   behavior that must be taught

MATERIALS
isopods   tape
paper towel   scissors
black paper   wall clock
petri dish and cover*   stick, plastic

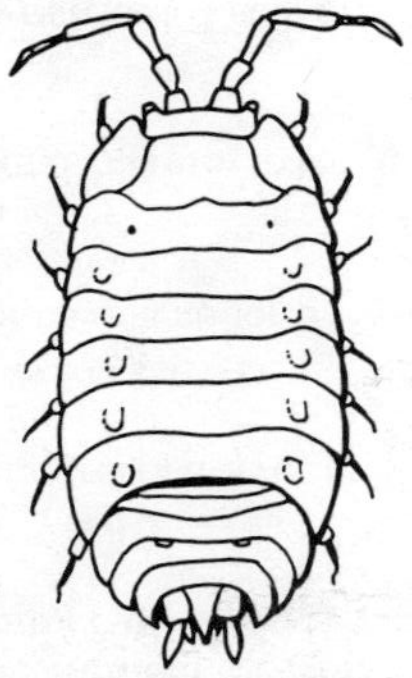

PROCEDURE
Part A. Do Isopods Prefer a Smooth or a Rough Surface?

1. Place the bottom of a round dish (petri dish) onto a paper towel.

*May substitute any dishes that are suitable—plastic cottage cheese or butter cartons

Collect isopods locally or order from a biological supply house. Substitute sow bug, cricket, or cockroach.

FIGURE 1. Isopod

141

2. Trace the bottom of the dish onto the towel.
3. Use scissors to cut out the paper outline of the dish and fold it in half.
*4. Place the folded paper towel in the bottom of the dish. Tape the paper towel in place along all sides as shown in Figure 2.
5. Place six isopods into the center of the dish. Cover the dish.
6. Examine the dish after exactly 5 minutes. Count the number of isopods that moved onto the toweled side. Count the number of isopods that moved onto the untoweled side.
7. Record your data in Table 1. Use the space marked Trial 1.
8. Repeat steps 5–7 two more times. Total your data and figure and record averages for the three trials in Table 1.

If time is a problem, reduce each trial to 2 or 3 minutes.

Part B. Do Isopods Prefer Wet or Dry Conditions?

1. Repeat steps 1–3 from Part A.
2. Moisten the paper towel and place it into the dish so that it is next to the dry towel used in Part A. Examine Figure 3 as a guide.
3. Tape the new towel in place, as before.
4. Place six isopods into the center of the dish. Cover the dish.
**5. Examine the dish after exactly 5 minutes. Count the number of isopods that moved onto the wet side. Count the number of isopods that moved onto the dry side.
6. Record these two results in Table 1. Use the space marked Trial 1.
7. Repeat steps 4–6 two more times. Total your data and record averages for the three trials in Table 1.

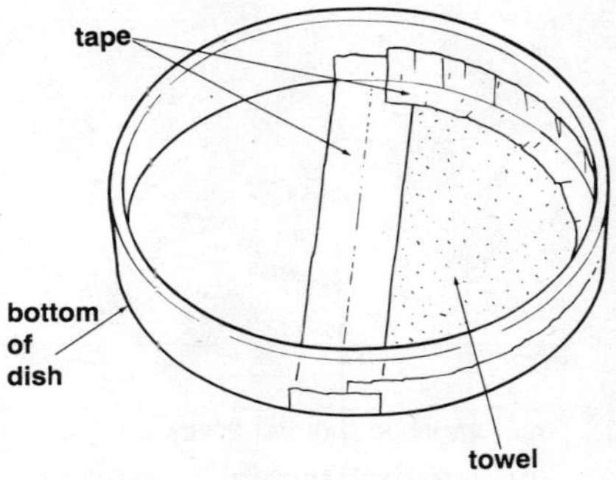

*Place the dish on a desktop or lab table that has a paper towel positioned below it. This will provide a background color that is similar on both sides.

**If time is a problem, reduce each trial to 2 or 3 minutes.

FIGURE 2.
FIGURE 3.
FIGURE 4.

142

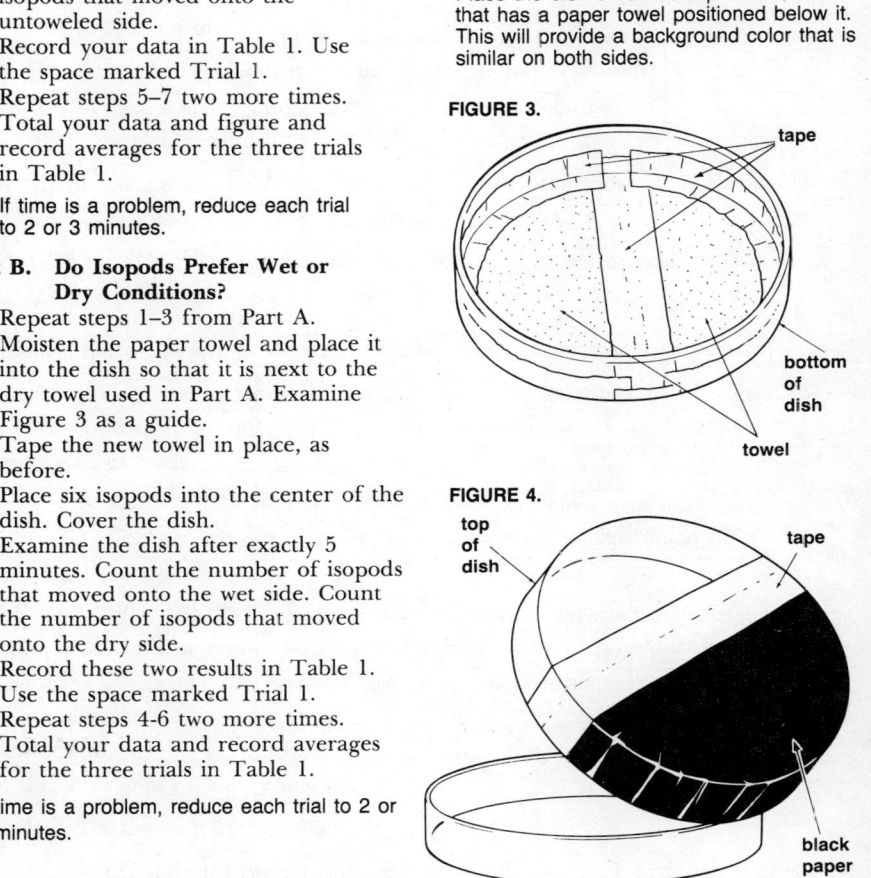

Name _____ Class _____ Period _____

Part C. Do Isopods Prefer Light or Dark?
1. Cover one-half of a new dish with black paper.
2. Tape the paper in place. Use Figure 4 as a guide.
3. Place six isopods into the center of the dish.
4. Cover the dish. Position a paper towel under the petri dish.
5. Examine the dish after exactly 5 minutes.* Count the number of isopods that moved into the dark side of the dish. Count the number of isopods that moved into the lighted side of the dish.
6. Record your data in Table 1. Use the space marked Trial 1.
7. Repeat steps 4–7 two more times. Total your data and record averages for the three trials in Table 1.
8. Return five of the animals to your teacher. Keep one for Part D.
*Reduce time for each trial to 2 or 3 minutes if students do not have time for 3 five-minute trials.

Part D. How Do Isopods Respond to Touch?
1. Place an isopod into a petri dish. Wait until the animal has uncurled itself before continuing on with the next step.
2. Lightly touch the animal with a plastic stick.
3. Record your observations in Table 2. Use the space marked Trial 1.
4. Repeat steps 2 and 3 for three more trials. Be sure to wait between trials until the animal has uncurled itself before touching it again.
5. Return the animal to your teacher when finished.

Table 1.

Trial	Rough Surface	Smooth Surface	Dry Condition	Wet Condition	Dark	Light
1	5	1	0	6	6	0
2	6	0	0	6	4	2
3	4	2	1	5	6	0
Total	15	3	1	17	16	2
Average	5	1	0.3	5.7	5.3	0.7

Table 2.

Trial	Observations
1	animal curls up into a ball
2	animal curls up into a ball
3	animal curls up into a ball
4	animal curls up into a ball

354

QUESTIONS

1. With your data on averages from Parts A–C of this experiment, do isopods:
   a. prefer a rough or a smooth surface? __rough__
   b. prefer a wet or a dry condition? __wet__
   c. prefer a light or a dark place? __dark__

2. Offer a reason why an isopod would choose the type of surface noted in Part A of the experiment. __A rough surface may be easier for the animal to walk on than one which is smooth. It may be better able to grasp a rough surface.__

3. Offer a reason why an isopod would choose the type of condition noted in Part B of the experiment. __Selecting a wet surface may prevent the animal from drying out and dying.__

4. Offer a reason why an isopod would choose the type of light noted in Part C of the experiment. __Selecting a dark place may prevent the animal from being seen by predators.__

5. Explain how the behavior seen in Part D may be protective to the animal. __Curling up into a ball will make it more difficult for a predator to attack. The ball shape may also be more difficult to see by a predator than when uncurled.__

6. a. Are the behaviors seen in the isopod learned or innate? __innate__
   b. How can you tell? __The animals do not learn how to react to these changes. In order for the isopod to survive, the behaviors must be innate.__

7. a. Might the results to the four experiments have been the same if you had used a different animal? __no__
   b. Why? __Each animal will respond differently to its surroundings.__
   c. How could you find out if another animal would behave in the same way that the isopod did? __Perform the same series of experiments.__

8. Explain why you used three trials rather than just one for each experiment. __to avoid chance results and to help establish an average which is more accurate than one trial__

144

Name _____ Class _____ Period _____

Before doing this lab, have students read section 17:1 in the text.

17-2 What Happens When You Learn By Trial and Error?

Suppose you were given a ring of keys and you were not told which key would open a door. How could you solve this problem? You could keep trying keys until you found the right one. Each time you made an error, you would try again. The process you used to find the correct key was trial and error. You learned by repeating over and over until you discovered the right key.

The next time you needed to use that same key, however, you would not have to go through the whole process again. You had already learned which key opens the door. This process of finding the key that fits is an example of learned behavior.

INVESTIGATION

Process Skills: observing, experimenting, predicting, recognizing and using spatial relationships

OBJECTIVES
In this exercise, you will:
a. solve the same paper puzzle several times.
b. run the same maze pattern several times.
c. see the effect of trial and error on learning how to solve puzzles.

The puzzle pieces fit together as shown here.

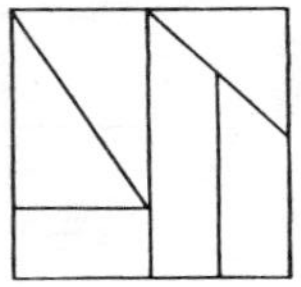

KEYWORDS
Define the following keywords:

innate behavior ___ a way of responding that does not require learning

learned behavior ___ a behavior that must be taught

maze ___ a network of paths from start to finish or a pathway puzzle having many "blind ends"

repeating or repetition ___ doing something over and over

trial and error ___ trying to solve something without knowing its solution in advance

MATERIALS
paper — ruler
file card — wall clock
scissors — folder
tape — manila folder or cardboard

PROCEDURE
Part A. Paper Puzzle Trial and Error

1. Trace the shapes shown in Figure 1 onto a piece of paper. You may want to photocopy the puzzle for students. Then omit the tracing step. Encourage students *not* to divulge answers to classmates.

FIGURE 1.

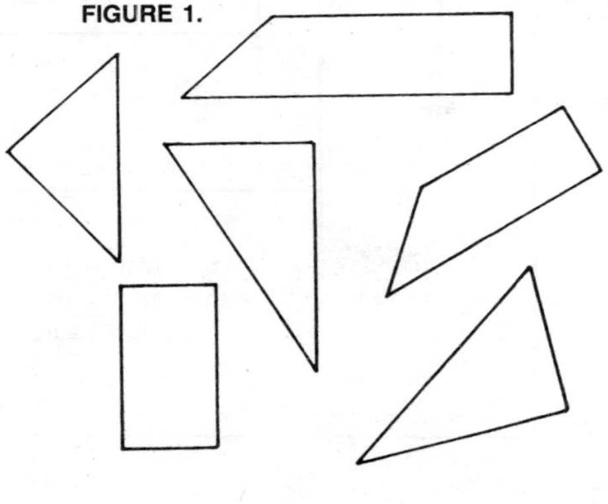

145

2. Tape the paper tracings onto a file card.
3. Cut out the six pieces with the scissors.
4. Put the six pieces together to form a single square. Time how long it takes. Point this out during the prelab.
5. Record in Table 1 the amount of time it took you to put the pieces together for trial 1.
6. Study how the pieces fit together to form a square.
7. Mix up the pieces of the square.
8. Repeat steps 4–7 three more times and enter your times in Table 1 as trials 2, 3, and 4.

Table 1. Student data will vary. This is sample data only.

Trial	Paper Puzzle Time	Maze Time	Number of Errors
1	6 minutes*	12 minutes	7
2	1 minute	6 minutes	4
3	20 seconds	1 minute	1
4	15 seconds	1 minute	0

You may have to set a maximum time limit so that students can complete Part B.

Part B. Maze Trial and Error

1. Make a hole 1 cm in diameter in the exact center of an opened folder.** Use Figure 2 as a guide.
2. Place the folder hole exactly over the letter "S" (for start) on the maze in Figure 3.
3. Keep the maze covered with the folder during the entire exercise. Move the hole to the letter "F" (for finish) by following the path lines. If you come to a "dead end," count that as one error and move the folder hole the other way.
4. Note your starting time. Record in Table 1 the total amount of time that it takes you to complete the maze. Record the time and number of errors as Trial 1.
5. Repeat steps 2–4 three more times and record your results as trials 2, 3, and 4.

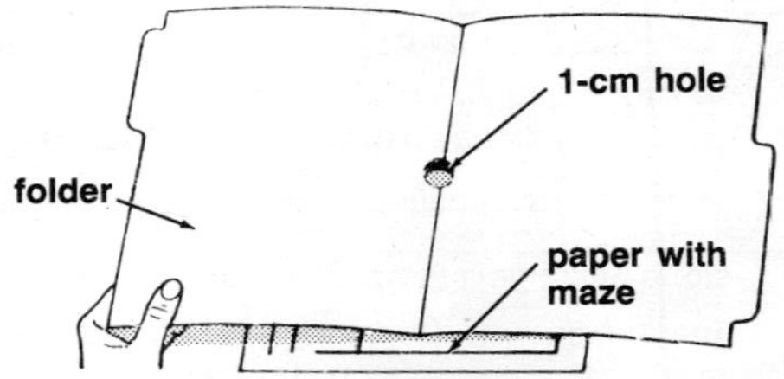

*If the majority of students cannot solve the problem, allow them to use the data from those students who did solve it.

**Use a cork borer for forming a 1-cm hole. Save the folders for use next year.

FIGURE 2. Holding folder over maze

146

Name _____ Class _____ Period _____

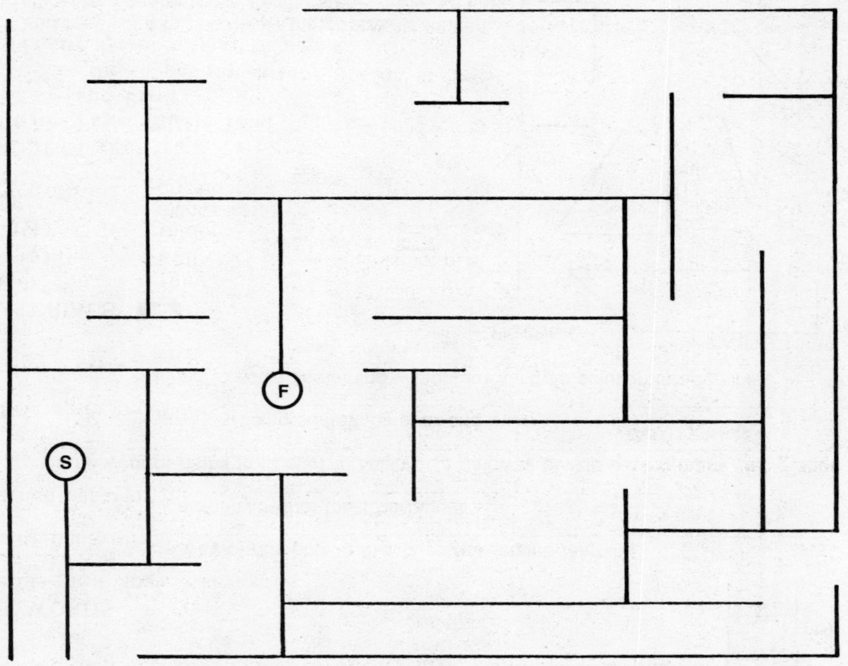

FIGURE 3.

QUESTIONS

1. a. Which trial in part A took the longest time to complete? __trial 1__
   b. Which trial in part A took the shortest time to complete? __trial 4__
2. Explain why the first trial in part A could be called trial and error learning. __Students did not know the solution in advance. Trial 1 was a series of errors.__
3. Explain how repeating the puzzle through repetition improved your time at solving it. What was happening to your behavior in terms of learning? __Students had solved the problem through repetition. They learned how the pieces fit together correctly. Future trials became "learned" rather than "trial and error."__
4. a. Explain why part A of the experiment was not an example of innate behavior. __Students were not born with the ability to solve the puzzle. They had to learn it.__

   b. How might your data have looked for part A if solving the paper puzzle was an example of innate behavior? __Each trial, including trial 1, might have taken only a few seconds.__
   c. Why? __Students would have already known the solution to the puzzle.__
5. a. Which trial in part B took the longest time to complete? __trial 1__
   b. Which trial in part B took the shortest time to complete? __trial 4__
   c. Which trial in part B had the most errors? __trial 1__
   d. Which trial in part B had the fewest errors? __trial 4__
6. Explain why your trial times and number of errors went down from trial 1 to trial 4. __Students were learning how to solve the problem or maze.__
7. Explain why your work with the maze is an example of learned behavior and not innate behavior. __Students did not know how to solve the maze at first. Many errors were first made through trial and error and each was corrected during learning.__
8. The following is a list of common behaviors. Place the word *innate* or *learned* after each behavior if you think it is an example of that type of behavior.
   a. dividing 7 by 24 __learned__ Why? __To do division, you must be taught numbers and simple math.__
   b. driving a car __learned__
   c. coughing __innate__ Why? __Coughing is performed without teaching.__
   d. sneezing __innate__
   e. using a skateboard __learned__
   f. blinking __innate__
   g. finding your way out of a building if lost in it __learned__
   h. memorizing the words to a song __learned__

Name _____ Class _____ Period _____

Before doing this lab, have students read section 18:3 in the text.

18-1 Which Antacid Works Best?

Antacids are a type of drug. The word *antacid* means against. These drugs are sold over-the-counter. They are used by people who have pain from too much acid in their stomach. The acid undergoes a chemical change with the antacid when this drug is taken. The antacid neutralizes the acid by changing it into water, a neutral liquid. The stomach pain disappears because there is no longer too much acid in the stomach.

Not all antacids are alike. Some may neutralize more acid than others. Television commercials are always telling us how much better one brand is than another. You can measure how good an antacid is by measuring how much acid it neutralizes.

Process Skills: experimenting, inferring, making and using tables

INVESTIGATION

OBJECTIVES

In this exercise, you will:
a. test water to find out if it is a good or a poor antacid.
b. test three different antacids in order to find out which neutralizes the most acid.
c. use your data to judge which antacid works best against stomach acid.
d. find out what chemicals are used as antacids.

Note: This lab could be performed as a teacher demonstration to conserve materials. Place beakers on the overhead projector for students to see color changes.

KEYWORDS

Define the following keywords:

acid _____ chemical, such as vinegar, that has a sour taste—can be neutralized by an antacid

antacid _____ a drug used against acid—it neutralizes acid

dosage _____ how much and how often to take a drug

neutralize _____ to change a liquid from acid or base to neutral

neutral _____ exactly between an acid and an antacid

sodium bicarbonate _____ baking soda—an antacid

Have students work in groups of four if lab is not performed as a demonstration.

MATERIALS

labels	graduated cylinder	Amounts given for antacid solutions are sufficient for four groups.
wooden sticks		
sodium bicarbonate liquid	¼ tsp. baking soda to 200 mL water.
red indicator chemical	Add 3 g Congo red to 100 mL distilled water. Stir to dissolved. Place in dropping bottles.
8 100 mL beakers	
water	If tablets, dissolve 1 in 200 mL water. If powder prepare as for baking soda.
antacids	
dropper	Add 35 mL concentrated hydrochloric acid to 500 mL water.
acid in dropper bottle	Place in dropping bottles. NOTE: DO NOT add water to acid.

149

PROCEDURE

Part A. Water as an Antacid

1. Label two small beakers, Beaker A and Beaker B.
2. Use a graduated cylinder to add 50 mL of water to each beaker.
3. Add 5 drops of red indicator chemical to each beaker as shown in Figure 1. Use a wooden stick to mix the red chemical in each beaker.
4. Record the color of each beaker in Table 1 under the column labeled Color before adding acid.
5. Read over steps 6 through 8 before starting step 6.
6. To *Beaker A only*, add acid one drop at a time. Use a stick to stir the liquid after adding each drop as shown in Figure 2. **CAUTION:** *Acid can burn skin or clothing. If acid is spilled, wash immediately with water and call your teacher.*
7. Count the number of drops of acid needed to change the color of the liquid to a deep red or purple color. Use Beaker B as a control for judging the original color of Beaker A. **NOTE:** *Once the color change has taken place in Beaker A, stop adding acid. At this point, you have neutralized all the acid that you can with the water in Beaker A.* The indicator chemical turns color because the water in Beaker A has become acid.
8. In Table 1, record the number of drops of acid used. Record the color of Beaker A and Beaker B in the column labeled Color after adding acid.

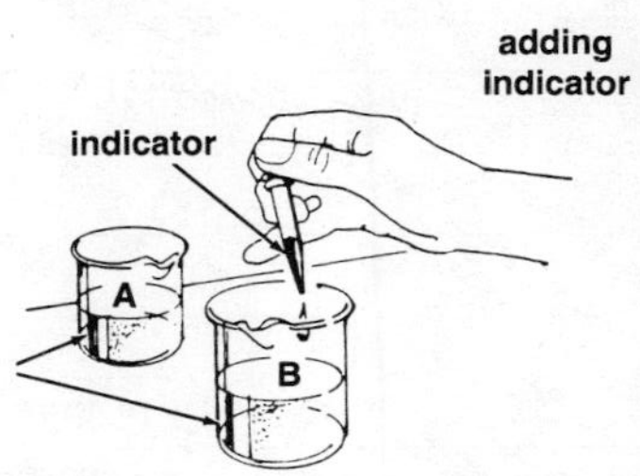

FIGURE 1. Adding red indicator chemical

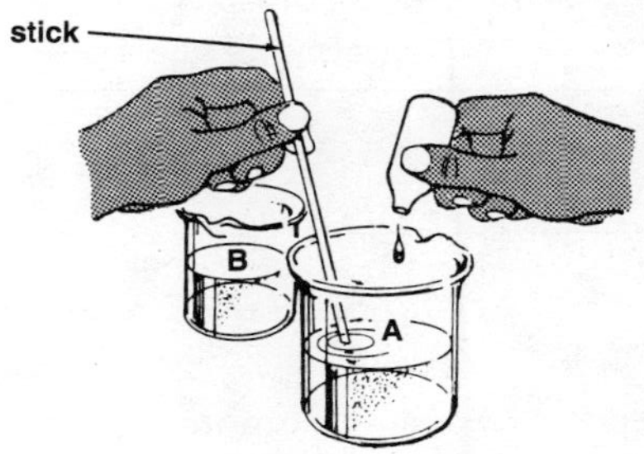

FIGURE 2. Adding acid

Table 1. Amount of Acid Neutralized by Antacid

Beaker	Color before adding acid	Number of drops of acid used	Color after adding acid
A	red	1 or 2 Answers may vary	deep red or purple
B	red	0	no acid added—still red

Part B. Testing Sodium Bicarbonate as an Antacid

1. Label two small beakers, Beaker C and Beaker D.
2. Use a graduated cylinder to fill each beaker with 50 mL of sodium bicarbonate liquid.

If time permits, use additional samples of antacids.

150

Name _____ Class _____ Period _____

3. Add 5 drops of red indicator chemical to each beaker. Use a clean wooden stick to mix the red chemical in each beaker.
4. To *Beaker C only*, add acid one drop at a time. Use a stick to stir the antacid after adding each drop. **CAUTION:** *Acid can burn skin or clothing. If acid is spilled, wash immediately with water and call your teacher.*
5. Count the number of drops of acid needed to change the color of the antacid from red to a deep red or purple color. Use Beaker D as a control for judging the original color of Beaker C. **NOTE:** *once the color change has taken place in Beaker C, stop adding acid. At this point, you have neutralized all the acid that you can with the antacid in Beaker C. The indicator chemical turns color because now the liquid in the beaker has turned acid.*
6. In Table 2, record the number of drops of acid added to Beaker C.

Student data will vary depending on antacid used. For example: one seltzer tablet in 200 mL of water can neutralize 55 drops of acid.

Table 2. Using an Antacid to Neutralize Acid

Beaker	Antacid name	Number of drops of acid added
C	sodium bicarbonate	60
D	sodium bicarbonate	0
E		will vary
F		0
G		will vary
H		0

Chewable antacid tablets can be used for E and G.

Part C. Testing Other Antacids

1. Repeat steps 2 through 8 in part B, using a different sample of antacid. This time, label the beakers E and F. Use Beaker F as a control for judging the original starting color, and record your results in Table 2. Complete the column labeled Antacid name. Your teacher will tell you the name of the antacid being tested.
2. Again repeat steps 2 through 8 in part B, using a different sample of another antacid. Your teacher will tell you the name of the antacid being tested. This time, label the beakers G and H. Use Beaker H as a control for judging the original starting color, and record your results in Table 2.

QUESTIONS

1. What color is an:
   a. antacid solution with red indicator? red
   b. acid solution with red indicator? deep red or purple
2. Explain how you were able to tell when the antacid had neutralized as much acid as it could? color changed from red to deep red or purple

151

3. The fewer drops a liquid takes before changing color, the poorer it is as an antacid. The more drops a liquid takes before changing color, the better it is as an antacid.
   a. How many drops of acid were used to neutralize water (Beaker A)? 1 or 2
   b. How many drops of acid were needed to neutralize sodium bicarbonate (Beaker C)? 60
   c. Is water a very good antacid compared to sodium bicarbonate? No. It took only 1 or 2 drops to change the color. Explain your answer.
4. a. Which antacid (Beaker C, E, or G) needed the most drops of acid before showing a color change? answer will vary
   b. Which antacid (Beaker C, E, or G) can neutralize the most acid? answers will vary
   c. Which antacid (Beaker C, E, or G) is the most efficient? answer will be same as 4b
5. a. Which antacid (Beaker C, E, or G) needed the fewest drops of acid before showing a color change? answer will vary
   b. Which antacid (Beaker C, E, or G) can neutralize the least acid? answer will be the same as 5a
   c. Which antacid (Beaker C, E, or G) is the least efficient? answer will be same as 5b

Have baking soda and antacid packages and their prices available for students.

6. Check the labels on the antacids used. What is the suggested dosage for: *You will need to look at the packages of the antacids used to answer questions 6, 7, 8 and 9.*
   a. antacid C? ½ teaspoon in a glass of water every 2 hours
   b. antacid E? answers will vary with type of antacid used
   c. antacid G? answers will vary with type of antacid used
7. What actual chemical is listed on the label as the antacid for:
   a. antacid C? sodium bicarbonate
   b. antacid E? answers will vary with sample used
   c. antacid G? answers will vary with sample used
8. a. Is antacid C purchased as a liquid, powder, or solid? powder
   b. Is antacid E purchased as a liquid, powder, or solid? answer will vary
   c. Is antacid G purchased as a liquid, powder, or solid? answers will vary
9. a. Compare the cost of a box of sodium bicarbonate with the other antacids used. Which antacid is the least expensive? sodium bicarbonate
   b. Does the most expensive antacid seem to neutralize the most acid? No.

152

Name _____ Class _____ Period _____
Before doing this lab, have students read section 18:1 in the text.

18-2 How Do Drugs Affect the Ability of Seeds to Grow into Young Plants?

Drugs change the way in which living things carry out every-day functions, such as growth and repair. It would be very hard to experiment on humans or other animals to show that this is true. It is possible to experiment with drugs and certain living things, such as plants. The living things used in this experiment are seeds.

INVESTIGATION

OBJECTIVES
In this exercise, you will:
a. test the effect of four drugs on seed growth.
b. compare the behavior of different seed types when tested with the same four drugs.

Process Skills: experimenting, separating and controlling variables, interpreting data, inferring

KEYWORDS
Define the following keywords:

caffeine _a drug found in foods such as coffee, cola, and chocolate_

drug _a chemical substance that has a physical effect on a living thing_

ethyl alcohol _a drug produced by yeast when no oxygen is present_

MATERIALS
5 small beakers
10 labels
cheesecloth
50 radish seeds
50 pinto bean seeds
spoon
scissors
5 petri dishes
2 baggie twist ties
graduated cylinder
4 cigarettes*
ground coffee (caffeinated)
ruler
paper towels
aspirin
ethyl alcohol**
water

*You may wish to buy loose tobacco or remove tobacco from cigarettes instead.
**denatured ethyl alcohol

PROCEDURE
Part A. Soaking Seeds
1. a. Label five small beakers as follows: water, alcohol, aspirin, nicotine, caffeine.
   b. Add your name to each beaker.
2. a. Use scissors to cut two pieces of cheesecloth so that each piece measures 12 cm by 12 cm.
   b. Remove the filter (if present) and paper from four cigarettes.
   c. Place the tobacco into the center of one cheesecloth square.
   d. Wrap the tobacco into the cheese cloth. Tie the ends of the bag together with a twist tie.

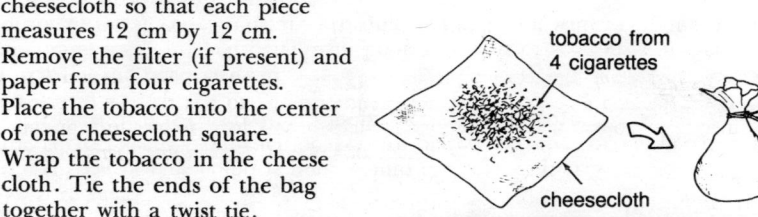

FIGURE 1. Making a cheesecloth bag

3. Prepare a second cheesecloth bag. In this bag, place one spoonful of ground coffee instead of tobacco.
4. Using a graduated cylinder to measure the different liquids, add the following to each beaker:
   a. water beaker, add 50 mL of water, 10 radish seeds, and 10 pinto bean seeds
   b. alcohol beaker, add 50 mL of ethyl alcohol, 10 radish seeds, 10 pinto been seeds
   c. aspirin beaker, add 50 mL of water, one aspirin tablet, 10 radish seeds, and 10 pinto bean seeds
   d. nicotine beaker, add 50 mL of water, cheesecloth bag with tobacco, 10 radish seeds, and 10 pinto bean seeds
   e. caffeine beaker, add 50 mL of water, cheesecloth bag with coffee, 10 radish seeds, and 10 pinto bean seeds
5. Allow the seeds to soak overnight in the beakers of liquid.

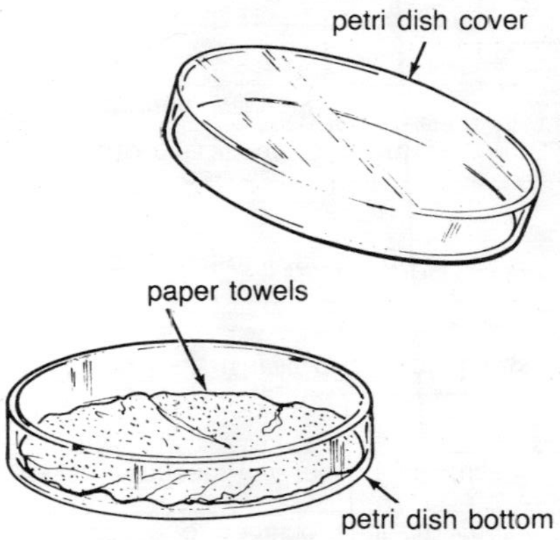

Part B. Preparing Chambers for Growing Seeds into Seedlings
1. Label the tops of five petri dishes as follows: water, alcohol, aspirin, nicotine, and caffeine. **NOTE:** *The top of the petri dish is wider than the bottom.*
2. Add your name to each label.
3. Prepare each dish to receive the soaked seeds by doing the following:
   a. Trace the bottom of a petri dish onto several thicknesses of paper toweling.
   b. Use scissors to cut out these circles of paper toweling.
   c. Place the circles of toweling into the bottom of each petri dish. Each dish should have at least five thicknesses of toweling (see Figure 2).

FIGURE 2. Making growth chambers

Part C. Adding seeds to Chambers
1. On the next day, get the beaker labeled *Water* and the petri dish labeled *Water*.
2. Pour about half the water that is in the beaker into the petri dish bottom that contains the paper toweling.
3. Remove the seeds from the beaker and sort them out by type.
4. Place the seeds in different halves of the petri dish as shown in Figure 3.

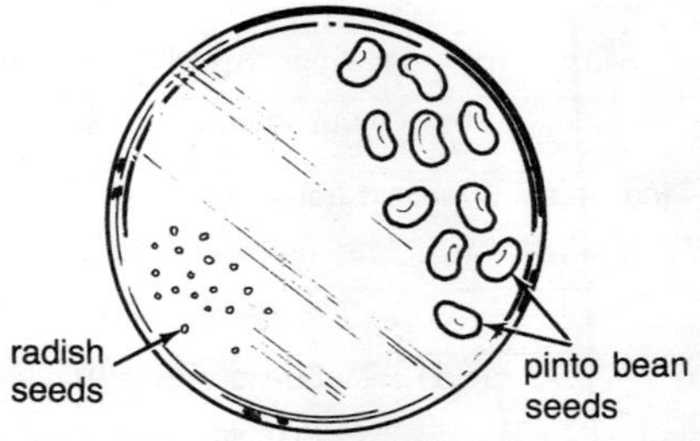

FIGURE 3. Adding seeds to growth chambers

Supervise students so that no errors are made in placing the wrong soaked seed type into the properly marked petri dish.

5. Cover the petri dish.
6. Repeat steps 1 through 5 for the remaining four beakers and petri dishes. Make sure that:
   a. the correct beaker liquid is poured into the correct petri dish.
   b. the correct seeds from each beaker are placed into the correct petri dish.
   c. each petri dish is covered with its properly labeled cover.

Part D. Observing Seed Growth

1. Examine all seeds in each petri dish. Look for signs of seed growth. See Figure 4 in order to tell if your seeds are growing. A small root will stick out from the seed if it is alive and growing.

Seed	Not Growing	Growing
Radish		→ root
Bean		→ root

FIGURE 4. How to tell if seeds are growing

2. Record in Table 1 how many seeds of each type and treatment are growing. Consider today as Day 1 in the experiment.
3. Place the petri dishes in a place indicated by your teacher. *No special temperature or light conditions are needed for seed storage. Room temperature conditions are suitable.*
4. Observe the seeds for four more days and record the number of seeds growing of each type in each petri dish under the correct day.

Table 1. Number of Seeds Growing in Each Petri Dish Under Each Treatment

Day	Water Radish	Water Bean	Aspirin Radish	Aspirin Bean	Alcohol Radish	Alcohol Bean	Nicotine Radish	Nicotine Bean	Caffeine Radish	Caffeine Bean
1	0	0	0	0	0	0	0	0	0	0
2	5	2	0	0	0	0	0	0	0	0
3	8	7	0	0	0	0	2	0	0	1
4	9	10	0	0	0	0	2	0	0	5
5	10	10	0	0	0	0	2	0	0	6

QUESTIONS

1. a. What different kinds of seeds were used in this experiment? __radish and pinto bean__
   b. What different drugs were used in this experiment? __aspirin, alcohol, nicotine, caffeine__
2. What was the purpose of this experiment? __to determine how different drugs will affect the growth of young seeds__

---

3. Using your results from Day 5, list the number of radish seeds that were growing in:
   a. water __10__ b. ethyl alcohol __0__ c. aspirin __0__
   d. nicotine __2__ e. caffeine __0__—student answers and data will vary

4. a. Do the different drugs used seem to affect the number of radish seeds that grew? __yes__
   b. Support your answer by comparing the number of radish seeds that grew in the drugs to the number that grew in water. __10 grew when placed in water, but only 2 grew in nicotine, while none (will vary) grew in the other three drugs__

5. Using your results from Day 5, List the number of pinto bean seeds that were growing in:
   a. water __10__ b. ethyl alcohol __0__ c. aspirin __0__
   d. nicotine __0__ e. caffeine __6__—student data and answers will vary

6. a. Do the different drugs used seem to affect the number of bean seeds that grew? __yes__
   b. Support your answer by comparing the number of pinto bean seeds that grew in the drugs to the number that grew in water. __10 grew when placed in water, but only 6 grew in caffeine (will vary), while none grew in the other three drugs__

7. Write a short paragraph that sums up your findings in this experiment. __Water does not prevent or affect the ability of radish and pinto seeds to grow. However, if these two seed types are soaked in alcohol, aspirin, nicotine, or caffeine, their growth is either slowed or prevented from occurring in five days. It appears that these four drugs do have a harmful effect on seed growth.__

8. a. Might all living things show the same type of results to these different drugs as did the seeds? __no__
   b. How could you find this out? __experiment with each type of living thing (student answers will vary)__
   c. Why might this type of experiment be difficult to carry out on animals? __injury or death to the animals might result (student answers will vary)__

You might wish to expand this experiment using other drugs that are easily available and/or using different seeds and/or by extending the time for observations from 5 to 10 days.

Name _____ Class _____ Period _____

Before doing this lab, have students read section 19:1 in the text.

19-1 What Do the Inside Parts of Leaves Look Like?

A leaf is one of the organs of a plant. The main function of a leaf is to make food for the rest of the plant. The food is carried to other plant parts. What parts of a leaf make and carry food?

Since leaves are organs, they are made of many different kinds of tissues. In leaves these tissues appear as layers of cells. Each tissue has a specific function and the shape of the cells in the tissue is related to its function. Some cells look like small boxes stacked side by side, while others are long and balloon-shaped. Certain cells are round and loosely packed while others look like small tubes stacked together.

EXPLORATION

Process Skills: observing, interpreting scientific illustrations, using a microscope, scientific terminology

OBJECTIVES
In this exercise, you will:
a. build a model of the structure of a leaf.
b. compare the model with the cells of a leaf as seen through the microscope.

KEYWORDS
Define the following keywords:

epidermis _outside layers of cells on a leaf_

palisade layer _a layer of long, closely packed cells on inside of leaf_

spongy layer _a layer of round, loosely packed cells on inside of leaf_

stoma _an opening between the guard cells in the epidermis—allows water and gases to enter and leave the leaf_

MATERIALS
scissors
1 sheet white paper
prepared slide of leaf
colored pencils: red, blue, purple, yellow, tan, light green, dark green
microscope
transparent tape

Photocopy Figure 1 on page 158 for each student.

PROCEDURE
Part A. Making a Model of a Leaf

1. Figure 1 shows the different cell layers of a leaf. Your teacher will provide you with a copy to color as follows:
   a. waxy layer—purple
   b. upper epidermis layer—yellow
   c. lower epidermis layer—tan
   d. spongy layer—light green
   e. palisade layer—dark green
   f. xylem—red
   g. phloem—blue
   h. guard cells—dark green

157

Make a copy of this page for each of your students.

FIGURE 1. Tissue layers of a leaf.

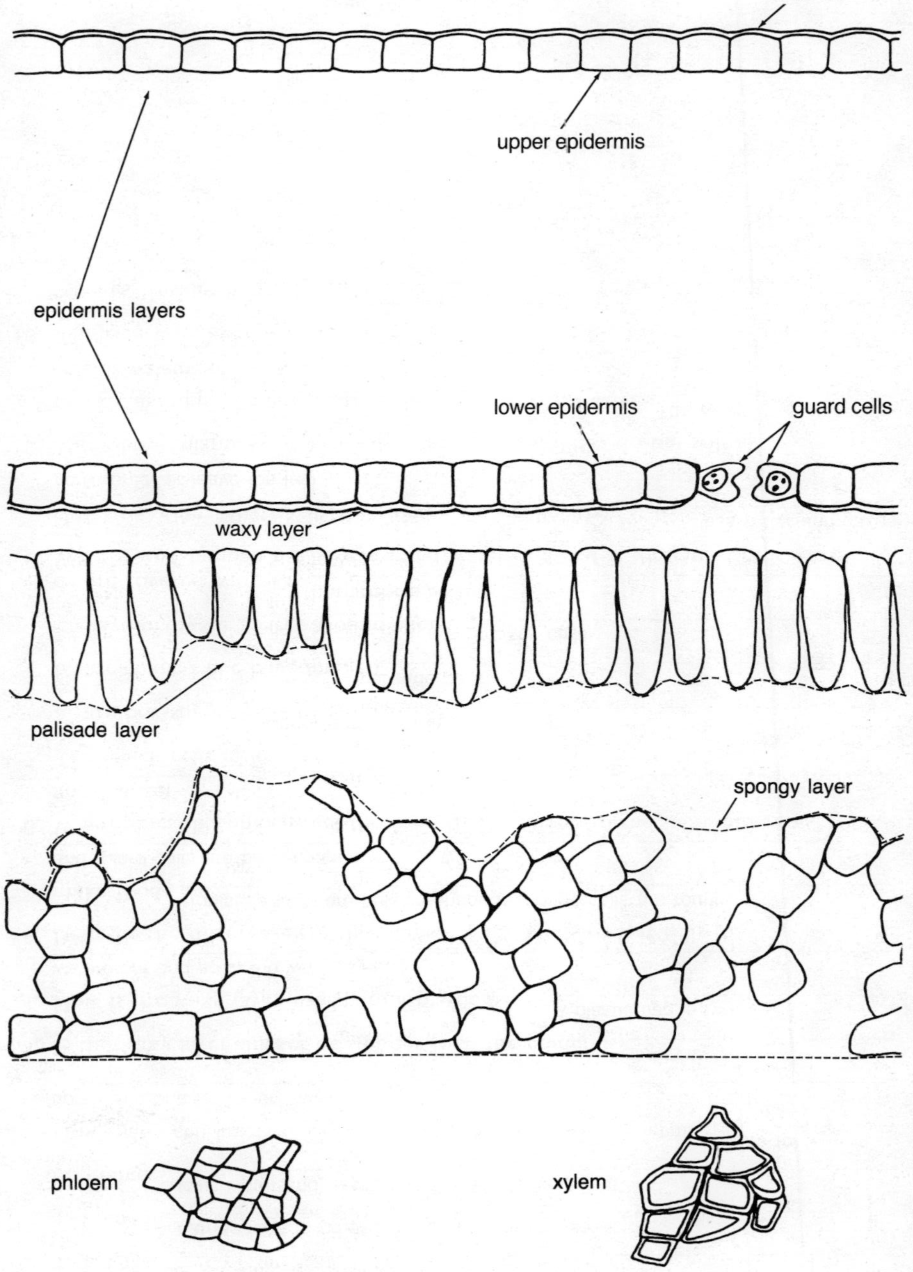

158

Name _____ Class _____ Period _____

2. Cut out the six layers of your colored copy of Figure 1. Cut along the dotted lines or follow the outlines of the cell layers.
3. Assemble your model of a leaf using Figure 19-5 on page 401 in your text.
4. Starting with the upper epidermis, tape each layer onto the blank sheet of paper. (HINT: The xylem and phloem fit together like parts of a puzzle.)
5. Label your leaf model by using the key in step 1.

Part B. Examining the Tissues of a Leaf

1. Examine the prepared slide of a leaf cross section on the low power of a microscope.
2. Locate and identify the six cell layers in the leaf section.
3. Draw a small section of the leaf that you see through the microscope on low power in the space provided. Label the parts of this leaf drawing by using the key in step 1.

FIGURE 2. Parts of a leaf

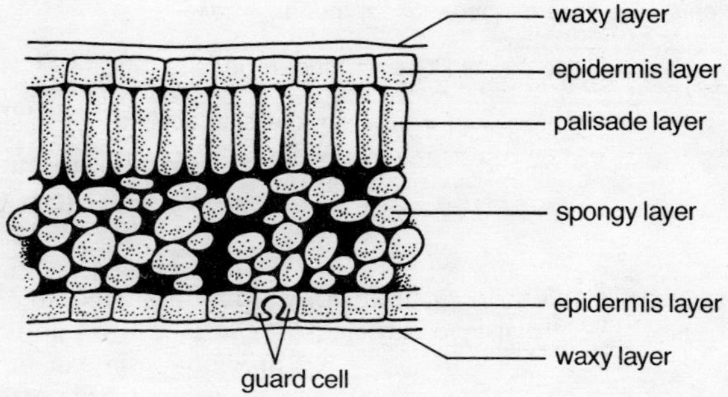

waxy layer
epidermis layer
palisade layer
spongy layer
epidermis layer
waxy layer
guard cell

Have students make a diagram of a small section of a leaf rather than trying to draw all they see on the slide.

QUESTIONS

1. List the job of each of the following leaf parts:
   a. waxy layer __protects the leaf from water loss and insects__
   b. upper and lower epidermis layers __protect the leaf from water loss and insects__
   c. guard cell __changes the size of the stoma__
   d. spongy layer __makes food for the plant__

159

   e. palisade layer __makes food for the plant__
   f. xylem __carries water to the leaf__
   g. phloem __carries food to other plant parts from the leaf__
   h. stoma __allows water and gases to enter and leave a leaf__

2. How many cell layers thick are the upper and lower epidermis? __one for upper and one for lower__
3. Is the waxy layer thicker or thinner than the epidermis? __thinner__
4. How is the waxy layer similar to a plastic bag? __A plastic bag is waterproof and so is the waxy layer.__
5. Describe the differences in the shapes of the kinds of leaf cells that make food. __Palisade cells are balloon-shaped, spongy cells are round, and guard cells are bean-shaped.__
6. Where does the phloem get the food that it carries to the stem and roots? __from the leaf__
7. Tell which cells in leaves
   a. are box-shaped. __upper and lower epidermis cells__
   b. are shaped like balloons. __palisade cells__
   c. have many air spaces around them. __spongy cells__
   d. are covered with wax. __epidermis cells__
8. What would happen to the gases and water in a leaf if the guard cells were closer together, that is, if the stoma were smaller than it is? __less water and gases could enter or leave the leaf__
9. Many houseplants have very thick, waxy leaves. They do not wilt as quickly as houseplants with thinner leaves. Explain why. __The waxy layer is waterproof.__
10. The celery stalks that you eat are leaf stalks. What kind of cells are the "strings" inside the celery? __xylem and phloem cells__

160

Name _____ Class _____ Period _____

Before doing this lab, have students read section 19:3 in the text.

19-2 Where in a Leaf Does Photosynthesis Take Place?

The entire leaf blade of most plants is green. Some leaves have white, red, and yellow parts as well as green parts. The green color is due to a chemical called chlorophyll. Chlorophyll is the green pigment found inside parts of plant cells called chloroplasts. Chloroplasts are found mainly in the leaves of plants.

Chlorophyll is needed by the plant to carry out photosynthesis. Chlorophyll traps the energy from light. This energy is used to change carbon dioxide and water to sugar. Sugar in many plants is then changed to starch. The starch is stored in other parts of the cell until needed by the plant.

Without light, plants cannot make food. If food is not made, cells of the plant can die. Sugar is made in the plant as long as light is present.

INVESTIGATION

Process Skills: experimenting, inferring, interpreting data, summarizing

OBJECTIVES

In this exercise, you will:
a. observe the parts of the leaf that make starch during photosynthesis.
b. decide what effect sunlight and lack of sunlight has on the making of starch in a plant.

For best results, place Coleus and bean plant in the dark for two days prior to starting Part A.

KEYWORDS

Define the following keywords:

carbon dioxide __a gas given off by living things during cellular respiration__

chlorophyll __a green pigment that helps the plant make food__

chloroplast __a cell part that contains chlorophyll__

photosynthesis __a process in which plants make food__

MATERIALS

1 500 mL beaker
2 250 mL beakers
scissors
50 mL rubbing alcohol
hot plate
bean plant, potted
4 per class

forceps
30 mL iodine solution
2 petri dishes
2 aluminum foil squares, 20 cm × 20 cm
Coleus plant, potted
1 per 2 students

Dissolve 1 g iodine crystals and 2 g potassium iodide in 300 mL water. Use Coleus that have red and yellow leaves. The red area hides the chlorophyll.

PROCEDURE

Part A. Setting Up the Experiment

1. Cut two pieces of foil, each 20 cm square.
2. Fold the pieces of foil in half as shown in Figure 1. Place one-half (bottom half) beneath a leaf of a bean plant as shown in Figure 2.
3. Fold the top half over the bottom half.*
4. Crimp three edges of the foil to keep out the light as shown in Figure 3. Be careful not to crush the leaf. Put a label with your name on the foil.
5. Repeat steps 2 through 4 with a Coleus plant.
6. Place both plants in the light for two days as shown in Figure 4.

*Make sure students completely wrap foil around the leaf to keep out the light.

Part B. Testing Leaves for Starch

1. Remove an uncovered leaf from each of the two plants. Cut off the stalks from each of your leaves with scissors.
2. Using forceps, carefully place your leaves into a beaker of boiling water that your teacher has set up on a hot plate as shown in Figure 5. This procedure ensures that the leaf is killed and cannot make more starch by photosynthesis. Remove the leaves with forceps after 3 minutes and place them in a petri dish.

You may set up hot plates at various stations around the room.

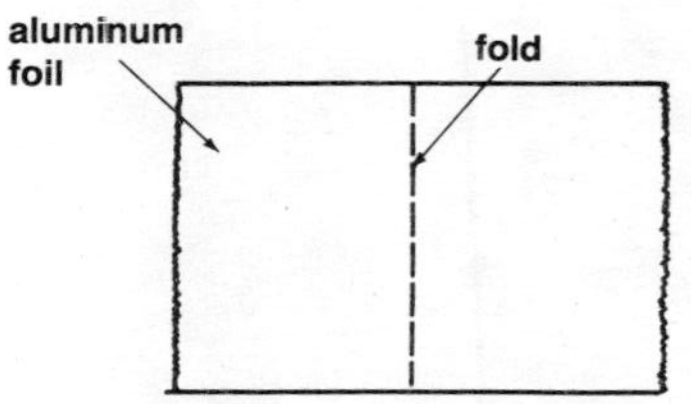

FIGURE 1. Folding aluminum foil

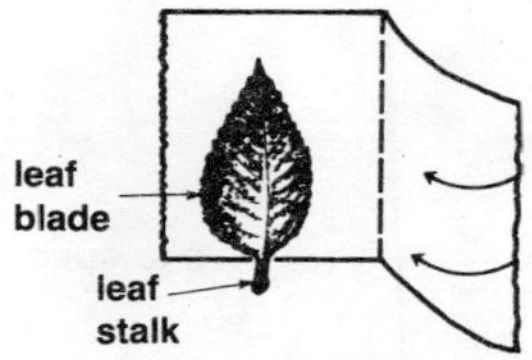

FIGURE 2. Covering a leaf

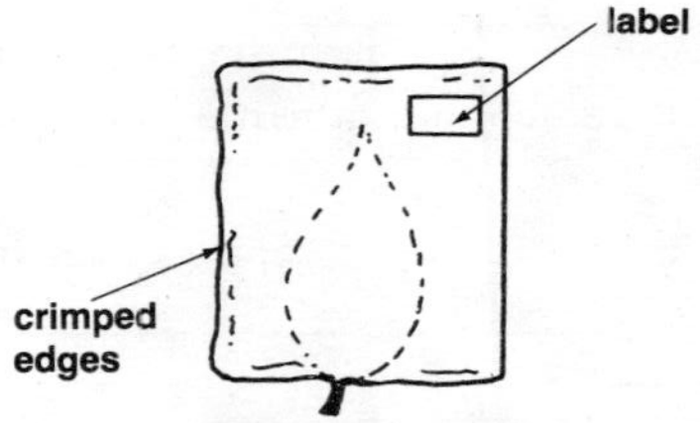

FIGURE 3. A covered leaf

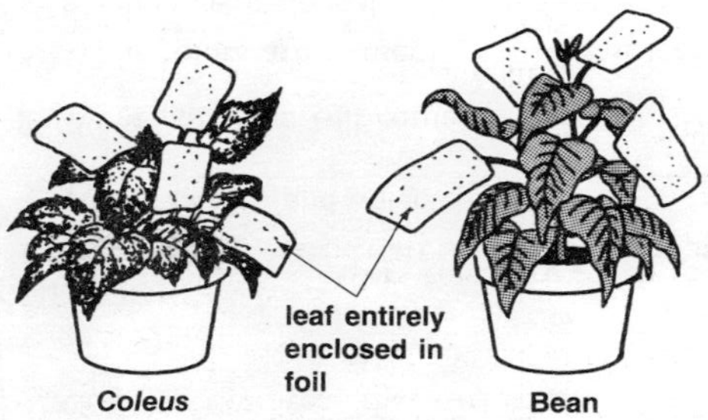

FIGURE 4. Placing plants in the light

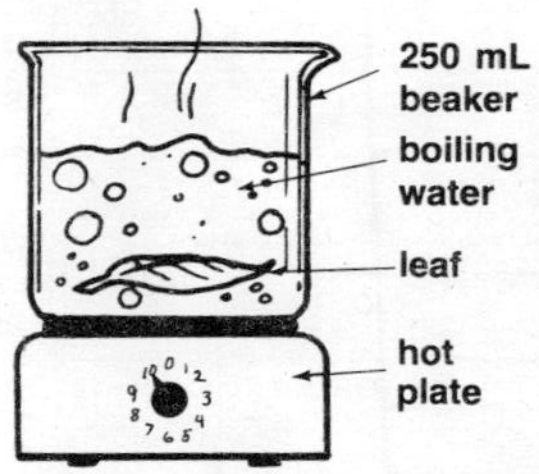

FIGURE 5. Killing leaves

Name _____ Class _____ Period _____

Do not heat the alcohol directly on the hot plate. You may wish to use a large pan in which to boil water to heat the alcohol. Several small beakers can be placed in the pan.

3. Give your leaves to your teacher who will place them into a small beaker of warm alcohol. This is prepared by placing the beaker of alcohol into a beaker of water heated on a hot plate as shown in Figure 6. The warm alcohol will remove the green pigment from the leaves. Your teacher will remove your leaves after 3 minutes.* To soften the leaves, use forceps to dip them one more time into the boiling water.
4. Put each leaf in a petri dish. Cover each leaf with iodine solution. Iodine turns starch a blue-black color. This is a positive test for starch.
5. Watch for any color change to appear in the leaves during the next 2 minutes.
6. Make simple outline drawings in Figure 7 to show the areas where starch is found on your uncovered leaves.

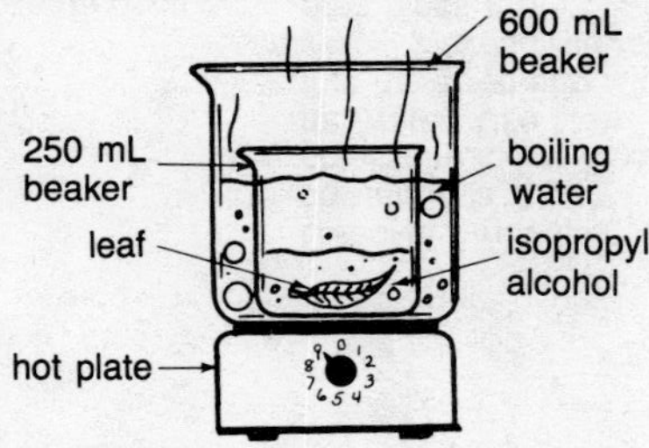

*Give each student one of each type of leaf after chlorophyll has been removed.

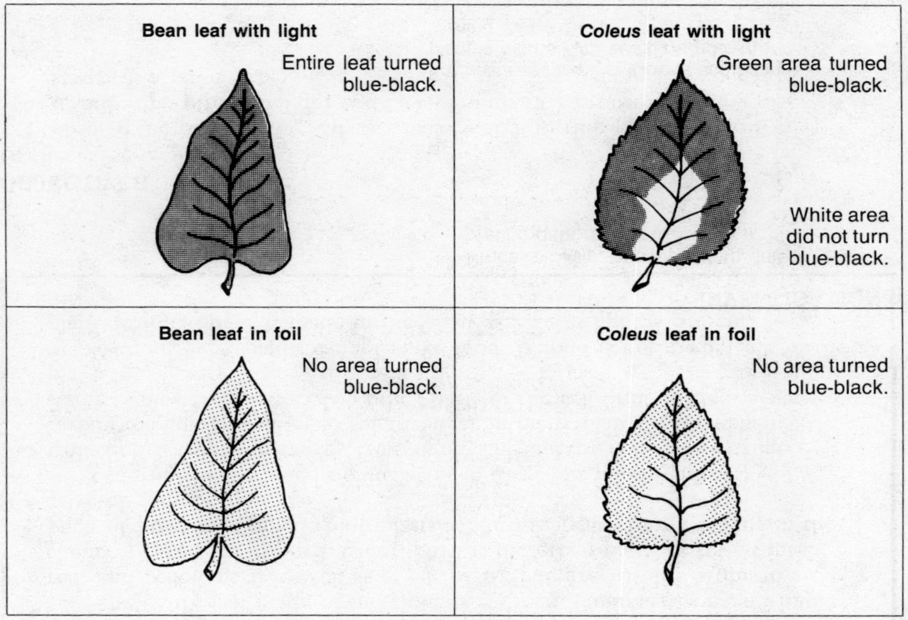

FIGURE 6. Removing chlorophyll

Bean leaf with light — Entire leaf turned blue-black.

Coleus leaf with light — Green area turned blue-black. White area did not turn blue-black.

Bean leaf in foil — No area turned blue-black.

Coleus leaf in foil — No area turned blue-black.

FIGURE 7. Areas of starch in leaves

---

8. Remove the foil-covered leaves from the two plants. Cut off the stalks from each of your leaves with scissors. Gently remove the foil from the leaves.
9. Repeat steps 2 through 6 with these two leaves and draw the areas of the leaves in Figure 7 where starch is found in these covered leaves.
10. Color in the areas that turned blue-black.

QUESTIONS

1. What two chemicals are needed by a plant for photosynthesis? carbon dioxide and water
2. What part of a cell contains chlorophyll? the chloroplast
3. What colors are present in the bean leaf and the *Coleus* leaf? green in the bean leaf; yellow, red, and green in the *Coleus* leaf
4. In what area of the uncovered bean leaf was starch made? the entire leaf
5. In what area of the uncovered *Coleus* leaf was starch made? the green area of the leaf
6. In what area of the foil-covered bean leaf was starch made? No starch was made.
7. In what area of the foil-covered *Coleus* leaf was starch made? No starch was made.
8. How can you explain the differences in the making of starch in the four leaves? The foil-covered leaves were not able to make starch because the light could not reach the chloroplasts. Chloroplasts need light in order to make starch.
9. Explain why photosynthesis takes place only in a certain area of a leaf. It takes place only where there is chlorophyll; the green parts of a leaf have chlorophyll.

Name _____ Class _____ Period _____

Before doing this lab, have students read section 20:1 in the text.

20–1 What Does a Woody Stem Look Like Inside?

Plant parts, like animal parts, are made of layers of cells called tissues. Each tissue is different in size and shape. The size and shape of each tissue is a clue to the job of the cells.

Woody stems have six kinds of tissues. The outer ring of cells, cork, is made of thick box-shaped cells. Cork protects the cells inside the stem. The thinner cells just inside the cork make up the cortex. These cells store food. Next to the cortex cells are phloem cells. Phloem cells are thin, tubelike cells that carry food. A thin ring of cells just inside the phloem is the cambium. These cells make new phloem and xylem cells. Next to the inside of cambium cells is the xylem. Xylem cells carry water inside the plant. The center of a woody stem is made of pith cells. Pith cells store food.

INTERPRETATION

Process Skills: interpreting scientific illustrations, scientific terminology, recognizing and using spatial relationships, inferring

OBJECTIVES
In this exercise, you will:
a. build a model of the structure of a woody stem.
b. compare the model to a diagram of a woody-stem cross section.

KEYWORDS
Define the following keywords:

bark   rough, outer covering of a woody stem

cork   an outer ring of cells on a woody stem

pith   cells in the center of a woody stem that store food

xylem   cells that carry water

MATERIALS
scissors   sheet of blank paper
transparent tape
colored pencils: purple, blue, red, green, orange, brown

Photocopy page 168 for each student.

PROCEDURE
1. Examine the woody-stem model parts on page 168. Obtain a photocopy of those model parts from your teacher. Color the different cell layers as follows:
pith—purple   phloem—blue
xylem—red   cambium—green
cortex—orange   cork—brown

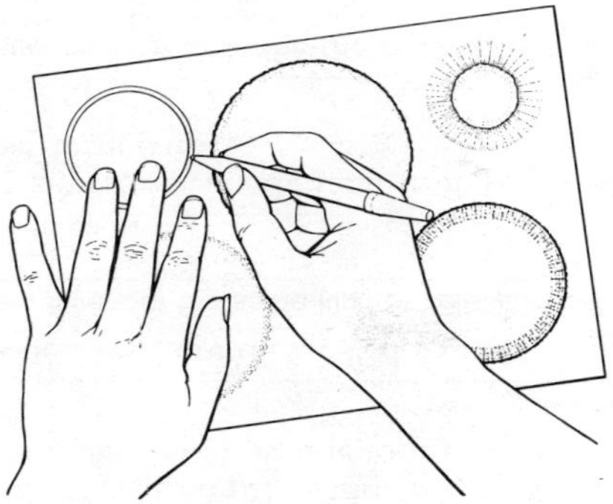

FIGURE 1. Coloring woody-stem layers

2. Cut out each layer of cells. Cut along the outer edge only.
3. Start assembling your model by placing one part on another. All parts fit on top of each other.
4. Tape your model on a blank sheet of paper. Label your model using step 1 as a guide.
5. Compare your woody-stem model with Figure 3, a drawing of the cross section of a woody stem.
6. Label the parts of the woody stem in Figures 3 and 4 by using your model as a guide. The diagram in Figure 4 shows what a woody stem looks like as seen through a microscope. Notice that the phloem is not an even layer as it is in your model. Also notice cells that make up each layer.

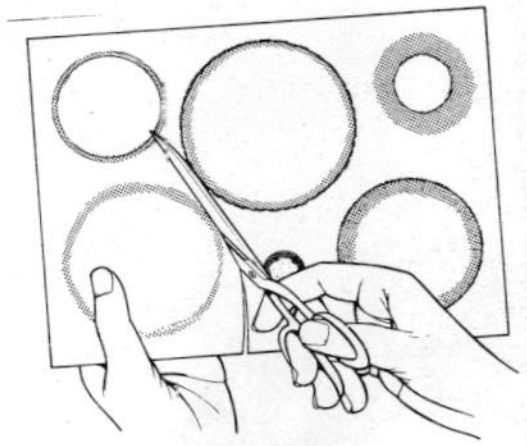

FIGURE 2. Cutting out woody-stem layers

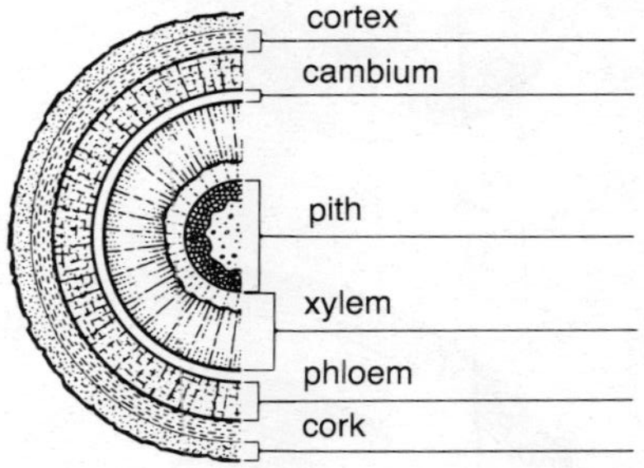

FIGURE 3. Cross section of a woody stem

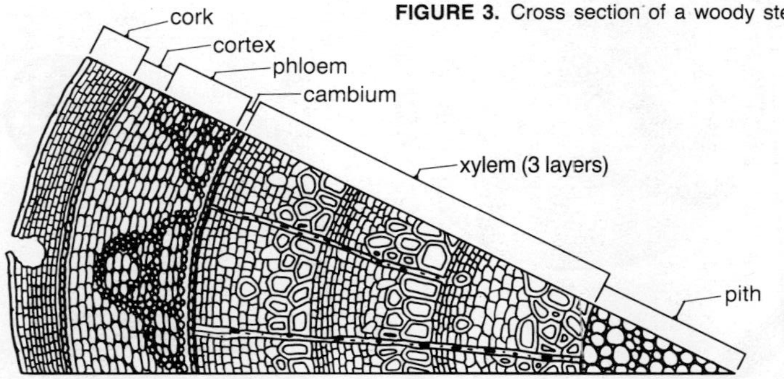

FIGURE 4. Layers of a woody stem as seen through a microscope

QUESTIONS
1. List the job of the stem parts in your model.

a. cork   protection

b. cambium forms new cells

c. xylem   carries water

d. phloem   carries food

e. cortex   stores food

f. pith   stores food

365

Name _____ Class _____ Period _____

2. Trace the path a beetle would follow as it burrows from the outside to the center of a tree stem. List the parts in correct order going from outside to inside.
cork, cortex, phloem, cambium, xylem, and pith

3. The cell layers in your model are called tissues. Why are they called tissues? Tissues are layers of the same kind of cells and the
(HINT: Use your text for help.)
cells in a woody stem are in layers.

4. Bark is the term used to describe the outer layers of a woody stem from cork through phloem. Name the tissue layers that form bark.
cork, cortex, and phloem

5. Use the drawing of the woody stem cross-section and describe the cell layers. Use a complete sentence.
   a. xylem   Xylem cells are thick, nearly circular cells that form in rings.
   b. pith   The pith cells are thin-walled cells in the center of a woody stem.
   c. phloem   The phloem cell is a small, thin-walled cell near the cortex.
   d. cork   Cork cells are thick, irregular-shaped cells on the outside of a woody stem.

6. The thickness of each xylem ring may differ each year. This may be the result of growing conditions during that year. A thick ring means that conditions such as rainfall were good and much growth took place. A thin ring means conditions were poor and little growth took place.
   a. Which ring, A or B, is the thinner ring? __A__
   b. Which ring, A or B, may have grown during a year when much rain fell? __B__   When little rain fell? __A__

Make a copy of this page for each of your students.

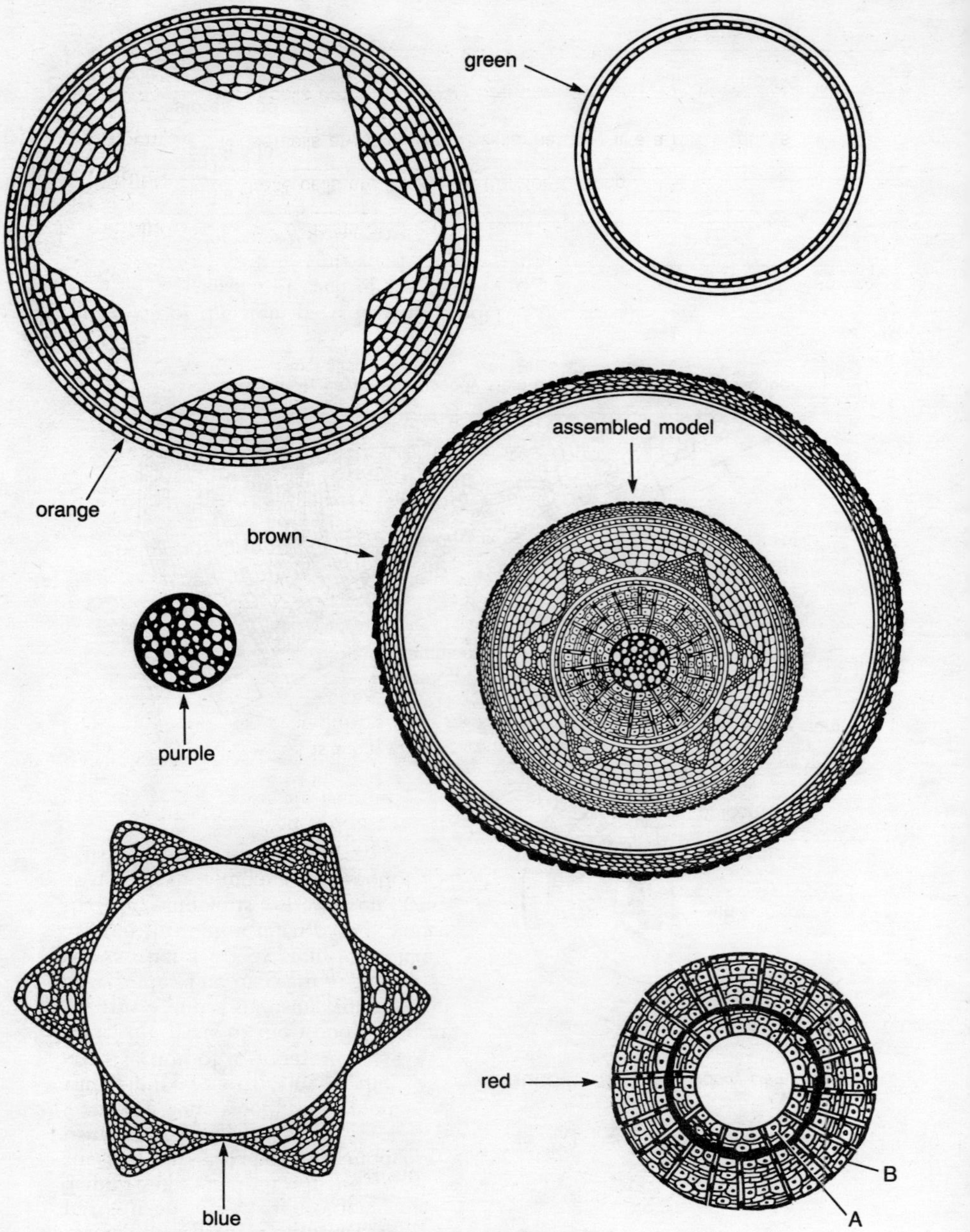

FIGURE 5. Woody-stem model parts

167

168

366

Name _____ Class _____ Period _____

Before doing this lab, have students read section 20:3 in the text.

20-2 What Do the Inside Parts of a Root Look Like?

Roots have five kinds of cells: epidermis, cortex, endodermis, xylem, and phloem. The cells do not look exactly the same as they do in the stem. The size and shape can be different. The way the cells are packed together can also be different.

The outer layer of cells is the epidermis, a layer of protective cells. These cells form a ring around the outside of the root. Certain long, thin cells of the epidermis that absorb water and nutrients are called root hairs. The layer of cells, or tissue, next to the epidermis is the cortex, a layer of food-storage cells. Within the cortex is a ring of cells called the endodermis. Endodermis cells protect the xylem and the phloem. Phloem cells are found inside and next to the endodermis cells. Phloem cells carry food to other cells. The center of the root is packed with xylem cells. Xylem cells carry water to all cells in the plant.

Building the root model enables students to not only see where root parts are but also how they fit together.

EXPLORATION

Process Skills: using a microscope, interpreting scientific illustrations, recognizing and using spatial relationships, inferring

OBJECTIVES

In this exercise, you will:
a. build a model of the structure of a root.
b. compare the model with the cells of a root as seen through a microscope.

KEYWORDS

Define the following keywords:

cortex ___ cells that store food

epidermis ___ cells that form a protective layer on the outside of the root

phloem ___ cells that carry food

root hairs ___ long, thin cells on the epidermis that absorb water

xylem ___ cells that carry water

MATERIALS

blank sheet of paper
scissors
light microscope
transparent tape
prepared slide of a dicot root
colored pencils: red, blue, green, brown, purple

Photocopy page 172 for each student.

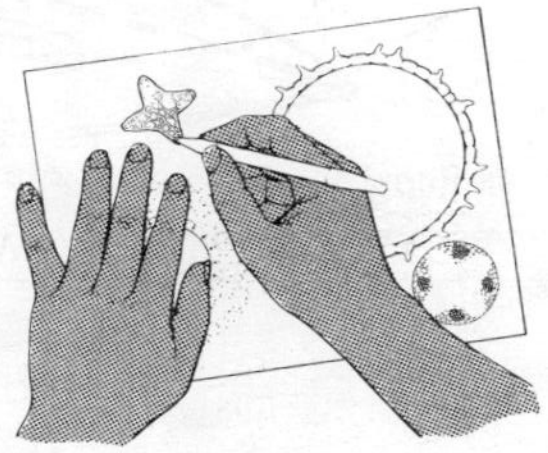

169

PROCEDURE

1. Examine the drawings of the root model parts in Figure 6 on page 160. Obtain a photocopy of the root model parts from your teacher. Color the different layers of cells in the root model parts as follows:
   a. epidermis—green
   b. cortex—purple
   c. endodermis—brown
   d. phloem—blue
   e. xylem—red

2. Cut out each layer of cells as shown in Figure 2. Cut along the outer edge only.

3. Assemble your model by placing one part on another. Start with the thick, outer layer of cells. All parts fit on top of each other.

4. Tape your completed model to a sheet of blank paper.

5. Label your root model parts by using step 1 as a guide.

6. Examine the prepared slide of the dicot root under low power, then high power, of a microscope. Locate each of the cell layers in the slide. Note how each layer looks.

7. Look at Figure 4. This is what the prepared root slide will look like. Use your model and root slide as guides to label the parts in Figure 4.

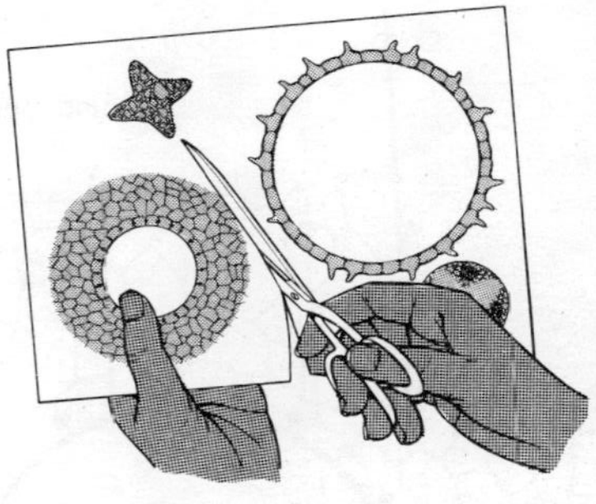

FIGURE 1. Coloring root model parts

FIGURE 2. Cutting out root model parts

FIGURE 3. Examining a root slide

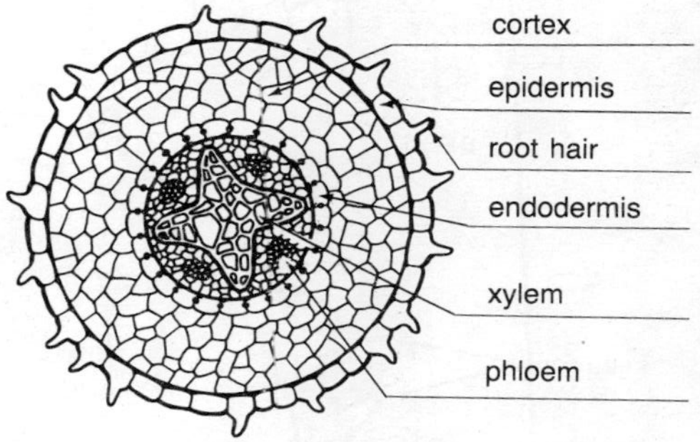

FIGURE 4. Cross section of root

labels: cortex, epidermis, root hair, endodermis, xylem, phloem

QUESTIONS

1. Write a description of each of the following cell layers as they looked through the microscope and in your model. Use complete sentences.

   a. cortex ___ The cortex consists of large, thin-walled cells inside the epidermis.
   b. endodermis ___ These cells are thick cells that form a ring.
   c. epidermis ___ These cells are thick-walled cells that look like a row of bricks.
   d. phloem ___ Phloem cells are small, tightly packed cells inside the endodermis.
   e. xylem ___ Xylem cells are cells that look like thick circles.

Name _____ Class _____ Period _____

Explain to students that they are looking at a longitudinal section or "lengthwise" cut of a carrot in Figure 5. The paper model they built is a cross section or a cut made through the middle part of the carrot.

2. Examine Figure 5, the diagram of a carrot cut lengthwise. Label the parts of the carrot by referring to your model.

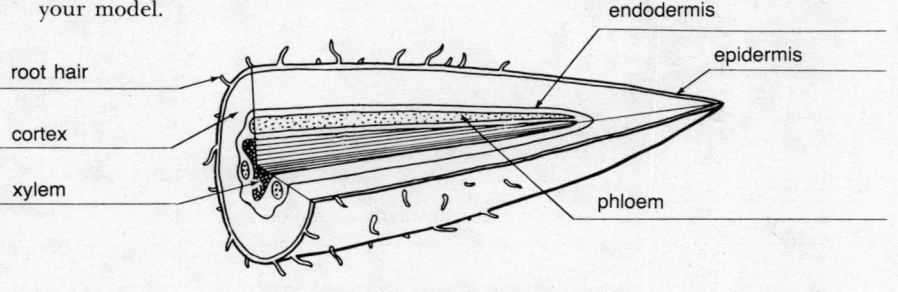

   — root hair
   — cortex
   — xylem
   — endodermis
   — epidermis
   — phloem

   FIGURE 5. Carrot root longitudinal section

3. Which of the three types of cells, xylem, phloem or cortex, has the largest diameter? __cortex__ Why? __because it stores food__

4. Which of the two types of cells, xylem or phloem, has the largest diameter? __xylem__ Why? __It carries water to all cells of the plant and plants need water for photosynthesis and to maintain cell "stiffness."__

5. If the phloem and xylem are tubelike cells, why do they look like circles and boxes in your models? __because you are looking at a cross section rather than a long section of the cells__

6. What is the advantage of making a model of the cell layers of a root? __to see how the layers of cells fit together__

7. Trace the path that a drop of water in the soil would follow as it moves into a root. List the parts in their correct order, going from outside to inside. Use your root model to help. (HINT: Water does not pass through the phloem.) __root hair of epidermis, cortex, endodermis, and xylem__

8. Many of the roots people eat are peeled first.
   a. What cell layer is peeled away? __epidermis and some cortex__
   b. Why do some people peel away the outer covering of carrots, turnips and beets before eating them? __because the outer covering is tough, and is made of thick epidermis cells__

9. Some roots grow thicker than others. What group of cells do you think are responsible for the thickness of a root? __cortex__ Why? __Some roots have a larger storage area than other roots.__

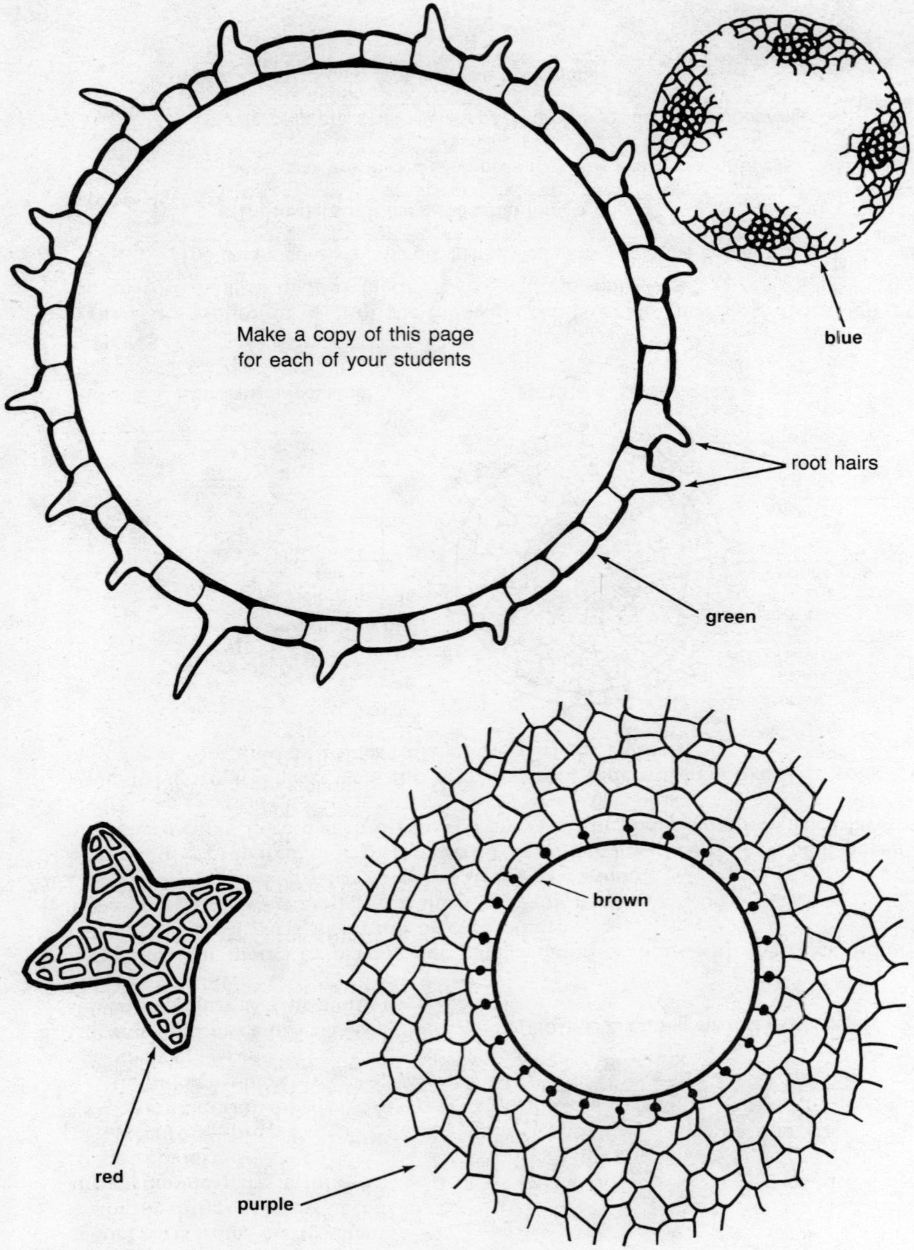

171

FIGURE 6. Root model parts

— blue
— green (root hairs)
— brown
— purple
— red

Make a copy of this page for each of your students

172

368

Name _____ Class _____ Period _____

Before doing this lab, have students read section 21:1 in the text.

21-1 What Tropisms Can Be Seen in Growing Plants?

Plants do not have a nervous system as animals do. But, a plant can respond to a stimulus in its environment. A response to a stimulus by a plant is called a tropism.

Most plant tropisms have something to do with growth movements. Unlike animals, plants are attached to the soil and are not able to move about. So, the movements made by a plant are made by a plant's parts, such as the stem, root, leaf, or flower. A growth movement in response to light is called a phototropism while a growth movement in response to touch is called a thigmotropism.

INVESTIGATION

Point out the difference between a plant's growth movement and how an animal uses muscles to move about.

OBJECTIVES

In this exercise, you will:
a. observe a growing bean plant's response to light.
b. observe a *Mimosa* plant as it responds to touch.
c. find out how much time it takes for a *Mimosa* plant to "recover" after being touched.

Process Skills: experimenting, predicting, inferring, interpreting data

KEYWORDS

Define the following keywords:

phototropism ___ the response of a plant to light

stimulus ___ something that causes a response

thigmotropism ___ the response of a plant to touch

tropism ___ the response of a plant to a stimulus

MATERIALS

for Part A
500 mL soil
2 plastic foam cups
* 4 germinated bean seedlings
metric ruler
cardboard box with slit

for Part B
potted *Mimosa* plant**
toothpick
watch or clock

* Dry kidney bean seeds from a grocery store grow better than lima beans.

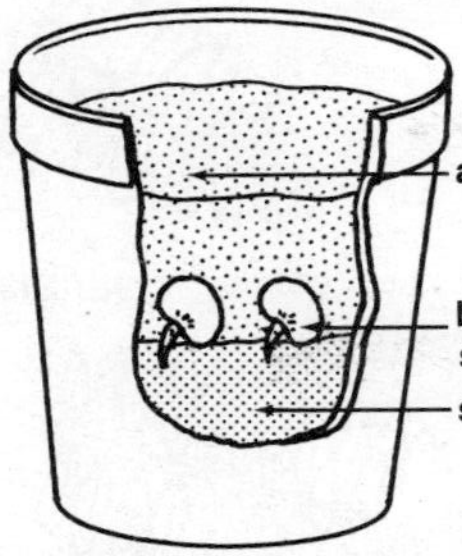

FIGURE 1. Planting bean seedlings

PROCEDURE
Part A. Phototropism

1. Punch four holes in the bottom of each cup. Label the cups A and B and put your name on each cup.
2. Fill the cups nearly full with soil.

Soak beans (pinto, kidney) for 12 hours. Then, place between moist paper towels. Seeds will germinate in about two days.
** *Mimosa* available from biological supply houses.

173

Seeds should be planted not more than 5 cm deep.

3. Put two bean seedlings on top of the soil in each cup as in Figure 1.
4. Fill the cups to the top with soil and add water to lightly dampen the soil.
5. Place the cup labeled A in an area of the room that gets even light from all sides.
* 6. Place the cup labeled B in a box that has been prepared for you. Note that a small strip has been cut out of the side of the box. Place the cup in the box as shown in Figure 2.
7. When a box is filled with cups, it should be closed so that light enters only through the cutout slot.
8. Examine the plants during each class period. Observe changes in the plants as they grow. Seedlings will break through the soil in 2-3 days after planting.
** 9. Measure the height of one of the bean plants in each cup. (Be sure to measure the same plant in each cup throughout this experiment.) Record this measurement for both plant A and plant B in the proper place in Table 1.
10. Make a small sketch of the plant in the proper place in Table 1. In the sketch you should show the direction, if any, in which the stem and leaves seem to be turning or bending.

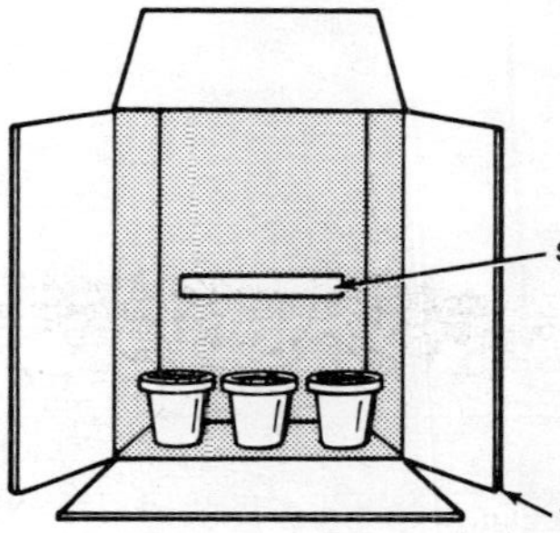

FIGURE 2. Placing bean seedlings in a cardboard box

* Tell students to put labeled side of cup B facing away from the slit in the box.

Have students label the parts they are measuring with self-sticking labels. Or, students can cut off the parts of the plants they will not be measuring.

Table 1. Height and Bending Response in Plants

	Plant A (in Light)		Plant B (in Dark)	
	Height in cm	Amount of Bending	Height in cm	Amount of Bending
Day 1	—	—	—	—
Day 2	—	—	—	—
Day 3	3		4	
Day 4	5.2		6.9	
Day 5	6.1		9.2	
Day 6	8.0		12.2	

(seeds not through soil)

Note which plant has a thinner stem at the end of this exercise. (Plant B)

Part B. Thigmotropism.

1. Observe the leaflets and stalk of the *Mimosa* plant without touching the plant.
2. Make a drawing in the space provided in Figure 3 to show what the leaflets and the stalk look like.
3. Put a toothpick in your fingers as shown in Figure 4.
** Have students begin measuring plants when they can see the hypocotyl arch above the soil. Students should measure the stem from the top of the soil to the tip of the stem. Have students run a string along the plant stem and use a ruler to measure the length of the string.

174

Name _____ Class _____ Period _____

4. Touch a leaflet by letting the toothpick spring forward as shown in Figure 4.
5. Observe the response of the leaflet to touch and note the time at which the response occurred. This is Trial 1. Record this time in the proper column in Table 2 under "Leaf."
6. Make a drawing, in the space provided in Figure 5, of what the leaflet looks like after it has been touched.
7. Record in Table 2 in the column marked "Time Recovered" the time at which the leaflet returns to its normal position.

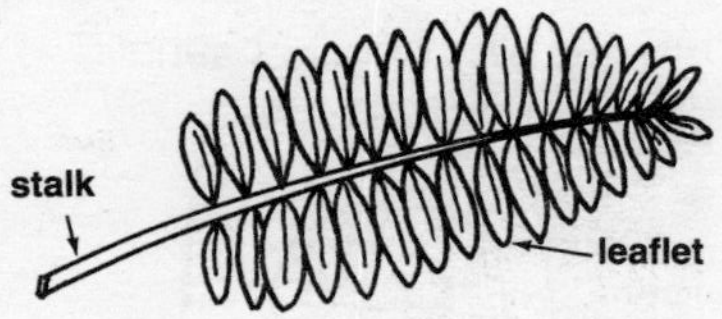

FIGURE 3. *Mimosa* plant

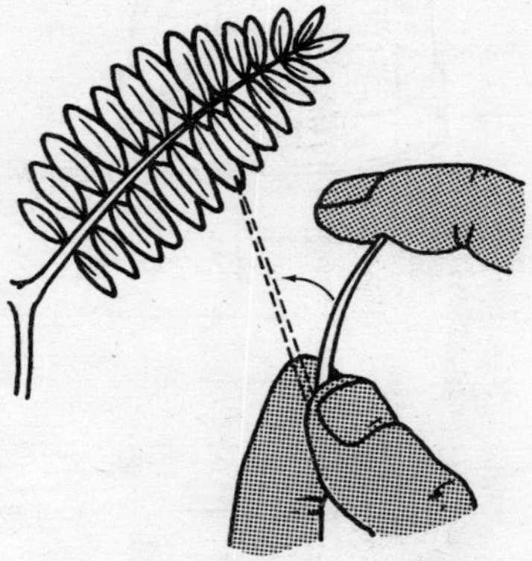

FIGURE 4. Touching *Mimosa* leaflet

FIGURE 5. *Mimosa* leaflet after touching

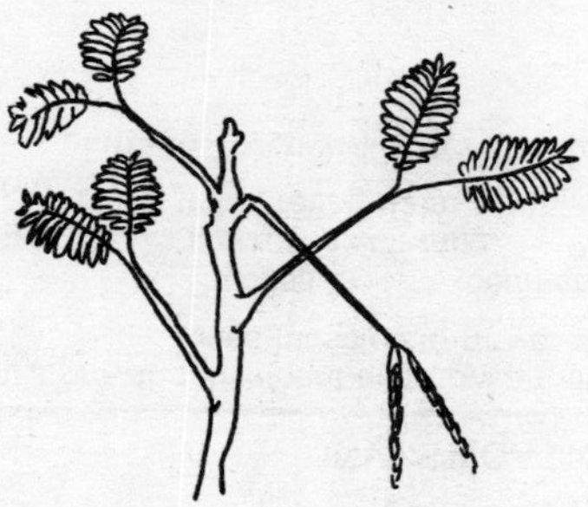

FIGURE 6. *Mimosa* stalk after touching

Table 2. Recovery Time for *Mimosa*

	Leaf			Stalk		
	Time Touched	Time Recovered	Amount of Time for Recovery	Time Touched	Time Recovered	Amount of Time for Recovery
Trial 1	10:00	10:05	5 minutes	10:30	10:35	5 minutes
Trial 2	10:07	10:12	5 minutes	10:37	10:42	5 minutes
Trial 3	10:14	10:12	5 minutes	10:43	10:48	5 minutes
Trial 4	10:20	10:25	5 minutes	10:49	10:54	5 minutes

Sample data are approximate. Students answers will vary.

*8. Repeat steps 4–7 at least three more times (Trials 2, 3, and 4) with a leaflet.
9. Repeat steps 4–7 at least four times (Trials 1, 2, 3, and 4) with a stalk. Record your results in the proper columns under "Stalk" and make a drawing of the stalk, after you have touched it, in the space provided in Figure 6.
10. For each of your trials calculate the "Amount of Time for Recovery" by subtracting the time in the "Time Touched" column from the time recorded in the "Time Recovered" column. Record your results in the proper columns in Table 2.

*Remind students not to start the next trial until the plant has fully recovered.

QUESTIONS

1. What is a tropism? _a plant's response to a stimulus_
2. What are the two types of tropisms being studied in this investigation? _phototropism and thigmotropism_
3. a. In which of the two types of plants were you able to see an immediate response? _the Mimosa_
   b. To what stimulus was it responding? _touch_
   c. Describe the response of this plant. _The leaflet folded and the stalk drooped quickly when touched._
4. a. In which of the two types of plants were you not able to see an immediate response? _the bean plant_
   b. To what stimulus was this plant responding? _light_
   c. Describe the response in this plant. _Over several days the plant stem bent toward the light as it grew (increased in height)._
5. a. If an insect landed on a *Mimosa* leaf, how do you think the leaflets would respond? _They would fold up quickly and the insect would fly away._
   b. What would be the advantage of this type of movement to the plant? _The insect might be prevented from damaging the plant._
6. What is one advantage of a plant's responding to light? _A plant needs to grow toward the light because it needs light for photosynthesis._
7. What would happen if a plant could not grow toward light? _It might not get enough light for photosynthesis and it would die._

Name _____ Class _____ Period _____

Before doing this lab, have students read section 21:2 in the text.

21-2 Is Light an Important Growth Requirement for Plants?

Plants need a certain number of hours of light each day during their growing season. This light is needed for photosynthesis and normal growth. Some kinds of plants need to get more light each day than other plants.

Did you ever notice that some plants grow best in the shade of other plants? Plants of this kind can grow in the shade of trees provide shade. Most mosses and ferns are plants that grow in the shade. Suppose you tried to grow corn or tomato plants in the shade. They probably would not grow well because they grow best in direct sun.

Start on a Monday if possible so students have 5 straight days of observations. Place newspapers on desks before handling soil and water.

Process Skills: experimenting, separating and controlling variables, inferring, interpreting data, making a line graph

OBJECTIVES

In this exercise, you will:
a. observe changes in height and color in two growing plants.
b. determine what effect sunlight has on the color and growth of plants.

KEYWORDS

Define the following keywords:

light — a growth requirement needed by plants

photosynthesis — the process by which plants make food

shade — a condition in which some plants can grow best

MATERIALS

Sprout 70–80 bean seeds in a container filled to a depth of 5 cm with soil. Prepare at least 5 days before starting this lab.

2 plastic foam cups   150 mL soil
metric ruler   water
2 bean seedlings   kidney beans available at grocery store
Discarded, washed, shampoo bottles work well as "watering cans."

INVESTIGATION

PROCEDURE

1. Punch four holes in the bottom of each cup.
2. Label the cups A and B and place your name on each cup.
3. Fill the cups nearly full with soil. Use Figure 1 as a guide.
4. Place a bean seedling in each cup. Try to choose two, nearly identical seedlings. Fill the cups to the top with soil. Water soil lightly.

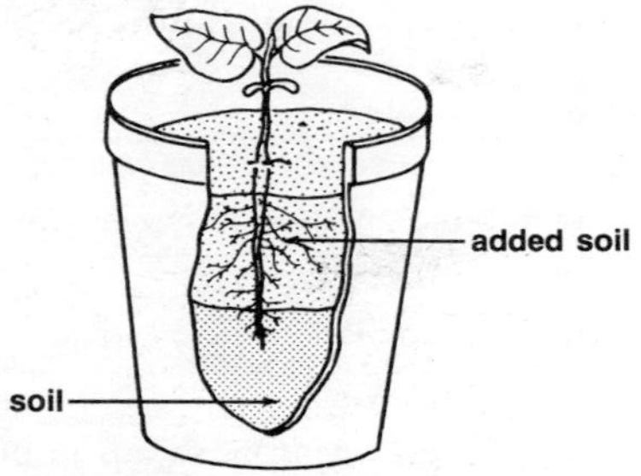

FIGURE 1. Planting bean seedlings

177

5. Measure the length of the stem (in cm) from the soil to the stem tip for both plants. Note the stem's color. Record your observations in the proper columns in Table 1 for Day 0.
6. Put cup A in a lighted area. Put cup B in a dark area.
7. Examine both plants at the same time each day for the next 4 days. Repeat steps 5 and 6 each day. On days 1 through 4, record leaf color. Keep the soil in the cups moist, but not wet, during the entire experiment.
8. On day 4, measure the length of one of the first true leaves from the base of the blade to the leaf tip for both plants. Record your observations in the proper columns in Table 1.
9. Make line graphs showing plant height for both plants using your data in Table 1.

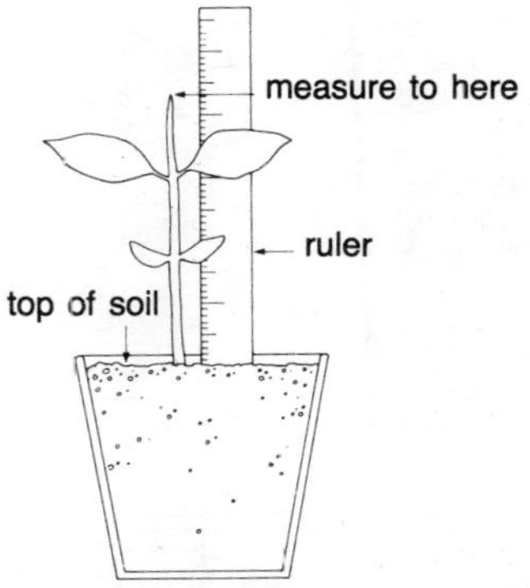

FIGURE 2. Measuring the stem of a bean seedling

Table 1.

Day	Plant A—Light			Plant B—Dark		
	Leaf* Color	Stem Color	Stem Length	Leaf Color	Stem Color	Stem Length
0	not visible	green	2.4 cm	not visible	green	2.4 cm
1	green	green	4.5 cm	green	green	4.6 cm
2	green	green	6.5 cm	yellow-green	yellow-green	8.0 cm
3	green	green	8.5 cm	yellow	yellow	11.2 cm
4	green	green	10.5 cm	yellow	yellow	13.6 cm

Leaf Length row: Day 4 Plant A = 8.6 cm; Plant B = 1.4 cm

*Have students record color of first true leaves when visible (not hidden in cotyledons).

QUESTIONS

1. a. What was the color of the leaves of both plants at the start of the experiment? __green__
   b. What was the height of both plant stems at the start of the experiment? __2.4 cm__
2. a. What was the color of the leaves at the end of the experiment for the plant kept in the light? __green__
   b. What was the color of the leaves at the end of the experiment for the plant kept in the dark? __yellow__

Name _____ Class _____ Period _____

3. a. What was the height of the stem at the end of the experiment for the plant kept in the light? __10.5 cm__
   b. What was the height of the stem at the end of the experiment for the plant kept in the dark? __13.6 cm__

4. How were the leaves and stems of the two plants at the end of the experiment different? __The dark-grown plant is yellowed, has thinner, flimsier leaves and a longer, thinner stem.__

5. Describe what you think is normal growth in a plant when it gets enough light. __The plant will have dark green leaves and stem. The stem will not droop.__

6. Describe what happens to plants that do not get enough light. __Green plants that do not get enough light lose their green color. They become yellow or yellow-white in color. They also grow taller and thinner than plants that get more light.__

7. Sometimes potatoes and carrots sprout while being kept in a refrigerator. Explain why they are usually yellow. __The new leaves do not get enough light and so they do not become green.__

8. A head of lettuce is mostly leaves. Explain why the leaves near the center of the head are yellow while the outer leaves are green. __The innermost leaves do not get enough light; the outer leaves do get enough light.__

FIGURE 3.

Height in Centimeters (y-axis, 1 to 14) vs Day (x-axis, 0 to 4)

● Plant A—in light
▲ Plant B—in dark

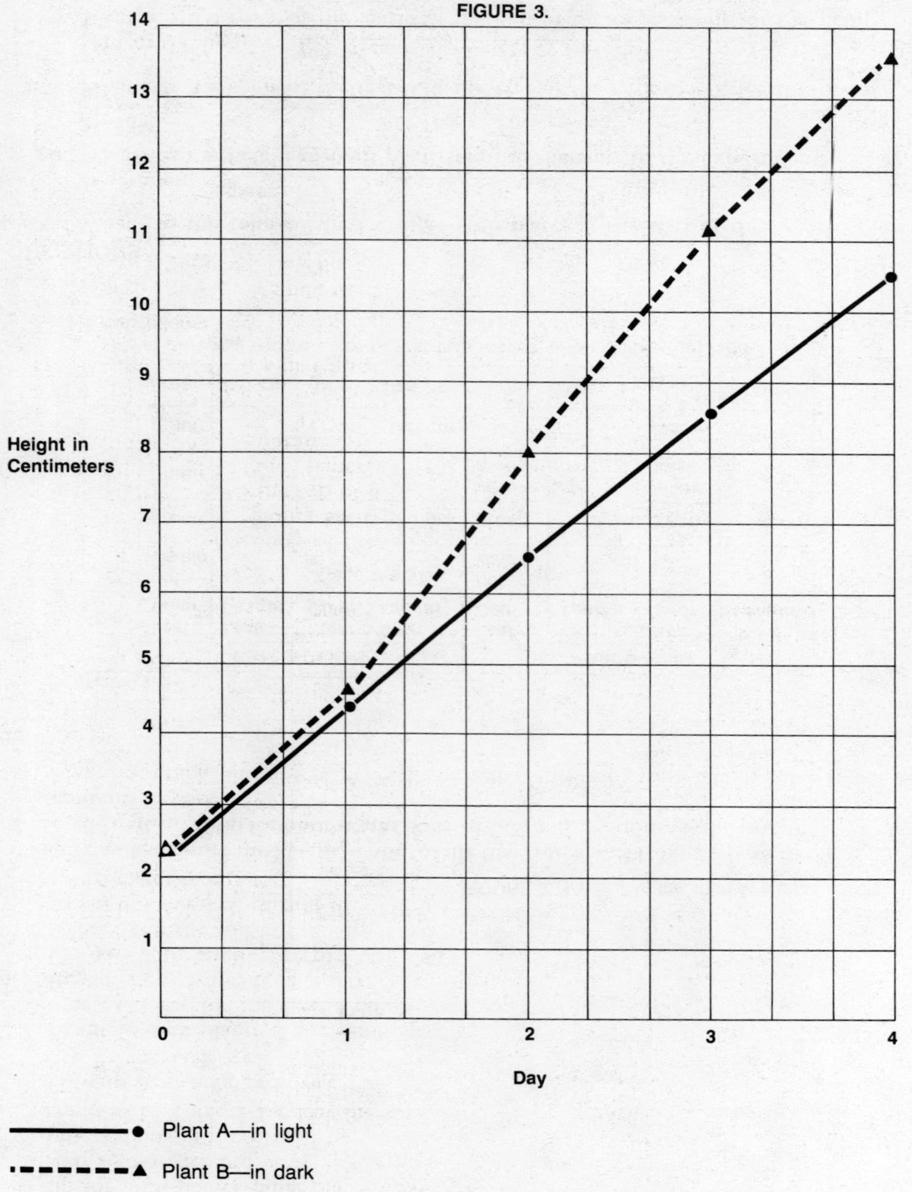

Name _____ Class _____ Period _____

Before doing this lab, have students read section 22:1 in the text.

22-1 What Happens When Cells Divide?

Cells form new cells by a process called cell division or mitosis. During mitosis, one cell divides in half to form two new cells. Suppose you could watch a cell divide. You could see that the cell parts called chromosomes move around the cell during mitosis. Because chromosomes move in particular ways, you could arrange the events of mitosis into several steps. Biologists have been able to arrange the events of mitosis into several steps. They examined many dividing cells in order to learn the steps. What are the steps of cell division? In what order do they occur?

INTERPRETATION

OBJECTIVES

In this exercise, you will:
a. build models of the steps of mitosis.
b. compare your models to the steps of animal-cell mitosis.

Have students work in teams of 2, 3, or 4 to construct one or two of the models. Have teams compare their models with those of other teams.

Process Skills: interpreting scientific illustrations, scientific terminology, sequencing, recognizing and using spatial relationships

KEYWORDS

Define the following keywords:

chromosome _____ cell part with information that determines the traits of a living thing

cytoplasm _____ clear, jellylike material that makes up most of a cell

nucleolus _____ cell part that helps make ribosomes

nucleus _____ cell part that controls cell activities

MATERIALS

4 pieces of different-colored construction paper
scissors   thread
glue   24 toothpicks
yarn   metric ruler

PROCEDURE

1. Using Figure 1 and your textbook, review the steps of mitosis.

FIGURE 1. Steps of mitosis

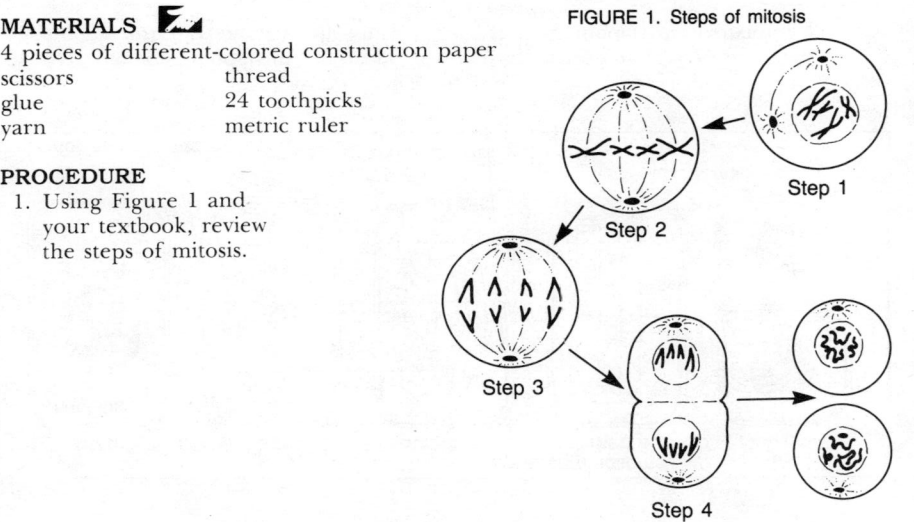

2. Use the materials listed in Table 1 to represent the cell parts. Cut the pieces of paper, yarn, and thread to the sizes given in Table 1.

Cell wall/membrane and cytoplasm sections can be cut prior to class by using a paper cutter. This will save time for the students.

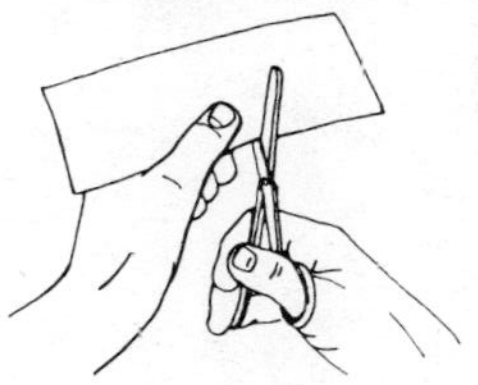

FIGURE 2. Making cell parts

Table 1. Making Cell Parts

Cell Part	Material to Use	Size	Number Needed
Cell wall and membrane	Dark-colored paper*	14 × 8 cm	5
Cytoplasm	Light-colored paper	13 × 7 cm	5
Nucleus	Dark-colored paper	5 cm circle	3
Nucleolus	Light-colored paper	1 cm circle	2
Chromosomes	Light-colored yarn	4 cm long	20
		10 cm long	2
Fibers	Toothpicks	full size	24
Cell wall between new cells	Dark-colored paper*	½ × 8 cm	1
Nuclei in new cells	Thread	½ m	2

*Use the same color

3. Begin building the models of the cell division steps by gluing each "cytoplasm" paper to the top of a "cell wall and membrane" paper. The cell wall and membrane should show on all sides. Use Figure 3 as a guide.
4. Following the diagrams in Figure 1, make each of the "cell wall-membrane-cytoplasm" pieces into a mitosis step. Use glue to attach the proper parts to the pieces. Be sure to study the diagrams so that you get the correct parts in each step.

FIGURE 3. Putting models together

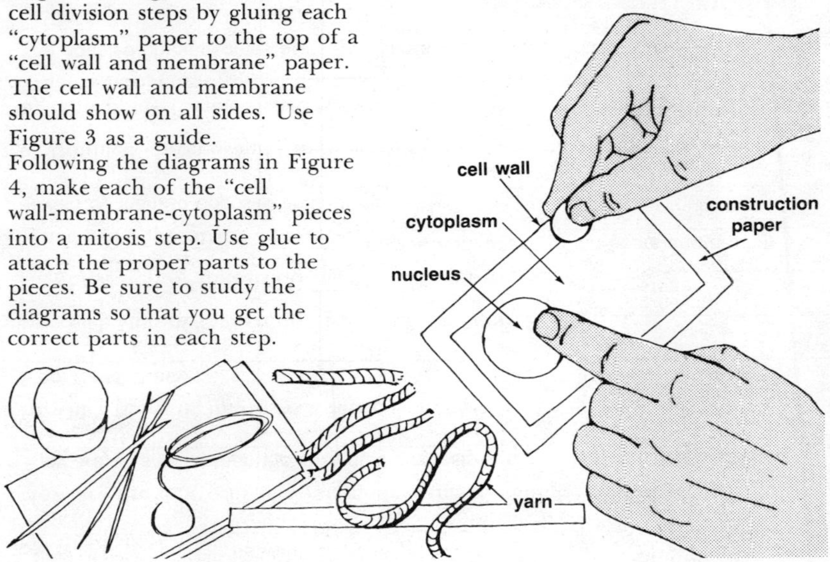

Name _____ Class _____ Period _____

FIGURE 4. Steps of plant cell mitosis

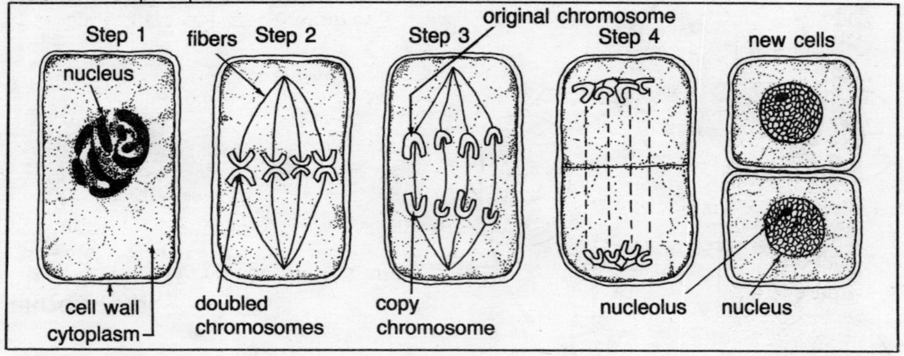

6. Arrange your models in the order in which mitosis occurs. Note how your models differ from those shown in Figure 1. Your models show dividing plant cells.
7. Compare your models with the drawings of the animal cells in Figure 5.
8. Write the number of the step of mitosis below each drawing of the animal cells.

FIGURE 5. Steps of animal cell mitosis

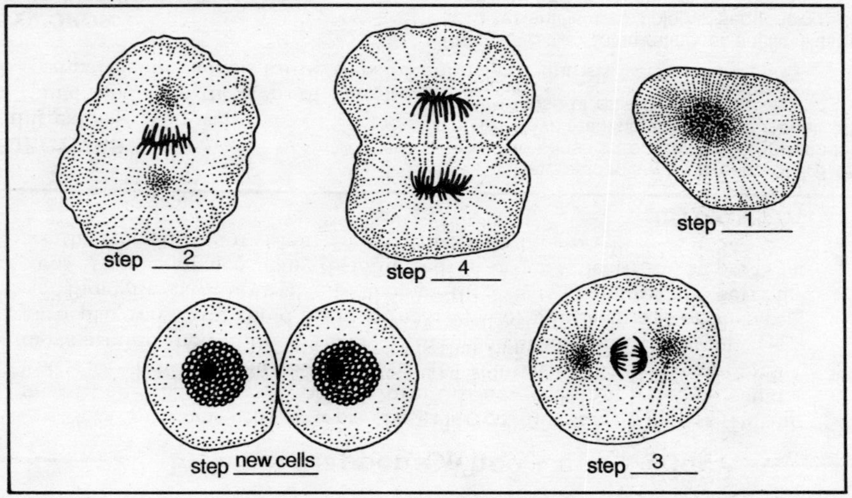

183

QUESTIONS

1. What part is present in plant cells but absent in animal cells? __cell wall__

2. How are the new cells of your models and the animal cells alike? __They are all smaller than the cells during mitosis and a distinct nucleus is present.__

3. In which steps of mitosis is a nucleus visible? __Step 1 and in new cells__

4. In which step of mitosis do you first see fibers? __step 2__

5. In which step of mitosis do the fibers begin to disappear? __step 4__

6. What is the job of the fibers? __Chromosomes attach to the fibers and are guided by them to opposite ends of the cell.__

7. a. Doubled chromosomes first become visible in which step of mitosis? __step 2__
   b. How many doubled chromosomes are visible in this step? __four (4)__

8. What is happening to the doubled chromosomes in steps 3 and 4? __The doubled chromosomes are separating and moving to opposite ends of the cell.__

9. How many cells does each dividing cell form? __two new cells__

10. What forms between the cells after the doubled chromosomes have pulled apart in plant cells? __a cell wall and membrane__

11. Match the following by writing the correct letter in the proper blank.

    __e__ Step 1   a. chromosomes move apart to the ends of each cell
    __d__ Step 2   b. nucleus reformed
    __c__ Step 3   c. doubled chromosomes separate
    __a__ Step 4   d. chromosomes become thick, dark, and doubled
    __b__ Step 5   e. membrane around nucleus disappearing

184

_____ Class _____ Period _____

ving this lab, have students read section 22:3 in the text.

22-2 Are There More Dividing Cells or Resting Cells in a Root Tip?

A plant grows in length at the tip of stem and root. In the stem and root tip there is a small group of cells that divide many times; however, not all cells in these parts may be dividing. A dividing cell may be next to several resting cells and a resting cell can be surrounded by several dividing cells.

Cells in mitosis are different from resting cells. Some parts of a cell are seen best only when a cell is dividing. These parts seem to disappear after a cell has divided.

Process Skills: using a microscope, using numbers, recognizing and using spatial relationships, interpreting data

OBJECTIVES
In this exercise, you will:
a. examine dividing and resting cells in an onion root.
b. compare the number of dividing cells to resting cells.

EXPLORATION

KEYWORDS
Define the following keywords:

dividing cell ____ a cell that is forming two cells from one

resting cell ____ a cell that is not forming new cells by cell division

root tip ____ the part of a root where cell division occurs

MATERIALS
prepared slide of an onion root tip, L.S.
light microscope

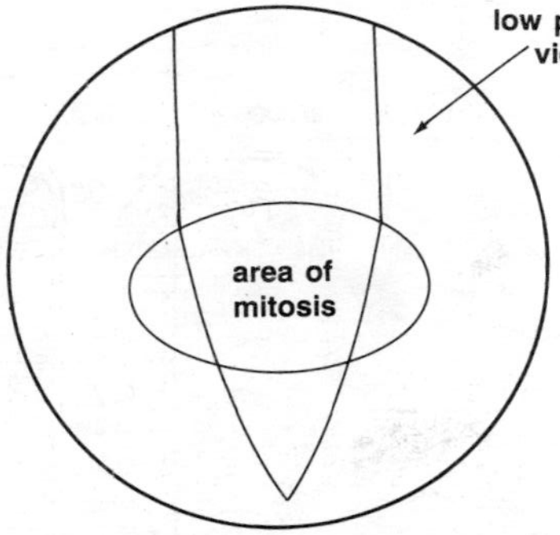

PROCEDURE
1. Obtain a microscope slide of an onion root tip.
2. Place the slide on the stage of the microscope and examine the root tip on low power. The tip of the root should be at the bottom of the field of view as in Figure 1.
3. Change the lens to high power. You will see several columns of cells.
4. Look for a cell that resembles Step 1 of mitosis, as shown in Figure 4 on page 187.

FIGURE 1. Root tip under a microscope

Have students use their models from Lab 22-1 as a comparison with the different steps of mitosis they will be viewing. This will allow students to see how well their models were made and will reinforce the steps of mitosis.

185

5. Draw the cell in the space provided and label the parts.
6. Repeat step 5 for the remaining steps of mitosis shown in Figure 4.
7. Count the cells in your field of view that resemble Step 1 of mitosis. Count the cells a column at a time going from left to right, as in Figure 2.
8. Record in Table 1 the number of cells that resemble Step 1.
9. Repeat steps 7 and 8 for the remaining steps of mitosis shown in Figure 4.
10. Compare your data with those of other students in your class.

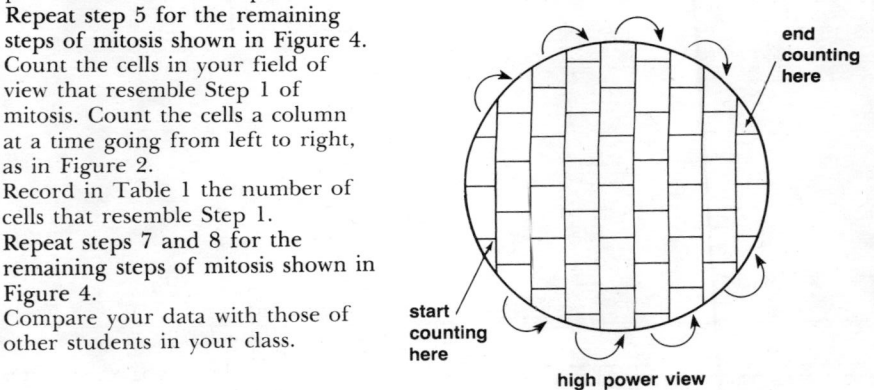

FIGURE 2. How to count cells

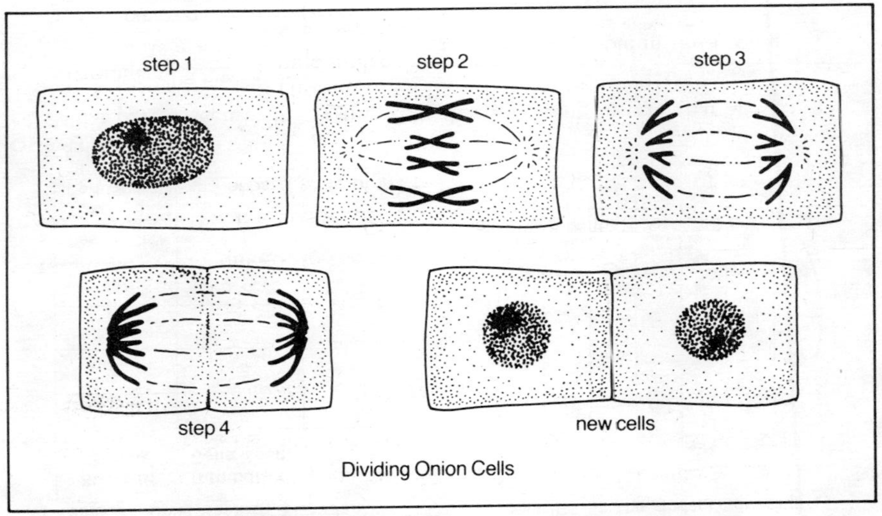

Dividing Onion Cells

FIGURE 3. Student drawings of the steps of mitosis

186

375

Name _____ Class _____ Period _____

Table 1.

Steps of Mitosis	Number of Cells Seen
1	17
2	7
3	2
4	1
Resting Cells	120
Total Cells Seen	157

FIGURE 4. Steps of mitosis in plant cells

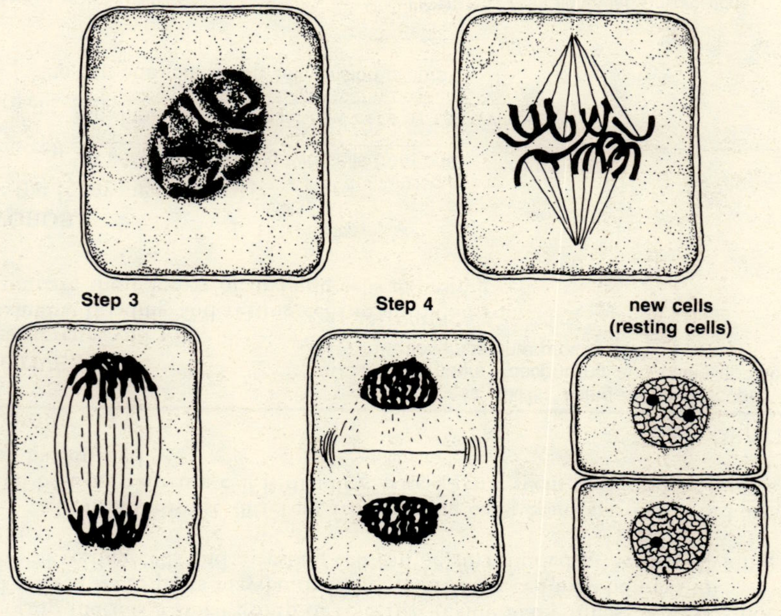

Step 1 • Step 2 • Step 3 • Step 4 • new cells (resting cells)

387

---

11. A biology student was looking at an onion root tip through the microscope and made a drawing of the cells she saw. Record on the chart how many cells you think she saw in each step of mitosis and the number of resting cells.

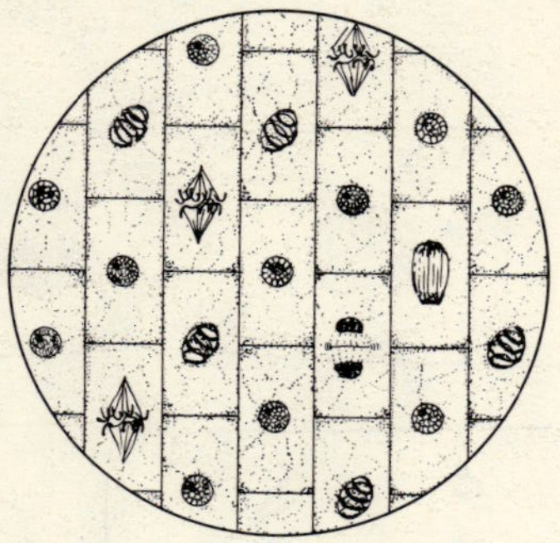

FIGURE 5. The student's drawing

Table 2. Cells Seen by a Student

Steps of Mitosis	Number of Cells Seen
1	5
2	3
3	1
4	1
Resting Cells	11
Total Cells Seen	21

QUESTIONS

1. What part is seen in the resting cells that is missing in cells that are dividing? __the nucleus__

2. What parts are seen in dividing cells that are not visible in the resting cells? __doubled chromosomes and fibers__

3. Look at Figure 4 again. Why do you think new cells are sometimes called resting cells? __They have not started to divide.__

4. Which cells did you see more of in the onion root, dividing cells or resting cells? __resting cells__

5. Which step of mitosis was most common in the onion root? __step 1__

6. Which step of mitosis was least common in the onion root? __step 4__

7. Suppose you examined another root tip and saw that half of the cells were dividing. Would this root be growing faster or slower than the one you examined in this exercise? Explain your answer. __It would be growing faster because there are more cells dividing and producing more new cells.__

188

Name _____ Class _____ Period _____

Before doing this lab, have students read section 23:2 in the text.

23-1 What Are the Parts of a Flower?

Many common plants you are familiar with form flowers at sometime during the year. The flower is the sexually reproductive part of a flowering plant. Certain flower parts are male, while other parts are female. Certain flower parts are neither male nor female. Which flower parts are male, female, or neither? How exactly do these parts help with sexual reproduction?

EXPLORATION

Process Skills: observing, interpreting scientific illustrations, scientific terminology, inferring

OBJECTIVES

In this exercise, you will:
a. dissect and examine the parts of a flower.
b. learn which parts are male, female, or neither male nor female.
c. find out how each flower part helps in reproduction.

KEYWORDS

Define the following keywords:

ovule   tiny round parts within an ovary that contain egg cells
pistil female reproductive organ of a flower
pollen male reproductive part that forms sperm cells
stamen male reproductive organ of a flower

Collect flowers locally if possible, use fresh iris or tulip, or use preserved *Lilium*, *Hemerocallis*, or *Amaryllis*.

MATERIALS

razor blade   colored pencils: red, blue, yellow, green, and purple
microscope   flower
1 coverslip   water
glass slide

PROCEDURE

Part A. Flower Parts That Are Neither Male nor Female

1. Examine the flower provided by your teacher.
2. Use Figure 1 to help locate the following two flower parts:
   a. sepals—small, green petallike parts forming the outside layer of the flower.

If flowers are preserved, colors will have faded.

   b. petals—large, brightly colored flower parts forming the layer inside the sepals.
3. Count the number of sepals and the number of petals present in your flower. Record these numbers in Table 1.
4. Write the functions of these two flower parts in Table 1.
5. Complete the last column of Table 1 for sepals and petals. (HINT: Read the title of this part of the activity.)
6. Color the sepals green and the petals yellow in Figure 1.

Refer students to their textbook.

Table 1. The Functions of the Parts of a Flower

Flower Part	Number of Parts	Function	Male, Female, or Neither
Sepals	3	protect the young flower in bud	Neither
Petals	3	protect the flower and attract insects	Neither
Stamens	6	male reproductive organs of the flower	Male
Anthers	6	form pollen grains	Male
Filaments	6	support the anther	Male
Pollen Grains	Approximately Thousands	reproductive parts that form sperm cells	Male
Pistil	1	female reproductive organ of the flower	Female
Stigma	1	sticky surface that traps pollen	Female
Style	1	supports stigma and connects to ovary	Female
Ovary	1	contains ovules and becomes the fruit	Female
Ovules	Approximately 1 to 100	reproductive parts that form egg cells and become seeds	Female

All numbers depend on the flower examined. Sample data is for a monocot such as a lily.

Part B. Male Flower Parts

1. Carefully remove first the sepals and then the petals. Try not to remove any smaller parts inside the petals.
2. Use Figure 1 to help locate the following male flower parts:
   a. stamen—long, slender parts made up of a top, often yellow, part and a lower stalklike part
   b. anther—top part of the stamen, often yellow in color
   c. filament—thin stalk that supports the anther
3. Count the stamens from your flower. Record this number in Table 1.
4. Write the function of these three flower parts in Table 1.*
5. Complete the last column of Table 1 for the stamen, anther, and filament.
6. Color the stamen in Figure 1 red.
7. Remove one stamen and place the anther on a glass slide.

*Refer students to their textbooks.

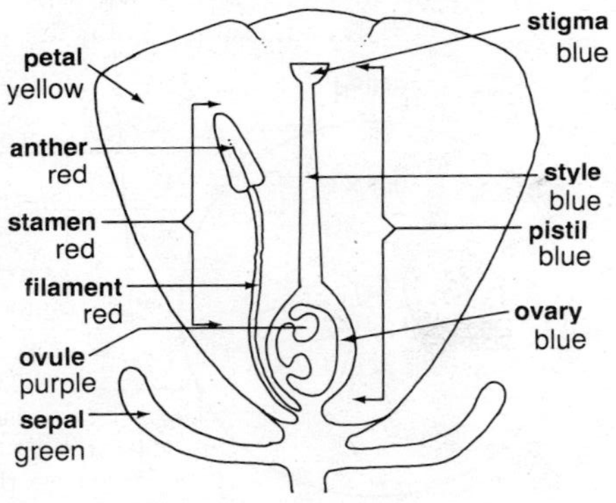

FIGURE 1. Parts of a flower

Labels:
- stigma — blue
- style — blue
- pistil — blue
- ovary — blue
- petal — yellow
- anther — red
- stamen — red
- filament — red
- ovule — purple
- sepal — green

189

190

377

Name _____ Class _____ Period _____

8. Crush the anther with the eraser end of a pencil. Add two drops of water and a cover slip.*
9. Examine the slide under low power, then high power, of your microscope.
10. The small round parts you see are pollen grains. Using high power, draw one or two pollen grains in the circular area provided here.
11. Complete Table 1 for pollen cells.

*Not all anthers may be mature—if not, no pollen will be present.

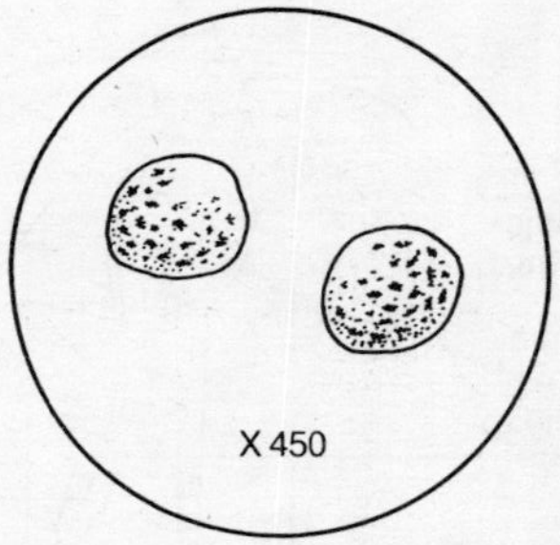

X 450

Pollen

Part C. Female Flower Parts

1. Use Figure 1 to help locate the following female flower parts:
   a. pistil—long, slender stalk with a round base in the center of the flower
   b. stigma—tip of the pistil
   c. style—slender stalk part of the pistil
   d. ovary—rounded, swollen bottom part of the pistil
2. Count the number of each of these flower parts examined in step 1. Record these numbers in Table 1.
3. Write the function of each of these parts in Table 1.
4. Complete the last column of Table 1 for these four parts.
5. Color the pistil blue in Figure 1.
6. Remove the pistil from the flower. Use a razor blade to cut down through the length of the ovary as shown in Figure 2.
   CAUTION: *Always use extreme care with the razor blade.*
7. Note the small, round seedlike parts inside the ovary. These are the ovules.
8. Complete Table 1 for the ovules. Color the ovules in Figure 1 purple.

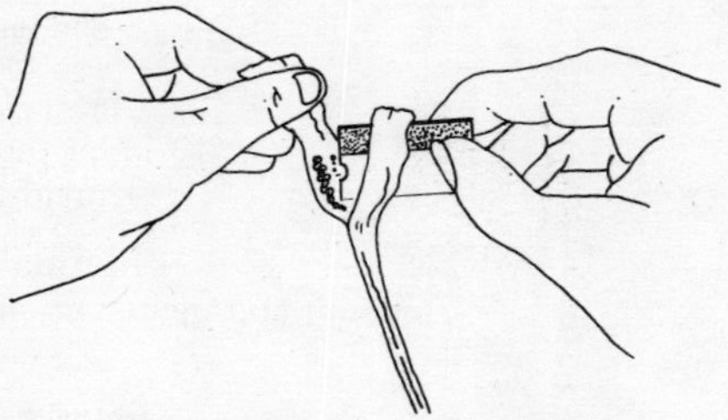

FIGURE 2. Cutting through the ovary

QUESTIONS

1. Group these flower parts—filament, ovary, ovule, petal, pollen grain, anther, stamen, pistil, stigma, sepal, style—under the following three headings:

Male Parts	Female Parts	Neither Male nor Female
filament stamen anther pollen	pistil ovary stigma ovule style	petal sepal

191

Refer students to their textbook.

2. a. What happens to pollen grains during pollination? They are carried from the anther to the stigma of a pistil.
   b. How might their small size help them for this job? They are easily carried because of their light weight.
   c. How might the fact that there are so many pollen cells help pollination occur? It increases the chances that at least some pollen will land on a stigma.

3. a. What flower part contains egg cells? ovule
   b. How do pollen grains on a stigma cause the fertilization of egg cells?* Pollen grows a tube down through the style to the ovary, and sperm cells pass along this tube to the egg cells in the ovule.

4. Look at Figure 3 and label the flower parts that are listed in question 1. Included in this diagram are close up views of the inside of an anther and an ovary. Locate and label a pollen grain and an egg cell.

5. Not all flowers have all the parts you have studied. Examine Figure 4 that shows sections of three different kinds of flowers. Label the parts that are present in each flower. Complete the chart that follows.

*Refer students to their textbook.

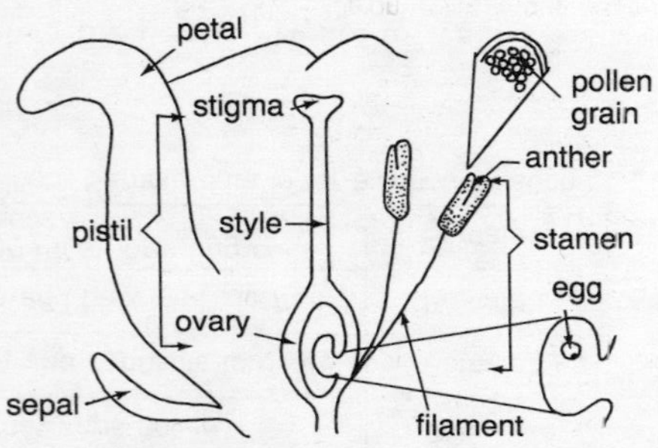

FIGURE 3. Reproductive parts of a flower

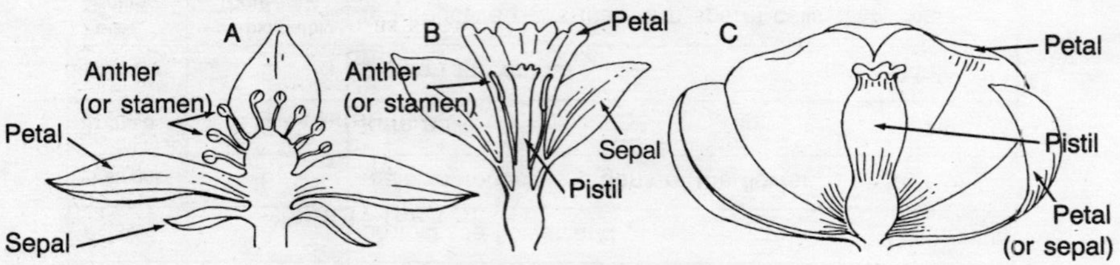

FIGURE 4. Parts of three different flowers

	Flower A	Flower B	Flower C
a. What are the missing flower parts?	pistil	none	stamens
b. Can flower make pollen? Why or why not?	yes—it has 8 stamens	yes—it has 2 stamens	no—it has no male parts
c. Can flower make eggs? Why or why not?	no—it has no female parts	yes—it has a pistil and ovary	yes—it has a pistil and ovary
d. Can flower self-pollinate? Why or why not?	no—stamens only present	yes—stamens and pistils both present	no—only pistil present

192

378

Name _____ Class _____ Period _____

Before doing this lab, have students read section 23:3 in the text.

23-2 What Plant Part Are You Eating?

Plants store food to provide food for the start of the next year's growth (for example in roots and stems) or for the start of the next generation (for example in seeds or bulbs). Both asexual and sexual reproduction often involve the production of plant parts that store large quantities of food.

We use many of these plant parts for food. Have you ever asked yourself "What part of a plant am I really eating?" You may be eating a plant's root, stem, leaf, flower, fruit, or seed. Sometimes it is difficult to recognize what plant part a particular food may be.

EXPLORATION

Process Skills: observing, classifying, using a microscope, inferring

Have students work in groups.

OBJECTIVES
In this exercise, you will:
a. examine identified parts of various food plants.
b. note the special characteristics of these plant parts.
c. examine some additional food plants and identify them as either root, stem, leaf, flower, fruit, or seed.

KEYWORDS
Define the following keywords:

asexual reproduction ___ a form of reproduction that uses one parent, no sex cells, and cell division by mitosis

embryo ___ a new, undeveloped living thing

meiosis ___ cell division that forms egg or sperm cells; part of sexual reproduction

mitosis ___ cell division that forms body cells; part of asexual reproduction

sexual reproduction ___ a form of reproduction that uses two parents, egg and sperm cells, and cell division by meiosis

MATERIALS
hand lens or stereomicroscope
glass slide
scalpel
dissecting needle
broccoli
* caper
* Available in bottles in the grocery store.

Use fresh or preserved samples. Presoak beans and peas overnight in water to soften the coats.

brussel sprout   lima bean
lettuce   pea
cherry pepper   sweet potato
okra   carrot
cucumber   onion
                 shallot

Reduce unknowns for Part B if cost becomes a limiting factor.

193

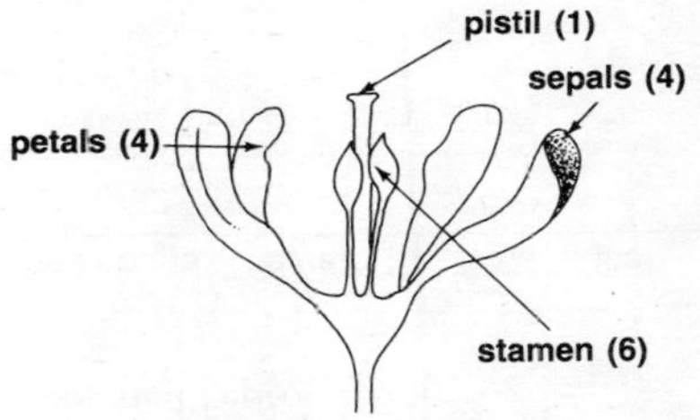

FIGURE 1. Broccoli flower (pistil (1), sepals (4), stamen (6), petals (4))

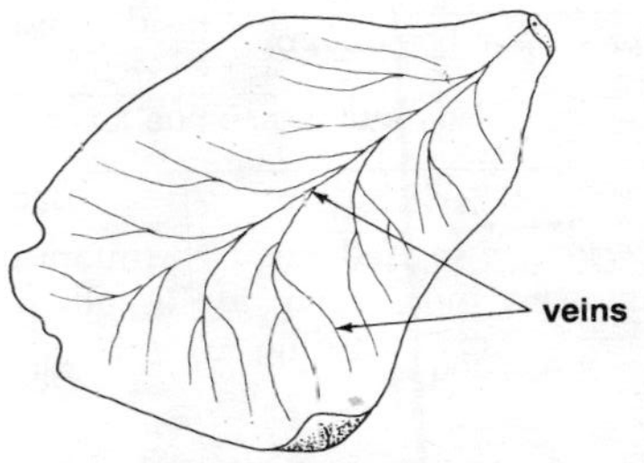

FIGURE 2. Lettuce leaf (veins)

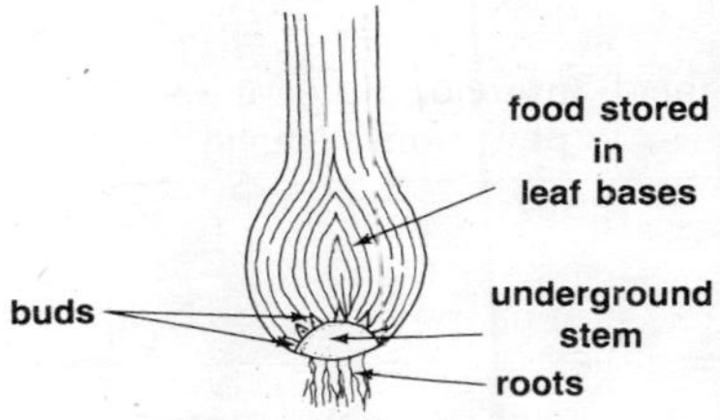

FIGURE 3. Onion underground stem (food stored in leaf bases, underground stem, roots, buds)

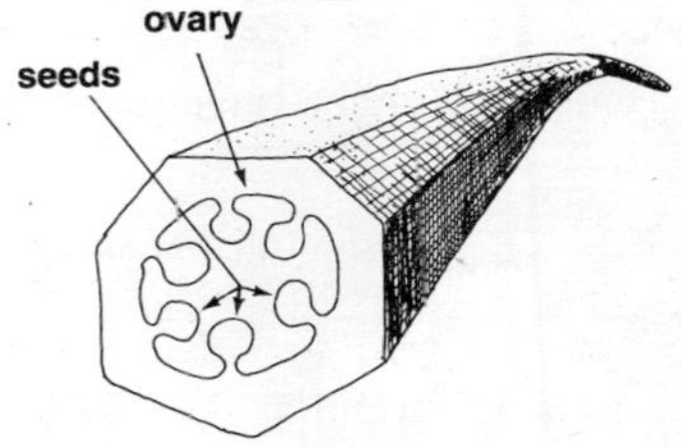

FIGURE 4. Okra fruit (ovary, seeds)

PROCEDURE
Part A. Examining Known Food Plant Parts
Flower The older the broccoli the better.
1. Examine the yellow tip ends of broccoli. These are tiny flowers.
2. Remove two or three of these flowers and place them on a glass slide.
3. Examine the broccoli flowers under a hand lens or stereomicroscope. Use a dissecting needle to spread and separate the flower parts.
4. Using Figure 1, identify the parts of the flower that you can see.

Leaf
5. Examine the flat green part of a piece of lettuce. This is a leaf.
6. Use Figure 2 to identify the veins on the lettuce leaf.

Underground Stem
7. Examine the lower, white part of an onion. This is a bulb.
8. Using a scalpel or a knife, make a section down through the bulb as shown in Figure 3. Notice how the bases of the leaves are thicker than higher up. Also notice how each leaf surrounds and is attached to the small underground stem. CAUTION: Take care when using a scalpel.
9. Peel away some of the layers of leaf bases that surround the bulb and look for the small buds on the surface of the stem inside each leaf base. Using Figure 3, identify the parts of a bulb.

Root
10. Examine the orange underground part of a carrot. This is a root.
11. Notice how the leaf stalks are attached to the top of the root.
12. Look at the small hairlike roots that come from the sides of the carrot.
13. Notice that the root does not have buds or leaves on its surface as in the underground stem of the bulb.

Fruit
14. Examine the green part of okra. This is a fruit. Use fresh or frozen.
15. Using a scalpel, make a cross section through the okra fruit as shown in Figure 4.
16. Notice the small seeds attached to the center part of the fruit.

194

Name _____ Class _____ Period _____

Seed Use dried or frozen.
17. Examine a lima bean. This is a seed.
18. Remove the outer coat and with your fingernails split the bean into its two halves.
19. Using Figure 5, identify the parts of a seed.

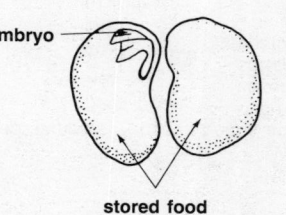

embryo
stored food

FIGURE 5. Lima bean seed

Part B. Identifying the Parts of Some Common Food Plants

1. Obtain a brussel sprout, caper, cherry pepper, pea, sweet potato, shallot, and cucumber.
2. Examine each food item.
3. From your observations in Part A, identify the plant parts of each of these foods. Are the parts we eat leaves, fruits, flowers, stems, seeds, or roots?
4. Record your results in Table 1.

Table 1. Identifying the Parts of Plants That We Eat

Food	Plant Part We Eat	Evidence That Tells You Which Plant Part We Eat
Pea	Seed	The inside contained an embryo.
Brussel Sprout	Leaf	Veins could be seen on the flat green parts.
Shallot	Leaf*	Each leaf base surrounds an underground stem and a bud.
Caper	Flower	Stamens and a pistal are inside petals.
Sweet Potato	Root	Is an underground part and does not have leaves or buds.
Cherry Pepper	Fruit	Seeds are attached to the inside.
Cucumber	Fruit	Seeds are attached to the inside.

If time and materials are available, other food plants you may wish students to examine are the stem tuber of the white potato (point out the leaf scars and buds that are known as "eyes"), the fruit of the pea or bean called the pod, the flower head of the artichoke (a member of the daisy family), the leaf stalk of the celery (point out the stem at the base of the celery bunch), and the seeds of the peanut plant (the tiny embryo can be seen still attached to the halves of roasted peanuts.

QUESTIONS

1. For each of the foods below, describe how you were able to identify what part of the plant it is.
    a. shallot __This is a bulb because it has thick leaf bases surrounding an underground stem.__
    b. cherry pepper __This is a fruit because seeds are found inside.__
    c. pea __This is a seed because it has a tiny embryo inside.__

*Students may report underground stem.

195

2. List two ways that an underground stem differs from a root. __a stem has buds or leaves on its surface and a root does not__
3. Using the information given to you in Figure 5:
    a. explain what the embryo may do with the stored food found in the seed. __use it for growth__
    b. explain why seeds are a good source of nutrients for humans. __they contain a high energy food source__
4. List several other examples (not studied in the experiment) commonly eaten by humans as food that are:
    a. roots __answers will vary__
    b. fruits __answers will vary__
    c. seeds __answers will vary__
5. Plants reproduce sexually and asexually. Identify the type of reproduction involved with the following plant parts:
    a. root __asexual__   b. flower __sexual__   c. bulb __asexual__
6. There are two types of cell reproduction that occur in plants, mitosis and meiosis. Name the type of cell division that occurs in each of the following plant parts:
    a. root __mitosis__   b. flower __meiosis__   c. bulb __mitosis__
7. There are two different types of cells in plants that are involved in reproduction: body cells and sex cells. Name the cell type involved in reproduction in each of the following plant parts:
    a. root __body cells__   b. flower __sex cells__   c. bulb __body cells__
8. A classmate tells you that tomatoes, pickles, and squash are fruits. Your classmate is correct. What feature should you look for to identify a plant part as a fruit? __seeds which are attached to the inside of the plant part__
9. What plant part will form fruit and seeds? __the ovary and ovules in a flower__
10. What is the main function of a flower's stamen and pistil? __to form egg and sperm cells needed for sexual reproduction__
11. Leaves and stems are plant parts that can also be involved in reproduction.
    a. Are these plant parts involved in sexual or asexual reproduction. __asexual__
    b. How? __No egg or sperm cells are formed by these parts. They are like roots in that they reproduce by mitosis using body cells.__

Name _____ Class _____ Period _____

Before doing this lab, have students read section 24:1 in the text.

24-1 How Do Some Animals Reproduce Asexually?

Animals, like plants, can reproduce both sexually and asexually. Two parents are needed for sexual reproduction. Only one parent is needed for asexual reproduction. A hydra can reproduce asexually by budding. Planarians are flatworms that can reproduce asexually by regeneration. Parts that are cut or broken off are replaced by new growth. When an animal is cut in half and the missing parts of each half grow back, the animal has reproduced by regeneration.

Asexual reproduction also occurs in the water flea. This is a small, shrimplike animal that lives in ponds or streams. A water flea produces eggs inside a transparent pouch. This is the brood pouch. Many eggs form at a time within this pouch. These eggs can develop into young water fleas without fertilization by sperm. When the young leave the body of the water flea, new eggs are produced.

EXPLORATION

OBJECTIVES

In this exercise, you will:
a. observe a planarian as it regenerates.
b. count the number of young produced by a hydra, planarian, and water flea.
c. compare asexual reproduction in a hydra, planarian, and a water flea.

Process Skills: observing, measuring in SI, experimenting, using a microscope

KEYWORDS

Define the following keywords:

asexual reproduction _reproduction by one parent, no sex cells, mitosis_

brood pouch _body part of a water flea that holds eggs_

budding _asexual reproduction in which a part of the parent body separates and forms a new organism_

egg _a female reproductive cell_

regeneration _asexual reproduction in which broken pieces of the parent body form new organisms_

MATERIALS

dropper	light microscope	planarian culture
glass slide	hand lens	2 small jars
paper towel	water flea culture	label
scalpel	*Euglena* culture	
metric ruler	1 L aged tap water	*Daphnia* can be bought from a pet store.

See 20T for preparation of cultures. Allow 1 L tap water to stand overnight.

197

PROCEDURE

Part A. Budding in Hydra

1. Look at Figure 1 of the budding hydra.
2. Count how many offspring buds of hydra grow from the parent in six days.

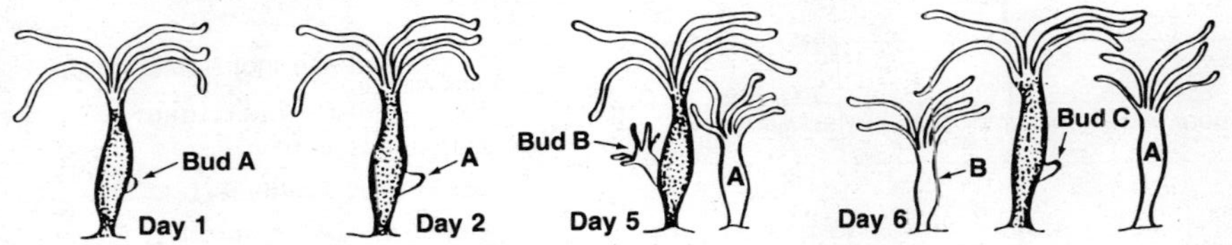

FIGURE 1. Budding hydra

Part B. Measuring Planarians

1. Look at Figure 2 of regenerating planarians.
2. Measure the length of each planarian.
3. Record your results in Table 1.

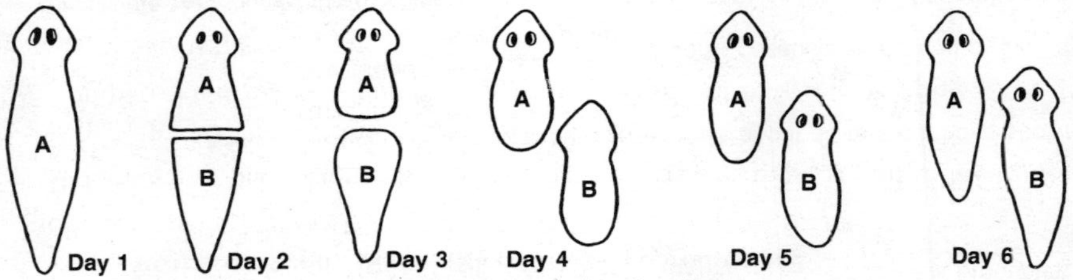

FIGURE 2. Regenerating planarians

The size of the planarian has been exaggerated for clarity.

Table 1. Measuring Planarians

Part	Length in millimeters on
	Day 1	Day 2	Day 3	Day 4	Day 5	Day 6
A	30 mm	14 mm	13 mm	16 mm	17.5 mm	22 mm
B	✕	16 mm	15 mm	17 mm	18 mm	23 mm

Part C. How Does a Planarian Reproduce?

1. Remove a planarian from the culture dish and place it on a glass slide with a drop of water.
2. Examine the planarian with a hand lens. Find the two eyespots on the head part.
3. Measure the total length in millimeters of the resting planarian. Figure 3 shows how. Record this length and make a drawing in Table 2.
4. Remove some of the water with a paper towel. Less water will cause the planarian to stretch out.
5. When the planarian is fully extended, cut it in half with a scalpel. **CAUTION:** *Use extreme care with the scalpel.*

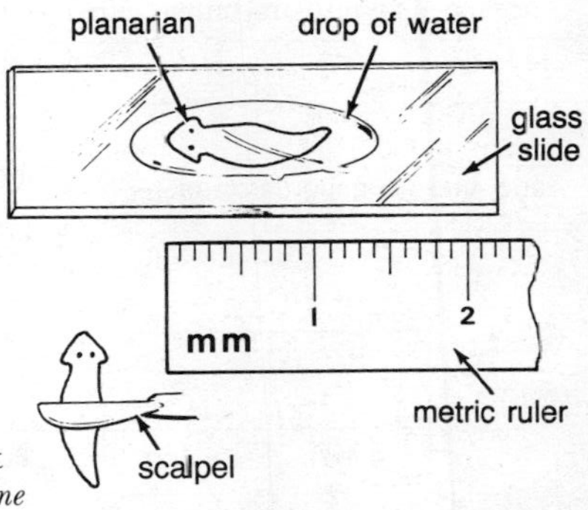

FIGURE 3. Measuring and cutting a planarian

198

6. Look at each cut area of the head and the body halves of the flatworm. Record the lengths and make drawings of the two halves in Table 2.
7. Use a dropper to move the halves of the planarian to a small jar filled with aged tap water. Put a label with your name on the jar.
8. Place the jar on a shelf in a cool, dark area of the classroom.
9. Add oxygen to the jar each day by squeezing air from a dropper through the water. The bubbles will contain oxygen.
10. Look at the planarian halves each day for the next five class days.
11. Record the changes in size and shape on Table 2.
12. Compare your results with those of your classmates.

Table 2. Regeneration in Planarians

Before Cutting		Day 1	Day 2	Day 3	Day 4	Day 5	Day 6
	Head Part						
	Length in mm	6	6	7	7	8	8
11 mm	Body Part						
	Length in mm	5	5	6	6	7	7

Part D. How Many Young Can a Water Flea Produce?

1. Study Figure 4 of the water flea. Locate the brood pouch and eggs.
2. Use a dropper to remove a water flea from the culture dish and place it on a glass slide.
3. Remove some of the water with a paper towel as shown, so the water flea cannot move very fast.
4. Examine the slide on low power of the microscope. Look at the brood pouch and eggs.
5. Fill a small jar with aged tap water and place four water fleas into the water.
6. Fill a dropper with *Euglena* culture and add this to the jar.
7. Repeat step 6 three more times. Add a dropper of *Euglena* culture to the jar each day for the next four days.

The *Euglena* culture serves as a food source for *Daphnia*.

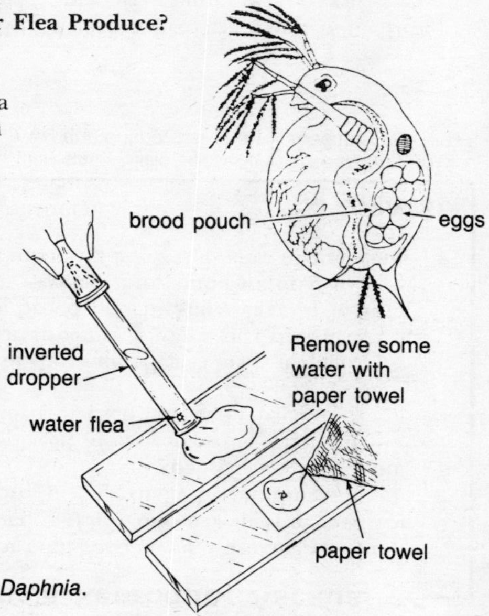

FIGURE 4. Water flea

8. Count the number of water fleas in the jar for five days. Record the data in Table 3.

Table 3. Water Flea Offspring

	Day 1	Day 2	Day 3	Day 4	Day 5
Number of water fleas	4	6	8	9	13

QUESTIONS

1. How many new hydra were reproduced in Part A. three hydra
2. Is reproduction by budding sexual or asexual? Why? asexual, because only one parent is involved
3. How does a planarian reproduce? by regeneration
4. How many new planarians were produced from the planarian that you cut in half? two planarians
5. What changes did you notice in the area where you cut the planarian? The cut area became darker in color shortly after it was cut.
6. What changes did you notice in the head part of the planarian over five days? The cut end of the head part began to form a new body.
7. What changes did you notice in the body part of the planarian over five days? The cut end of the body part began to form a new head.
8. Why can regeneration also be called asexual reproduction? because it produces offspring from one parent
9. Did your new planarians in Part C regenerate as fast as those shown in the figure in Part B? Answers will vary, but they should regenerate about as fast as those shown.
10. Compare Figure 1 of a budding hydra with Figure 2 of a regenerating planarian. Which reproduces faster? Explain. Hydra, because at the end of six days one hydra reproduced three offspring, whereas the planarian produced only two offspring during the same time.
11. What did the water flea eggs look like when they were in the brood pouch? They were tiny ovals and some of them had eyes like the parent.
12. What happens to the female water flea after a batch of young leave the brood pouch? The female produces more eggs.
13. How many young can a water flea produce? Answers will vary, but usually between 12-25 over a four-day period.

Name _____ Class _____ Period _____

Before doing this lab, have students read section 24:2 in the text.

24-2 How Do Internal and External Reproduction Compare?

Some animals that live in water gather in large groups during their breeding season. The females release their eggs into the water. The males of the species then release their sperm over the eggs. The sperm cells fertilize the eggs in the water. This is called external fertilization because it takes place outside the animal's body.

In other animals, the male releases sperm directly into the female's body where fertilization takes place. This is internal fertilization.

Process Skills: interpreting data, using numbers, predicting, inferring

INTERPRETATION

OBJECTIVES
In this exercise, you will:
a. interpret and chart data collected by a student.
b. determine if there is a relationship between the number of young produced, the type of fertilization, and the type of place where an animal lives.
c. determine how parental care helps the young.

KEYWORDS
Define the following keywords:

external fertilization ___ the joining of egg and sperm outside the body

internal fertilization ___ the joining of egg and sperm inside the body

parental care ___ behavior in adults of giving care to eggs and offspring

MATERIALS
pencil

This exercise helps students learn to organize data into a more meaningful format. At the same time, it gives students information about pond and land animals and fertilization.

PROCEDURE
Part A. Observing Data On Animal Reproduction
Study the data in Table 1 below that was collected by a student.

Table 1.

Animal	Where Animal Lives	Number Of Young Produced	Number Of Young That Survive	Parental Care Yes	Parental Care No
Horse	land	1	1	X	
Spider	land	100	10	X	
Oyster	water	750,000	100		X
Jellyfish	water	1,000	10		X
Tiger Salamander	land	100	8		X
Starfish	water	1,500	5		X
Garter Snake	land	35	3	X	
Falcon	land	5	2	X	
Humpback Whale	water	1	1	X	
Clam	water	1,000,000	30		X

201

Part B. Using the Table
Use the data from Table 1 in Part A to complete Table 2.
1. Look at Table 1 to find where each animal lives.
2. Look at the number of young produced by each animal in Table 1 and decide if fertilization is internal or external. Hint: If the young produced are greater than 150, then fertilization is usually external.

Table 2.

	Lives in Water		Lives on land	
	Fertilization		Fertilization	
Internal	External	Internal	External
Humpback Whale	Oyster	Horse	
	Jellyfish	Spider	
	Starfish	Tiger Salamander	
	Clam	Garter Snake	
		Falcon	

Part C. Applying The Use of Data
1. Look at Table 3. It gives data on reproduction for another list of animals.
2. Using what you have learned from Parts A and B, complete the information in Table 3 under the columns C and D.

Table 3.

A Animal	B Number Of Young Produced	C Fertilization Internal	C Fertilization External	D Where Animal Lives Land	D Where Animal Lives Water
Bluegill	15,000		X		X
Chameleon	8	X		X	
Leech	60	X			X
Lobster	8,500		X		X
Rainbow Trout	4,000		X		X
Crayfish	40	X			X
Lizard	10	X		X	
Cow	1	X		X	
Blue Crab	950,000				X
Toad	10,000		X	X	

202

Name _____ Class _____ Period _____

Part D. Using Data to Draw Conclusions

Using what you have learned from Parts A, B, and C, fill in Table 4 to show if an animal gives parental care or not.

Table 4.

Animal	Number Of Young Produced	Number Of Young That Survived	Parental Care Yes	Parental Care No
Opossum	10	8	X	
Grasshopper	300	14		X
Carp	12,000	22		X
Deer	1	1	X	
Pig	8	6	X	

QUESTIONS

1. What kind of fertilization do animals that live in water usually show? _external_
2. Where do most of the animals live that show internal fertilization? _on land_
3. What is an animal that has internal fertilization and lives in water? _the humpback whale_
4. Do animals that fertilize their eggs internally produce more or less young than animals that fertilize their eggs externally? _less young_
5. Which eggs do you think have a better chance of being fertilized, those that are fertilized internally or those fertilized externally? _those that are fertilized internally_
6. Freshwater clams produce 20 young a breeding season; some freshwater jellyfish produce 500 young. Which animal do you think has internal fertilization? Why? _clams—because they produce less than 150 young_
7. Do most of the animals that provide parental care have many or few young? _few young_
8. Where do most animals live that give no parental care for their young? _in water_
9. From Tables 1, 3, and 4 how can you tell which young animals have a better chance of surviving? _Those animals with a higher number of young that survive give more parental care._
10. In Table 3 of Part C, which of the animals do not fit the pattern between numbers of offspring and kind of fertilization as seen in Part A? Explain. _The blue crab and the lobster produce more than 150 young and have internal fertilization._

203

---

11. Which of the animals do you think have the best chance of survival? Why? _the young of the horse, cow, and deer because only one offspring is produced and there is parental care_
12. The bluegill sunfish cares for its eggs and young while the rainbow trout does not. Which would probably have a greater number of young survive to reproduce? Explain. _the bluegill sunfish because the parent protects the young_
13. Write a short paragraph that explains each of the following points:
    a. The type of fertilization of an animal is different between animals that live on land and those that live in water. _Most animals that live on land fertilize their eggs internally while most animals that live in water fertilize their eggs externally._
    b. The number of eggs produced by an animal is different between animals that have internal fertilization and those that have external fertilization. _Animals that fertilize eggs internally produce less eggs than those animals that fertilize their eggs externally. External fertilization is mostly by chance._
    c. The number of young that survive differs between animals that have internal fertilization and those that have external fertilization. _There is a higher number of survivors among animals that fertilize their eggs internally than those that fertilize their eggs externally._
    d. The number of young that survive differs between animals that give parental care and those that do not. _There are more young that survive when parental care is given to the young. Parents that do not give parental care to eggs or young produce more eggs._

204

Name _____ Class _____ Period _____
Before doing this lab, have students read sections 24:3 in the text.

24-3 What are the Stages in the Menstrual Cycle?

The human female reproductive system has several different roles to perform. It produces a new egg each month for fertilization. It must prepare the lining of the uterus for a new embryo if fertilization of an egg does occur. This is accomplished by a thickening of the uterus lining.

It must shed the egg and thickened uterus lining if fertilization does not occur. All of these events occur in a cyclic pattern each month in a sexually mature female.

INTERPRETATION

Process Skills: interpreting data, inferring, classifying, communicating

OBJECTIVES

In this exercise, you will:
a. review the organs that form the human female reproductive system.
b. prepare a calendar that shows the changes occurring during the human menstrual cycle if no fertilization occurs.
c. prepare a calendar that shows the changes occurring during the human menstrual cycle if fertilization occurs.

KEYWORDS

Define the following keywords:

fertilization ___ joining of egg and sperm cell

menstrual cycle ___ monthly changes that take place in female reproductive system

ovary ___ female sex organ that produces eggs

oviduct ___ tubelike organ that connect the ovaries to uterus

uterus ___ muscular organ in which fertilized egg develops

MATERIALS
scissors   You will need to photocopy Figures 2 and 4 for students.
tape   You may wish to have students work in groups.

PROCEDURE

Part A. Review of the Female Reproductive System

1. Use the following parts and their description for help in properly labeling Figure 1.
NOTE: The diagram is 1/3 natural size.
a. ovary—two are present, round in shape
b. egg—small cells present within ovary
c. uterus—large muscle, V-shaped, largest part of reproductive system
d. oviduct—thin tube connecting each ovary to uterus
e. uterus lining—inner wall or lining of uterus

205

FIGURE 1.

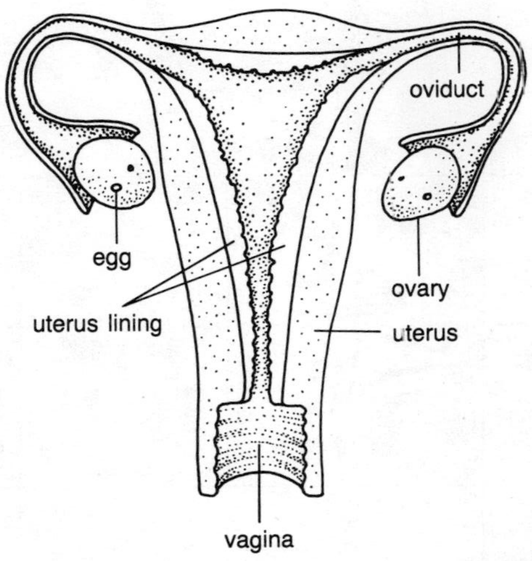

Part B. Changes in the Menstrual Cycle; No Fertilization of Egg

1. Obtain a copy of Figure 2 from your teacher.
2. Use scissors to cut out the square diagrams in Figure 2. These diagrams show the different stages that occur during the menstrual cycle if fertilization does not occur.
3. Look over the calendar marked Figure 3. It describes a series of events that take place in the female reproductive system if fertilization does not take place.
4. Match the diagrams that you cut out with the events being described in the calendar.
5. When all diagrams have been properly matched, tape them onto the calendar in their proper location to the right of the brackets describing the events.
Be sure students use Figure 2 models with Figure 3.

Part C. Changes in the Menstrual Cycle; Fertilized Does Occur

1. Obtain a copy of Figure 4 from your teacher.
2. Use scissors to cut out the square diagrams in Figure 4. These diagrams show the different stages that occur during the menstrual cycle if fertilization does occur.
3. Look over the calendar marked Figure 5. It describes a series of events that take place in the female reproductive system if fertilization does take place.
4. Match the diagrams that you cut out with the events being described in the calendar.
5. When all diagrams have been properly matched, tape them onto the calendar in their proper location to the right of the brackets describing the events.
Be sure students use Figure 4 models with Figure 5.

Advise students that Figures 2 and 4 are not drawn in their proper sequence. Students will have to determine the proper order of events based on information provided on the calendars. You may wish to check calendars before students tape models in place to verify that models are positioned in correct sequence.

206

385

Name _____ Class _____ Period _____

Provide copies of this page for students.

FIGURE 2. No fertilization of egg

FIGURE 4. Fertilization of egg

Name _____ Class _____ Period _____

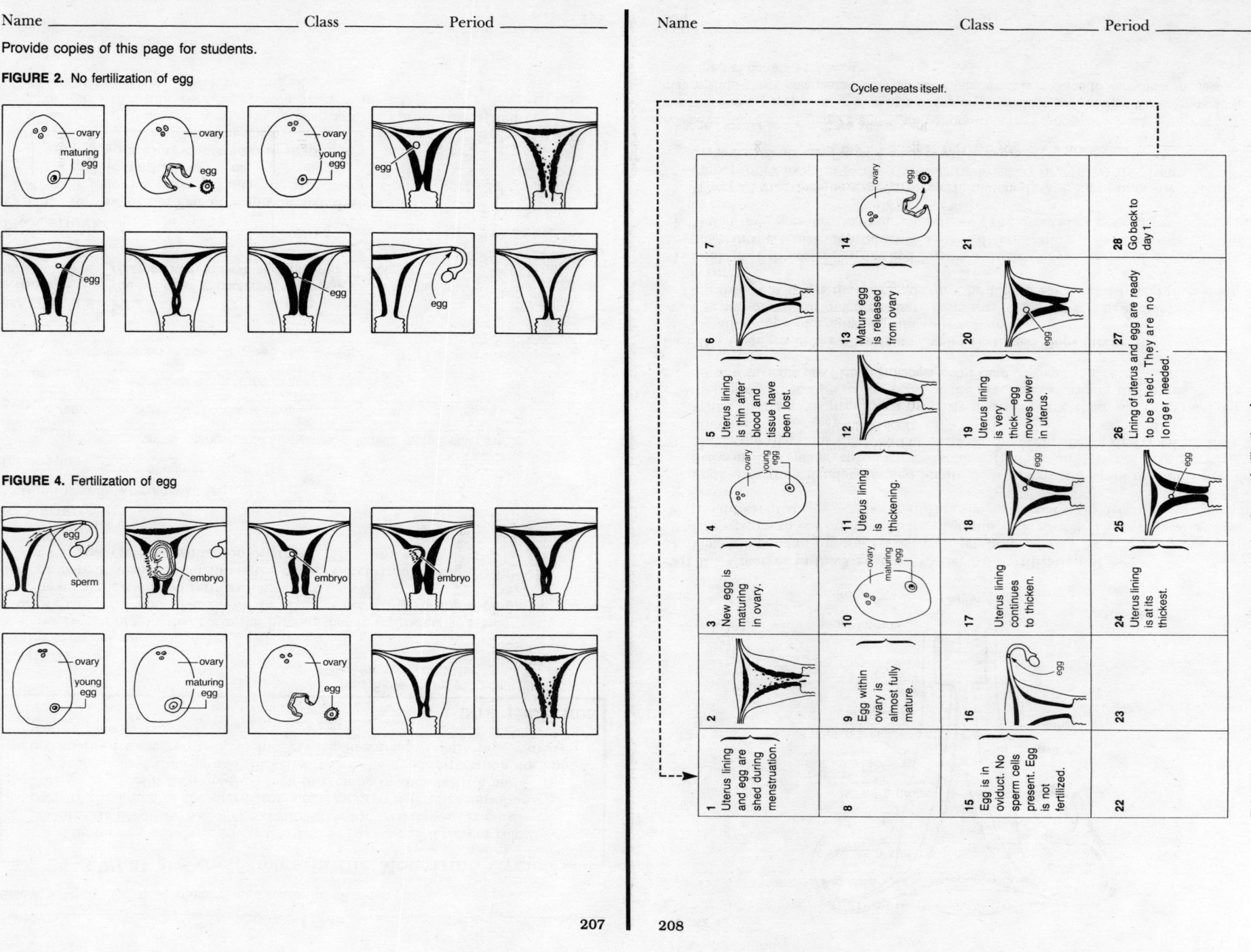

1 Uterus lining and egg are shed during menstruation.	2	3 New egg is maturing in ovary.	4	5 Uterus lining is thin after blood and tissue have been lost.	6	7
8 Egg within ovary is almost fully mature.	9	10	11 Uterus lining is thickening.	12	13 Mature egg is released from ovary.	14
15 Egg is in oviduct. No sperm cells present. Egg is not fertilized.	16	17 Uterus lining continues to thicken.	18	19 Uterus lining is very thick—egg moves lower in uterus.	20	21
22 Uterus lining is at its thickest.	23	24	25 Lining of uterus and egg are ready to be shed. They are no longer needed.	26	27 Go back to day 1.	28

Cycle repeats itself.

FIGURE 3. Day by day changes in the menstrual cycle—no fertilization of egg

Name _____ Class _____ Period _____

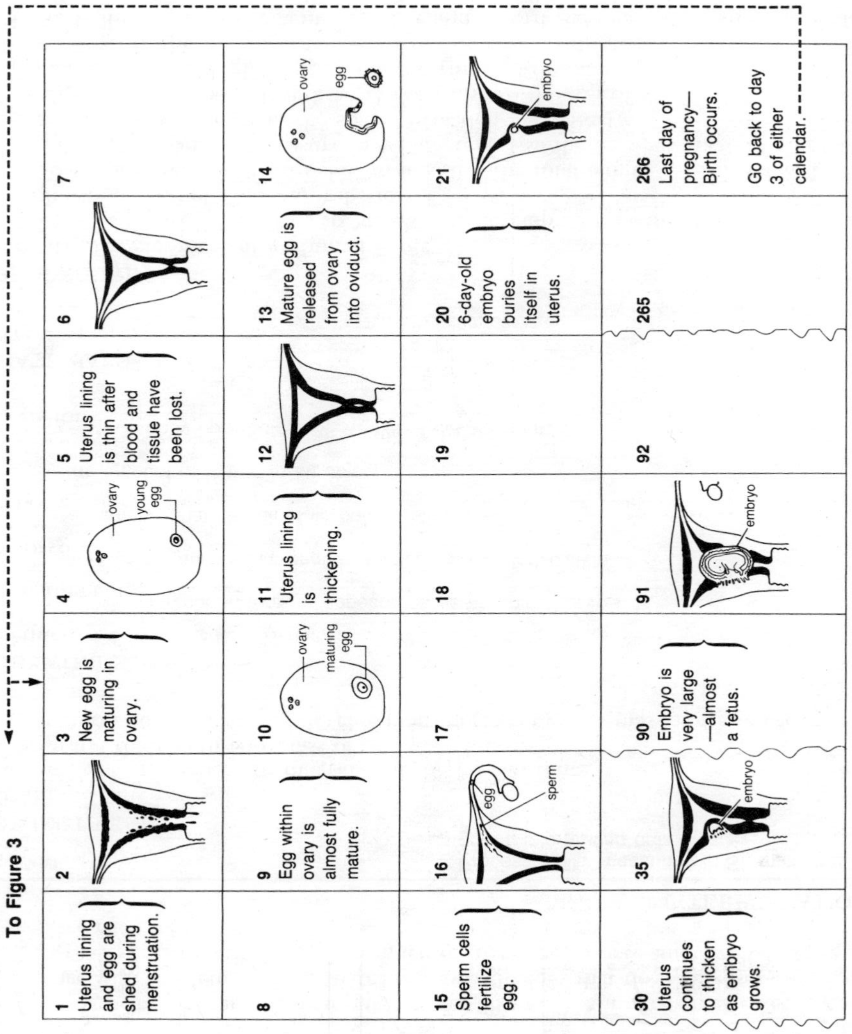

QUESTIONS

1. Describe the role or function of the following:
   a. ovary   produces egg cells
   b. uterus lining   serves as a surface upon which young embryo can attach
   c. uterus muscle   serves as a protective chamber for developing embryo; aids in pushing out fetus at birth
   d. oviduct   if sperm are present, allows egg to move from ovary to uterus, site of fertilization of egg

2. An average menstrual cycle with no fertilization takes how many days?   28

3. Describe the changes that take place during the menstrual cycle from day 1-4 to the following:
   a. unfertilized egg   lost from body through vagina
   b. uterus lining   lost from body through vagina
   c. egg in ovary   begins to mature

4. Describe the changes that take place during the menstrual cycle from day 5-13 to the following:
   a. uterus lining   thickens
   b. egg in ovary   continues to mature

5. Describe what happens to the egg during the menstrual cycle on day 14. it is released from ovary into oviduct

6. Describe the changes that take place to an egg
   a. from day 15-28 if no fertilization occurs.   moves from oviduct to uterus
   b. from day 15-21 if fertilization does occur.   moves from oviduct to lining of uterus where it attaches
   c. from day 21-266 if fertilization does occur.   changes from a fertilized egg to embryo to fetus

7. Explain why the female
   a. needs a thick uterus lining if fertilization does occur.   the embryo will attach to the lining and remain there during entire pregnancy
   b. no longer needs a thick lining if fertilization does not occur.   no need for a site for embryo attachment

210

FIGURE 5. Day by day changes in the menstrual cycle—fertilization of egg

7 Uterus lining is thin after blood and tissue have been lost.	**14** Mature egg is released from ovary into oviduct.	**21** 6-day-old embryo buries itself in uterus.	**266** Last day of pregnancy—Birth occurs. Go back to day 3 of either calendar.
6	**13**	**20**	**265**
5	**12** Uterus lining is thickening.	**19**	**92**
4 New egg is maturing in ovary.	**11**	**18**	**91** embryo
3	**10** maturing egg, ovary	**17**	**90** Embryo is very large—almost a fetus. embryo
2 young egg, ovary	**9** Egg within ovary is almost fully mature.	**16** Sperm cells fertilize egg. egg, sperm	**35** embryo
1 Uterus lining and egg are shed during menstruation.	**8**	**15**	**30** Uterus continues to thicken as embryo grows.

To Figure 3

209

387

Name _____ Class _____ Period _____

Before doing this lab, have students read section 25:1 in the text.

25–1 How Does a Human Fetus Change During Development?

INTERPRETATION

Development in a human takes about 38 weeks. Many changes take place with the fetus during that time. Two changes that do occur are increases in size and mass. How much of a change in mass and size takes place each week?

Process Skills: measuring in SI, making a line graph, interpreting data

OBJECTIVES
In this activity, you will:
a. measure the length of diagrams of the human fetus.
b. graph the length and mass of a human fetus.
c. determine when during development most changes in mass and size occur.

KEYWORDS
Define the following keywords:

development _change in form or appearance as an animal grows_

embryo _a young animal that begins developing after fertilization_

fetus _an embryo that has all of its body systems_

mass _the amount of matter in an object_

premature _a baby born before 38 weeks of development_

MATERIALS
metric ruler

PROCEDURE
Part A. Development of a Human Fetus

1. Look at Figure 1. It shows six stages of a developing human fetus. They are shown at 40% of their natural size.
2. Follow the steps outlined below to measure the total length of each stage. Use the metric ruler and measure in millimeters. Use the 38-week stage as a guide and record your data in the spaces provided in Table 1.
   a. Measure the body length from the rump to the top of the head.
   b. Measure the thigh length from the rump to the knee.
   c. Measure the length of the leg from the knee to the foot.
4. Add all three measurements together and record the total in the space provided in Table 1.
5. Multiply the total by 2.5 to give a figure that is close to the actual size of the fetus at each stage.
6. Record this actual size in the table.

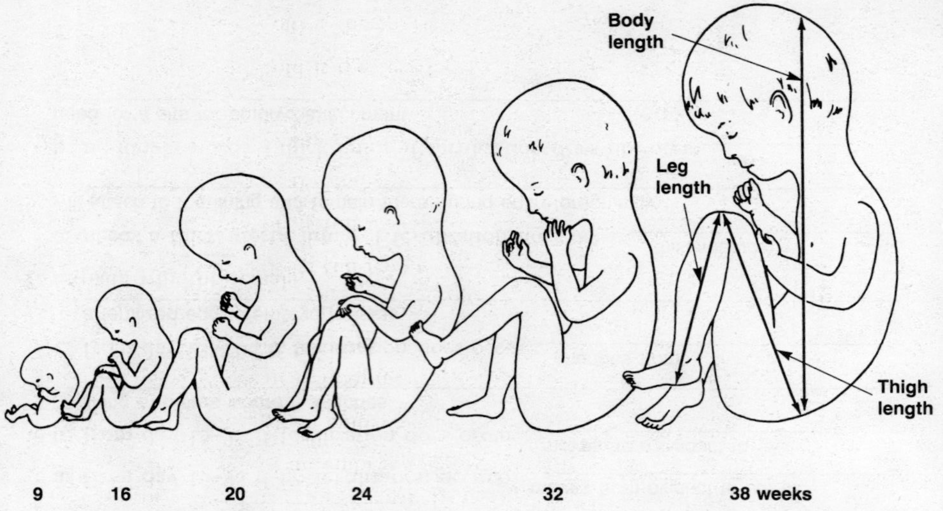

388

FIGURE 1. Stages in the development of a human fetus

(9, 16, 20, 24, 32, 38 weeks)

Body length, Thigh length, Leg length

Table 1. Lengths of a Developing Fetus

Age of fetus in weeks	Body length	+	Thigh length	+	Leg length	=	Total length	× 2.5 =	Actual length
2									2 mm
9	13		4		4		21		52.5 mm
16	28		12		10		50		125 mm
20	48		19		16		83		207.5 mm
24	57		22		22		101		252.5 mm
32	65		28		25		118		295 mm
38	78		37		33		148		370 mm

Part B. Plotting Length of a Developing Fetus

1. Plot the data from Table 1 onto the graph in Figure 2.
2. Plot the actual fetal length against the age of the fetus.

211

Name _____ Class _____ Period _____

FIGURE 2. Length of a developing fetus

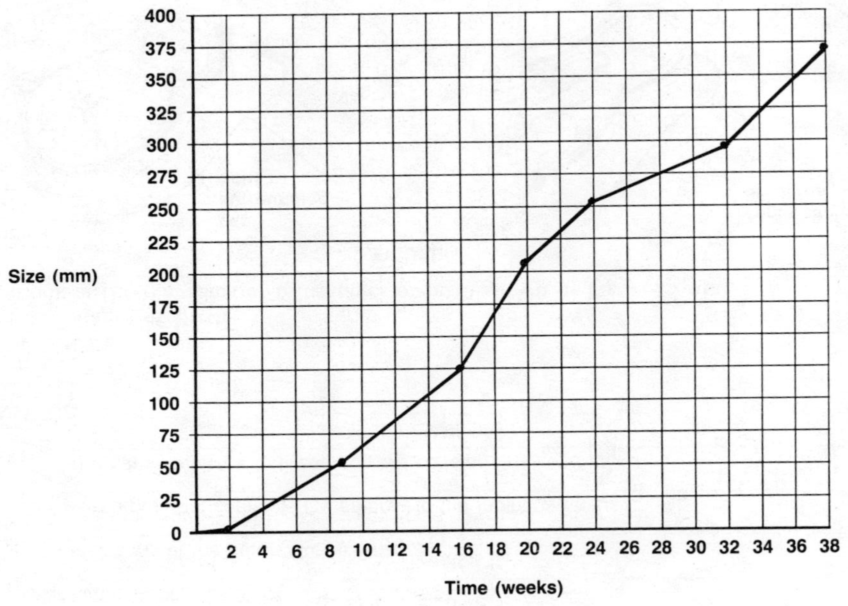

Part C. Plotting Mass of a Developing Fetus
1. Look at the data supplied in Table 2.
2. Plot the data of the developing fetus from Table 2 onto the graph in Figure 3.
3. Plot the mass of the fetus against the age of the fetus.

Table 2. Mass of a Developing Fetus

Time (weeks)	Mass (grams)	Time (weeks)	Mass (grams)
4	0.5	24	650
8	1	28	1100
12	15	32	1700
16	100	36	2400
20	300	38	3300

213

FIGURE 3. Mass of a developing fetus

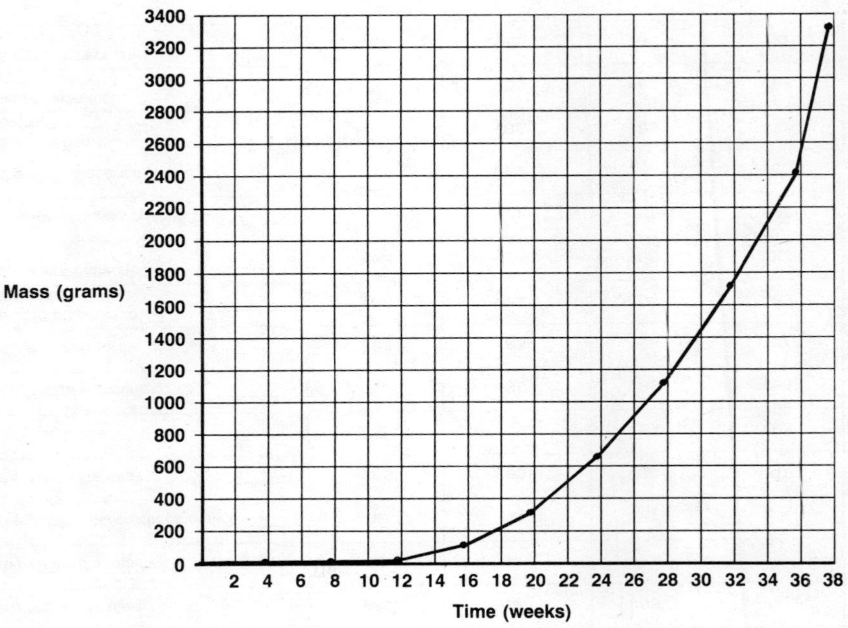

QUESTIONS
1. During what weeks of development is the human baby called an embryo? _____ It has grown to 45 mm. from 0 to 8 weeks
2. What is the length of an embryo during this time? _____
3. How much mass does an embryo gain during this time? 1 gram
4. During what weeks of development is the human baby called a fetus? _____ at the start of the third month or the 9th week
5. Look at Figures 2 and 3 for the halfway point in development at week 19.
   a. Is the fetus half of its full length at this time? yes
   b. Is the fetus half of its full mass at this time? no
6. a. At what week does the fetus reach half its full length? week 19
   b. At what week does the fetus reach half its full mass? week 31
7. If a premature baby is born with a mass of
   a. 2200 grams, how old is the fetus? 35 weeks old
   b. 1800 grams, how old is the fetus? 32 weeks old (Students may say 33 weeks.)

214

Name _____ Class _____ Period _____

Before doing this lab, have students read section 25:1 in the text.

25-2 What Changes Occur During Birth?

INTERPRETATION

A human baby develops for about 38 weeks inside the mother. Then labor begins and the baby is born. What changes take place during and after birth? Why must a caesarean delivery may be needed. c. compare a delivery through the birth canal with a caesarean delivery to help in delivery?

Process Skills: interpreting scientific illustrations, interpreting data, scientific terminology, summarizing

OBJECTIVES
In this activity, you will:
a. compare the changes that occur during birth.
b. learn why a caesarean delivery may be needed.
c. compare a delivery through the birth canal with a caesarean delivery.

KEYWORDS
Define the following keywords:

caesarean __the birth of a baby in which the uterus must be cut open__

contractions __muscles of the uterus shortening during birth__

fetus __an embryo that has all of its body systems__

labor __contractions of the uterus during birth__

placenta __an organ that connects the embryo to the mother's uterus__

uterus __an organ in which a fertilized egg will develop__

MATERIALS
metric ruler

PROCEDURE
Part A. Stages of Birth
1. Look at the diagrams of four stages of birth shown in Figures 1 and 2.

FIGURE 1.

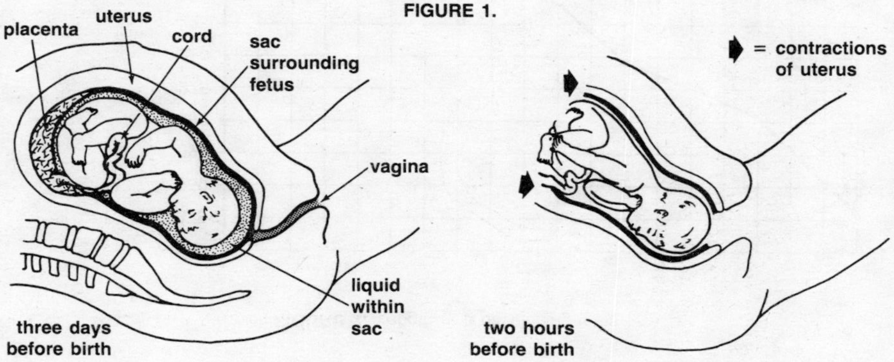

three days before birth

two hours before birth

FIGURE 2.

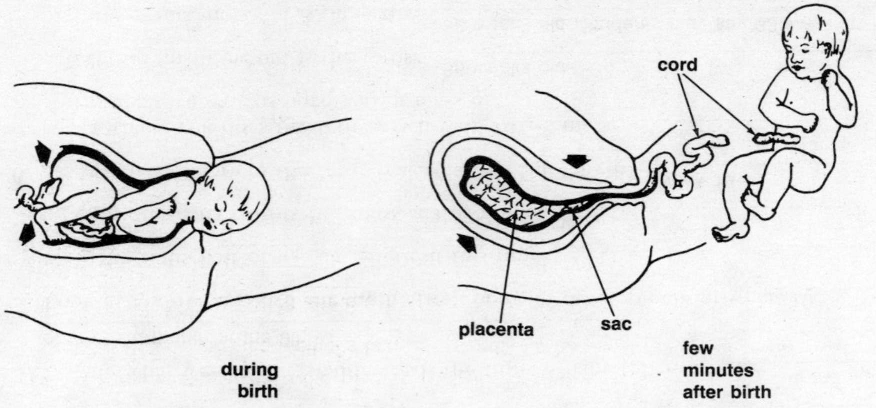

during birth

few minutes after birth

2. Answer *yes* or *no* to each of the following questions in Table 1.

Table 1. Stages During Birth

	Three days before birth	Two hours before birth	During birth	Few minutes after birth
Is baby inside the uterus?	yes	yes	yes	no
Is baby inside the vagina?	no	no	yes	no
Is baby outside the mother's body?	no	no	no	yes
Is baby inside the sac?	yes	yes	yes	no
Has the sac broken?	no	yes	yes	yes
Are contractions occurring?	no	yes	yes	yes
Is baby attached to the cord?	yes	yes	yes	yes
Is the cord attached to the placenta?	yes	yes	yes	no
Is the placenta attached to the uterus?	yes	yes	no	no
Is the placenta being pushed out?	no	no	no	yes
Has the vagina opened?	no	yes	yes	yes
Is baby attached to the mother?	yes	yes	yes	no
Has liquid been lost from the sac?	no	no	yes	yes
Is baby still dependent on the mother?	yes	yes	yes	no

Name _____ Class _____ Period _____

Part B. What Is a Caesarean Birth?

1. Look at the diagram in Figure 3 that shows the outline of the pelvis and the head of a fetus just before the time of birth.
2. Note carefully that the head must be able to pass through the opening in the pelvis during birth.
3. Measure line a. This represents the width of the opening in the pelvis.
4. Measure line b. This represents the width of the head of the fetus.
5. Record your data here:
   a. width of pelvis opening   31 mm
   b. width of fetus head   34 mm
6. Notice that this fetus would not be able to pass through this pelvis opening.
7. A caesarean operation must be done to deliver the baby.
8. Look at how a caesarean birth is done in Figure 4. This is usually done before the mother goes into labor.
9. To compare a birth canal delivery with a caesarean delivery, answer the questions in Table 2.

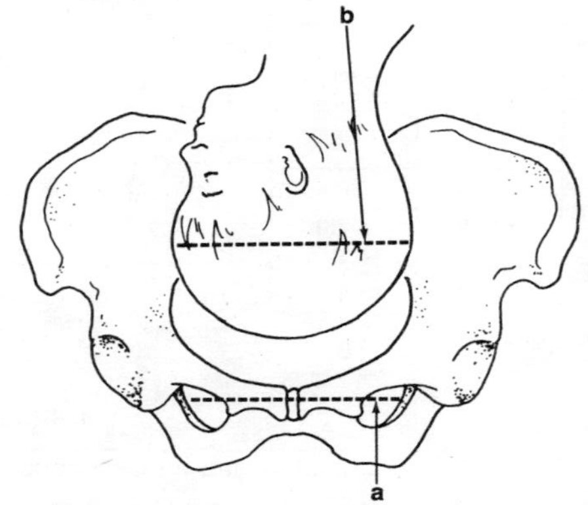

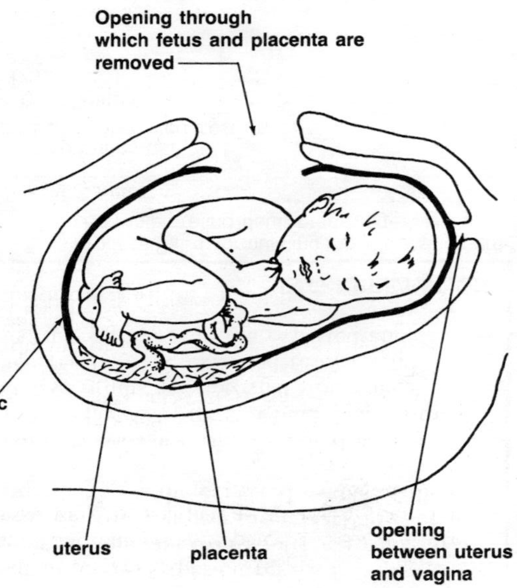

FIGURE 3. Sizes of pelvis and head of fetus

FIGURE 4. Caesarean birth

Table 2. Comparing a Caesarean Delivery With a Birth Canal Delivery

Trait	Birth canal	Caesarean
Does the fetus pass through opening in the pelvis?	yes	no
Does the fetus pass through the vagina?	yes	no
Does the placenta move through the vagina?	yes	no
Is the fetus lifted from the uterus?	no	yes
Is the uterus cut open?	no	yes
Is the sac cut open?	no	yes
Must the cord be cut to separate the fetus from the placenta?	yes	yes
Do contractions occur?	yes	no

QUESTIONS

1. What two body parts surround and protect the fetus as it develops? the mother's uterus and the fluid-filled sac
2. What is the job of the placenta? carries food and oxygen to embryo and carries waste from the embryo
3. What is the job of the cord? connects embryo to placenta
4. What is meant by the word *labor*? contractions of the muscles of the uterus during birth
5. The placenta is sometimes called the *afterbirth*. Why is this a good name for this part? The placenta is "born" or leaves the uterus after the baby is born.
6. List several changes that take place several hours before birth. (1) Contractions are occurring. (2) The vagina has opened. (3) The liquid is lost from the sac.
7. List several changes that take place a few minutes after birth. (1) The baby is out of the mother's body. (2) The baby is out of the sac. (3) The baby is not attached to the cord or the mother. (4) The placenta is not attached to the uterus.

Name _____ Class _____ Period _____

Before doing this lab, have students read section 26:1 in the text.

26-1 How Can the Genes of Offspring Be Predicted?

The Punnett square can be used to predict expected results from a genetic cross. In mice, black coat color is dominant over white coat color. If two heterozygous parents are crossed, we would expect three black to every one white offspring. The Punnett square shows the expected results of the $Bb \times Bb$ cross.

You know observed results do not always agree with expected results. Four offspring of the $Bb \times Bb$ parents may really all be white. That is, the observed results may all be white. These results are not what would be expected. Of what good are expected results? The expected results help us determine what the observed results will be. When do the expected and observed results agree? When are they the same?

INTERPRETATION

Process Skills: formulating models, using numbers, interpreting data, predicting

OBJECTIVES

In this exercise, you will:
a. set up a model to compare expected and observed results.
b. find when expected and observed results agree.
c. predict some genes in offspring of mice.

KEYWORDS

Define the following keywords:

expected results ___ results that can be predicted

observed results ___ traits or results that are actually seen

Punnett square ___ a way to show which genes will combine when an egg and a sperm join

pure dominant ___ a combination of two dominant genes for a trait

MATERIALS

2 pennies   masking tape

PROCEDURE

Part A. Calculating Expected Results

Assume that a female mouse has several litters of young in one year. She is heterozygous (*Bb*) for coat color and mates with a male that is also heterozygous (*Bb*) for coat color.

	B	b
B	BB black	Bb black
b	Bb black	bb white

FIGURE 1. Punnett square

219

Remind students that B = black; b = white.
You can predict what kind of offspring she will have by constructing a Punnett square as shown in Figure 1. Results from mating two mice can be shown by tossing and reading coins.

1. Place two coins in your cupped hands and shake the coins. Drop the coins on a desktop.
2. Examine the coins and determine whether you have two heads, two tails or a head and a tail.
3. Make a mark (/) in Table 1 under the correct combination of genes.
4. Repeat shaking and reading the coins for a total of 40 times. These 40 shakes will represent the combination of genes you might have observed in the offspring of several litters.
5. Count the marks for each gene combination and write the total observed in Table 1.
6. Calculate the expected number for each gene combination by using the Punnett square in Figure 1.
   a. First divide the number of *BB* squares in Figure 1 by 4; multiply that number (a percentage) by 40. Record this number as the expected number for the gene combination of *BB* genes in Table 1.
   b. Repeat these calculations for *Bb* and *bb* squares. Record your values for each gene combination in Table 1.
   c. Record the colors in the last row of Table 1.

Table 1. Results of Coin Tosses (Coat Color in Mice Bb × Bb)

Coin combinations	Head-Head	Head-Tail	Tail-Tail
Gene combinations	BB	Bb	bb
Observed results	︎ ﾊﾞ ///	ﾊﾞ ﾊﾞ ﾊﾞ ﾊﾞ /	ﾊﾞ ﾊﾞ /
Total observed in 40 tosses	8*	21*	11*
Total expected in 40 tosses	10	20	10
Coat color in mice	black	black	white

*Answers will vary.

Part B. Predicting Mouse Offspring

Suppose you mate a female mouse that is heterozygous (*Bb*) with a male that is pure recessive (*bb*). Predict what kind of offspring she will have by completing a Punnett square like that show in Figure 2.

1. Place tape on both sides of the two coins and mark both sides of one coin with a *b*. Mark one side of the other coin with a *b* and the other side with a *B*.

	B	b
b	Bb black	bb white
b	Bb black	bb white

FIGURE 2. Punnett square

220

Name _____ Class _____ Period _____

2. Place the two coins in your cupped hands and shake the coins. Drop the coins on a desktop.
3. Examine the coins and determine whether the offspring are heterozygous (**Bb**) or pure recessive (**bb**).
4. Make a mark (/) in Table 2 under the correct combination of genes.
5. Repeat shaking and reading the coins a total of 40 times. The 40 shakes will show the combination of genes you observed in the offspring of several litters.
6. Count the marks for each gene combination and write the total observed in Table 2.
7. Use the method of calculating expected numbers for each gene combination as in Part A step 6 by using the Punnett square in Figure 2.
8. Determine the coat color of the offspring by using the Punnett square in Figure 2.
2. Record the color in the proper part of Table 2.

Table 2. Results of Coin Tosses (Coat Color in Mice Bb × bb)

Coin combinations	B–b	b–b
Gene combinations	Bb	bb
Total observed in 40 tosses	19* ++++ ++++ ++++ ++++	21* ++++ ++++ ++++ ++++ +
Total expected in 40 tosses	20	20
Coat color in mice	black	white

*Answers will vary.

QUESTIONS

1. How often do you expect a tossed coin to land on heads? __one time out of two__
2. How often do you expect a tossed coin to land on tails? __one time out of two__
3. When two coins are tossed, how often do you expect to get the following combinations:
   a. heads/heads? __one out of four times__
   b. heads/tails? __two out of four times__
   c. tails/tails? __one out of four times__
4. What gene combinations and features (coat color) do the following coin tosses produce:

gene combinations	features
a. heads/heads? BB	black
b. heads/tails? Bb	black
c. tails/tails? bb	white

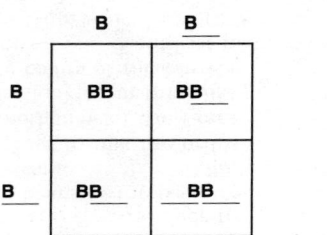

5. What gene combinations and features (coat color) do the following marked coin tosses produce:

gene combinations	features
a. B/b? Bb	black
b. b/b? bb	white

6. How close were your observed results to what you expected? __Answers will vary. They should be close.__
7. When will the observed results be close to the expected? __When the sample size or the number of coin flips is larger in number.__
8. What is the expected result of a cross **BB** × **bb**? Use the Punnett square.

	b	b
B	Bb	Bb
B	Bb	Bb

   BB 0
   Bb 4
   bb 0

9. Suppose you mated two other mice and expected one black mouse for each white mouse. How many of the following mice would you expect to observe out of 100 offspring:
   a. black? __50__   b. white? __50__
10. A scientist made a cross between two black mice. The cross was repeated between the same two mice several times. The data chart showed the color of all 42 offspring to be black. Use the Punnett squares below to show what you think the gene combinations were of both parents.

	B	B
B	BB	BB
B	BB	BB

	B	b
B	BB	Bb
B	BB	Bb

Name _____ Class _____ Period _____

Before doing this lab, have students read section 26:2 in the text.

26-2 What Is a Test Cross?

An organism with a pure dominant trait has two dominant genes for that trait. An organism with a heterozygous trait has only one dominant gene for the trait; the other gene is recessive. Most of the time, both organisms look alike. You cannot tell them apart by looking at them. How can you tell if an organism is pure dominant or heterozygous for a trait? How can you tell what genes the parent with the dominant trait has?

There is a way to tell if an organism is pure dominant or heterozygous for a trait. If the organism is mated to a pure recessive and features of the offspring are examined, you can tell the parent's genes. The cross to a pure organism to determine unknown genes is called a test cross.

Here an example: In dogs, rust red fur color (R) is dominant over brown fur color (r). A rust red dog can have either RR or Rr genes. To determine which genes a rust red dog has, it can be mated to a brown dog (rr) and the offspring can be examined. If all the offspring are red (Rr), the parent dog is pure dominant for fur color and has RR genes. If about half of the offspring are rust red and about half are brown, the unknown parent is heterozygous for fur color (Rr).

INTERPRETATION

Process Skills: formulating models, using numbers, interpreting data, predicting

OBJECTIVES
In this exercise, you will:
a. set up a model to show a test cross.
b. determine if a parent is pure dominant or heterozygous for a trait.

KEYWORDS
Define the following keywords:

cross _____ a mating

heterozygous _____ a trait controlled by a dominant gene and a recessive gene

offspring _____ young produced from reproduction

pure recessive _____ a trait controlled by two recessive genes

test cross _____ a cross between an unknown and an organism pure recessive for a trait

MATERIALS
marking pen
3 prepared bags of beans

Set up bags as follows: A—10 brown and 10 red beans. B—20 brown beans. C—20 red beans. Mark the bag with the color of parent and fill with beans prior to class. Use red kidney beans and brown pinto beans. Both can be obtained in a grocery store.

PROCEDURE
Part A. Test Cross 1

1. Obtain two bags marked "Parent A" and "Parent B" from your teacher. Do not look inside. These bags represent the parent dogs. The beans inside represent the genes.

223

---

2. "Parent A" is a red dog. It has one R gene and one gene that may be R or r. Therefore, it has genes R___. "___" means that the second gene of the dog is not known.

3. "Parent B" is a brown dog. It has rr genes.

4. Remove one bean from each bag without looking into the bags. Place the bean in front of the bag from which it was removed. Record the color of the beans in Table 1. The beans represent the genes of a puppy.

5. Return the beans to the bags from which they came.

6. Shake the bags to mix the beans. Remove another bean from each bag without looking in the bag. Record the colors of the beans in the table.

7. Repeat Steps 5 and 6 until 20 pairs have been recorded. Always return the beans to the bags from which they came.

8. Complete the table by writing in the genes the beans represent and the colors of the offspring. Total the offspring colors and complete the boxes below the table.

9. Make a Punnett square to show what both genes are for each parent. Then fill in the unknown gene of Parent A in the title of Table 1.

FIGURE 1.

Table 1. Test Cross 1: Parent A (R_r_) X Parent B (rr)   Sample data given.

Trial	Bean colors	Genes	Offspring color	Trial	Bean colors	Genes	Offspring color
1	brown, brown	rr	brown	11			
2	red, brown	Rr	red	12			
3	red, brown	Rr	red	13			
4	red, brown	Rr	red	14			
5	brown, brown	rr	brown	15			
6				16			
7				17			
8				18			
9				19			
10				20			

No RR will appear. About ½ the offspring will be brown and ½ will be red.
Total brown offspring = 10*   **Total red offspring** = 10*
*Answers will vary. In Part A, that about ½ the offspring are brown indicates that the unknown parent had to have the r gene or, in other words, had to be Rr.

224

Name _____ Class _____ Period _____

Part B. Test Cross 2

1. Obtain another bag, "Parent C," from your teacher. C is a red dog. You will determine if Parent C is **RR** or **Rr**. The genes are **R** ___.

2. Repeat Steps 4 to 8 of Part A using Parents C and B. This time you are crossing Parent C with pure recessive Parent B. Recall that using the pure recessive as a parent is a test cross.

3. Record your results in Table 2. Complete the table as before. Total the results.

4. Make a Punnet square to determine what both genes are for each parent. Fill in Parent C's genes at the top of Table 2.

FIGURE 2.

Table 2. Test Cross 2: Parent C (R _R_) X Parent B (rr)

Trial	Bean colors	Genes	Offspring color	Trial	Bean colors	Genes	Offspring color
1	brown, red	Rr	red	11			
2	brown, red	Rr	red	12			
3	brown, red	Rr	red	13			
4	brown, red	Rr	red	14			
5	brown, red	Rr	red	15			
6				16			
7				17			
8				18			
9				19			
10				20			

Sample data given.

There should be no brown offspring, rr. All offspring should be red.
Total brown offspring = [0]   **Total red offspring =** [20]

That no rr offspring appear indicates that the unknown parent must be RR. If it had the r gene, one would expect some of the offspring to be brown.

QUESTIONS

1. Before the test crosses, what possible genes could Parent A have had? **RR or Rr**

2. Before the test crosses, what possible genes could Parent C have had? **RR or Rr**

3. Could Parents A and B have had any brown offspring? **Yes** Explain. Results indicate that about half of the offspring are brown; thus, Parent A must be Rr.*

4. Could Parents C and B have had any brown offspring? **No** Explain. Results indicate that all offspring are red; thus, Parent C is probably RR.*

*The Punnett squares in question 5 will verify these crosses for students.

5. Complete the Punnett squares for the test crosses shown below for dog fur color.

	R	R
r	Rr red	Rr red
r	Rr red	Rr red

	R	r
r	Rr red	rr brown
r	Rr red	rr brown

6. Suppose long fur (**L**) is dominant over short fur (**l**) in guinea pigs. How can you determine whether a particular long-furred guinea pig has **LL** or **Ll** genes? You can cross the long-furred guinea pig to a short-furred guinea pig and examine the offspring for fur length.

7. Suppose a long-furred guinea pig is crossed to a guinea pig with short fur. They have 4 offspring. Two of the offspring have short fur. The other two have long fur. What genes does the long-furred parent have, **LL** or **Ll**? **Ll** In the space below draw a Punnett square that shows the parents and the four offspring.

Long-furred parent

	L	l
l	Ll long	ll short
l	Ll long	ll short

Short-furred parent

Name _____ Class _____ Period _____

Before doing this lab, have students read section 27:2 in the text.

27-1 What Do Normal and Sickled Cells Look Like?

EXPLORATION

Sickle-cell anemia is a disorder in which red blood cells are sickle-shaped rather than round. Sickle-cell anemia is a genetic disorder. The sickled-cells do not carry as much oxygen as normal red blood cells. A person with this disorder does not get enough oxygen to the body cells.

Process Skills: using a microscope, recognizing and using spatial relationships, interpreting scientific illustrations, predicting

OBJECTIVES

In this exercise, you will:
a. examine and compare slides of normal and sickled red blood cells.
b. learn that there are two different conditions to this disease.
c. solve genetic problems involving sickle-cell anemia.

KEYWORDS

Define the following keywords:

genetic disorder ___ a health problem that occurs in humans and is caused by genes ___

hemoglobin ___ protein in red blood cells that carries oxygen around the body ___

sickle-cell anemia ___ a disorder in which all red blood cells are sickle-shaped* ___

sickle-cell trait ___ trait is carried by a person but doesn't show up in that person ___
*In cases of oxygen deprivation (high altitude or strenuous exercise), a person with sickle-cell anemia will show severe sickling of red blood cells.

MATERIALS

compound light microscope   red pencil
prepared slides of normal and sickled red blood cells

Slides are available from biological supply houses or local hospitals.

PROCEDURE
Part A. Observation of Normal and Sickled Red Blood Cells

1. Examine a slide of normal blood cells under the microscope. Locate the cells first under low power, then high power.
2. Draw two or three red blood cells in the space provided marked "normal." Draw the cells to scale. (Note: Red cells are round, have no nucleus, and are very pale pink in color. Cells that appear dark blue in color are stained white blood cells.)
3. Label the following cell parts: cell membrane, cytoplasm.
4. Shade, with red pencil, the parts of the cell in which hemoglobin is found.

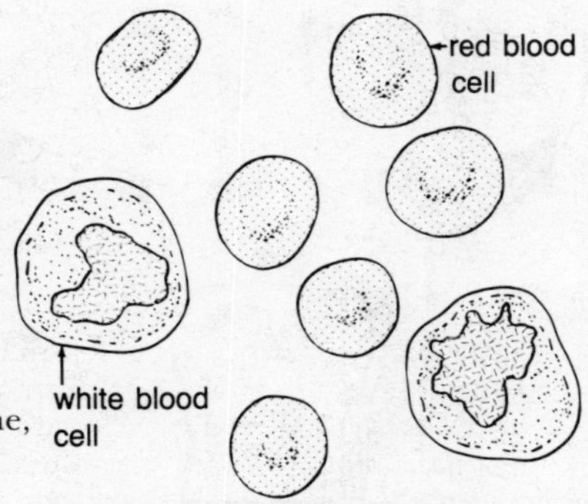

FIGURE 1. Normal blood cells

5. Examine a slide of sickled red blood cells under the microscope. Locate the cells first under low power, then high power.
6. Draw two or three sickled red blood cells in the space provided marked "sickled." (Note: Sickled cells are irregular in shape and may look like crescent moons or teardrops.)
7. Label the following cell parts: cell membrane, cytoplasm.
8. Shade, with red pencil, the parts of the cell in which hemoglobin is found.

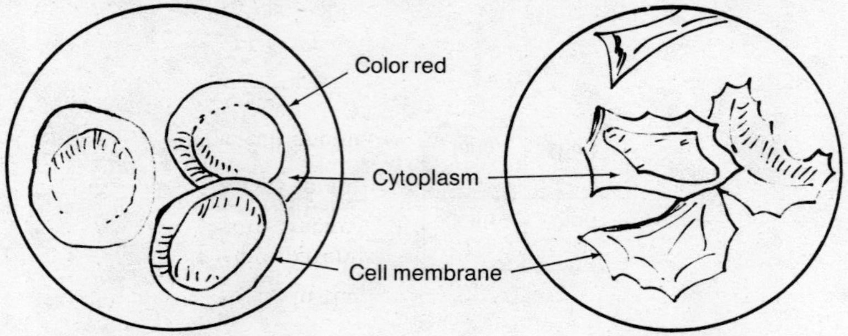

FIGURE 2. Normal and sickled cells

Part B. Comparison of Sickle-cell Trait with Sickle-cell Anemia

1. Examine Table 1. Note that not all people that possess the sickle-cell gene have sickle-cell anemia. There are two different sickle-cell conditions that are determined by the genes that are received from the parents. Sickle-cell trait is less of a problem than sickle-cell anemia. For more information read the material in section 27:5 of your text.

Table 1. Gene Combinations

Gene combination	Blood cell shape	Name of disease
RR	All Round	Normal
RS	Half Round Half Sickled	Sickle-Cell Trait
SS	All Sickled	Sickle-Cell Anemia

2. Examine Figure 3. These drawings represent blood samples from three different people. Count the number of normal and sickled cells in all three samples. For each sample record your totals in Table 2.

Table 2. Number of Normal and Sickled Cells Seen

Blood sample	Number of normal cells	Number of sickled cells
A	38	36
B	67	0
C	0	67

Name _____ Class _____ Period _____

Point out that the homozygous condition, in which both alleles for sickled cells are present, is fatal.

3. Examine your data and Table 1. Determine the blood condition for each of the samples in Figure 3 and write the correct condition ("normal," "sickle-cell trait," or "sickle-cell anemia") below each blood sample.

4. Examine your data and Table 1. Determine the correct gene combination (*RR*, *RS* or *SS*) that produced each sample in Figure 3 and write them below each blood sample also.

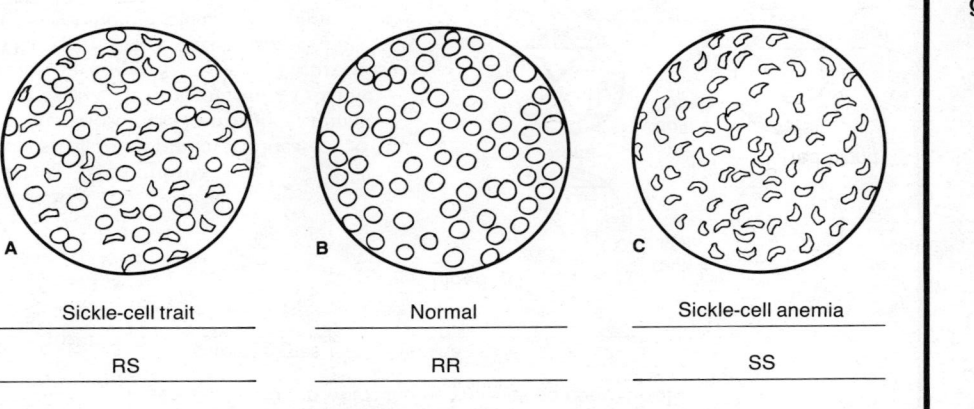

Sickle-cell trait	Normal	Sickle-cell anemia
RS	RR	SS

FIGURE 3. Blood samples

Part C. Genetics Problems

Construct and use a Punnett square for each of the following problems. Record your answers in the spaces provided.

1. Two parents have the following genes for blood cell shape: *RS* and *RR*. What kind of blood might their children have?

	R	R
R	RR	RR
S	RS	RS

	number of children
have normal blood	2
have sickle-cell trait	2
have sickle-cell anemia	0

2. Two parents have the following genes for blood cell shape: *RS* and *RS*. What kind of blood might their children have?

	R	S
R	RR	RS
S	RS	SS

	number of children
have normal blood	1
have sickle-cell trait	2
have sickle-cell anemia	1

229

3. Two parents have the following genes for blood cell shape: *RS* and *SS*. What kind of blood might their children have?

	S	S
R	RS	RS
S	SS	SS

	number of children
have normal blood	0
have sickle-cell trait	2
have sickle-cell anemia	2

QUESTIONS

1. Describe the shape of normal red blood cells. __They are round.__

2. Describe the shape of sickled cells. __irregular shape—pointed, jagged, and elongated__

3. Explain how the number of normal and sickled red blood cells differ in a person with sickle-cell trait and sickle-cell anemia. __Sickle-cell trait—½ of cells are sickled; sickle-cell anemia—all cells are sickled.__

4. Examine the drawings of the red blood cells in Figure 1 again. Which type of cell, normal or sickled, probably contains more hemoglobin? __The normal cell—it is larger.__

5. The less hemoglobin a person has, the more difficult it is for the cells of their body to get enough oxygen. With too little oxygen, a person will tire easily and cannot do a lot of exercising. Would a person with sickle-cell trait be able to do more or less exercising than a person with sickle-cell anemia? __more__ Why? __That person has more normal red blood cells.__

6. Which condition is more of a problem for a person, sickle-cell trait or sickle-cell anemia? __sickle-cell anemia__ . Why? __They cannot get as much oxygen to their cells as a person with sickle-cell trait.__

7. A scientist gathered the following information while studying people with sickle-cell anemia.
   Number of children born in the United States with sickle-cell anemia:
   to white parents—less than 1/100,000 births
   to Afro-American parents—about 200/100,000 births
   Which racial group seems to suffer most from this genetic disorder? __black__

8. Why do we call sickle-cell anemia a genetic disorder? __because the disorder is caused by inherited genes__

9. How does this disorder serve as an example of lack of dominance? __The heterozygous person (RS) shows both normal cells (R) and sickled cells (S)__

230

Name _____ Class _____ Period _____

Before doing this lab, have students read section 27:2 in the text.

27-2 How Are Traits on Sex Chromosomes Inherited?

Hemophilia is a disorder in which the person's blood will not clot. The disorder is inherited. If you have the dominant gene H, you will have normal blood. If you have only the recessive gene h, your blood will not clot.

Color blindness is also a genetic disorder. In this disorder, the person does not see certain colors, such as red and green. This person will see green as a grey color and red as a yellow color. If you have at least one dominant gene C, you see all colors. If you have only recessive genes, you cannot see red and green.

INTERPRETATION

Process Skills: predicting, recognizing and using spatial relationships, using numbers

OBJECTIVES
In this exercise, you will:
a. toss coins to show children born in five families.
b. see how hemophilia and color blindness are inherited in several families.
c. solve genetic problems involving hemophilia and color blindness in some families.

KEYWORDS
Define the following keywords:

color blindness   a condition in which some colors are not seen as they should be

hemophilia   a genetic disorder in which the person's blood doesn't clot

sex chromosomes   chromosomes that determine sex

MATERIALS
8 coins   tape   pen

If you make up the coins in advance there will be less chance for error on the part of the student. The coins can be put in small envelopes and stored for future use. Seven coins are all that are needed because coins 2 and 4 are alike.

PROCEDURE
Part A. Hemophilia

Genes for hemophilia are located on the sex chromosomes. Remember, females have two X chromosomes (XX) while males have one X and one Y chromosome (XY). Only the X chromosomes have the genes for hemophilia. A female can be $X^H X^H$, $X^H X^h$, or $X^h X^h$ for the clotting trait. A male can be $X^H Y$ or $X^h Y$.

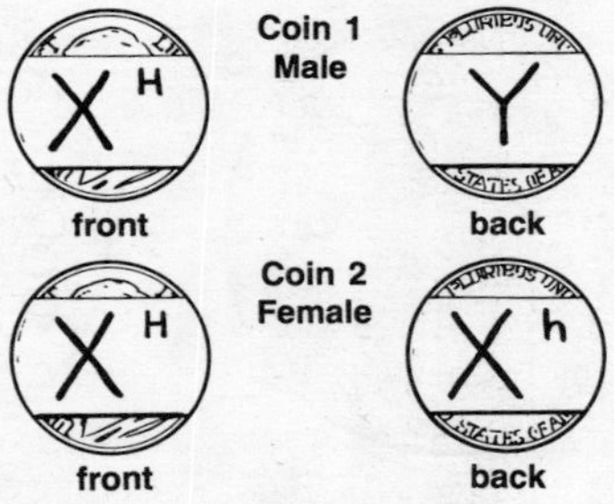

Family 1. Offspring of parents who are normal; the mother is heterozygous.
1. Place the tape on both sides of two coins.
2. Mark the coins as shown in Figure 1. These coins represent the genes of the parents. The coin with the Y chromosome is the father. The coin with an X on each side is the mother.

FIGURE 1. Marking coins for family 1

231

3. Place both coins in your cupped hands. Shake the coins, and then drop them on your desktop.
4. Read the combination of letters that appears. This combination represents the result that might appear in an offspring of these parents.
5. Make a mark (/) in Table 1 beside the correct gene combination in the column marked "Offspring Observed."
6. Repeat shaking and reading the coins for a total of 40 times.
7. Figure the total marks for each gene combination in Table 1 and write these totals in the proper space in the table.

Table 1. Offspring of $X^H Y$ Father and $X^H X^h$ Mother

Gene combinations	Offspring observed	Total
$X^H X^H$	|||| |||||	9
$X^H X^h$	|||| ||||| ||	12
$X^h X^h$		0
$X^H Y$	|||| |||	8
$X^h Y$	|||| ||||| /	11

Answers will vary.

Family 2. Offspring of a father who has hemophilia and a heterozygous mother.
1. Place tape on two coins and mark them as shown in Figure 2.
2. Place the coins in your hands and shake. Read the results and make a proper mark in Table 2.
3. Repeat step 2 for a total of 40 times. Total your marks in Table 2.

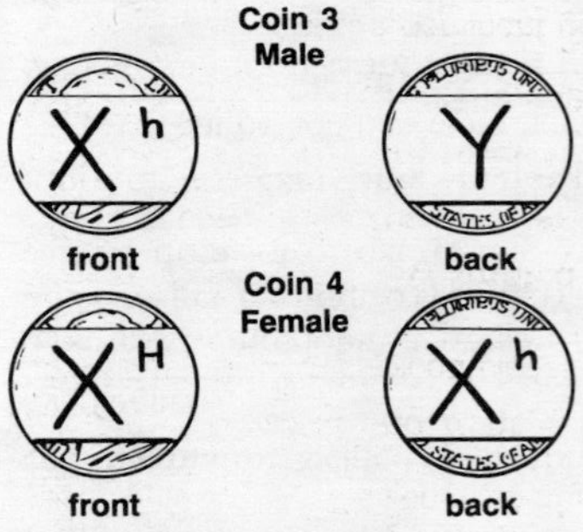

Table 2. Offspring of $X^h Y$ Father and $X^H X^h$ Mother

Gene combinations	Offspring observed	Total
$X^H X^h$		0
$X^h X^h$	|||| |||	8
$X^H Y$	|||| |||||	10
$X^h Y$	|||| |||| ||||	14
									8

FIGURE 2. Marking coins for family 2

Part B. Color Blindness

The genes for color blindness are also located on the sex chromosomes. For the genes controlling color blindness a female can be $X^C X^C$, $X^C X^c$, or $X^c X^c$. A male can be either $X^C Y$ or $X^c Y$.

Family 3. Offspring of a father who is color-blind and a mother who has two dominant genes.
1. Place tape on two coins and mark them as shown in Figure 3.
2. Shake the coins and read the results. Place a proper mark in Table 3.
3. Repeat step 2 for a total of 40 times. Total your marks in Table 3.

Family 4. Offspring of parents who are normal but the mother is heterozygous.
1. Place tape on two coins and mark them as shown in Figure 4.
2. Shake the coins and read the results. Place a proper mark in Table 4.

232

Name _____ Class _____ Period _____

3. Repeat step 2 for a total of 40 times. Total your marks in Table 4.

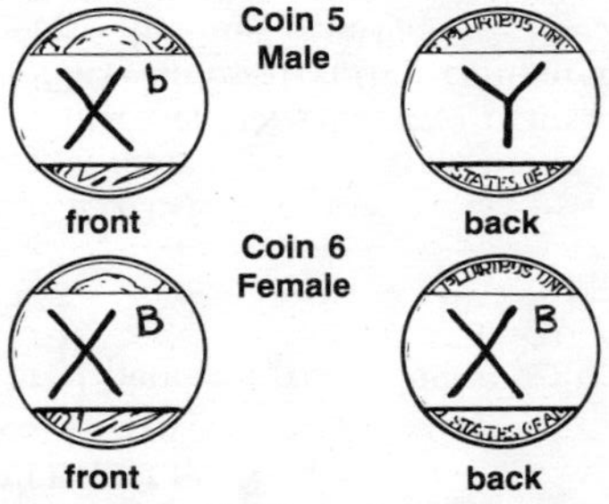

Table 3. Offspring of X^B Y Father and X^B X^b Mother

Gene combinations	Offspring Observed	Total
X^B X^B		0
X^B X^b	++++ ++++ ++++ ++++ I	21
X^b X^b		0
X^B Y	++++ ++++ ++++ ++++	19
X^b Y		0

Coin 5 Male — front: X^b / back: Y
Coin 6 Female — front: X^B / back: X^B

FIGURE 3. Marking coins for family 3

Table 4. Offspring of X^a Y Father and X^B X^b Mother

Gene combinations	Offspring observed	Total
X^B X^B	++++ III	8
X^B X^b	++++ ++++ I	11
X^b X^b		0
X^B Y	++++ IIII	9
X^b Y	++++ ++++ II	12

Coin 7 Male — front: X^B / back: Y
Coin 8 Female — front: X^B / back: X^b

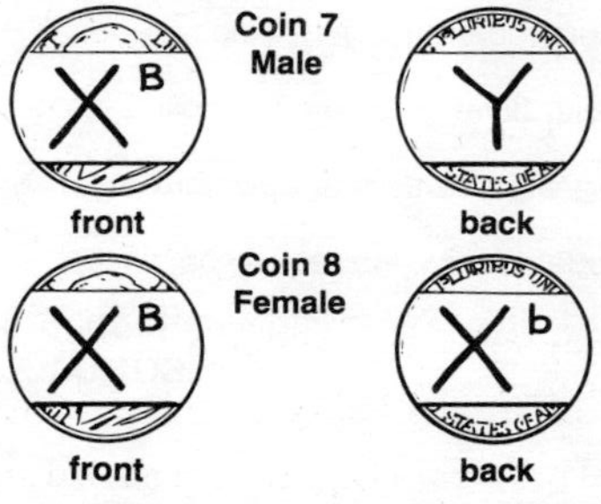

FIGURE 4. Marking coins for family 4

Part C. Problems

For each of the following problems construct and use a Punnett square. Record your answers in the spaces provided. Traits are sex-linked recessive.

1. Two parents have the following genes for hemophilia: $X^H X^h$ and $X^H Y$. What kind of blood could their children have?

	X^H	X^h
X^H	$X^H X^H$	$X^H X^h$
Y	$X^H Y$	$X^h Y$

Children | Number of males | Number of females
have normal clotting | 1 | 2
have hemophilia | 1 | 0

2. Two parents have the following genes for color blindness: $X^B X^b$ and $X^B Y$. What kind of color vision could their children have?

	X^B	X^b
X^B	$X^B X^B$	$X^B X^b$
Y	$X^B Y$	$X^b Y$

Children | Number of males | Number of females
have normal vision | 2 | 2
have color blindness | 0 | 0

3. Two parents have the following genes for color blindness: $X^b X^b$ and $X^b Y$. What type of color vision could their children have?

	X^b	X^b
X^b	$X^b X^b$	$X^b X^b$
Y	$X^b Y$	$X^b Y$

Children | Number of males | Number of females
have normal vision | 1 | 1
have color blindness | 1 | 1

QUESTIONS

1. What sex chromosomes do female offspring have? __XX__
2. What sex chromosomes do male offspring have? __XY__
3. How many genes do females have:
   a. for blood clotting? __two__   b. for color blindness? __two__
4. How many genes do males have
   a. for blood clotting? __one__   b. for color blindness? __one__
5. Why is there a difference in the number of genes for blood clotting and color blindness in males and females? __The male has a Y chromosome and there are no genes for the traits of blood clotting and color blindness on the Y chromosome.__
6. Which of the two traits studied in this exercise are genetic disorders? __Both are genetic disorders because they are caused by inherited traits.__
7. In Problem 2, why are there no color-blind children even though one of the parents is color blind? __The trait is recessive and the children that inherit the color-blind gene also have a dominant gene for normal color vision.__
8. Which of the parents give the trait of hemophilia to their son? __The mother gives the trait to the son because the father only gives his son a Y chromosome.__
9. Which of the parents give the trait of hemophilia to their daughter? __Both the mother and father give a gene for hemophilia to their daughter because both parents give an X chromosome.__

Name _____ Class _____ Period _____

Before doing this lab, have students read section 28:1 in the text.

28-1 How Does DNA Make Protein?

DNA directs your cells to make certain proteins. How does DNA make proteins? DNA is a model for making a molecule called messenger RNA (mRNA). Messenger RNA is much like DNA. RNA is made of substances, called nitrogen bases, that must match up with the nitrogen bases in DNA. These nitrogen bases will only match up with the nitrogen bases in the nucleus.

After it is formed, mRNA leaves the nucleus and attaches to a ribosome in the cytoplasm of the cell. Other RNA molecules, called transfer RNA (tRNA), bring protein parts to the mRNA on the ribosome. The two types of RNA molecules match up, join protein parts together, and make a protein. Figure 1 shows the steps involved in making a protein. DNA determines what proteins are produced.

Process Skills: interpreting scientific illustrations, sequencing, scientific terminology, formulating models

OBJECTIVES

In this exercise, you will:
a. use models to show how DNA makes mRNA.
b. use models to show how mRNA leaves the nucleus and causes tRNA to make proteins.

KEYWORDS

Define the following keywords:

DNA _a molecule that makes up genes and controls traits of organisms_

mRNA _a molecule that carries the DNA code from the nucleus to the ribosome_

protein _building material of living things_

tRNA _a molecule that carries protein parts to the mRNA on the ribosome_

MATERIALS

scissors
colored pencils: red, blue and green

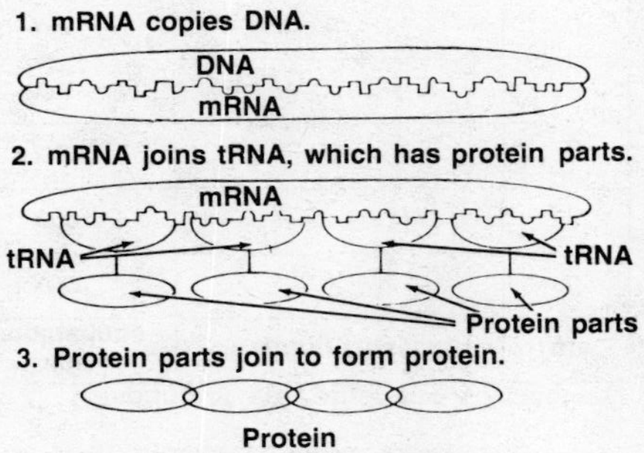

PROCEDURE

1. Examine Figure 2, a model of a DNA molecule. DNA has two main sides. These sides are often compared with the upright sides of a ladder. The squares in the model represent sugar molecules. The nitrogen bases A, C, G, and T join to connect the two sides.

Make copies of Figures 2, 3, 4, and 5 for students to make models.

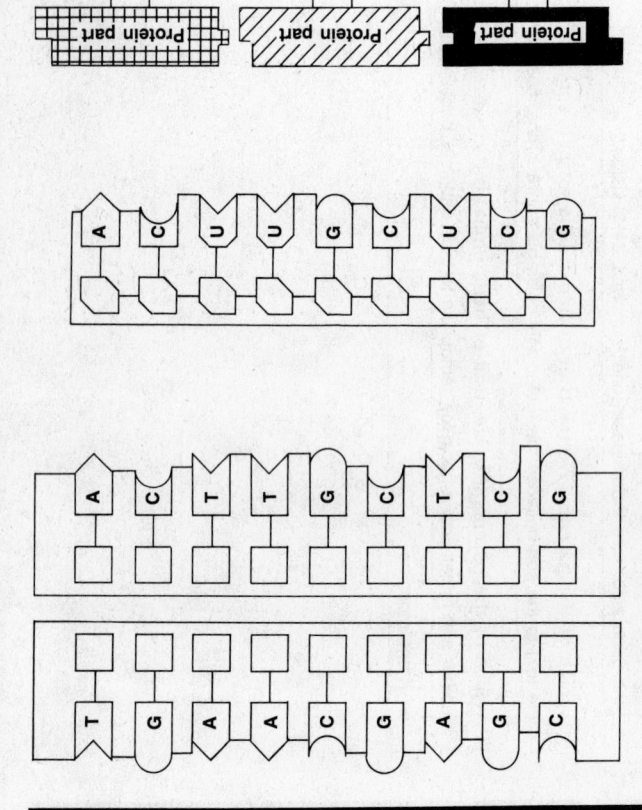

INTERPRETATION

1. mRNA copies DNA.

2. mRNA joins tRNA, which has protein parts.

3. Protein parts join to form protein.

FIGURE 1. Formation of protein

235

FIGURE 2. DNA molecule

FIGURE 3. Messenger RNA

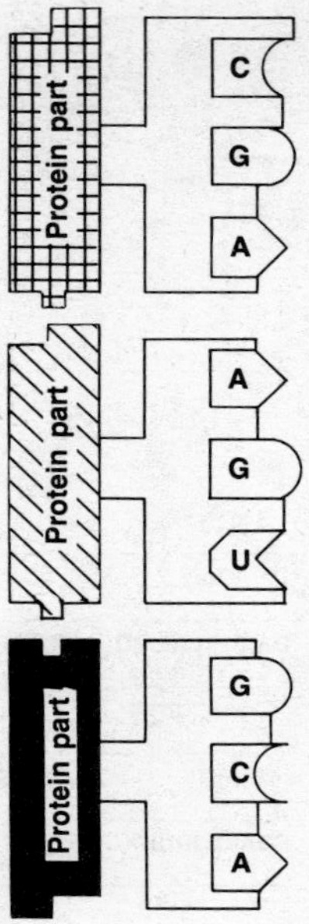

FIGURE 4. Transfer RNA

2. Cut out the two sides of the DNA model in Figure 2.
3. Color the two sides red.
4. Put the two sides together so that they fit together like the pieces of a puzzle. Note that nitrogen base A only binds with T and base G only with C.
5. Examine Figure 5, a model of a cell. The nucleus is in the upper left corner. Place the model of DNA in the nucleus. DNA carries the code for making cell proteins. That code is the order in which the nitrogen bases appear.
6. Cut out the model of the mRNA molecule in Figure 3. This molecule has only one side.
7. Color this model blue. Observe that the sugar in the mRNA molecule is different from the sugar in DNA. Also, the nitrogen base U is present instead of T.
8. Open the two sides of the DNA model.
9. Place the mRNA molecule along one side of the DNA model. Note that its bases will fit only one side of the DNA. In an actual cell, the mRNA is assembled from small molecules to fit exactly along one side of the DNA. The nitrogen bases can only fit certain other bases because of their shape. mRNA copies the code of DNA.
10. Move the mRNA molecule out of the nucleus to the cytoplasm by following the dotted line as a path. This shows that mRNA carries the code of the DNA to the ribosomes.
11. Move the mRNA to the cell part called the ribosome. Place it on the dashed lines at the ribosome.

236

Name _____ Class _____ Period _____

12. Put the DNA model sides back together.
13. Cut out the three tRNA molecules shown in Figure 4. Using a green pencil, color only the lower parts (that contain the letters A, U, C, and G). This type of RNA is different from mRNA in two ways. First, each tRNA molecule has only three nitrogen bases and second, a certain protein part is attached to it. Transfer RNA is found in the cytoplasm of the cell. The top of each tRNA has a specific protein part attached to it.
14. Fit the tRNA molecules to the mRNA molecule again, so the bases fit together tightly. Observe which bases of tRNA bind with which bases of mRNA (A with U, G with C).
15. With the tRNA molecules in place on the mRNA molecule, the protein parts can now join with each other. The linked protein parts carried by the tRNA make a chain. This chain separates from the tRNA molecules and leaves the ribosome to become a protein. The code of the DNA molecule directs certain steps in a cell for the process of forming a certain protein.

QUESTIONS

1. What do the letters DNA stand for? __deoxyribonucleic acid__
2. In DNA, what nitrogen base always binds with A? __T__   G? __C__
3. How is mRNA different from DNA? __It has the nitrogen base U instead of T. It has only one side.__
4. In mRNA, what nitrogen base binds with the DNA base
   A? __U__   G? __C__   T? __A__
5. Where in the cell is mRNA made? __the nucleus__
6. To what cell part does mRNA attach? __the ribosome__
7. What carries the protein parts to the ribosome and the mRNA? __tRNA__
8. How are mRNA and tRNA alike? __They both have the same nitrogen bases, A, U, C, and G. They both have only one side.__
9. What does tRNA have that mRNA does not have? __protein parts attached to it__
10. Where in the cell are proteins made? __ribosomes__
11. What determines which proteins are produced? __DNA__

237

FIGURE 5. Model of a cell

238

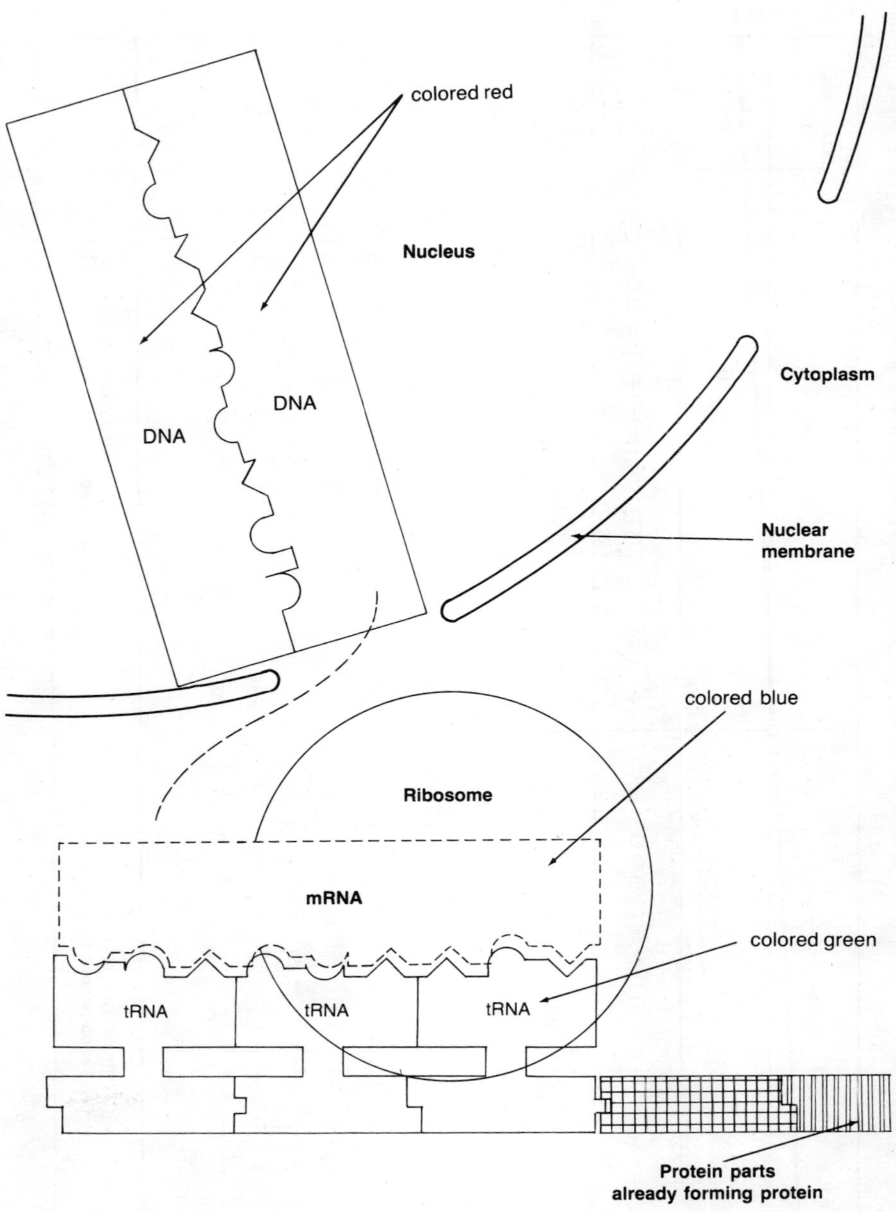

Name _____ Class _____ Period _____

Before doing this lab, have students read section 28:2 in the text.

28-2 How Can a Mutation in DNA Affect an Organism?

Sometimes the DNA code that makes up a gene has an error in it. This error is called a mutation. When the DNA contains an error, the mRNA it makes will copy that error. When the mRNA contains an error, it will code for incorrect tRNAs and produce an incorrect protein.

Sickle-cell anemia is a disorder that gets its name from the sickle shape of the red blood cells. The sickled red blood cells are caused by a mutation in the hemoglobin of the person with the disorder. Hemoglobin is the main protein in red blood cells. Each hemoglobin molecule carries oxygen from the lungs to all other parts of the body.

Process Skills: interpreting scientific illustrations, recognizing and using spatial relationships, sequencing, inferring

INTERPRETATION

OBJECTIVES
In this exercise, you will:
a. examine the coding errors produced in mRNA and tRNA when there is a mutation in the DNA.
b. examine the effect of a mutation in the gene that codes for blood hemoglobin.

KEYWORDS
Define the following keywords:

gene _____ a chromosome part that determines the trait of an organism

hemoglobin _____ the protein in red blood cells that carries oxygen around the body

mutation _____ when the DNA code makes up a gene that has an error in it

sickle-cell anemia _____ disorder in which the red blood cells are sickle-shaped

MATERIALS
colored pencil

PROCEDURE
1. Examine Table 1. The two columns show a section of normal DNA and a section of DNA that has a mutation in it. The mutation is called *sickle hemoglobin*.

Table 1. Comparing Normal With Sickle Mutation DNA

	This section codes for normal hemoglobin	This section codes for "sickle" hemoglobin
DNA code	G G G C T T C T T T T T	G G G C T T A A C T T T
mRNA code	C C C G A A G A A A A A	C C C G A A U U G A A A
tRNA code	G G G C U U C U U U U U	G G G C U U A A C U U U
Order of protein parts	A B C	A X B C
Shape of blood cells	normal	sickled

Name _____ Class _____ Period _____

2. In Table 1, in the row marked *mRNA code*, write in the correct letters that will match with the nitrogen base letters of DNA given in the row above. Do this for both columns. Remember that A matches with U, T matches with A, C matches with G, and G matches with C.

3. In the row marked *tRNA code*, write in the correct letters that will match with the nitrogen base letters of mRNA in the row above. Remember that A matches U, U matches with A, C matches with G, and G matches with C.

4. Examine Table 2. This table shows which protein parts are coded for by specific sets of nitrogen bases (three per set) of the mRNA molecule. For example, the mRNA sequence CCC codes for protein part A.

Table 2. Nitrogen Bases of Protein Parts

Protein part	mRNA
A	CCC
B	GAA
C	AAA
X	GUU

5. In Table 1, in the row marked *Order of protein Parts*, write in the correct order of protein parts coded for by the mRNA. Do this for both normal and sickle hemoglobin.

6. In the row marked *Shape of blood cells*, draw in what you think will be the correct shape of blood cells for the kind of protein found in the row above. Use the diagrams in Figure 1 for reference.

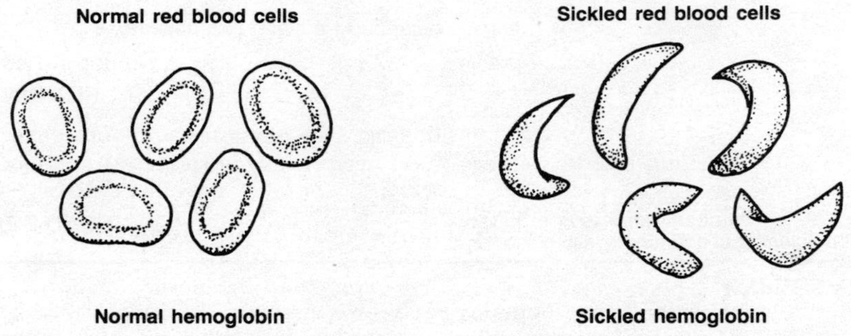

Normal red blood cells   Sickled red blood cells

Normal hemoglobin   Sickled hemoglobin

FIGURE 1. Shapes of blood cells

7. In the column marked *This section codes for sickle hemoglobin*, locate the two nitrogen bases that are different in DNA, mRNA, and tRNA from those in the column for normal hemoglobin. Color those bases that are mutations with the colored pencil.

241

QUESTIONS

1. Look at the two DNA molecules in Table 1. What nitrogen bases in the sickle mutation DNA are different from those of the normal DNA?

   The sickle mutation DNA has CAA instead of CTT. _____

2. If every three nitrogen bases on DNA represent a gene, how many genes are shown on

   a. the section of normal DNA? __4__

   b. the section of sickle hemoglobin DNA? __4__

3. List the nitrogen bases (examined in Table 1) for

   a. the normal genes of hemoglobin __GGG, CTT, CTT, and TTT__

   b. the sickle hemoglobin genes __GGG, CAA, CTT, and TTT__

4. How many genes are different in sickle hemoglobin DNA compared with normal hemoglobin DNA? __1__

5. How many protein parts are different in sickle hemoglobin compared with normal hemoglobin? __1__

6. How many genes are needed to code one protein part into a protein such as hemoglobin? __1__

7. Define the word *mutation* _____

   a. by using the word "gene." _____ A mutation takes place when a gene codes for the wrong protein.

   b. by using the phrase "DNA code." _____ A mutation is a change that takes place when the DNA code has an error in it.

8. It is possible to move genes from one molecule of DNA to another. A normal gene could be put in the place of a gene with a mutation.

   a. If the DNA with a mutation were corrected in this way, what would happen to the mRNA that DNA makes? _____ If the DNA were corrected, the mRNA it makes would be corrected. _____

   b. What would happen to the protein formed by this mRNA? _____ If the mRNA were corrected, the protein that formed from the mRNA would be corrected. _____

242

Name _____ Class _____ Period _____

Before doing this lab, have students read section 29:1 in the text.

29–1 How Do Some Living Things Vary?

EXPLORATION

All the living things that are within one species are not alike. The differences in a trait are called variations (ver ee AY shunz.) For example, we see variations in height, eye color, and ear shape.

Variations in traits may be helpful. They may help survival. A living thing with a variation may have an advantage (ud VANT ihj) over a living thing without that variation. We say the variation is an adaptation.

Process Skills: measuring in SI, using numbers, making a line graph, inferring

OBJECTIVES

In this activity, you will:
a. observe variations in leaves, bean pods, bean seeds, and humans.
b. determine how variations may be helpful.

KEYWORDS

Define the following keywords:

adaptation _a variation that helps a living thing survive_

variation _a difference in a trait_

MATERIALS

conifer twig
metric ruler
10 pinto bean seeds
10 opened bean pods

Pine, spruce, or fir twigs are suitable for this activity. Pine leaves occur in bundles of two, three or five. Leaves of spruce or fir occur singly on the twig.

PROCEDURE

Part A. Variation in Leaf Length

1. Examine the conifer twig. Each needlelike structure is a leaf.
2. Remove 10 leaves from the twig as shown in Figure 1.
3. Measure the length of each leaf.
4. Record your results in Table 1 by making a check below the millimeter length for each leaf.
5. Count the marks and enter the number of leaves for each length.

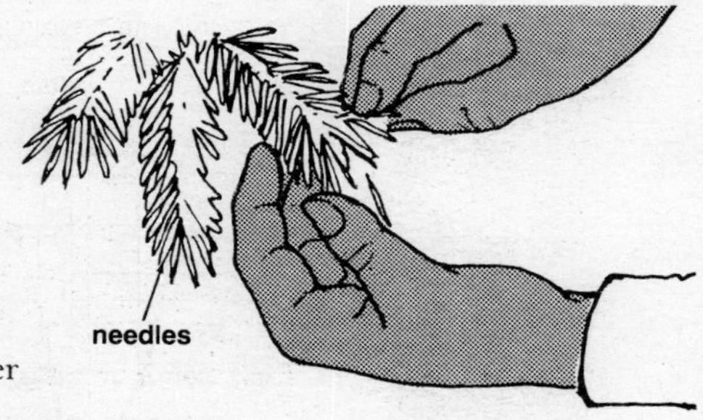

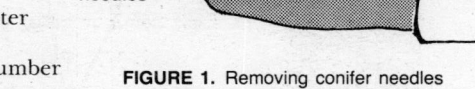

FIGURE 1. Removing conifer needles

Table 1. Variation in Leaf Lengths

Measurement (mm)	5	6	7	8	9	10	11	12	13	14	15	16	17	18	19	20	21	22	23	24
Checks							/	///	/	////	/									
Number of leaves							1	3	1	4	1									

Answers will vary. Sample data are given.

243

Part B. Variation in the Number of Seeds in Bean Pods

1. Examine 10 opened bean pods. Each pod has seeds inside.
2. Count the number of seeds in each pod.
3. Record your results in Table 2 by making a mark below the number of seeds for each pod.
4. Count the marks and enter the number of pods for each seed number.

Table 2. Variation in Seeds in Pods

Number of seeds	5	6	7	8	9	10	11	12
Checks	/	////	//					
Number of pods	1	5	2					

Answers will vary. Sample data are given.

Part C. Variation in Seed Coats

Pinto beans are three main colors.

1. Examine 10 pinto seeds for variations in color: light, medium, and dark brown.
2. Group the seeds according to their color.
3. Count the number of seeds in each group and record this in the table.

Table 3. Variation in Seed Coat Color

Color	Light brown	Medium brown	Dark brown
Number of seeds	6	1	3

Answers will vary. Sample data are given.

Part D. Variation in Seed Size

1. Measure the length of each pinto bean seed as shown in Figure 2.
2. Record your results in Table 4 by making a mark below the length for each seed.

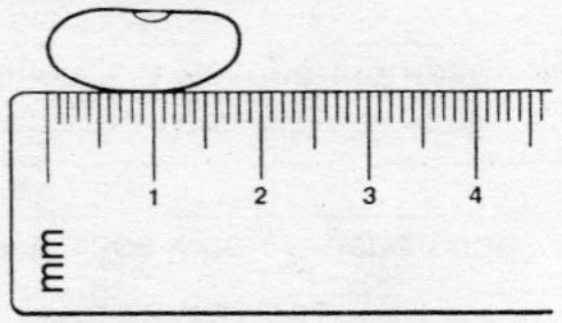

FIGURE 2. Measuring bean

Table 4. Variation in Seed Size

Measurement (mm)	5	6	7	8	9	10	11	12	13	14	15	16	17	18	19	20
Checks							/	/	////	//	/					
Number of beans							1	1	4	2	1					

Answers will vary. Sample data are given.

Part E. Hand Spread

1. Stretch the fingers of your right hand out flat on your desk top.
2. Measure the distance from the tip of your thumb to the tip of your little finger in cm (Figure 3).
3. Write the measurement on the chalkboard.
4. Record the measurements of the class in Table 5.

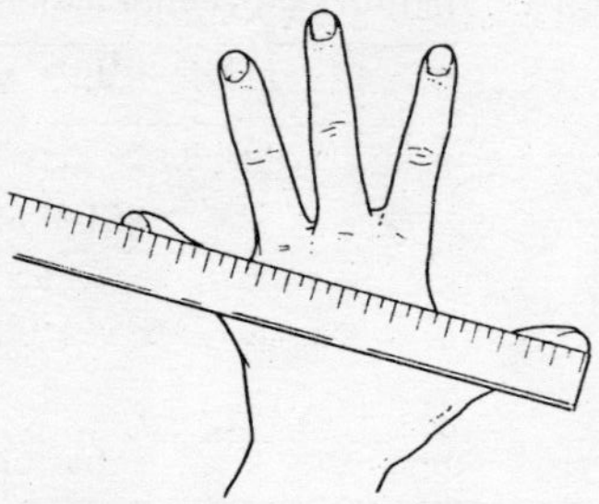

FIGURE 3. Measuring hand spread

244

404

Name _____ Class _____ Period _____

Table 5. Variation in Hand Spread

Measurement (cm)	13	14	15	16	17	18	19	20	21	22	23	24	25	26	27	28
Number of hands	0	1	1	3	6	6	5	4	2	1	1	1	0	0	0	0

Answers will vary. Sample data are given.

5. Plot the information from Table 5 onto the graph in Figure 4.

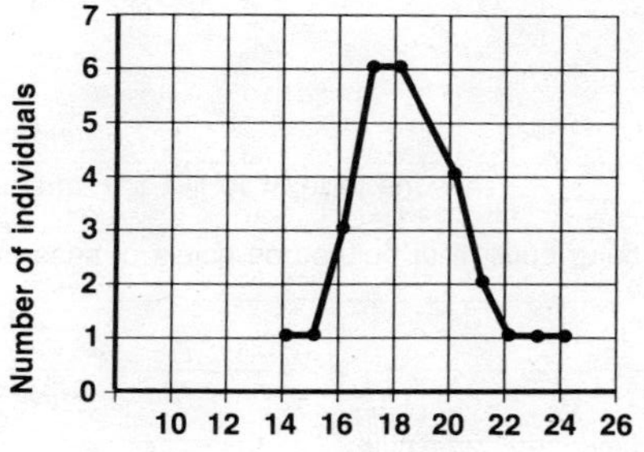

[Graph: Number of Individuals (y-axis, 0-7) vs Hand spread (cm) (x-axis, 10-26)]

FIGURE 4. Variation in hand sizes within a class of students

QUESTIONS Answers will vary. Students should give data they have recorded in this activity.

1. Describe the variation that you saw in
   a. leaf length. _The needles varied in length from 11 mm to 16 mm._
   b. seed number in each pod. _The number of seeds varied from 5 to 7 in each pod._
   c. seed coat color. _The color of seed coats varied from light, to medium, to dark brown._
   d. seed size. _The seeds varied in size from 10 mm to 15 mm._
   e. hand spread. _The hand size varied from 15 cm to 28 cm wide._

2. The living things that you observed varied in ways other than those you noted in this activity. In Table 6, list another trait for which the living thing varies. You may want to look at the samples again. The table is started for you.

Table 6. Other Traits of Living Things

Sample	Trait examined	Other traits
leaves	length	color, width
pods	seed number	length, width
seeds	color	size, mass, shape
hands	size	lengths of fingers, lines

3. In your leaf sample what is:
   a. the length of most of the coniferous needles? _15 mm_
   b. the length of the shortest needle? _11 mm_
   c. the length of the longest needle? _16 mm_

4. In your bean sample:
   a. how many seeds do most of the bean pods have? _6 seeds_
   b. what color are most of the bean seeds? _light brown_
   c. what size are most of the bean seeds? _13 mm_

5. How wide do most of the hand spreads measure? _17 to 19 cm_

6. In which width range are the fewest hand spreads? _14, 15, 22, 23, 24 cm_

7. What is the general shape of the graph of hand spread in Part E? _The shape of the graph is a bell._

8. What does the shape of the graph tell you about the width of the hand spreads in your class? _The shape shows that although the widths vary, most of the widths are near the average with fewer samples at both extremes._

9. What advantage could the following traits be to the organism:
   a. longer tree leaves? (HINT: What is the main job of leaves?) _Longer leaves have more cells for making food._
   b. more seeds in a pod? _More seeds could produce more plants._
   c. larger bean seeds? _Larger seeds have more stored food._
   d. larger hand spreads? _Larger hands hold more and reach farther._

Name _____ Class _____ Period _____

Before doing this lab, have students read section 29:2 in the text.

29-2 How Do Fossils Show Change?

Most organisms live, die, and decompose. They leave no traces of having lived. Under certain conditions, an organism's remains or tracks may be preserved as a fossil. Fossils give clues about how an organism looked and where it lived. They are often used by scientists as evidence of change.

A fossil is any remains of a once-living thing. Fossils may only be the outline of some plant, animal, or other organism that is preserved in rock. Sometimes, entire skeletons of animals that lived millions of years ago are found.

INTERPRETATION

Process Skills: interpreting scientific illustrations, recognizing and using spatial relationships, measuring in SI, interpreting data

OBJECTIVES

In this activity, you will:
a. examine diagrams of fossil horses and present-day horses shown in their surroundings.
b. examine diagrams of the structure of the front foot of fossil horses and present-day horses.
c. note the changes in horses that have taken place over time.

KEYWORDS

Define the following keywords:

adaptation ___ a variation that helps a living thing survive _____

Equus ___ present-day horse _____

fossil ___ remains of a once-living thing _____

Hyracotherium ___ fossil horse _____

natural selection ___ the process in which something in a living thing's surroundings

___ determines if it will or will not survive _____

MATERIALS

metric ruler
colored pencils: red, blue, green, and yellow

PROCEDURE

Part A. Change in Size With Time

1. Examine the diagrams in Figure 1 of *Hyracotherium*, *Miohippus*, *Merychippus*, and *Equus*.
2. Use the diagrams to fill in Table 1.

Table 1. Evolution in the Horse

Horse	*Hyracotherium*	*Miohippus*	*Merychippus*	*Equus*
Size	38 cm	65 cm	100 cm	140 cm
Type of surroundings	swamp with dense forest	dry with trees and shrubs	grassland	grassland

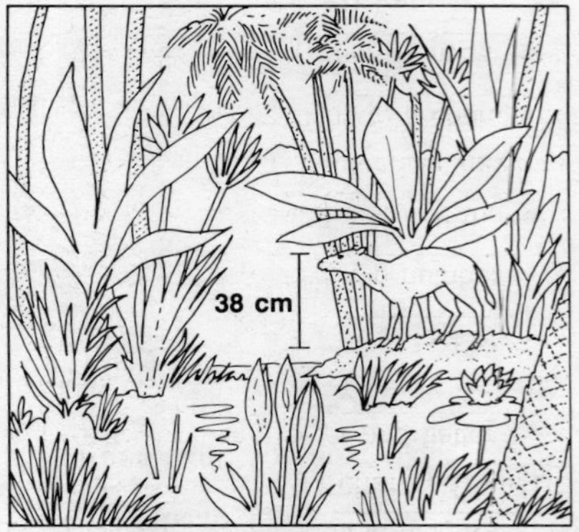

Hyracotherium
55 million years ago

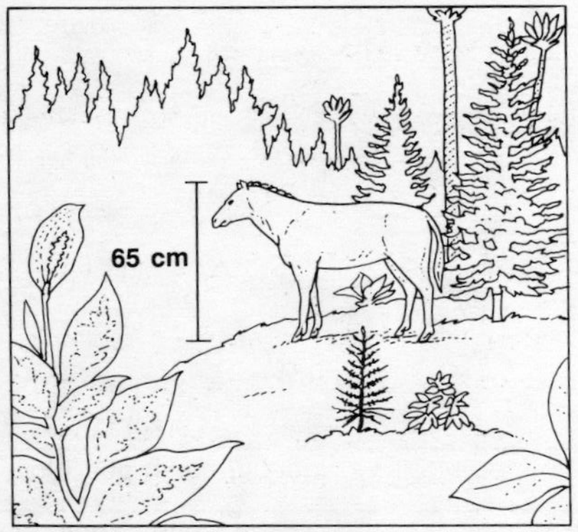

Miohippus
30 million years ago

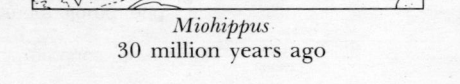

Merychippus
13 million years ago

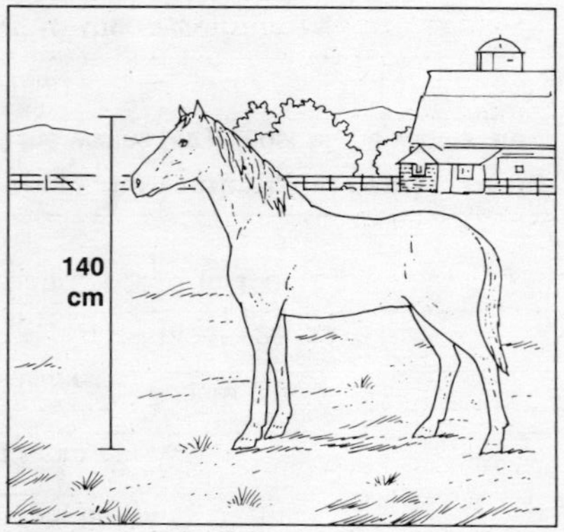

Equus
Today

FIGURE 1. Evolution of the horse

Name _____ Class _____ Period _____

Part B. Changes in Bone Structures With Time

The changes in horses over the last 55 million years have been shown by studies of large numbers of fossils. The earliest kind of horse was small and had teeth that were adapted to browsing on young shoots of trees and shrubs. The present-day horse is much larger and has larger teeth that are adapted to grazing on the tough leaves of grasses. Early horses were adapted to living in wooded, swampy areas where more toes were an advantage. The single-hoofed toes of the present-day horse allow it to travel fast in the plains.

1. Examine the diagrams in Figure 2. They show fossils of the front foot bones and the teeth of horses. The foot bones at the upper right of each diagram indicate the relative bone sizes of each kind of horse.

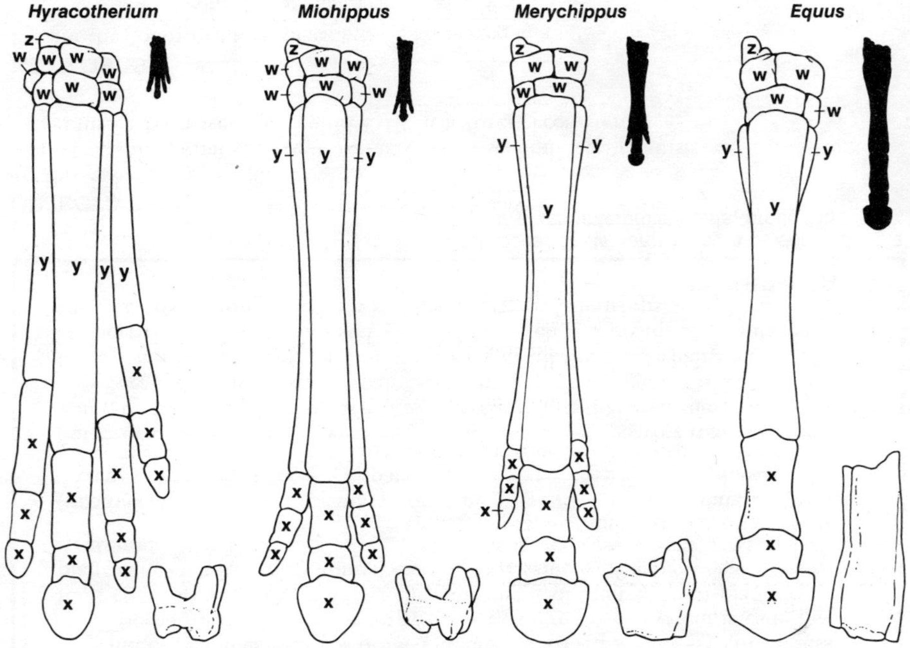

FIGURE 2. Forefoot bones and teeth of horses

2. Look for and color the following kinds of bones for each fossil horse.
   a. Color the toe bones red. These are marked for you with an *x*.
   b. Color the foot bones blue. These are marked with a *y*.
   c. Color the ankle bones green. These are marked with a *w*.
   d. Color the heel bones yellow. These are marked with a *z*.
3. Using the diagrams in Figure 2, make measurements to fill in Table 2.

Table 2. Evolution of the Horse

Kind of horse	Hyracotherium	Miohippus	Merychippus	Equus
Number of toes	4	3	3	1
Number of toe bones	12	9	9	3
Number of foot bones	4	3	3	3
Number of ankle bones	7	6	4	4
Number of heel bones	1	1	1	1
Total number of foot bones	24	19	17	11
Length of foot (measure inset diagrams) (mm)	9 mm	14 mm	24 mm	35 mm
Height of teeth (mm)	15 mm	15 mm	19 mm	37 mm

QUESTIONS
Answers will vary. Possible answers are given.

1. What changes occurred in the surroundings of horses from *Hyracotherium* to *Equus*? The land became drier. The forests and heavy vegetation were replaced by grasses and plains.

2. What change occurred in the shape of the horse from *Hyracotherium* to *Equus*? The early horses were small and had curved backs. The present-day horses are much larger and have straight backs.

3. What changes occurred in the size of the horse from *Hyracotherium* to *Equus*? The horse has increased in height from 38 cm to 140 cm.

4. As the surroundings changed, what happened to the teeth of the horse? The teeth became larger and flatter as an adaptation to grazing.

5. Describe the overall changes in foot length, number of toes, and size of toes in the horse over time. The foot of the horse over time increased in length, the number of ankle bones decreased, the number of toes decreased, the size of the side toes became shorter than the central toe, and the central toe became larger.

6. How would natural selection have caused changes in the size, feet, and teeth of the horse? Adaptations to a change in the environment would have given an advantage to the horses that had them. These horses would have had a greater chance of surviving than horses that were not adapted.

Name _____ Class _____ Period _____

Before doing this lab, have students read section 30:3 in the text.

30-1 What Are Some Parts of a Food Chain and a Food Web?

Plants use light energy of the sun to make food. The food is stored in the cells of the plant. Plants are called producers because they make food. Some of the stored energy in the food that plants make is passed on to the animals that eat the plants. Plant-eating animals are called primary consumers. Some of the energy is passed on to the animals that eat primary consumers. Animals that eat other animals are called secondary consumers.

The pathway that food energy takes through an ecosystem is called a food chain. A food chain shows the movement of energy from plants to plant eaters and then to animal eaters. An example of a food chain can be written as follows:

(producer)   (primary consumer)   (secondary consumer)
seeds   →   sparrow   →   hawk

Some of the food energy in the seeds moves to the sparrow that eats them. Some of the food energy then moves to the hawk that eats the sparrow. Because a hawk eats animals other than sparrows, you could make a food chain for each animal the hawk eats. If all the food chains were connected, the result is a food web. A food web is a group of connected food chains. A food web shows many energy relationships.

INTERPRETATION

Process Skills: formulating a model, making a bar graph, interpreting data, predicting

OBJECTIVES
In this exercise, you will:
a. determine what different animals eat in several food chains.
b. build a food web that could exist in a forest ecosystem.

KEYWORDS
Define the following keywords:

consumer   a living thing that eats other living things

food chain   pathway of food through an ecosystem

food energy   the energy an organism gets from the food it eats

food web   interconnecting food chains in an ecosystem

producer   living things that make their own food by photosynthesis

MATERIALS
colored pencils   metric ruler

408

PROCEDURE
Part A. Examining Food Chains

1. Read the introduction and examine the food chains given below.

(producer)	(primary consumer)	(secondary consumers)
plant roots	→ rabbit	→ fox
plant seeds	→ mouse	→ fox
plant leaves	→ earthworm	→ robin → snake
plant leaves	→ rabbit	→ snake
plant leaves	→ cricket	→ robin → fox
plant stems	→ earthworm	→ snake → hawk → fox
plant stems	→ rabbit	→ hawk
plant stems	→ small insects	→ mouse → owl
plant leaves	→ rabbit	→ owl
plant leaves	→ cricket	→ mouse → hawk
plant fruits	→ mouse	→ snake
plant fruits	→ small insects	→ robin → snake

2. Answer the questions that follow:

a. List the organisms that you think are producers.   plants

b. Why are they called producers?   They make their food by photosynthesis.

c. List the organisms that you think are primary consumers.   rabbit, mouse, earthworm, cricket, small insects

d. Why are they called primary consumers?   They eat plants.

e. List the organisms that you think are secondary consumers.   fox, robin, snake, hawk, mouse, owl

f. Why are they called secondary consumers?   They eat other animals.

g. Herbivores are organisms that eat plants. List the herbivores in the food chains.   rabbit, mouse, earthworm, cricket, small insects

h. How does your list of herbivores compare with your list in question c?   They are the same organisms.

i. Carnivores are organisms that eat other animals. List the carnivores in the food chains.   fox, robin, snake, hawk, mouse, owl

j. How does your list of carnivores compare with your list in question e?   They are the same.

k. Make two food chains using animals not listed in the above food chains.

plant parts → millipede → spider

plant parts → squirrel → cat

(Answers will vary.)

Name _____ Class _____ Period _____

Part B. Making a Food Web

1. Use the information in Part A on the previous page to complete Figure 1.
2. Draw lines from each organism to other organisms that eat it.
3. Show which organism gets the energy by making an arrow pointing in the direction of energy flow from producers to primary consumers, to secondary consumers. One food chain has already been done for you.
4. Draw your lines with different colored pencils for different food chains. To make it easier to read when finished, do not draw through the circles.

FIGURE 1. A food web in a forest ecosystem

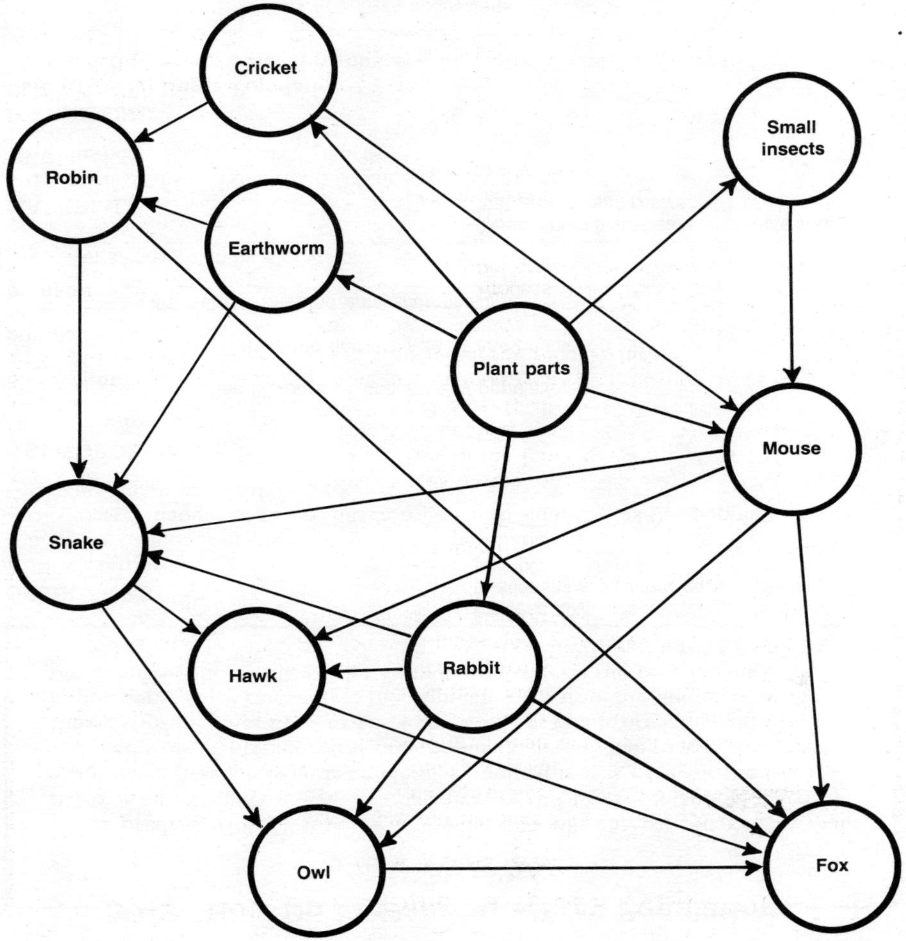

253

QUESTIONS

1. How many of the food chains you made in Figure 1 include the following animals? hawk __3__ earthworm __2__ fox __5__ small insects __2__ owl __3__ snake __5__
2. How many of the food chains include plant parts? __12 (all of them)__
3. Give the names of the producers that are in the food web. __plants__
4. Give the names of the consumers that eat both plants and animals. __mouse__
5. What would happen to the food web if all the plants were removed? __Everything would die.__
   Explain your answer. __Plants are involved in all of the food chains and are the base of the food chain in all ecosystems.__
6. What might happen to the owl population if there were less rabbits, mice, and snakes in a certain year? __There would be less owls because of the lack of food.__
7. What organisms will be affected if crickets, small insects, and earthworms are killed by pesticides? __Robins will be most affected because these three organisms are their only food in the food web. Other organisms that eat these three organisms also feed on other organisms in the food web.__
8. Draw three food chains below that can be connected in a food web. Show producers and consumers that you might see in your backyard or on your way to school.

   Answers will vary depending on the organisms the students see near their house.

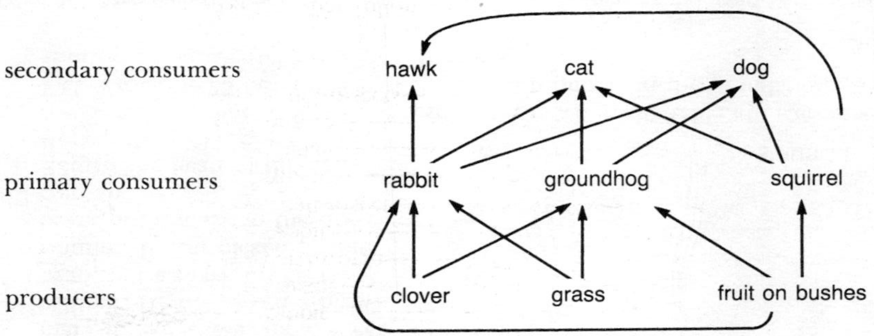

   secondary consumers

   primary consumers

   producers

254

Name _____ Class _____ Period _____

Before doing this lab, have students read section 30:4 in the text.

30-2 How Do Predator and Prey Populations Change?

A predator (PRED ut ur) is an animal that kills and eats another animal. A fox is an example of a predator. The prey (PRAY) is the animal killed by a predator. A rabbit is an example of an animal that is prey for the fox.

The sizes of predator and prey populations can change with the seasons. Biologists sometimes need to know the sizes of certain predator and prey populations. They can sample the population by trapping and/or counting the animals. The results of the samplings change as the populations change.

INTERPRETATION

OBJECTIVES
In this exercise, you will:
a. set up a model of predator and prey populations.
b. observe changes in the results you get from sampling as the populations change.
c. construct a graph showing your results.

Process Skills: recognizing and using spatial relationships, sequencing, formulating models, predicting

KEYWORDS
Define the following keywords:

population change __an increase or decrease in population size__

population sampling __a way to find the size of any population__

predator __an animal that kills and eats another animal__

prey __an animal eaten by a predator__

MATERIALS
101 brown beans* small paper bag graph paper
13 white beans* colored pencils

*Use brown pinto beans and white soup beans that are the same size.

PROCEDURE
Part A. Sampling a Population

1. Read this report about the animals on the abandoned James Hyde farm.

The James Hyde farm has not had people living on it since June of 1979. An interstate highway was put through the middle of the farm. Now there are only 100 acres of land left on this farm. In April of 1986, two biologists wanted to find out how the fox and rabbit populations were changing on the farm. They counted rabbits by trapping them and releasing them. They counted foxes by looking for them with field glasses because the foxes would not go near the traps. They trapped and released 23 rabbits. They saw two foxes.
This sample represented numbers in the actual populations.

255

---

2. Put 92 brown beans and 8 white beans into a bag. Assume brown beans are rabbits and white beans are foxes. This number of beans is four times the sample size in the example above. This will represent the numbers in the actual populations of rabbits and foxes.

3. Shake the beans in the bag.
Pick a bean without looking as shown in Figure 1. Put a strike mark in Table 1 in the correct column. If you picked a brown bean, put a mark in the rabbit column. If you picked a white bean, put a mark in the fox column.

4. Return the bean to the bag. Repeat the picking, returning the bean each time. Record the result by a mark in the table after each selection. Pick a total of 25 beans (25% of the actual numbers in the population). Total your results in the table.

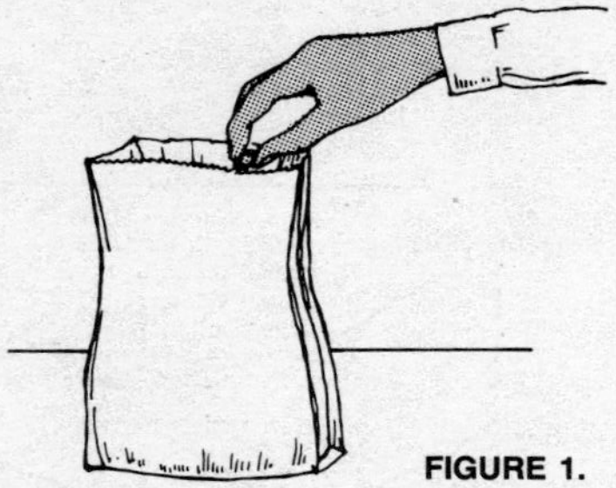

FIGURE 1.

Table 1. Recording Data in a Table   Answers will vary. Sample data given.

Data	Rabbits (brown beans)		Foxes (white beans)	
	Marks	Total	Marks	Total
April 1986	ℋℋ ℋℋ ℋℋ ℋℋ //	22	///	3

Part B. Recording Changes in Populations

1. Examine Table 2 that explains how to change your numbers of beans to show how the rabbit and fox populations changed at later dates.

Table 2. Population Changes

Sampling date	Rabbit population	Fox population
October 1986	Remove 10 brown beans. (Winter was harsh and food was low. Many rabbits died.)	Add 2 white beans. (Foxes also ate pheasants. Fox numbers increased.)
October 1987	Add 15 brown beans. (Food was plentiful. More rabbits moved into the area.)	Add 2 white beans. (Foxes had larger litters than usual.)
April 1988	Remove 8 brown beans. (Many rabbits died from disease.)	Remove 3 white beans. (Food was low. Some foxes left the area.)
October 1988	Add 12 brown beans. (Spring came early. Rabbits could breed earlier.)	Remove 4 white beans. (Rabbits were fewer from disease. Foxes decreased.)
April 1989	No change.	Add 8 white beans. (Food was plentiful. Foxes moved into the area.)
October 1989	Remove 14 brown beans. (Hunters killed pheasants. Foxes ate more rabbits.)	Remove 2 white beans. (Hunters shot some foxes.)

256

Name _____ Class _____ Period _____

2. Using Table 2 and the sampling method in Part A, sample the populations of rabbits and foxes nine more times to fill in the data for Table 3.
   a. Compare the dates in Tables 2 and 3. For each date in Table 3, sample beans 25 times. Make marks and fill in the totals of brown and white beans.
   b. When you come to a date in Table 2 that indicates a change in population size, follow the directions as to adding or removing beans from the bag. Record this data in the same date listed in Table 3.

Table 3. Population Sampling — Answers will vary. Sample data given.

Date	Rabbits (brown beans) Marks	Total	Foxes (white beans) Marks	Total
October 1986	ℋℋ ℋℋ ℋℋ ℋℋ	20	ℋℋ	5
April 1987	ℋℋ ℋℋ ℋℋ ℋℋ I	21	IIII	4
October 1987	ℋℋ ℋℋ ℋℋ ℋℋ II	22	III	3
April 1988	ℋℋ ℋℋ ℋℋ ℋℋ III	23	II	2
October 1988	ℋℋ ℋℋ ℋℋ ℋℋ IIII	24	I	1
April 1989	ℋℋ ℋℋ ℋℋ IIII	19	ℋℋ I	6
October 1989	ℋℋ ℋℋ ℋℋ ℋℋ	20	ℋℋ	5
April 1990	ℋℋ ℋℋ ℋℋ ℋℋ II	22	III	3
October 1990	ℋℋ ℋℋ ℋℋ IIII	19	ℋℋ I	6

Part C. Graphing the Data

1. Use a sheet of graph paper and make a graph like Figure 2. The number of animals are listed up the side, and the dates of sampling are along the bottom.
2. Fill in your graph using the data in Table 3 for your sampling of populations. Use different colors to color in the blocks for each animal. The first one is done for you with the biologists' data. (If you have trouble graphing, ask your teacher for help.)

Figure 2. Changes in rabbit and fox populations over 4 years

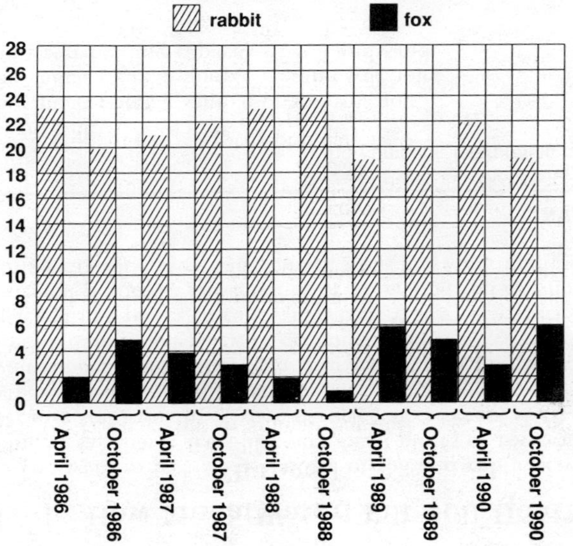

▨ rabbit ■ fox

Graphs will vary depending on each student's data. Sample graph given.

QUESTIONS

1. Which animal was the predator and which was the prey? _The fox was the predator and the rabbit was the prey._

2. How did your sampling in Part A compare with those of the two biologists in April 1986? _Answers will vary. They should be similar to the 2 foxes and 23 rabbits of the biologists' sample._

3. Give three factors that caused a decrease in the rabbit population. _harsh weather, disease, predators_

4. Give two factors that caused an increase in the rabbit population. _migration from other areas, warmer weather, birth_

5. Give three factors that caused a decrease in the fox population. _lack of food, hunters, migration to other areas_

6. Give three factors that caused an increase in the fox population. _good food supply, larger litters, migration into area_

7. How would the presence of pheasants affect the fox population? _Pheasants would mean more food for foxes and could increase the fox population._

8. What will happen to the rabbits when there is a decrease in the pheasant population? _More rabbits will be eaten by foxes so the rabbit population will decrease._

9. In some areas rewards are given to humans for killing certain animals. Animals such as coyotes and foxes are, therefore, hunted for the rewards. Farmers and ranchers often claim that these animals are bad because they kill farm animals. Biologists think these animals are important to the areas where they are found. Write a short paragraph explaining what some of the things that animals such as coyotes and foxes do that make them important. What could happen if these animals are all removed from their natural environments? _Coyotes and foxes eat other animals, thus controlling population sizes of other animals. They usually prey on sick or lame animals. Many animals that humans consider pests, such as mice and moles, are also eaten by these predators._

Name _____ Class _____ Period _____

Before doing this lab, have students read section 31:1 in the text.

31-1 How Much Water Will Soil Hold?

Suppose you were to pick up a handful of soil and rub it between your hands. You might say that it feels like very small pieces of rock. You would be correct because soil is made of rock particles of different sizes and different shapes that are mixed together.

When you rubbed the soil between your hands, you may have noticed that it felt wet or damp. Soil particles do not fit tightly together. Rather they are loosely packed like pieces of candy in a bag. Water is in spaces between and within the soil particles. The amount of water that soil can hold is called its water-holding capacity. How can you tell what the water-holding capacity of soil is?

INVESTIGATION

OBJECTIVES
In this exercise, you will:
a. find out how much water a sample of soil will hold.
b. find out how much water a sample of sand will hold.
c. compare the water-holding capacity of soil and sand.

Process Skills: experimenting, measuring in SI, using numbers, interpreting data

KEYWORDS
Define the following keywords:

balance   an instrument used to measure mass

mass   the amount of matter in something

soil   a mixture of mineral particles, living matter, and dead matter

water-holding capacity   the amount of water that soil can hold

MATERIALS
2 paper cups   dry soil*   pencil   large container
2 paper towels   dry sand**   balance   graduated cylinder

*Students can bring soil samples from home and air dry them for several days in the classroom before the experiment.

**can be purchased from a garden center or a hardware store

PROCEDURE
Part A. Comparing Water-Holding Capacity of Different Soils
1. Mark the two cups with the numbers 1 and 2.
2. Using a pencil, punch several holes in the bottom of each paper cup as shown in Figure 1. CAUTION: *Be very careful and work slowly as you punch the holes.*
3. Line each cup with a soaked paper towel as shown in Figure 1.

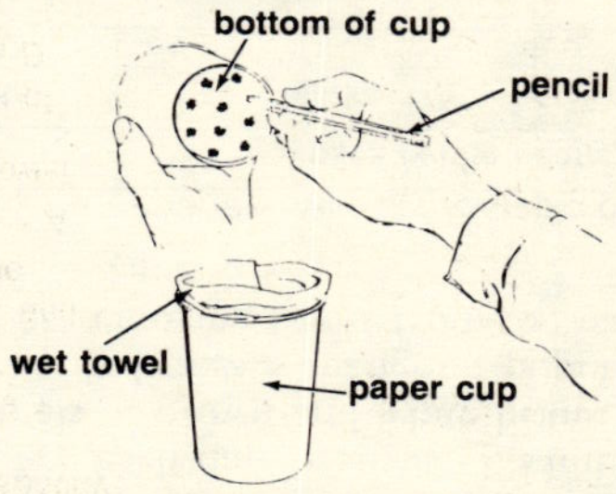

FIGURE 1. Preparing cups

4. Follow the procedure as shown in Figure 2.
   a. Find the mass of cup 1 with its towel using a balance. Record this mass in Table 1.
   b. Fill cup 1 three-fourths full with soil.
   c. Find the mass of cup 1 with the towel and soil. Record this mass in Table 1.
   d. Hold the cup over the large container. Slowly add 500 mL of water to the cup of soil. Let the water drain through the soil into the container.
   e. When no more water drips through the bottom of the cup, find the mass of the cup of wet soil. Record this mass in Table 1.
   f. Repeat steps 4a to 4e using cup 2 and sand this time in place of soil.
   g. Record your masses in Table 1.

FIGURE 2. Preparing soil samples

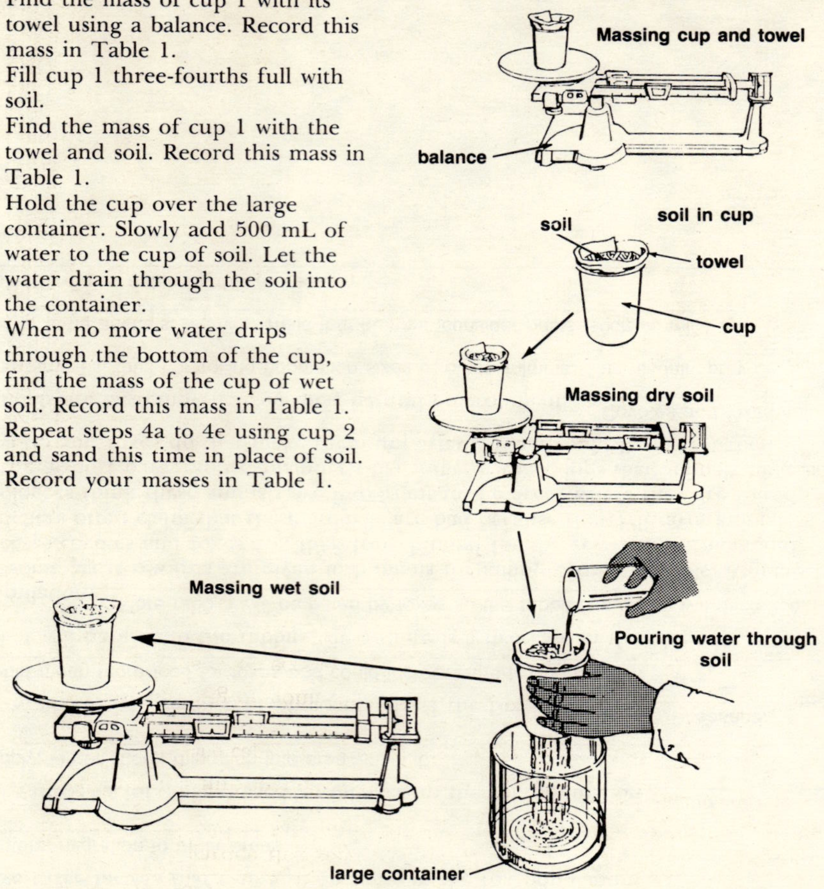

Table 1. Masses of Samples
Answers will vary. Sample data given.

	Mass of sample	
	Cup 1 (Soil)	Cup 2 (Sand)
A. Cup and towel	4.8 g	4.8 g
B. Cup, towel, and dry sample	32.1 g	67.6 g
C. Cup, towel, and wet sample	63.5 g	92.5 g

Part B. Figuring Water-Holding Capacity of Soils
1. Using the data recorded in Table 1, complete the boxes in Table 2.
2. Follow the steps set out in Table 2 to calculate the water-holding capacity of each soil sample.

Name _____ Class _____ Period _____

Answers will vary.
Sample data given.

Table 2. Procedure for Calculating Water-Holding Capacity

	Cup 1 (soil)	Using Table 1	Cup 2 (sand)
	32.1	Mass B of cup, towel, and dry sample	67.6
−	4.8	Subtract mass A of cup and towel	− 4.8
	27.3	Gives mass of dry sample D	62.8
	63.5	Mass C of wet sample	92.5
−	27.3	Subtract mass D of dry sample from above	− 62.8
=	36.2	Gives gain in mass of soil E	29.7
=	3620	Multiply E by 100 for percentage gain F in mass of soil	= 2970
	3620 ÷	Enter F again	2970 ÷
	63.5	Enter here the mass C of wet sample	92.5
=	57%	For percentage of water in the sample divide F by C	= 32.1%

QUESTIONS

1. a. What percent of the soil sample is water? **57%**
   b. What percent of the sand sample is water? **32.1%**
   c. Which holds more water, soil or sand? **soil**

2. What determines whether soil can hold more or less water? **How closely packed the soil particles are determines if more or less water can be held.**

3. What does it mean when a soil sample has a mass of 40 g and a 50% water-holding capacity? **If a wet soil sample has a mass of 40 g, then 20 g of that mass is water.**
   Explain your answer. **Soil that is tightly packed or has much rock or sand present will hold less water than other soil.**

---

5. Suppose soil particles become tightly packed together. What will happen to the water-holding capacity of the soil? **The water-holding capacity of the soil will decrease.**
   Explain your answer. **Tightly packed soil holds less water than loosely packed soil because there is less space available for water between tightly packed soil particles.**

6. Suppose soil particles become more loosely packed. What will happen to the water-holding capacity of the soil? **The water-holding capacity will increase.**
   Explain your answer. **Loosely packed soil holds more water than tightly packed soil because there is more space available for water between the loosely packed soil particles.**

7. Why do gardeners use a hoe to break up soil clumps in a garden? **to increase the air spaces in the soil**

8. Suppose you have a garden that stays too wet most of the time. Your friend tells you to dump a truckload of sand onto the garden and to mix it with the garden soil. Explain why this information could be good advice. **Sand added would decrease the water-holding capacity of the soil because sandy soil holds less water.**

9. Are soil and water living or nonliving parts of an ecosystem? **Soil and water are nonliving parts of the ecosystem.**

Name _____ Class _____ Period _____

Before doing this lab, have students read section 31:1 in the text.

31-2 How Can a Nonliving Part of an Ecosystem Harm Living Things?

An ecosystem (EE koh sihs tum) is a place with its living and nonliving things. The living parts are the biotic (bi AHT ihk) parts. The nonliving parts are the abiotic (ay bi AHT ihk) parts.

The biotic parts of an ecosystem need the abiotic parts. Air and water are examples of abiotic parts that are needed by living things. The abiotic parts must be present in the right amounts.

One abiotic part that living things need is mineral salts. Mineral salts are chemicals that contain several different elements. Some mineral salts, for example, are made of nitrogen, calcium, sulfur, and phosphorus. Living things get their mineral salts from soil or water. Often the soil or water contains too much or too little mineral salts for living things.

INVESTIGATION

OBJECTIVES
In this exercise, you will:
a. observe what happens when different salt solutions are added to paramecia.
b. determine which solution is harmful to paramecia.

Tardigrades (water bears) may also be used to show interesting effects of salt solutions on an organism.

Process Skills: experimenting, separating and controlling variables, using a microscope, interpreting data

KEYWORDS
Define the following keywords:

abiotic ____ the nonliving parts of an ecosystem

biotic ____ the living parts of an ecosystem

calcium ____ a mineral salt needed by living things in an ecosystem

mineral salt ____ a chemical made of several different elements

MATERIALS
See page 20T for preparation of culture and page 19T for preparation of salt solutions.

Paramecium culture   cotton ball   paper towel
3 glass slides   3 droppers   3 coverslips
                      light microscope   0.5%, 2%, and 5% mineral salt solutions
                                          Label these 1, 2, and 3.

PROCEDURE
1. Pull a small piece of cotton from the cotton ball and put it on the microscope slide. Add a drop of the *Paramecium* culture to the cotton on the slide. The strands of cotton will help keep the paramecia in the field of view.
2. Add a coverslip to the drop. Examine the slide under low power and then high power of the microscope.
3. Compare the paramecium with that shown in Figure 1.

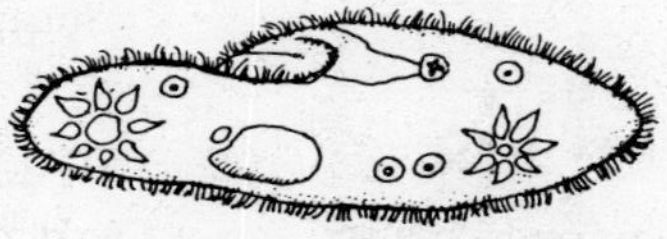

FIGURE 1. A paramecium

263

---

4. Make a drawing of your paramecium in Table 1. Draw a series of arrowed lines in the table to show how the paramecium moves.
5. Add a drop of salt solution 1 to the slide at the edge of the coverslip as shown in Figure 2. Be careful not to get the solution on top of the coverslip.
6. Place the edge of a paper towel on the other side of the coverslip so that it touches the solution. This draws the solution across the slide.
7. Use the microscope to look for the paramecia again. Record your observations of appearance and movement in Table 1.
8. Repeat steps 1 to 7 using solutions 2 and 3. Be sure to use a clean slide, coverslip, and dropper each time.

FIGURE 2. Preparing a paramecium slide

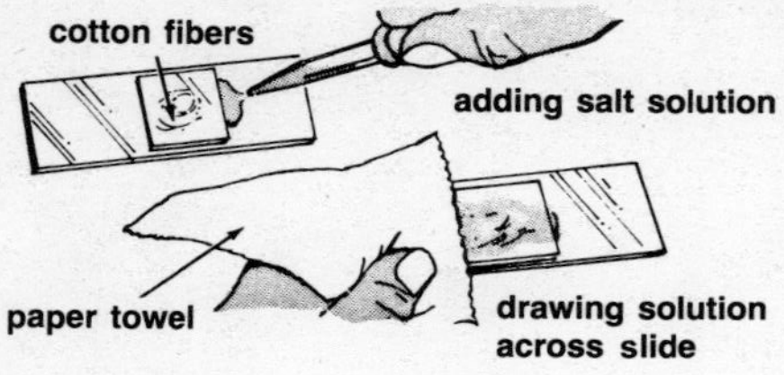

Table 1. Effects of Salt Solutions on Paramecia

Observations	No salt	Solution 1 (0.5%)	Solution 2 (2%)	Solution 3 (5%)
Appearance		no change	thinner, appears dead	thinner, needle-like pattern on edges
Movement	moves normally	moves normally	does not move	does not move

QUESTIONS

1. Which salt solutions caused a shape change in the paramecium? 2% and 5%
2. Which solutions caused a change in the movement of the paramecium? 2% and 5%
3. Which solution did not change the shape and movement of paramecium? 5%
4. Which solution is the strongest salt solution? 5%
5. Which solution do you think was most harmful to the paramecium? 5% salt solution
   Explain your answer. The paramecia were killed more quickly in 5% salt solution
6. Sometimes chemicals are added to streams and ponds to kill unwanted animals or plants. Not all animals die when chemicals are added to a stream. How could you explain this? Those that live are able to tolerate the amounts of chemicals. Those that die cannot tolerate the amounts of chemicals.

Name _____ Class _____ Period _____

Before doing this lab, have students read section 31:2 in the text.

31-3 How Can a Nonliving Part of an Ecosystem Help Living Things?

In the last exercise, you found out that an abiotic part can harm living things. Abiotic parts of a ecosystem help a living thing grow and reproduce. Duckweed is a small plant that lives in aquatic ecosystems such as ponds, lakes, or streams. The oval-shaped parts that float on the water are leaflike, but are in fact tiny flattened stems. Below the water hangs a single root. Minerals and water are taken in by the root and used for plant growth. One sign of growth is an increase in the number of leaflike parts, which will be called leaves here, and an increase in new plants. One plant can produce two or three new plants every 24 hours. The right amount of minerals can cause this good growth. Fertilizers contain large amounts of minerals that can be used by plants for growth.

INVESTIGATION

Process Skills: experimenting, separating and controlling variables, interpreting data, inferring

OBJECTIVES
In this exercise, you will:
a. find out how abiotic parts help a living thing.
b. determine which solution has the proper amount of minerals to cause growth in duckweed.

KEYWORDS
Define the following keywords:

duckweed __a small plant that lives in an aquatic ecosystem__

ecosystem __community interacting with the environment__

fertilizer __nutrients needed by plants for growth__

MATERIALS
9 duckweed plants *Lemna*** **Buy from biological supply houses or collect locally.
100 mL cow manure solution*   100 mL distilled water
100 mL 0.1% liquid plant food solution   3 small jars
small culture dish   hand lens
Add 20 drops of liquid plant food to 1 liter distilled water. Stir until dissolved.

FIGURE 1. Duckweed plants

The plant body of duckweed is reduced to a simple thallus. There is no distinction between stems and leaves. Some duckweed has roots, some do not.

PROCEDURE
1. Half fill a culture dish with water and put a duckweed plant in the dish. Look at the duckweed plant with a hand lens.
2. Look at the leaflike part and the root. Compare your duckweed plant with those shown in Figure 1.

* Dry cow manure and liquid plant food can be purchased in the garden section of hardware or variety stores. Add 50 mL dried (or fresh) cow manure to 1 liter distilled water. Allow to sit overnight. Drain off and use supernatant.

265

3. Label three small jars with the numbers 1, 2, and 3. Nearly fill each jar with the solutions as follows:
   jar 1—cow manure solution
   jar 2—0.1% liquid plant food solution
   jar 3—distilled water
4. Put three duckweed plants into each of the jars. Place the jars in a lighted area. Count the total number of leaves that you see in each jar. Record this number in Table 1 under Day 1.
5. Record the total number of leaves each day for the next four days in each jar.
6. At the same time look at the color of the leaves. If they are dark green, record this in Table 1 with a D. If they are pale green, record this with a P.

Table 1. Growth in Duckweed Plants

Jar number	Number and color (D or P) of leaves each day				
	Day 1	Day 2	Day 3	Day 4	Day 5
1	3 (D)	5 (D)	5 (D)	6 (D)	6 (D)
2	3 (D)	3 (D)	3 (D)	4 (D)	5 (D)
3	3 (D)	3 (D)	3 (P)	4 (P)	4 (P)

QUESTIONS
1. Which jar had the most duckweed leaves after Day 5? __jar 1__
2. Which jar had the least duckweed leaves after Day 5? __jar 3__
3. Dark green leaves tell us that these plants are growing best. In which jar did the best growth take place? __jar 1__
4. Pale green leaves are a sign of poor growth. In which jar did the poorest growth take place? __jar 3__
5. Are your answers to questions 1 and 3 the same? __yes__
   Explain your answer. __These plants had the best abiotic factors for growth.__
6. Are your answers to questions 2 and 4 the same? __yes__
   Explain your answer. __These plants had the poorest abiotic factors for growth.__
7. What was the biotic part that was a limiting factor in this exercise? __lack of proper minerals__
8. Explain how an abiotic part of an ecosystem can help a living thing. __Abiotic factors can cause proper growth and reproduction.__

266

Name _____ Class _____ Period _____
Before doing this lab, have students read section 32:2 in the text.

32-1 How Do Chemical Pollutants Affect Living Things?

Chemical pollutants get into water and soil from farms and factories. Sometimes factories dump their chemical wastes into holding ponds. If it rains before these chemicals can be hauled away, the ponds may overflow. The chemicals are then carried to fields and streams by the rainwater. Farmers spray chemicals on their crops to kill bacteria, fungi, insects, and weeds. They put fertilizer on the newly seeded fields to help plants grow and to produce a good crop. Chemical sprays and fertilizer are washed into streams during rainy weather. These chemicals are called pollutants because they can kill or harm living things. Unclean or discolored water is called polluted. Living things that live in the water are limited to where they can live because of the pH of the water. Some can live in water that is slightly acidic or slightly basic. Very few can live in water that is strongly acidic or strongly basic. Chemicals in the rain or soil enter freshwater ponds and streams. Sometimes these chemicals change the pH of the water to that of an acid. The rain falling on most of the United States has a pH in the acid range.

INVESTIGATION

Process Skills: experimenting, separating and controlling data, making and using tables, interpreting data

OBJECTIVES

In this exercise, you will:
a. compare the effects of two chemical pollutants on *Euglena*.
b. determine how different amounts of chemical pollutants affect *Euglena*.
c. determine if an acid or base added to the water can affect *Daphnia*.

KEYWORDS

Define the following keywords:

chemical wastes ___ chemicals no longer needed by humans

pH ___ a scale used to tell if a substance is an acid or base

pollutant ___ anything that causes pollution

Euglena and *Daphnia* culture techniques are found on pages 20T-21T.
* See page 19T for preparation of solutions.

MATERIALS

Euglena culture
10 mL fertilizer solution*
10 mL chemical solution*
10 test tubes
test-tube rack
hand lens

Daphnia culture
3 droppers
4 small jars
aged tap water
marking pencil
plastic spoon

acid solution*
base solution*
pH paper with pH color chart
applicator stick
forceps

pH paper on plastic strips can be reused several times during an experiment. Tell students not to touch the pH "paper" directly but rather hold on to the plastic end.

416

PROCEDURE

Part A. Observing the Effect of Pollution on *Euglena*

1. Use the marking pencil to label ten test tubes 1 to 10.
2. Two-thirds fill each tube with *Euglena* culture as shown in Figure 1.
3. Notice that the color of the culture in each tube is green. This color is caused by the green color of the euglenas that are present throughout the water in each tube.
4. Look at Table 1. Add the amounts of fertilizer and chemical solutions to the numbered test tubes as listed in Table 1.
5. Place the test tubes in the light and leave them undisturbed.
6. After 2 days observe the test tubes. Compare the colors of tubes 2 through 5 with tube 1. Describe the colors in Table 1. Note if they are dark green, pale green, or colorless. If the euglenas have reproduced, the color of the solution will appear a darker green than in the culture solution. If the euglenas have died, they will have settled on the bottom giving the solution a colorless appearance.
7. In the same way compare the colors of tubes 7 through 10 with tube 6.

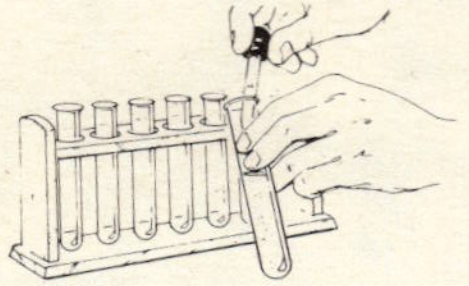

FIGURE 1. Adding *Euglena* culture to a test tube

Table 1. Differences in Color of *Euglena* Cultures

Test tube	Amount of fertilizer	Color at end of two days
1	0 drops	light green
2	5 drops	dark green
3	10 drops	dark green
4	15 drops	darker green
5	20 drops	darker green

Test tube	Amount of chemical	Color at end of two days
6	0 drops	light green
7	5 drops	clear
8	10 drops	clear
9	15 drops	clear
10	20 drops	clear

Part B. Studying the Effect of pH on *Daphnia*

1. Label four small jars 1 to 4. Nearly fill each jar with aged tap water.
2. Use pH paper to find the pH of the water. To do this place a drop of water on a small section of the pH paper as shown in Figure 2.
3. Look at the pH color chart and compare and match the color of your pH paper with one on the chart. This will give the pH of the water.
4. Adjust the pH of each of the four jars of water by adding drops of acid or base solutions until they match the following:
   jar 1—pH 10 (base)
   jar 2—pH 8 (base)
   jar 3—pH 6 (acid)
   jar 4—pH 4 (acid)

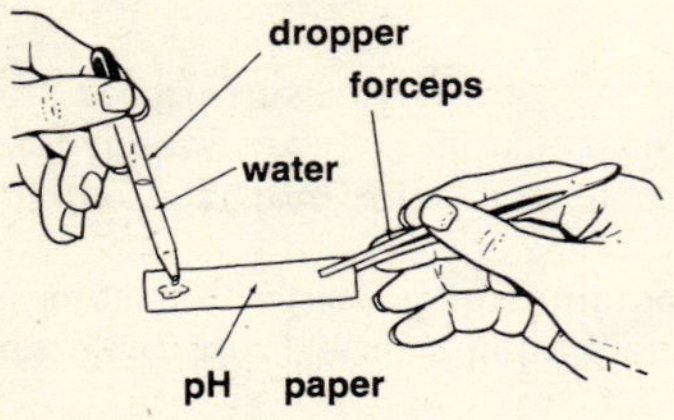

FIGURE 2. Testing pH

Name _____ Class _____ Period _____

5. To make the water in jars 1 and 2 more basic, add drops of base solution as shown in Figure 3. To make the water in jars 3 and 4 more acidic, add drops of acid solution.
6. Add a few drops of acid or base at a time and stir with a stick. Test the pH of the water in the jars after each stir as shown in Figure 2. Continue this procedure until the correct pH for each jar is reached.
7. Add 6 *Daphnia* to each jar as shown in Figure 4.
   a. First take a spoonful of culture.
   b. With an inverted dropper pick up the *Daphnia* from the spoon.
   c. Carefully drop them into the jar.
   d. Record this starting number of *Daphnia* in Table 2.
8. Add 3 droppersful of *Euglena* culture to each jar. This will provide the *Daphnia* with food.
9. Place the jars on a shelf in the lab.
10. Add 3 droppersful of *Euglena* culture to each jar for the next 2 days.
11. After three days observe each jar and record numbers of living *Daphnia* in Table 2.

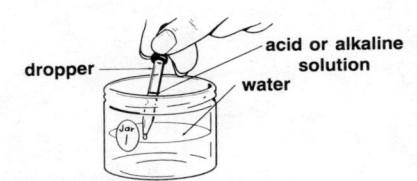

FIGURE 3. Changing pH

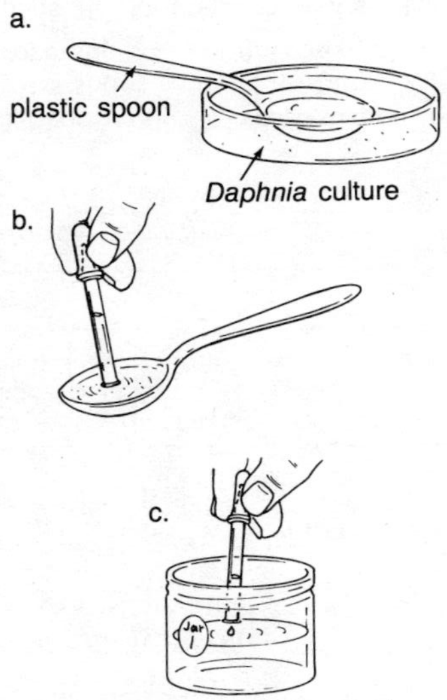

FIGURE 4. Transferring *Daphnia*

Table 2. Changes in Numbers of *Daphnia*

Jar	Numbers of *Daphnia* at start	pH of water	Numbers of living *Daphnia* after 3 days
1	6	10	5
2	6	8	7
3	6	6	8
4	6	4	0

269

QUESTIONS

1. In part A what did the fertilizer solution do to the euglenas? __It caused the euglenas to grow in numbers.__
2. What did the chemical solution do to the euglenas? __The euglenas were killed by the chemical solution.__
3. Why were fertilizer and chemical solutions not added to test tubes 1 or 6? __These tubes were controls for the experiment.__
4. Why did some of the test tubes of euglenas in part A become a darker green? __Fertilizer contains chemicals used for growth by euglenas.__
5. Which tubes had pollutants in them? __All tubes contain pollutants.__ These chemicals can be called chemical wastes.
6. A pond is light green in color for several weeks. Several days after a rain the water begins to turn dark green. How can you explain this? __Rainwater washed chemicals (fertilizer) into the pond; euglenas and algae use fertilizer for growth.__
7. In part B which jar had the most *Daphnia* at the end of three days? __jar 3__
8. What was the pH of this jar? Is this pH acidic or basic? __slightly acidic pH 6__
9. Why were there more *Daphnia* in this jar than in other jars? __Daphnia are able to live in water at a pH of 6.__
10. Which jar had the least *Daphnia* at the end of three days? __jar 4__
11. What was the pH of this jar? Is this pH acidic or basic? __4 (very acidic)__
12. Why were there less *Daphnia* in this jar than in other jars? __The water is very acidic and Daphnia cannot live in highly acidic water.__
13. Is there a limiting factor for *Daphnia*? __yes__
    Explain your answer. __Acid water because it limits the growth of Daphnia.__
14. Farmers put fertilizer on their soil, which can make the soil acidic. Before they plant crops, farmers usually add lime to the soil. Lime is a base. Why do farmers add lime to the soil? __Lime is added to make the soil less acidic. Seeds will not sprout or grow if soil is too acid.__
15. A chemical factory cleaned all of its large storage tanks on a weekend. Explain why many living things in a nearby stream died within a few days. __Many living things cannot live in streams where chemical wastes are present.__

270

Name _____ Class _____ Period _____

Before doing this lab, have students read section 32:2 in the text.

32–2 How Does Thermal Pollution Affect Living Things?

Sometimes the environment becomes too warm for living things. Thermal pollution is heat that is discharged into the soil, water, or air of a biological community. This heat can harm or kill living things.

Some industries heat water during the process of cooling their electric generators. While still warm, the water is sometimes dumped into small streams or ponds. Many of the organisms that make up the food chains and food webs in these water biomes may be killed.

INVESTIGATION

OBJECTIVES

In this exercise, you will:
a. find out if heated water can kill or stop the growth of living things.
b. determine if yeast cells can live in heated water for a short time.

Process Skills: experimenting, making a bar graph, interpreting data, using numbers

KEYWORDS

Define the following keywords:

community _____ living things interacting in an area

environment _____ the surroundings of living things

food chain _____ pathway of food through an ecosystem

food web _____ interconnecting food chains

thermal pollution _____ unneeded heat discharged into a biological community

MATERIALS

4 test tubes   toothpick   yeast suspension*
test-tube rack   glass beaker   blue stain*
test-tube holder   hot plate   microscope
clock with second hand marking pen

*See page 19T for preparation of solutions.

PROCEDURE

Part A. The Effect of Heat on Yeast

1. Label four test tubes 1 to 4. Place the test tubes in the rack.
2. Fill each test tube 1/3 full with tap water.
3. Add five drops of yeast suspension to each test tube as shown in Figure 1.
4. Shake the tubes back and forth to mix the yeast cells in the water.

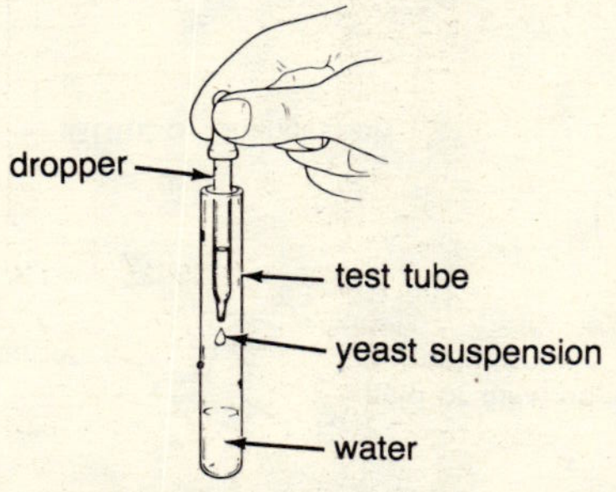

FIGURE 1. Adding yeast suspension

(labels: dropper, test tube, yeast suspension, water)

271

5. Heat a beaker of water to boiling over a hot plate.
6. Attach a test-tube holder to tube 2 and hold it in the boiling water for 20 seconds as shown in Figure 2. Then return it to the rack. **CAUTION:** *Always use the test-tube holder when placing the test tubes in or out of the boiling water.*
7. Repeat this process for test tube 3 but keep the test tube in the water for 40 seconds.
8. Repeat this process for test-tube 4 for 60 seconds.
9. Stir up the yeast cells in test tube 1 by filling a dropper with the yeast solution and squirting it back into the tube three times.
10. Place one drop of yeast solution from test tube 1 on a clean slide.
11. Using a clean dropper, add a drop of blue stain to the drop of yeast.
12. Use a toothpick to mix the stain with the drop of yeast solution.
13. Add a coverslip. Locate the yeast cells on low power and then turn to high power. Yeast cells will appear as small dots on low power. Look at Figure 3 to see their appearance at high power.

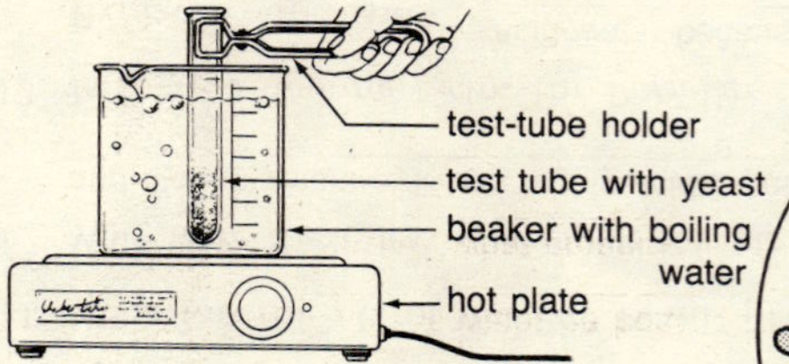

FIGURE 2. Heating yeast

(labels: test-tube holder, test tube with yeast, beaker with boiling water, hot plate)

Yeast suspension will yield 50-70 cells per field of view.

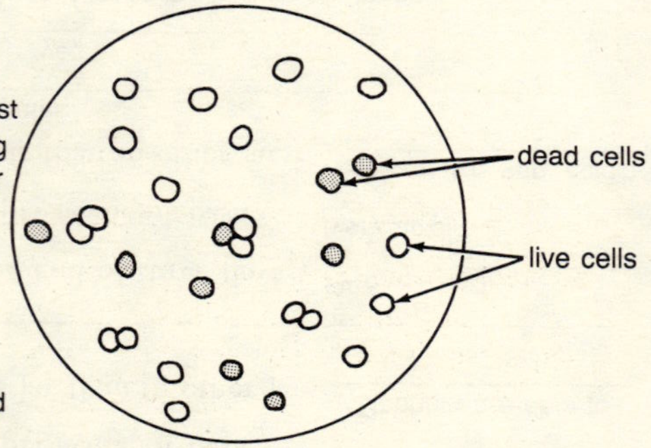

FIGURE 3. Yeast cells

(labels: dead cells, live cells)

14. Look for live yeast cells. These will appear very light blue in color. Look for dead yeast cells. These will have the same dark blue color as the stain on the slide.
15. For each yeast cell in one field of view, make a mark in Table 1 to show if it is alive or dead. Continue counting until 50 cells have been recorded. If there are less than 50 cells in one field of view, move to another area of the slide and continue counting until 50 has been reached.
16. Repeat steps 9 to 15 with test tubes 2, 3, and 4.

Table 1. Number of Yeast Cells

Answers will vary. Sample data given.

Test tube	Time in boiling water	Number of live yeast cells	Number of dead yeast cells	Total number of cells counted
1	0 seconds	⦀⦀ ⦀⦀ ⦀⦀ ⦀⦀ ⦀⦀ ⦀⦀ ⦀⦀ ⦀⦀ ⦀⦀ ⦀⦀		50
2	20 seconds	⦀⦀ ⦀⦀ ⦀⦀ ⦀⦀ ⦀⦀ ⦀⦀ ⦀⦀ ⦀⦀ ⦀⦀ ⦀⦀ ⦀⦀⦀⦀	⦀⦀ /	50
3	40 seconds	⦀⦀ ⦀⦀ ⦀⦀ ⦀⦀ ⦀⦀ ⦀⦀ ⦀⦀ /	⦀⦀ ⦀⦀ ⦀⦀ ⦀⦀ ⦀⦀ ⦀⦀ ⦀⦀ ⦀⦀ //	50
4	60 seconds		⦀⦀ ⦀⦀ ⦀⦀ ⦀⦀ ⦀⦀ ⦀⦀ ⦀⦀ ⦀⦀ ⦀⦀ ⦀⦀	50

272

Name _____ Class _____ Period _____

Part B. Plotting the Data

Using the data recorded in Table 1, plot two bar graphs in Figure 4 to show the number of live and dead cells in each tube.

Answers will vary. Sample data given.

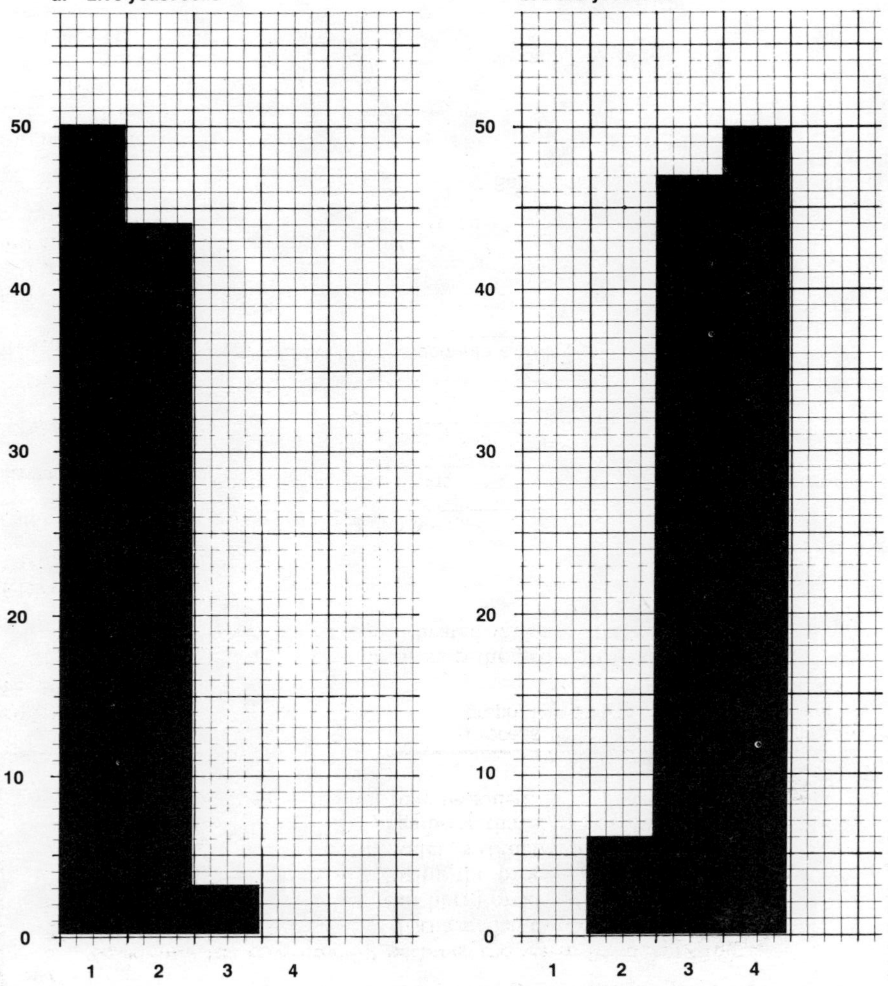

a. Live yeast cells

b. Dead yeast cells

FIGURE 4. Bar graphs

QUESTIONS

1. Which tube contained the most live cells? **tube 1** It was not heated so no cells were killed by excess heat.

2. Why were so many cells in this tube alive? _____

3. Which tube contained the most dead cells? **tube 4**

4. Why were so many dead cells in this tube? **The cells in this tube were in boiling water longer.**

5. Why was test tube 1 not placed in boiling water? **It was a control tube to show what the yeast cells would look like if unheated.**

6. Using your results, write three sentences that explain what thermal pollution is. Thermal pollution means too much heat in a community. Yeast cells and other living things cannot live very long in heated water. The longer they are in heated water, the more dead cells there will be.

7. Suppose that algae living in a stream react the same way as yeast cells did in this exercise. What would happen to food chains in the stream if thermal pollution occurred? **Algae are producers and the beginning of the food chain; most herbivores would die or move to another part of the stream.**

8. A new industry wants to move to your town. This industry wants to use water from the local river for its production line. What questions should the townspeople ask the new industry about the water? **How much water will they be using per day? Will this water be heated and returned to the river at its original temperature? If heated, what organisms will it kill?**